Lecture Notes in Computer Science   8246

Commenced Publication in 1973
Founding and Former Series Editors:
Gerhard Goos, Juris Hartmanis, and Jan van Leeuwen

T0236527

Gregory Gutin   Stefan Szeider (Eds.)

Parameterized and Exact Computation

8th International Symposium, IPEC 2013
Sophia Antipolis, France, September 4-6, 2013
Revised Selected Papers

 Springer

Volume Editors

Gregory Gutin
Royal Holloway, University of London
Egham Hill, Egham, TW20 0EX, UK
E-mail: g.gutin@cs.rhul.ac.uk

Stefan Szeider
Vienna University of Technology
Institute of Information Systems (184/3)
Favoritenstraße 9-11, 1040 Vienna, Austria
E-mail: stefan@szeider.net

ISSN 0302-9743   e-ISSN 1611-3349
ISBN 978-3-319-03897-1   e-ISBN 978-3-319-03898-8
DOI 10.1007/978-3-319-03898-8
Springer Cham Heidelberg New York Dordrecht London

Library of Congress Control Number: 2013954800

CR Subject Classification (1998): F.2.1-3, G.1-2, G.2.3, I.3.5, G.4, E.1, I.2.8

LNCS Sublibrary: SL 1 – Theoretical Computer Science and General Issues

Typesetting: Camera-ready by author, data conversion by Scientific Publishing Services, Chennai, India

Printed on acid-free paper

Springer is part of Springer Science+Business Media (www.springer.com)

Preface

The International Symposium on Parameterized and Exact Computation (IPEC, formerly IWPEC) is an international symposium series that covers research in all aspects of parameterized and exact algorithms and complexity. Started in 2004 as a biennial workshop, it became an annual event in 2008. This volume contains the papers presented at IPEC 2013: the 8th International Symposium on Parameterized and Exact Computation held during September 4–6, 2013, in Sophia Antipolis, France. The symposium was part of ALGO 2013, which also hosted six other workshops and symposia, including the Annual European Symposium on Algorithms (ESA 2013). The seven previous meetings of the IPEC/IWPEC series were held in Bergen, Norway (2004), Zürich, Switzerland (2006), Victoria, Canada (2008), Copenhagen, Denmark (2009), Chennai, India (2010), Saarbrücken, Germany (2011), and Ljubljana, Slovenia (2012).

The IPEC 2013 invited plenary keynote talk was given by Ramamohan Paturi (University of California, San Diego) on "Exact Complexity and Satisfiability" who won, together with his coauthors Chris Calabro (Google Inc.), Russell Impagliazzo (University of California, San Diego), Valentine Kabanets (Simon Fraser University), and Francis Zane (Alcatel Lucent), the *EATCS-IPEC Nerode Prize 2013*. The prize was delivered during the symposium. These proceedings contain an extended abstract of this talk. We had two additional invited tutorial speakers: Marek Cygan (University of Warsaw) speaking on treewidth and Daniel Lokshtanov (University of Bergen) speaking on representative sets. We thank the speakers for accepting our invitation.

In response to the call for papers, 58 papers were submitted. Each submission was reviewed by at least three reviewers. The reviewers were either Program Committee members or invited external reviewers. The Program Committee held electronic meetings using the EasyChair system, went through extensive discussions, and selected 29 of the submissions for presentation at the symposium and inclusion in this LNCS volume. The numbers of submitted and accepted papers are the highest in the history of the IPEC/IWPEC series. The Program Committee decided to award three Excellent Student Paper Awards: The winners are Bart M.P. Jansen (Utrecht University) for "On Sparsification for Treewidth Computations," Lukas Mach (Warwick University) and Tomas Toufar (Charles University) for "Amalgam Width of Matroids," and Mateus de Oliveira Oliveira (KTH Royal Institute of Technology) for "Subgraphs Satisfying MSO Properties on z-Topologically Orderable Digraphs." We thank Frances Rosamond for sponsoring the award.

We are very grateful to the Program Committee, and the external reviewers they called on, for the hard work and expertise which they brought to the difficult selection process. We also wish to thank all the authors who submitted their work for our consideration. Last but not least we would like to thank the local organizers of ALGO, in particular to Monique Teillaud.

September 2013

Gregory Gutin
Stefan Szeider

Organization

Program Committee

Faisal Abu-Khzam	Lebanese American University, Lebanon
Andreas Björklund	Lund University, Sweden
Rod Downey	Victoria University of Wellington, New Zealand
Mike Fellows	Charles Darwin University, Australia
Serge Gaspers	University of New South Wales, Australia
Gregory Gutin *(Co-chair)*	Royal Holloway University of London, UK
Pinar Heggernes	University of Bergen, Norway
Eun Jung Kim	Paris Dauphine University, France
Christian Komusiewicz	TU Berlin, Germany
Daniel Lokshtanov	University of Bergen, Norway
Venkatesh Raman	Institute of Mathematical Sciences, India
Peter Rossmanith	RWTH Aachen, Germany
Stefan Szeider *(Co-chair)*	Vienna University of Technology, Austria
Anders Yeo	University of Johannesburg, South Africa

IPEC Steering Committee

Dániel Marx	Hungarian Academy of Sciences, Hungary
Gregory Gutin	Royal Holloway University of London, UK
Gerhard Woeginger	Eindhoven University of Technology, The Netherlands
Hans Bodlaender *(Chair)*	Utrecht University, The Netherlands
Rolf Niedermeier	TU Berlin, Germany
Peter Rossmanith	RWTH Aachen, Germany
Dimitrios Thilikos	National and Kapodistrian University of Athens,Greece
Stefan Szeider	Vienna University of Technology, Austria
Thore Husfeldt	Lund University, Sweden

Additional Reviewers

Balasundaram, Baski	Chen, Yijia
Belmonte, Rémy	Courcelle, Bruno
Bliznets, Ivan	Crowston, Robert
Bousquet, Nicolas	Cygan, Marek
Bui-Xuan, Binh-Minh	Dawar, Anuj
Böcker, Sebastian	Dorn, Frederic
Cao, Yixin	Drucker, Andrew

Fan, Jia-Hao
Feldman, Vitaly
Fertin, Guillaume
Ganian, Robert
Golovach, Petr
Golovnev, Alexander
Gu, Qianping
Guillemot, Sylvain
Guo, Jiong
Hermelin, Danny
Hlineny, Petr
Hüffner, Falk
Jansen, Bart
Johnson, Matthew
Kanj, Iyad
Kaski, Petteri
Koivisto, Mikko
Kolay, Sudeshna
Koutis, Yiannis
Kral, Damiel
Kratsch, Dieter
Langer, Alexander
Lu, Yue
Makowsky, Janos
Marx, Dániel
Mathieson, Luke
Mayhew, Dillon
Meister, Daniel
Misra, Neeldhara
Mouawad, Amer
Muciaccia, Gabriele
Nederlof, Jesper
Nichterlein, André
Niedermeier, Rolf

Nishimura, Naomi
Obdrzalek, Jan
Ordyniak, Sebastian
Oum, Sang-Il
Panolan, Fahad
Paul, Christophe
Philip, Geevarghese
Pilipczuk, Marcin
Rai, Ashutosh
Ramanujan, M.S.
Rao B.V., Raghavendra
Regan, Kenneth
Reidl, Felix
Sarma, Jayalal
Sau, Ignasi
Saurabh, Saket
Schweitzer, Pascal
Shachnai, Hadas
Sharmin, Sadia
Shaw, Peter
Sikdar, Somnath
Simjour, Narges
Sorge, Manuel
Suchy, Ondrej
Telle, Jan Arne
Van 'T Hof, Pim
van Leeuwen, Erik Jan
Van Rooij, Johan M.M.
Villanger, Yngve
Weller, Mathias
Whittle, Geoff
Williams, Ryan
Xia, Ge
Xiao, Mingyu

Table of Contents

Exact Complexity and Satisfiability

(Invited Talk)

Russell Impagliazzo* and Ramamohan Paturi*

Department of Computer Science and Engineering
University of California, San Diego
La Jolla, CA 92093-0404, USA
{russell,paturi}@cs.ucsd.edu

All **NP**-complete problems are equivalent as far as polynomial time solvability is concerned. However, their exact complexities (worst-case complexity of algorithms that solve every instance correctly and exactly) differ widely. Starting with Bellman [1], Tarjan and Trojanowski [9], Karp [5], and Monien and Speckenmeyer [7], the design of improved exponential time algorithms for **NP**-complete problems has been a tremendously fruitful endeavor, and one that has recently been accelerating both in terms of the number of results and the increasing sophistication of algorithmic techniques. There are a vast variety of problems where progress has been made, e.g., k-SAT, k-COLORABILITY, MAXIMUM INDEPENDENT SET, HAMILTONIAN PATH, CHROMATIC NUMBER, and CIRCUIT SAT for limited classes of circuits. The "current champion" algorithms for these problems deploy a vast variety of algorithmic techniques, e.g., back-tracking search (with its many refinements and novel analysis techniques), divide-and-conquer, dynamic programming, randomized search, algebraic transforms, inclusion-exclusion, color coding, split and list, and algebraic sieving. In many ways, this is analogous to the diversity of approximation ratios and approximation algorithms known for different **NP**-complete problems. In view of this diversity, it is tempting to focus on the distinctions between problems rather than the interconnections between them. However, over the past two decades, there has been a wave of research showing that such connections do exist. Furthermore, progress on the exact complexity of **NP**-complete problems is linked to other fundamental questions in computational complexity, such as circuit lower bounds, parameterized complexity, data structures, and the precise complexity of problems within **P**. We are honored that our work helped to catalyze this wave of research, and humbled by the extent to which later researchers went far beyond our dreams of what might be possible.

We are severely constrained by space restrictions here, so we will not be able to give due credit or even describe many of the significant results. We will confine ourselves to informally defining the questions to be investigated, describing the initial steps to answer them, and giving some open problems.

The basic issue is to understand the extent to which "exhaustive search" can be beaten. However, for a decision problem, there might be several ways to

* This research is supported by NSF grant CCF-1213151 from the Division of Computing and Communication Foundations.

G. Gutin and S. Szeider (Eds.): IPEC 2013, LNCS 8246, pp. 1–3, 2013.

associate a search problem, and hence a trivial algorithm. We decided to finesse this issue by utilizing parameterized complexity [2], introducing a *complexity parameter* in addition to the usual size parameter. If the number of bits required to describe solutions is bounded by the complexity parameter, then we use the label "brute-force" to describe an algorithm that is exponential in the complexity parameter and polynomial in the size. For example, for the k-SAT problem, we will typically use the number of variables as a complexity parameter. However, we could also consider another parameterization for the k-SAT problem where the number of clauses is the complexity parameter. In either case, the size of the instance, in terms of bits used to represent it, will typically be significantly larger than this parameter.

For **NP**-hard problems with input size m and complexity parameter n, we call an algorithm *brute-force* if it runs in time $\texttt{poly}(m)2^n$. We say that algorithm is an *improvement* if its running time is bounded by $\texttt{poly}(m)2^{\mu n}$ for some $\mu <$ 1. An *exponential* improvement is one where $\mu \in 1 - \Omega(1)$, and a *nontrivial* improvement is one where $\mu = 1 - \omega(\log m/n)$. If $\mu = o(1)$, then the algorithm is *sub-exponential*.

Many **NP**-complete problems have exponential improvements, and a few such as chromatic number or Hamiltonian path for planar graphs have subexponential improvements. On the other hand, for problems like CIRCUIT SAT, we do not even know of any nontrivial improvements. It is not clear whether the series of improvements for problems such as k-SAT and k-COLORABILITY will lead to subexponential algorithms. For what problems, can we expect improved algorithms? Is progress on various problems connected? Can we give complexity-theoretic reasons why improvements or further improvements might not be possible for some problems?

To understand the difficulty of connecting the exact complexities, consider the standard reduction from k-SAT to k-COLORABILITY. It maps a k-SAT instance F of size m and n variables to a graph on $O(m+n)$ variables and $O(m+n)$ vertices. Since $m = O(n^k)$, a subexponential time algorithm for k-COLORABILITY would not a priori let us conclude anything useful about k-SAT.

In [4], progress was made to resolve this issue where it is shown subexponentital time solvability of both the problems is equivalent. The key new ideas include the notion of subexponential time Turing reductions and the Sparsification Lemma, which states that any k-CNF can be expressed as the subexponential disjunction of linear (in the number of variables) size k-CNF in subexponential time. From this lemma, we can derive a way to convert a subexponential time algorithm for k-COLORABILITY into a subexponential time algorithm for k-SAT. More generally, in [4], it is shown that all the **SNP**-complete problems (under subexponential time Turing reductions), are equivalent as far as subexponential time complexities are concerned. **SNP** is a subclass of **NP** which includes k-SAT and k-COLORABILITY, and is defined as the class of problems expressible by second order existential quantifiers followed by a first order universal quantifiers followed by a basic formula, introduced by Papadimitriou and Yannakakis [8]. The result also extends to size-constrained **SNP** where the second order quantified variables are restricted in size.

The result provides evidence that k-SAT (equivalently, 3-SAT) may not have subexponential time algorithms since otherwise the entire *logically* defined class **SNP** would have subexponential time algorithms. While we are not in a position to resolve this issue, it is interesting to explore the state of affairs assuming the likelihood. Let $s_k = \inf\{\delta | \exists\ 2^{\delta n}$ algorithm for k-SAT$\}$. We define the *Exponential Time Hypothesis* (**ETH**) to be the statement: $s_3 > 0$. We call it a hypothesis rather than a conjecture, because we are agnostic about whether it is actually true, but think that its truth value has interesting ramifications either way.

We understand very little about exponential time algorithms. **ETH** will be useful if it helps factor out the essential difficulty of dealing with exponential time algorithms for **NP**-complete problems. More concretely, the usefulness of **ETH** is its explanatory value regarding the exact complexities of various **NP**-complete problems, ideally, by providing lower bounds that match the best known upper bounds.

One of the first nontrivial consequences of **ETH** is that the exponential complexities s_k of k-SAT strictly increase infinitely often as k increases [3]. This result tempts one to posit the *Strong Exponential Time Hypothesis* (**SETH**) that says $s_\infty := \lim_{k \to \infty} s_k = 1$. There have been numerous lower bounds on the exact complexities of **NP**-complete problems based on **ETH** and **SETH**. Lokshtanov, Marx, and Saurabh provide a systematic summary of these in results in [6].

However, a number of questions remain regarding the explanatory power of **ETH**. Assuming **ETH** or other well known complexity assumption, can we obtain a positive constant lower bound on s_3? Can we prove that **ETH** implies **SETH**? Assuming **SETH**, can we prove a $2^{(n-o(n))}$ lower bound for CHROMATIC NUMBER?

References

1. Bellman, R.: Dynamic programming treatment of the travelling salesman problem. Journal of the ACM 9(1), 61–63 (1962)
2. Downey, R.G., Fellows, M.R.: Parameterized Complexity. Springer, New York (1999)
3. Impagliazzo, R., Paturi, R.: The complexity of k-SAT. Journal of Computer and Systems Sciences 62(2), 367–375 (2001); Preliminary version in 14th Annual IEEE Conference on Computational Complexity, pp. 237–240 (1999)
4. Impagliazzo, R., Paturi, R., Zane, F.: Which problems have strongly exponential complexity? Journal of Computer and System Sciences 63, 512–530 (1998); Preliminary version. In: 39th Annual IEEE Symposium on Foundations of Computer Science, pp. 653–662 (1998)
5. Karp, R.M.: Dynamic programming meets the principle of inclusion and exclusion. Operations Research Letters 1, 49–51 (1982)
6. Lokshtanov, D., Marx, D., Saurabh, S.: Lower bounds based on the exponential time hypothesis. Bulletin of the EATCS 105, 41–72 (2011)
7. Monien, B., Speckenmeyer, E.: Solving satisfiability in less than 2^n steps. Discrete Applied Mathematics 10, 287–295 (1985)
8. Papadimitriou, C., Yannakakis, M.: Optimization, approximation, and complexity classes. Journal of Computer and System Sciences 43, 425–440 (1991)
9. Tarjan, R., Trojanowski, A.: Finding a maximum independent set. SIAM Journal of Computing 6, 537–546 (1977)

The Parameterized Complexity of Fixpoint Free Elements and Bases in Permutation Groups

Vikraman Arvind

The Institute of Mathematical Sciences
Chennai 600 113, India
arvind@imsc.res.in

Abstract. In this paper we study the parameterized complexity of two well-known permutation group problems which are NP-complete.

1. Given a permutation group $G = \langle S \rangle \leq S_n$ and a parameter k, find a permutation $\pi \in G$ such that $|\{i \in [n] \mid \pi(i) \neq i\}| \geq k$. This generalizes the NP-complete problem of finding a fixed-point free permutation in G [7,14] (this is the case when $k = n$). We show that this problem with parameter k is fixed-parameter tractable. In the process, we give a simple deterministic polynomial-time algorithm for finding a fixed point free element in a transitive permutation group, answering an open question of Cameron [8,7].

2. A *base* for G is a subset $B \subseteq [n]$ such that the subgroup of G that fixes B pointwise is trivial. We consider the parameterized complexity of checking if a given permutation group $G = \langle S \rangle \leq S_n$ has a base of size k, where k is the parameter for the problem. This problem is known to be NP-complete [4]. We show that it is fixed-parameter tractable for cyclic permutation groups and for permutation groups of constant orbit size. For more general classes of permutation groups we do not know whether the problem is in FPT or is W[1]-hard.

1 Introduction

Let S_n denote the group of all permutations on a set of size n. The group S_n is also called the symmetric group of degree n. We refer to a subgroup G of S_n, denoted by $G \leq S_n$, as a permutation group (of degree n). Let $S \subseteq S_n$ be a subset of permutations. The permutation group *generated* by S, denoted by $\langle S \rangle$, is the smallest subgroup of S_n containing S. A subset $S \subseteq G$ of a permutation group G is *a generating set* for G if $G = \langle S \rangle$. It is easy to see that every finite group G has a generating set of size at most $\log_2 |G|$.

Let $G = \langle S \rangle \leq S_n$ be a subgroup of the symmetric group S_n, where G is given as input by a generating set S of permutations. Each permutation $\psi \in S$ can be described as a list of n ordered pairs $\langle i, j \rangle \in [n] \times [n]$.

Algorithmic problems on permutation groups, where the groups are given as input by their generating sets are well studied in the literature (e.g. see [19,12,15,17]). Some of them have efficient algorithms, some others are NP-complete, and yet others have a status similar to Graph Isomorphism: they

G. Gutin and S. Szeider (Eds.): IPEC 2013, LNCS 8246, pp. 4–15, 2013.

are neither known to be in polynomial time and unlikely to be NP-complete (unless the Polynomial-Time Hierarchy collapses). The reader may refer to [15] for different examples. Efficient permutation group algorithms have played an important role in the design of algorithms for the Graph Isomorphism problem [1,2]. In fact, the algorithm with the best running time bound for general Graph Isomorphism is group-theoretic [3,21].

We recall some group-theoretic definitions and notation. An excellent modern reference on permutation groups is Cameron's book [5]. Algorithmic permutation group problems are well treated in [15,17]. More details can be found in these references.

The symmetric group S_n is a group under permutation composition. We evaluate the composition gh of permutations $g, h \in S_n$ from left to right, following the standard convention in permutation groups. That is to say, given a point i in the domain $[n]$ we will apply g first and then h. The notation i^g will denote the g-image of i. This is more convenient for a left to right evaluation; notice that $i^{gh} = (i^g)^h$. We denote the identity permutation in S_n by id.

Let $\pi \in S_n$ be a permutation. A *fixpoint* of π is a point $i \in [n]$ such that $i^\pi = i$ and a permutation π is *fixpoint free* if $i^\pi \neq i$ for all $i \in [n]$.

Let $G \leq S_n$ and $\Delta \subseteq [n]$ be a subset of the domain. The subgroup of G that is the *pointwise stabilizer* of Δ is $G_\Delta = \{g \in G \mid i^g = i \text{ for all } i \in \Delta\}$.

For groups G and H, the expression $H < G$ means that H is a subgroup of G (not necessarily a proper subgroup). For $\pi \in G$ the subset $H\pi = \{h\pi : \pi \in H\}$ of G is a *right coset* of H in G. Any two right cosets of H in G are either disjoint or identical. Thus, the right cosets of H in G form a partition of G and we can write

$$G = H\pi_1 \cup H\pi_2 \cup \ldots \cup H\pi_m.$$

Each coset of H has exactly $|H|$ elements and $\{\pi_1, \pi_2, \ldots, \pi_m\}$ is a complete set of distinct *coset representatives* for H in G. For each $i \in [n]$, $G_{[i]} = \{g \in G : \forall j \in [i], j^g = j\}$ is the pointwise stabilizer of the subset $[i]$.

As developed by Sims [19], pointwise stabilizers are fundamental in the design of algorithms for permutation group problems. The structure used is the tower of stabilizers subgroups in G:

$$\{id\} = G_{[n-1]} < G_{[n-2]} < \ldots < G_{[1]} < G_{[0]} = G.$$

The union of the right coset representative sets T_i for the groups $G_{[i]}$ in $G_{[i-1]}, 1 \leq i \leq n-1$, forms a generator set for G called a *strong generating set* for G [19,20].

Theorem 1 (Schreier-Sims). [19] *Let $G < S_n$ be a permutation group input by some generating set. There is a polynomial-time algorithm for computing a strong generating set $\bigcup_{i=1}^{n-1} T_i$, where T_i is the set of coset representatives for $G^{(i)}$ in $G^{(i-1)}$, $1 \leq i \leq n-1$ with the following properties:*

1. *Each $\pi \in G$ can be uniquely written as a product $\pi = \pi_1 \pi_2 \ldots \pi_{n-1}$ with $\pi_i \in T_i$,*
2. *Membership in G of a given permutation can be tested in polynomial time.*

A subset $B \subseteq [n]$ is called a *base* for G if the pointwise stabilizer subgroup G_B is trivial. Thus, if B is a base for G then each element of G is uniquely determined by its action on B. The problem of computing a base of minimum cardinality is known to be computationally very useful. Important algorithmic problems on permutation groups, like membership testing, have nearly linear time algorithms in the case of small-base groups (e.g. see [17]). We will discuss the parameterized complexity of the minimum base problem in Section 3.

Basic definitions and results on parameterized complexity can be found in Downey and Fellows' classic text on the subject [9]. Another, more recent, reference is [11].

2   Fixpoint Free Elements

The starting point is the Orbit-Counting lemma. Our discussion will follow Cameron's book [5]. For each permutation $g \in S_n$ let fix(g) denote the number of points fixed by g. More precisely,

$$\text{fix}(g) = |\{i \in [n] \mid i^g = i\}|.$$

A permutation group $G \leq S_n$ induces, by its action, an equivalence relation on the domain $[n]$: i and j are in the same equivalence class if $i^g = j$ for some $g \in G$. Each equivalence class is an *orbit* of G. G is said to be *transitive* if there is exactly one G-orbit. Let orb(G) denote the number of G-orbits in the domain $[n]$.

Lemma 1 (Orbit Counting Lemma). [6] *Let $G \leq S_n$ be a permutation group. Then*

$$\text{orb}(G) = \frac{1}{|G|} \sum_{g \in G} \text{fix}(g). \tag{1}$$

In other words, the number of G orbits is the average number of fixpoints over all elements of G.

Proof. It is useful to recall a proof sketch. Define a $|G| \times n$ matrix with rows indexed by elements of G and columns by points in $[n]$. The $(g, i)^{th}$ entry is defined to be 1 if $i^g = i$ and 0 otherwise. Clearly, the gth row has fix(g) many 1's in it. Let G_i denote the subgroup of G that fixes i. The ith column clearly has $|G_i|$ many 1's. Counting the number of 1's in the rows and columns and equating them, keeping in mind that, by the Orbit-Stabilizer lemma [5], $|G|/|G_i|$ is the size of the orbit containing i yields the lemma.

We now recall a theorem of Jordan on permutation groups [13]. See [18,8] for very interesting accounts of it. A permutation group $G \leq S_n$ is *transitive* if it has exactly one orbit.

Theorem 2 (Jordan's theorem). *If $G \leq S_n$ is transitive then G has a fixpoint free element.*

The theorem follows directly from the Orbit counting lemma. Notice that the left side of Equation 1 equals 1. The right side of the equation is the average over all fix(g). Now, the identity element id fixes all n elements. Thus there is at least one element $g \in G$ such that fix$(g) = 0$. Cameron and Cohen [6] do a more careful counting and show the following strengthening.

Theorem 3. [6] *If $G \leq S_n$ is transitive then there are at least $|G|/n$ elements of G that are fixpoint free.*

We discuss their proof, because we will build on it to obtain our results. If G is transitive, the orbit counting lemma implies

$$|G| = \sum_{g \in G} \text{fix}(g).$$

Take any point $\alpha \in [n]$. We can write the above equation as

$$|G| = \sum_{g \in G_\alpha} \text{fix}(g) + \sum_{g \in G \setminus G_\alpha} \text{fix}(g).$$

By the orbit counting lemma applied to the group G_α we have

$$\sum_{g \in G_\alpha} \text{fix}(g) = \text{orb}(G_\alpha) \cdot |G_\alpha|.$$

Let $A \subset G$ be the set of all fixpoint free elements of G. Clearly, the sum $\sum_{g \in G \setminus G_\alpha} \text{fix}(g) \geq |G \setminus A|$ as $A \subseteq G \setminus G_\alpha$ and each element of $G \setminus A$ fixes at least one element. Combining with the previous equation we get

$$|A| \geq \text{orb}(G_\alpha) \cdot |G_\alpha| = \text{orb}(G_\alpha) \cdot \frac{|G|}{n} \geq \frac{|G|}{n}.$$

2.1   The Algorithmic Problem

We now turn to the problem of computing a fixpoint free element in a permutation group $G \leq S_n$ and a natural parameterized version.

As observed by Cameron and Wu in [7], the result of [6] gives a simple randomized algorithm to find a fixpoint free element in a transitive permutation group $G \leq S_n$, where G is given by a generating set S: Using the Schreier-Sims polynomial-time algorithm [19] we can compute a *strong generating set S'* for G in polynomial time. And using S' we can sample uniformly at random from G. Clearly, in $O(n)$ sampling trials we will succeed in finding a fixpoint free element with constant probability. We will show in the next section that this algorithm can be *derandomized* to obtain a deterministic polynomial time algorithm (without using CFSG). This answers an open problem of Cameron discussed in [7,8].

This result is to be contrasted with the fact that computing fixed point free elements in nontransitive groups $G \leq S_n$ is NP-hard. The decision problem is shown NP-complete in [7]. This is quite similar to Lubiw's result [14] that checking if a graph X has a fixpoint free automorphism is NP-complete.

We will now introduce the parameterized version of the problem of computing fixpoint free elements in permutation groups. First we introduce some terminology. We say that a permutation π *moves* a point $i \in [n]$ if $i^\pi \neq i$.

k-MOVE Problem

INPUT: A permutation group $G = \langle S \rangle \leq S_n$ given by generators and a number k.

PROBLEM: Is there an element $g \in G$ that moves at least k points?

For $k = n$ notice that k-MOVE is precisely the problem of checking if there is a fixpoint free element in G. The parameterized version of the problem is to treat k as parameter. We will show that this problem is fixed-parameter tractable.

Let move(g) denote the number of points moved by g. We define two numbers fix(G) and move(G):

$$\text{fix}(G) = |\{i \in [n] \mid i^g = i \text{ for all } g \in G\}|$$
$$\text{move}(G) = |\{i \in [n] \mid i^g \neq i \text{ for some } g \in G\}|$$

That is to say, fix(G) is the number of points fixed by all of G and move(G) is the number of points moved by some element of G. Clearly, for all $g \in G$, move$(g) = n - \text{fix}(g)$ and move$(G) = n - \text{fix}(G)$. Furthermore, notice that orb$(G) \leq \text{fix}(G) + \text{move}(G)/2$, and we have $n - \text{orb}(G) \geq \text{move}(G)/2$. Let $G = \langle S \rangle \leq S_n$ be an input instance for the k-MOVE problem. Substituting $n - \text{move}(g)$ for fix(g) in Equation 1 and rearranging terms we obtain

$$\text{move}(G)/2 \leq n - \text{orb}(G) = \frac{1}{|G|} \sum_{g \in G} \text{move}(g) = \mathbb{E}_{g \in G}[\text{move}(g)], \qquad (2)$$

where the expectation is computed for g picked uniformly at random from G.

We will show there is a deterministic polynomial time algorithm that on input $G = \langle S \rangle \leq S_n$ outputs a permutation $g \in G$ such that move$(g) \geq n - \text{orb}(G) \geq \text{move}(G)/2$. Using this algorithm we will obtain an FPT algorithm for the k-MOVE problem. We require the following useful lemma about computing the average number of points moved by uniformly distributed elements from a *coset* contained in S_n.

Lemma 2. *Let $G\pi \subseteq S_n$ be a coset of a permutation group $G = \langle S \rangle \leq S_n$, where $\pi \in S_n$. There is a deterministic algorithm that computes $\mathbb{E}_{g \in G}[\text{move}(g\pi)]$ in time polynomial in $|S|$ and n.*

Proof. We again use a double counting argument. We define a matrix whose rows are indexed by $g\pi, g \in G$ and columns by $i \in [n]$. The $(g\pi, i)^{th}$ entry of this matrix is 1 if $i^{g\pi} \neq i$ and 0 otherwise. Thus, the number of 1's in the ith column of the matrix is $|G| - |\{g \in G \mid i^{g\pi} = i\}|$. Now, $|\{g \in G \mid i^{g\pi} = i\}| = |\{g \in G \mid i^g = i^{\pi^{-1}}\}|$, which is zero if $i^{\pi^{-1}}$ and i are in different G-orbits and is $|G_i|$ if they are in the same G-orbit. In polynomial time we can compute the orbits of G from its given generating set [19,12] and check this condition. Also, the number $|G| - |\{g \in G \mid i^{g\pi} = i\}|$ is computable in polynomial time since $|G_i|$ is computable in polynomial time. Call this number N_i. It follows that the total number of 1's in the matrix is $\sum_{i=1}^{n} N_i$, which is computable in polynomial time. Since $\sum_{i=1}^{n} N_i = \sum_{g \in G} \text{move}(g\pi)$, it follows that $\frac{1}{|G|} \sum_{g \in G} \text{move}(g\pi) = \mathbb{E}_{g \in G}[\text{move}(g\pi)]$ can be computed exactly in polynomial time.

Theorem 4. *There is a deterministic polynomial-time algorithm that takes as input a permutation group $G = \langle S \rangle \leq S_n$ given by generating set S and a permutation $\pi \in S_n$ and computes an element $g \in G$ such that $\text{move}(g\pi) \geq \mathbb{E}_{g \in G}[\text{move}(g\pi)]$.*

Proof. By Lemma 2 we can compute in polynomial time the following quantity:

$$\mu = \frac{1}{|G|} \sum_{g \in G} \text{move}(g\pi) = \mathbb{E}_{g \in G}[\text{move}(g\pi)].$$

Now, we can write G as a disjoint union of cosets $G = \bigcup_{i=1}^{r} G_1 g_i$, where G_1 is the subgroup of G that fixes 1 and g_i are the coset representatives, where the number of cosets $r \leq n$. Using Schreier-Sims algorithm [19] we can compute all coset representatives g_i and a generating set for G_1 from the input in polynomial time.

Now, we can write the summation $\frac{1}{|G|} \sum_{g \in G} \text{move}(g\pi)$ as a sum over the cosets $G_1 g_i \pi$ of G_1:

$$\frac{1}{|G|} \sum_{g \in G} \text{move}(g\pi) = \frac{1}{|G|} \sum_{i=1}^{r} \sum_{g \in G_1} \text{move}(gg_i\pi).$$

For $1 \leq i \leq r$ let

$$\mu_i = \frac{1}{|G_1|} \sum_{g \in G_1} \text{move}(gg_i\pi).$$

Since $|G|/|G_1| = r$, it follows that $\mu = \frac{1}{r} \sum_{i=1}^{r} \mu_i$ is an average of the μ_i. Let μ_t denote $\max_{1 \leq i \leq r} \mu_i$. Clearly, $\mu \leq \mu_t$ and therefore there is some $g \in G_1 g_t \pi$ such that $\text{move}(g) \geq \mu_t \geq \mu$ and we can continue the search in the coset $G_1 g_t$ since we can compute all the μ_i in polynomial time by Lemma 2. Continuing thus for $n-1$ steps, in polynomial time we will obtain a coset $G_{n-1}\tau$ containing a single element τ such that $\text{move}(\tau) \geq \mu$. This completes the proof.

Cameron, in [7] and in the lecture notes [8], raises the question whether the randomized algorithm, based on uniform sampling, for finding a fixpoint free element in a transitive permutation group (given by generators) can be derandomized. In [7] a deterministic algorithm (based on the classification of finite simple groups, CFSG) is outlined. The algorithm does a detailed case analysis based on the CFSG and is not easy to verify. Here we show that the randomized algorithm can be easily derandomized yielding a polynomial-time algorithm. The derandomization is essentially a simple application of the "method of conditional probabilities" [10,16].

Corollary 1. *Given a transitive permutation group* $G = \langle S \rangle \le S_n$ *by a generating set* S, *we can compute a fixpoint free element of* G *in deterministic polynomial time.*

Proof. Notice that $\mathbb{E}_{g \in G}[\text{move}(g)] = n - 1$ to begin with. However, since G_1 has at least two orbits, we have by orbit counting lemma that $\mathbb{E}_{g \in G_1}[\text{move}(g)] \le n-2$. Hence, for some coset $G_1 g_i$ of G_1 in G we must have $\mathbb{E}_{g \in G_1 g_i}[\text{move}(gg_i)] > n-1$. The polynomial-time algorithm of Theorem 4 applied to G will therefore continue the search in cosets where the expected value is strictly more than $n - 1$ which means that it will finally compute a fixpoint free element of G.

Given $G = \langle S \rangle \le S_n$ by its generating set S there is a trivial exponential time algorithm for finding a fixpoint free element in G: compute a strong generating set for G in polynomial time [19]. Then enumerate G in time $|G|.n^{O(1)}$ using the strong generating set, checking for a fixpoint free element. This algorithm could have running time $n!$ for large G. We next describe a $2^n n^{O(1)}$ time algorithm for finding a fixpoint free element based on inclusion-exclusion and coset intersection.

Theorem 5. *Given a permutation group* $G = \langle S \rangle \le S_n$ *and* $\pi \in S_n$ *there is a* $2^{(n+O(\sqrt{n}\lg n))} n^{O(1)}$ *time algorithm to test if the coset* $G\pi$ *has a fixpoint free element and if so compute it.*

Proof. For each subset $\Delta \subseteq [n]$, using the Schreier-Sims algorithm [19,12] we can compute in polynomial time the pointwise stabilizer subgroup G_Δ. This will take time $2^n n^{O(1)}$ overall. For each $i \in [n]$, let $(G\pi)_i$ denote the subcoset of $G\pi$ that fixes i. Indeed,

$$(G\pi)_i = \{g\pi \mid g \in G, i^{g\pi} = i\} = G_i \tau_i \pi,$$

if there is a $\tau_i \in G$ such that $i^{\tau_i} = i^{\pi^{-1}}$ and $(G\pi)_i = \emptyset$ otherwise.

Clearly, $G\pi$ has a fixpoint free element if and only if the union $\bigcup_{i=1}^{n}(G\pi)_i$ is a *proper* subset of $G\pi$. Hence we need to check if $|\bigcup_{i=1}^{n}(G\pi)_i| < |G\pi| = |G|$. Now, $|\bigcup_{i=1}^{n}(G\pi)_i|$ can be computed in $2^{(n+O(\sqrt{n}\lg n))} n^{O(1)}$ time using the inclusion exclusion principle: there are 2^n terms in the inclusion-exclusion formula. Each term is the cardinality of a coset intersection of the form $\bigcap_{i \in I}(G\pi)_i$, for some subset of indices $I \subseteq [n]$, which can be computed in $n^{O(\sqrt{n})}$ time [2]. Hence, we can decide in $2^{(n+O(\sqrt{n}\lg n))} n^{O(1)}$ time whether or not $G\pi$ has a fixpoint free element. Notice that this fixpoint free element must be in one of the $n - 1$ subcosets of $G\pi$ that maps 1 to j for $j \in \{2,3,\ldots,n\}$. The subcoset of $G\pi$ mapping 1 to j can be computed in polynomial time [19]. Then we can apply the inclusion-exclusion principle to each of these subcosets, as explained above, to check if it contains a fixpoint free element and continue the search in such a subcoset. Proceeding thus for $n - 1$ steps we will obtain a fixpoint free element in $G\pi$, if it exists, in $2^{(n+O(\sqrt{n}\lg n))}n^{O(1)}$ time.

We now prove the main result of this section.

Theorem 6. *There is a deterministic* $3^{(2k+O(\sqrt{k}\lg k))}k^{O(1)} + (n|S|)^{O(1)}$ *time algorithm for the k-MOVE problem and hence the problem is fixed-parameter tractable. Furthermore, if $G = \langle S \rangle \leq S_n$ is a "yes" instance the algorithm computes a $g \in G$ such that* $\text{move}(g) \geq k$.

Proof. Let $G = \langle S \rangle \leq S_n$ be an input instance of k-MOVE with parameter k. By Equation 2 we know that $\mathbb{E}_{g \in G}[\text{move}(g)] \geq \text{move}(G)/2$. We first compute $\text{move}(G)$ in $(n|S|)^{O(1)}$ time by computing the orbits of G. If $\text{move}(G) \geq 2k$ then the input is a "yes" instance to the problem and we can apply Theorem 4 to compute a $g \in G$ such that $\text{move}(g) \geq k$ in $(n|S|)^{O(1)}$ time. Otherwise, $\text{move}(G) \leq 2k$. In that case, the group G is effectively a permutation group on a set $\Omega \subseteq [n]$ of size at most $2k$. For each subset $\Delta \subseteq \Omega$ of size at most k, we compute the pointwise stabilizer subgroup G_Δ of G in polynomial time [19]. Note that effectively the permutations of G_Δ have support only in $\Omega \setminus \Delta$. Now, if the input is a "yes" instance of k-MOVE, some G_Δ will contain an element that is fixpoint free in $\Omega \setminus \Delta$. We can apply the algorithm of Theorem 5 to compute this element in time $2^{(|\Omega \setminus \Delta| + O(\sqrt{k}\lg k))}k^{O(1)}$. Since $1 \leq |\Delta| = d \leq k$, the overall running time of these steps can be bounded by $2^{O(\sqrt{k}\lg k)}k^{O(1)}\sum_{1 \leq d \leq k}\binom{2k}{d}2^{2k-d} \leq 3^{2k}2^{O(\sqrt{k}\lg k)}k^{O(1)}$.

Remark 1. We note from the first few lines in the proof of Theorem 6 that the application of Theorem 4 is actually a polynomial time reduction from the given k-MOVE instance to an instance for which $\text{move}(G) \leq 2k$. Given $G = \langle S \rangle \leq S_n$ such that $\text{move}(G) \leq 2k$, note that G is effectively a subgroup of S_{2k}. We can apply the Schreier-Sims algorithm [19] to compute from S a generating set of size $O(k^2)$ for G, therefore yielding a polynomial time computable, $k^{O(1)}$ size kernel (see [11] for definition) for the k-MOVE problem.

3   The Parameterized Minimum Base Problem

In this section we turn to another basic algorithmic problem on permutation groups.

Definition 1. *Let $G \leq S_n$ be a permutation group. A subset of points $B \subseteq [n]$ is called a base if the pointwise stabilizer subgroup G_B of G (subgroup of G that fixes B pointwise) is the identity.*

Since permutation groups with a small base have fast algorithms for various problems [17], computing a minimum cardinality base for G is very useful. The decision problem is NP-complete. On the other hand, the optimization problem has a $\lg \lg n$ factor approximation algorithm [4].

In this section we study the parameterized version of the problem with base size as parameter. We are unable to resolve if the general case is FPT or not, we give FPT algorithms in the case of cyclic permutation groups and for permutation groups with orbits of size bounded by a constant.

k-BASE Problem

INPUT: A permutation group $G = \langle S \rangle \leq S_n$ given by generators and a number k.

PROBLEM: Is there a base of size at most k for G?

A trivial $n^{k+O(1)}$ algorithm would cycle through all candidate subsets B of size at most k checking if G_B is the identity.

Remark 2. If the elements of the group $G \leq S_n$ are explicitly listed, then the k-BASE problem is essentially a hitting set problem, where the hitting set B has to intersect, for each $g \in G$, the subset of points moved by g. However, the group structure makes it different from the general hitting set problem and we do not know how to exploit it algorithmically in the general case.

3.1 Cyclic Permutation Groups

We give an FPT algorithm for the special case when the input permutation group $G = \langle S \rangle$ is cyclic. While this is only a special case, we note that the minimum base problem is NP-hard even for cyclic permutation groups [4, Theorem 3.1].

Theorem 7. *The k–BASE problem is fixed-parameter tractable for cyclic permutation groups.*

Proof. Let $G = \langle S \rangle \leq S_n$ be a cyclic permutation group as instance for k-BASE. First, using the Schreier-Sims algorithm we compute $|G|$. We can assume $|G| \leq n^k$, Otherwise, G does not have a size k base and the algorithm can reject the instance. Then we check that G is abelian in polynomial time by checking if $gh = hg$ for all $g, h \in S$. Suppose $|G| = p_1^{r_1} p_2^{r_2} \ldots p_\ell^{r_\ell}$ where the p_i are distinct primes. Using known polynomial-time algorithms [19,15] we can compute a decomposition of G into a direct product of cyclic groups:

$$G = \langle g_1 \rangle \times \langle g_2 \rangle \times \ldots \times \langle g_\ell \rangle,$$

where g_i has order $p_i^{r_i}, 1 \leq i \leq \ell$. We write g_i as a product of disjoint cycles. The length of each such cycle is a power of p_i that divides $p_i^{r_i}$, and there is at least one cycle of length $p_i^{r_i}$. Any base for G must include at least one point of a $p_i^{r_i}$-cycle (i.e. cycle of length $p_i^{r_i}$) of g_i. Otherwise, the subgroup $\langle g_i \rangle$ will not become identity when the points in the base are fixed. For each index $i : 1 \leq i \leq \ell$, let $S_i = \{ \alpha \in [n] \mid \alpha$ is in some $p_i^{r_i}$ cycle of $g_i \}$.

Claim. Let $B \subseteq [n]$ be a subset of size k. Then B is a base for G if and only if B is a hitting set for the collection of sets $\{S_1, S_2, \ldots, S_\ell\}$.

Proof of Claim. Clearly, it is a necessary condition. Conversely, suppose $|B| = k$ and $B \cap S_i \neq \emptyset$ for each i. Consider the partition of $[n]$ into the orbits of G:

$$[n] = \Omega_1 \cup \Omega_2 \cup \cdots \cup \Omega_r.$$

For each g_i, a cycle of length $p_i^{r_i}$ in g_i is wholly contained in some orbit of G. Indeed, each orbit of G must be a union of a subset of cycles of g_i. Since $B \cap S_i \neq \emptyset$, some $p_i^{r_i}$-cycle C_i of g_i will intersect B.

Assume, contrary to the claim, that there is a $g \in G_B$ such that $g \neq$ id. We can write $g = g_1^{a_1} g_2^{a_2} \ldots g_\ell^{a_\ell}$ for nonnegative integers $a_i < p_i^{r_i}$. Suppose $g_j^{a_j} \neq$ id. Then raising both sides of the equation $g = g_1^{a_1} g_2^{a_2} \ldots g_\ell^{a_\ell}$ to the power $\frac{|G|}{p_j^{r_j}}$, we have

$$g' = g^{\frac{|G|}{p_j^{r_j}}} = g_j^{\beta_j},$$

where $\beta_j < p_j^{r_j}$. Moreover, $\beta_j = \frac{|G|a_j}{p_j^{r_j}}(mod\ p_j^{r_j})$ is nonzero because $a_j \neq 0(mod\ p_j^{r_j})$ and $|G|/p_j^{r_j}$ does not have p_j as factor.

By assumption, some $p_j^{r_j}$-cycle C_j of g_j intersects B. Since β_j is nonzero and strictly smaller than $p_j^{r_j}$, none of the points of C_j are fixed by $g_j^{\beta_j}$ which contradicts the assumption that g and hence g' is in G_B. This proves the claim.

We now describe the FPT algorithm. First compute $|G|$ using the Schreier-Sims algorithm. If $|G| > n^k$ then there is no base of size k. Hence we can assume $|G| \leq n^k$. If $k \lg k > \lg \lg n$ then $n^k \leq 2^{k^{k+1}}$. Hence the brute-force search algorithm for a base of size k is already an FPT algorithm with running time $2^{k^{k+1}}$.

Hence, we can assume $k \lg k \leq \lg \lg n$. By the above claim, it suffices to solve the k-hitting set problem for a collection of ℓ sets $\{S_1, S_2, \ldots, S_\ell\}$. This is a problem of k-coloring the indices $\{1, 2, \ldots, \ell\}$ such that for each color class I we have $\cap_{i \in I} S_i \neq \emptyset$ and we can pick any one point in the hitting set for each such intersection. Notice that there are at most k^ℓ many colorings. Since

$$(\ell/e)^\ell \leq \ell! \leq p_1 p_2 \cdots p_\ell \leq n^k,$$

it follows that $\ell \lg \ell = O(k \lg n)$ and hence $\lg \ell = O(\lg k + \lg \lg n)$. Since we are considering the case when $k \lg k \leq \lg \lg n$, we have $\lg \ell = O(\lg \lg n)$, which implies $\ell = O(\frac{k \lg n}{\lg \lg n})$.

It follows that the total number of k-colorings $k^\ell = 2^{O(\frac{k \lg k \lg n}{\lg \lg n})} = 2^{O(\lg n)} = n^{O(1)}$. Hence, we can cycle through all these k-colorings in polynomial time and find a good k-coloring if it exists.

3.2   Bounded Orbit Permutation Groups

We give an FPT algorithm for another special case of the k-BASE problem: Let $G = \langle S \rangle \leq S_n$ such that G has orbits of size bounded by a fixed constant b. More precisely, $[n] = \biguplus_{i=1}^{m} \Omega_i$, where Ω_i are the G-orbits and $|\Omega_i| \leq b$ for each i. This is again an interesting special case as the minimum base problem is NP-hard even for orbits of size bounded by 8 [4, Theorem 3.2].

Suppose G has a base $B = \{i_1, i_2, \ldots, i_k\}$ of size k. Then G has a pointwise stabilizer tower $G = G_0 \geq G_1 \geq \ldots \geq G_k = \{1\}$ obtained by successively fixing the points of B. More precisely, G_j is the subgroup of G that pointwise fixes $\{i_1, i_2, \ldots, i_j\}$. Now, $\frac{|G_{j-1}|}{|G_j|}$ is the orbit size of the point i_j in the group G_{j-1}. Furthermore, b is also a bound on this orbit size. Therefore, $|G| \leq b^k$. Hence in $b^k n^{O(1)}$ time we can list all elements of G. Let $G = \{g_1, g_2, \ldots, g_N\}$, where $N \leq b^k$, where g_1 is the identity element.

For each $g_i \in G, i \geq 2$, let $S_i = \{j \in [n] \mid j^{g_i} \neq j\}$ denote the nonempty subset of points not fixed by g_i. Then a subset $B \subset [n]$ of size k is a base for G if and only if B is a hitting set for the collection $S_2, S_3, \ldots, S_N$. The next claim is straightforward.

Claim. There is a size k hitting set contained in $[n]$ for the sets $\{S_2, S_3 \ldots, S_N\}$ if and only if there is a partition of $\{2, 3, \ldots, N\}$ into k parts $I_1, I_2, \ldots, I_k$ such that $\cap_{j \in I_r} S_j \neq \emptyset$ for each $r = 1, 2, \ldots, k$.

As $N \leq b^k$, the total number of k-partitions of $\{2, 3, \ldots, N\}$ is bounded by $k^N \leq k^{b^k}$. We can generate them and check if any one of them yields a hitting set of size k by checking the condition in the above claim. The overall time taken by the algorithm is given by the FPT time bound $k^{b^k} n^{O(1)}$. We have shown the following result.

Theorem 8. Let $G = \langle S \rangle \leq S_n$ such that G has orbits of size bounded by b, be an instance for the k-BASE problem with k as parameter. Then the problem has an FPT algorithm of running time $k^{b^k} n^{O(1)}$.

4   Concluding Remarks

The impact of parameterized complexity on algorithmic graph theory research, especially its interplay with graph minor theory, has been very fruitful in the last two decades. This motivates the study of parameterized complexity questions in other algorithmic problem domains like, for example, group-theoretic computation. To this end, we considered parameterized versions of two well-known classical problems on permutation groups. We believe that a similar study of other permutation group problems can be a worthwhile direction.

Acknowledgments.   I thank the referees for valuable remarks and suggestions.

References

1. Babai, L.: Monte-Carlo algorithms in graph isomorphism testing. Technical Report 79–10, Universit de Montral (1979)
2. Babai, L., Kantor, W.M., Luks, E.M.: Computational complexity and the classification of finite simple groups. In: Proc. 24th IEEE Symposium on Foundations of Computer Science, pp. 162–171 (1983)
3. Babai, L., Luks, E.M.: Canonical labeling of graphs. In: 15th Annual ACM Symposium on Theory of Computing, pp. 171–183 (1983)
4. Blaha, K.D.: Minimal bases for permutation groups: the greedy approximation algorithm. Journal of Algorithms 13, 297–306 (1992)
5. Cameron, P.J.: Permutation Groups. London Mathematical Society, Student Texts 45. Cambridge Univ. Press, Cambridge (1999)
6. Cameron, P.J., Cohen, A.M.: On the number of fixed point free elements of a permutation group. Discrete Mathematics 106/107, 135–138 (1992)
7. Cameron, P.J., Wu, T.: The complexity of the weight problem for permutation and matrix groups. Discrete Mathematics 310, 408–416 (2010)
8. Cameron, P.J.: Lectures on derangements. In: Pretty Structures Conference in Paris (2011), http://www.lix.polytechnique.fr/Labo/Leo.Liberti/pretty-structures/pdf/pcameron-ps11.pdf
9. Downey, R.G., Fellows, M.R.: Parameterized Complexity. Springer (1999)
10. Erdös, P., Selfridge, J.L.: On a combinatorial game. Journal of Combinatorial Theory, Series A 14(3), 298–301 (1973)
11. Flum, J., Grohe, M.: Parameterized Complexity Theory. Texts in Theoretical Computer Science. An EATCS Series. Springer, Berlin (2006)
12. Furst, M., Hopcroft, J.E., Luks, E.M.: Polynomial-time algorithms for permutation groups. Technical report, Cornell University, 10 (1980)
13. Jordan, C.: Recherches sur les substitutions. J. Math. Pures Appl. (Liouille) 17, 351–387 (1872)
14. Lubiw, A.: Some NP-Complete Problems Similar to Graph Isomorphism. SIAM J. Comput. 10(1), 11–21 (1981)
15. Luks, E.M.: Permutation groups and polynomial-time computation. In: Finkelstein, L., Kantor, W.M. (eds.) Groups and Computation. DIMACS series, vol. 11, pp. 139–175. AMS (1993)
16. Raghavan, P.: Probabilistic construction of deterministic algorithms: approximating packing integer programs. Journal of Computer and System Sciences 37(2), 130–143 (1988)
17. Seress, A.: Permutation Group Algorithms. Cambridge University Press (2003)
18. Serre, J.P.: On a theorem of Jordan. Bull. Amer. Math. Soc. 40, 429–440 (2003)
19. Sims, C.C.: Computational methods in the study of permutation groups. In: Leech, J. (ed.) Proc. Conf. on Computational Problems in Abstract Algebra, pp. 169–183. Pergamon Press, Oxford (1967, 1970)
20. Sims, C.C.: Some group theoretic algorithms. In: Newman, M.F., Richardson, J.S. (eds.) Topics in Algebra. Lecture Notes in Mathematics, vol. 697, pp. 108–124. Springer, Heidelberg (1978)
21. Zemlyachenko, V.N., Kornienko, N.M., Tyshkevich, R.I.: Graph isomorphism problem. Zapiski Nauchnykh Seminarov Leningradskogo Otdeleniya Matematicheskogo Instituta 118, 83–158 (1982)

Parameterized Complexity of Two Edge Contraction Problems with Degree Constraints*

Rémy Belmonte[1], Petr A. Golovach[1], Pim van 't Hof[1], and Daniël Paulusma[2]

[1] Department of Informatics, University of Bergen, Norway
{remy.belmonte,petr.golovach,pim.vanthof}@ii.uib.no
[2] School of Engineering and Computing Sciences, Durham University, UK
daniel.paulusma@durham.ac.uk

Abstract. Motivated by recent results of Mathieson and Szeider (J. Comput. Syst. Sci. 78(1): 179–191, 2012), we study two graph modification problems where the goal is to obtain a graph whose vertices satisfy certain degree constraints. The REGULAR CONTRACTION problem takes as input a graph G and two integers d and k, and the task is to decide whether G can be modified into a d-regular graph using at most k edge contractions. The BOUNDED DEGREE CONTRACTION problem is defined similarly, but here the objective is to modify G into a graph with maximum degree at most d. We observe that both problems are fixed-parameter tractable when parameterized jointly by k and d. We show that when only k is chosen as the parameter, REGULAR CONTRACTION becomes W[1]-hard, while BOUNDED DEGREE CONTRACTION becomes W[2]-hard even when restricted to split graphs. We also prove both problems to be NP-complete for any fixed $d \geq 2$. On the positive side, we show that the problem of deciding whether a graph can be modified into a cycle using at most k edge contractions, which is equivalent to REGULAR CONTRACTION when $d = 2$, admits an $O(k)$ vertex kernel. This complements recent results stating that the same holds when the target is a path, but that the problem admits no polynomial kernel when the target is a tree, unless NP $\subseteq$ coNP/poly (Heggernes et al., IPEC 2011).

1 Introduction

Graph modification problems play an important role in algorithmic graph theory due to the fact that they naturally appear in numerous practical and theoretical settings. Typically, a graph modification problem takes as input a graph G and an integer k, and the task is to decide whether a graph with certain desirable structural properties can be obtained from G by applying at most k graph operations, such as vertex deletions, edge deletions, edge additions, or a combination of these. The problems VERTEX COVER, FEEDBACK VERTEX SET, MINIMUM FILL-IN and CLUSTER EDITING are just a few famous examples of problems

---

* This work has been supported by the Research Council of Norway (197548/F20), the European Research Council (267959), EPSRC (EP/G043434/1) and the Royal Society (JP100692).

G. Gutin and S. Szeider (Eds.): IPEC 2013, LNCS 8246, pp. 16–27, 2013.

that fall into this framework. Graph modification problems have received a huge amount of interest in the literature for many decades, and due to the fact that the vast majority of such problems turn out to be NP-hard [11, 15], the area has also been intensively studied from a parameterized complexity point of view.

Moser and Thilikos [14] studied the parameterized complexity of the problem of deciding, given a graph G and an integer k, whether there is a subset of at most k vertices in G whose deletion yields an r-regular graph, where r is a fixed constant. They showed that, for every value of r, this problem is fixed-parameter tractable when parameterized by k, and admits a kernel of size $O(kr(k+r)^2)$. On the other hand, they showed that the problem becomes W[1]-hard for every fixed $r \geq 0$ with respect to the dual parameter $|V(G)| - k$. Mathieson and Szeider [13] showed that the aforementioned positive result by Moser and Thilikos crucially depends on the fact that r is a fixed constant, as they proved the problem to be W[1]-hard when r is given as part of the input. This result by Mathieson and Szeider is a particular case of a much more general result in [13] on graph modification problems involving degree constraints. We refer to [13] for more details, and only mention here that the Classification Theorem in [13] shows that the behavior of the investigated graph modification problems heavily depends on the graph operations that are allowed.

Motivated by the results of Moser and Thilikos [14] and Mathieson and Szeider [13], we study the parameterized complexity of two graph modification problems involving degree constraints when *edge contraction* is the only allowed operation. The parameterized study of graph modification problems with respect to this operation has only recently been initiated, but has already proved to be very fruitful [7–10]. In general, for every graph class $\mathcal{H}$, the $\mathcal{H}$-CONTRACTION problem takes as input a graph G and an integer k, and asks whether there exists a graph $H \in \mathcal{H}$ such that G is k-*contractible* to $\mathcal{H}$, i.e., such that H can be obtained from G by contracting at most k edges. A general result by Asano and Hirata [1] shows that this problem is NP-complete for many natural graph classes $\mathcal{H}$. On the positive side, when parameterized by k, the problem is known to be fixed-parameter tractable when $\mathcal{H}$ is the class of paths or trees [9], bipartite graphs [10, 12], or planar graphs [7]. Interestingly, the problem admits a linear vertex kernel when $\mathcal{H}$ is the class of paths, but does not admit a polynomial kernel when $\mathcal{H}$ is the class of trees, unless NP $\subseteq$ coNP/poly [9].

Before we formally define the two problems studied in this paper and state our results, let us mention one more recent paper that formed a direct motivation for this paper. For any integer $d \geq 1$, let $\mathcal{H}_{\geq d}$ denote the class of graphs with minimum degree at least d. Golovach et al. [8] studied the DEGREE CONTRACTIBILITY problem, which takes as input a graph G and two integers d and k, and asks whether there exists a graph $H \in \mathcal{H}_{\geq d}$ such that G is k-contractible to H. They proved that this problem is fixed-parameter tractable when parameterized jointly by d and k, but becomes W[1]-hard when only k is the parameter. They also showed that the problem is para-NP-complete when parameterized by d by proving the problem to be NP-complete for every fixed value of $d \geq 14$. These results by Golovach et al. [8] raise the question what happens to the complexity of the problem when the objective is not to increase the minimum degree of the input graph, but to decrease the maximum degree instead.

Our Contribution. For any integer $d \geq 1$, let $\mathcal{H}_{\leq d}$ denote the class of graphs that have maximum degree at most d, and let $\mathcal{H}_{=d}$ denote the class of d-regular graphs. In this paper, we study the complexity of different parameterizations of the following two decision problems:

BOUNDED DEGREE CONTRACTION
Instance: A graph G and two integers d and k.
Question: Is there a graph $H \in \mathcal{H}_{\leq d}$ such that G is k-contractible to H?

REGULAR CONTRACTION
Instance: A graph G and two integers d and k.
Question: Is there a graph $H \in \mathcal{H}_{=d}$ such that G is k-contractible to H?

Throughout the paper, we will use n and m to denote the number of vertices and edges, respectively, of the input graph G. Moreover, since edge contractions leave the number of connected components of a graph unchanged, we assume throughout the paper that the input graph G in each of our problems is connected.

In Section 2, we first observe that both problems can be solved in $O((d + k)^{2k} \cdot (n + m))$ time using a simple branching algorithm. This implies that both problems are fixed-parameter tractable when parameterized jointly by d and k, and that both problems are in XP when parameterized by k only. This naturally raises the following two questions:

1. Are the two problems fixed-parameter tractable when parameterized by k?
2. Are the two problems in XP when parameterized by d?

In the remainder of Section 2, we provide strong evidence that the answer to both these questions is "no". We first show that REGULAR CONTRACTION is W[1]-hard when parameterized by k, before proving that BOUNDED DEGREE CONTRACTION is W[2]-hard with the same parameter, even when restricted to the class of split graphs. This implies that neither of the two problems is in FPT, assuming that FPT $\neq$ W[1] and FPT $\neq$ W[2], respectively. The negative answer to question 2, this time under the assumption that P $\neq$ NP, is given in Theorem 3, where we show that both problems are NP-complete for every fixed value of $d \geq 2$, and hence para-NP-complete when parameterized by d. Note that both problems are trivially solvable in polynomial time when $d = 1$. The results of Section 2 are summarized in Table 1.

To complement our hardness results, we show in Section 3 that REGULAR CONTRACTION admits a kernel with at most $6k + 6$ vertices when $d = 2$. Equivalently, we show that the $\mathcal{H}$-CONTRACTION problem admits a linear vertex kernel when $\mathcal{H}$ is the class of cycles. We point out that this problem is para-NP-complete with respect to the dual parameter $|V(G)| - k$, i.e., when parameterized by the length of the obtained cycle, since the problem of deciding whether or not a graph can be contracted to the cycle C_ℓ is NP-complete for every fixed $\ell \geq 4$ [2].

Table 1. An overview of the results presented in Section 2

Parameter	REGULAR CONTRACTION	BOUNDED DEGREE CONTRACTION
d, k	FPT	FPT
k	W[1]-hard	W[2]-hard on split graphs
d	para-NP-complete	para-NP-complete

Our kernelization result complements the aforementioned known results stating that $\mathcal{H}$-CONTRACTION admits a linear vertex kernel when $\mathcal{H}$ is the class of paths, but admits no polynomial kernel when $\mathcal{H}$ is the class of trees, unless NP $\subseteq$ coNP/poly [9].

Preliminaries. All graphs considered in this paper are finite, undirected and simple. We refer to the textbook by Diestel [4] for graph terminology and notation not defined below. For a thorough background on parameterized complexity, we refer to the monographs by Downey and Fellows [5].

Let $G = (V, E)$ be a graph and let U be a subset of V. We write $G[U]$ to denote the subgraph of G induced by U. We write $G - U = G[V \setminus U]$, or simply $G - u$ if $U = \{u\}$. We say that two disjoint subsets $U \subseteq V$ and $W \subseteq V$ are *adjacent* if there exist two vertices $u \in U$ and $w \in W$ such that $uw \in E$. The *contraction* of edge uv in G removes u and v from G, and replaces them by a new vertex made adjacent to precisely those vertices that were adjacent to u or v in G. Instead of speaking of the contraction of edge uv, we sometimes say that a vertex u is *contracted onto* v, in which case we use v to denote the new vertex resulting from the contraction. For a set $S \subseteq E$, we write G/S to denote the graph obtained from G by repeatedly contracting an edge from S until no such edge remains. Note that, by definition, edge contractions create neither self-loops nor multiple edges.

Let H be a graph. We say that H is a *contraction* of G if H can be obtained from G by a sequence of edge contractions. We say that G is *k-contractible* to H if H can be obtained from G by at most k edge contractions. An *H-witness structure* $\mathcal{W}$ is a partition of $V(G)$ into $|V(H)|$ nonempty sets $W(x)$, one for each $x \in V(H)$, called *H-witness sets*, such that each $W(x)$ induces a connected subgraph of G, and for all $x, y \in V(H)$ with $x \neq y$, the sets $W(x)$ and $W(y)$ are adjacent in G if and only if x and y are adjacent in H. Clearly, H is a contraction of G if and only if G has an H-witness structure; H can be obtained by contracting each witness set into a single vertex.

2 Contracting to Graphs with Degree Constraints

We start by observing that the problems BOUNDED DEGREE CONTRACTION and REGULAR CONTRACTION are FPT when parameterized jointly by k and d.

Theorem 1. *The problems* BOUNDED DEGREE CONTRACTION *and* REGULAR CONTRACTION *can be solved in time* $O((d + k)^{2k} \cdot (n + m))$.

Proof. We first present an algorithm for BOUNDED DEGREE CONTRACTION, and then describe how it can be modified to solve REGULAR CONTRACTION in the same running time.

Let (G, d, k) be an instance of BOUNDED DEGREE CONTRACTION. We first check if G has a vertex of degree at least $d + k + 1$. If so, then (G, d, k) is a trivial no-instance, since the contraction of any edge in G decreases the degree of each vertex in G by at most 1. Hence we output "no" in this case. Suppose every vertex in G has degree at most $d + k$, but G has a vertex v such that $d_G(v) \geq d + 1$. In order to contract G to a graph of maximum degree at most d, we must either contract v onto one of its neighbors, or contract all the edges of a path between two of the neighbors of v. In either case, we must contract an edge e incident with a neighbor of v. Since $\Delta(G) \leq d + k$, there are at most $(d + k)^2$ such edges e. We branch on each of them, calling our algorithm recursively for $G' = G/e$ with parameter $k' = k - 1$. Since the parameter decreases by 1 at every step, this branching algorithm runs in time $O((d + k)^{2k} \cdot (n + m))$.

We can also obtain an algorithm for REGULAR CONTRACTION with same running time by replacing the branching rule with the following one: if there is a vertex v with $d_G(v) \neq d$, then we branch over all the edges e that are incident with a vertex in $N_G(v)$. For each branch, we contract the edge e and decrease k by 1. The correctness of this branching rule follows from the observation that if we contract any edge e' that is not incident with a neighbor of v, then the degree of v before and after the contraction is the same. □

We now show that REGULAR CONTRACTION becomes W[1]-hard when only k is chosen as the parameter. In the proof of Theorem 2 below, we will reduce from the following problem:

REGULAR MULTICOLORED CLIQUE
Instance: A regular graph G, an integer k, and a partition $X_1, \ldots, X_k$ of
    $V(G)$ into k independent sets of size p each.
Question: Does G have a clique $K \subseteq V(G)$ such that $|K \cap X_i| = 1$
    for every $i \in \{1, \ldots, k\}$?

It is well-known that the CLIQUE problem, asking whether a given graph has a clique of size k, is W[1]-hard when parameterized by k [5]. Cai [3] proved that this remains true on regular graphs. Using this fact and the standard parameterized reduction from CLIQUE to MULTICOLORED CLIQUE due to Fellows et al. [6], we obtain the following result.

Lemma 1. (★)[1] *The* REGULAR MULTICOLORED CLIQUE *problem is* W[1]-*hard when parameterized by* k *for* d-*regular graphs when* $k < d < p$.

We now use the above lemma to prove our first hardness result.

Theorem 2. *The* REGULAR CONTRACTION *problem is* W[1]-*hard when parameterized by* k.

---

[1] Proofs marked with a star have been omitted due to page restrictions.

Proof. We reduce from the restricted version of the REGULAR MULTICOLORED CLIQUE problem described in Lemma 1. Let $(G, k, X_1, \ldots, X_k)$ be an instance of this problem where G is a d-regular graph, $p = |X_1| = \ldots = |X_k|$, and $k < d < p$. We construct an instance (G', d', k) of REGULAR CONTRACTION as follows:

- construct a copy of G with the corresponding partition $X_1, \ldots, X_k$ of the vertex set;
- for each $i \in \{1, \ldots, k\}$, construct a vertex x_i and then make the set $X_i \cup \{x_i\}$ into a clique by adding edges;
- make the set $\{x_1, \ldots, x_k\}$ into a clique by adding edges.

Let G' denote the obtained graph, and let $d' = d + p - 1$.

Suppose that G has a clique $K = \{y_1, \ldots, y_k\}$ such that $y_i \in X_i$ for $i \in \{1, \ldots, k\}$. It is straightforward to verify that contracting the edges $x_i y_i$ for $i \in \{1, \ldots, k\}$ in G' results in a d'-regular graph.

Assume now that (G', d', k) is a YES-instance of REGULAR CONTRACTION, i.e., there is a set S of at most k edges such that G'/S is a d'-regular graph. Notice that each x_i in G' has degree $p + k - 1 < p + d - 1 = d'$. Therefore, for each $i \in \{1, \ldots, k\}$, S contains at least one edge incident to x_i. Suppose that S contains an edge $x_i x_j$ for $1 \leq i < j \leq k$. Let G'' be the graph obtained from G' by the contraction of $x_i x_j$, and denote by z the vertex obtained from x_i, x_j. The degree of z in G'' is $2p + k - 2 > p + d + k - 2 = d' + k - 1$. It means that we have to contract at least k edges to obtain a vertex of degree d' from z. It contradicts the assumption that $|S| \leq k$. Hence, for each $i \in \{1, \ldots, k\}$, S contains an edge $x_i y_i$ for $y_i \in X_i$. Since $|S| \leq k$, $S = \{x_1 y_1, \ldots, x_k y_k\}$. We claim that $\{y_1, \ldots, y_k\}$ is a clique in G. To see this, assume that some y_i, y_j are not adjacent in G. Then y_i, y_j are not adjacent in G' but are adjacent in G'/S, and the degree of the vertex obtained from x_i and y_i in G'/S is at least $d + p > d'$. This contradiction to the assumption that G'/S is d'-regular completes the proof of Theorem 2. □

We expect that the arguments in the proof of Theorem 2 can also be used to show that BOUNDED DEGREE CONTRACTION is W[1]-hard when parameterized by k. However, we obtain a stronger result below by proving that BOUNDED DEGREE CONTRACTION is W[2]-hard when parameterized by k, even when restricted to split graphs. This result will be a corollary of the following lemma.

Lemma 2. *The problem of deciding whether the maximum degree of a split graph can be reduced by at least 1 using at most k edge contractions is W[2]-hard when parameterized by k.*

Proof. We give a reduction from the problem RED-BLUE DOMINATING SET, which takes as input a bipartite graph $G = (R \cup B, E)$ and an integer k, and asks whether there exists a *red-blue dominating set* of size at most k, i.e., a subset $D \subseteq B$ of at most k vertices such that every vertex in R has at least one neighbor in D. This problem, which is equivalent to SET COVER and HITTING SET, is well-known to be W[2]-complete when parameterized by k [5].

Let (G, k) be an instance of RED-BLUE DOMINATING SET, where $G = (R \cup B, E)$ is a bipartite graph with partition classes R and B. We assume that every vertex in G has degree at least 1. We create a split graph G' from G by making the vertices of R pairwise adjacent, and by adding, for each vertex $u \in R$, $(\Delta(G) - d_G(u)) + k + 2$ new vertices that are made adjacent to u only. Let $B' = V(G') \setminus (R \cup B)$ be the set of all vertices of degree 1 that were added to G this way. Clearly, G' is a split graph, since its vertex set can be partitioned into the clique R and the independent set $B \cup B'$. Observe that each vertex in R has degree $\Delta := \Delta(G') > k + 2$.

We claim that G has a red-blue dominating set of size k if and only if G' can be contracted to a split graph of maximum degree at most $\Delta - 1$ using at most k edge contractions.

First, suppose there is a red-blue dominating set $D \subseteq B$ such that $|D| \leq k$. For every $v \in D$, we choose an arbitrary neighbor w of v in R and contract v onto w. Note that contracting v onto w is equivalent to deleting v from the graph due to the fact that $N_{G'}[v] \subseteq N_{G'}[w]$. Since every vertex in R is adjacent to at least one vertex in D, these $|D| \leq k$ edge contractions decrease the degree of every vertex in R by at least 1. Since the degree of each vertex in $B \cup B'$ in G' was already at most $\Delta(G) \leq \Delta - 1$, the obtained graph has maximum degree at most $\Delta - 1$.

For the reverse direction, suppose there exists a set $S \subseteq E(G')$ of at most k edges such that G'/S has maximum degree at most $\Delta - 1$. We claim that S does not contain any edge whose endpoints both belong to R. To see this, observe that contracting any such edge would create a vertex of degree at least $\Delta + k + 1$, and the degree of such a vertex cannot be decreased to Δ by contracting at most $k - 1$ other edges. Suppose S contains an edge uv such that $u \in R$ and $v \in B'$, and let w be an arbitrary neighbor of u in B. Note that contracting v onto u decreases the degree of u by 1 but leaves the degrees of all other vertices in R unchanged, whereas contracting w onto u decreases the degree of every neighbor of w in R, and of u in particular. Hence we may assume, without loss of generality, that every edge in S is incident with one vertex of R and one vertex of B'. Since the degree of every vertex in R decreases by at least 1 when we contract the edges in S, every vertex in R must be incident with at least one edge in S. This implies that $D := V(S) \cap B$ is a red-blue dominating set of G, where $V(S)$ denotes the set of endpoints of the edges in S. The observation that $|D| \leq |S| \leq k$ completes the proof. $\square$

Since an instance (G, k) of the problem defined in Lemma 2 is a yes-instance if and only if $(G, \Delta(G) - 1, k)$ is a yes-instance of BOUNDED DEGREE CONTRACTION, we immediately obtain the following result.

Corollary 1. *The* BOUNDED DEGREE CONTRACTION *problem is* W[2]-*hard on split graphs when parameterized by* k.

To conclude this section, we also consider the complexity of our two problems when we take only d to be the parameter. The following result shows that both our problems are para-NP-complete with respect to this parameter. Note that both problems can trivially be solved in polynomial time when $d = 1$.

Theorem 3. (★) *The problems* REGULAR CONTRACTION *and* BOUNDED DE-GREE CONTRACTION *are* NP-*complete for any fixed* $d \geq 2$.

3 A Linear Vertex Kernel

In this section, we show that the problem REGULAR CONTRACTION admits a kernel with at most $6k + 6$ vertices in case $d = 2$. Since this problem is equivalent to the $\mathcal{H}$-CONTRACTION problem when $\mathcal{H}$ is the class of cycles, we will refer to the problem as CYCLE CONTRACTION throughout this section.

We first introduce some additional terminology. Let G and H be two graphs, and suppose that there exists an H-witness structure $\mathcal{W}$ of G. If a witness set of $\mathcal{W}$ contains more than one vertex of G, then we call it a *big* witness set; a witness set consisting of a single vertex of G is called *small*.

Observation 1 ([9]) *If a graph G is k-contractible to a graph H, then any H-witness structure $\mathcal{W}$ of G satisfies the following three properties:*

- *no witness set of $\mathcal{W}$ contains more than $k + 1$ vertices;*
- *$\mathcal{W}$ has at most k big witness sets;*
- *all the big witness sets of $\mathcal{W}$ together contain at most $2k$ vertices.*

Let G be a graph. A cycle C is *optimal for G* if G can be contracted to C but cannot be contracted to any cycle longer than C. Note that if G is a connected graph that is not a tree, then an optimal cycle for G always exists. The following structural lemma will be used in the correctness proof of our kernelization algorithm.

Lemma 3. *Let (G, k) be a yes-instance of CYCLE CONTRACTION, let C be an optimal cycle for G, and let $\mathcal{W}$ be a C-witness structure of G. If G is 2-connected and G contains two vertices u and v such that $d_G(u) = d_G(v) = 2$ and $G - \{u, v\}$ has exactly two connected components G_1 and G_2, then the following three statements hold:*

- *(i) either $\{u\}$ and $\{v\}$ are small witness sets of $\mathcal{W}$, or u and v belong to the same big witness set of $\mathcal{W}$;*
- *(ii) if u and v belong to the same big witness set $W \in \mathcal{W}$, then W contains all the vertices of G_1 or all the vertices of G_2;*
- *(iii) if G_1 and G_2 contain at least $k + 1$ vertices each, then $\{u\}$ and $\{v\}$ are small witness sets of $\mathcal{W}$.*

Proof. Suppose G is 2-connected and contains two vertices u and v such that $d_G(u) = d_G(v) = 2$ and $G - \{u, v\}$ has exactly two connected components G_1 and G_2. Let p and q denote the two neighbors of u, and let x and y denote the two neighbors of v. Without loss of generality, suppose $p, x \in V(G_1)$ and $q, y \in V(G_2)$.

To prove statement (i), suppose, for contradiction, that u belongs to a big witness set $W \in \mathcal{W}$ and $v \notin W$. Let $W_1 = (W \setminus \{u\}) \cap V(G_1)$ and $W_2 = (W \setminus \{u\}) \cap V(G_2)$. Since u has degree 2 in G and $G[W]$ is connected by the definition of a witness set, the graphs $G[W_1]$ and $G[W_2]$ are connected. Moreover, by the definition of G_1 and G_2, there is no edge between W_1 and W_2 in G. Let $\mathcal{W}'$ be the C'-witness structure of G obtained from $\mathcal{W}$ by replacing W with the sets W_1, $\{u\}$, and W_2. Then C' is a cycle that has two more vertices than C. This contradicts the assumption that C is an optimal cycle for G.

We now prove statement (ii). Suppose u and v both belong to the same witness set $W \in \mathcal{W}$. Note that $V(G) \setminus W$ induces a connected subgraph of G, and assume, without loss of generality, that $(V(G) \setminus W) \subseteq V(G_1)$. Then we must have $V(G_2) \subseteq W$.

To prove statement (iii), suppose $|V(G_1)| \geq k + 1$ and $|V(G_2)| \geq k + 1$. Suppose, for contradiction, that u and v belong to the same big witness set of $\mathcal{W}$. Then W contains all the vertices of either G_1 or G_2 by statement (ii). This implies that W contains at least $k + 3$ vertices, contradicting the fact that every big witness set of $\mathcal{W}$ contains at most $k + 1$ vertices due to Observation 1.   □

We now describe four reduction rules that will be used in our kernelization algorithm for CYCLE CONTRACTION. Each of the reduction rules below takes as input an instance (G, k) of CYCLE CONTRACTION and outputs a reduced instance (G', k') of the same problem, and the rule is said to be *safe* if the two instances (G, k) and (G', k') are either both yes-instances or both no-instances.

Rule 1 *If G is 3-connected and $|V(G)| \geq 2k + 4$, then return a trivial no-instance.*

Lemma 4. *Rule 1 is safe.*

Proof. Let G be a 3-connected graph on at least $2k+4$ vertices. We show that G is not k-contractible to a cycle. For contradiction, suppose G is k-contractible to a cycle C. Let $\mathcal{W}$ be a C-witness structure. Then $\mathcal{W}$ has at most three small witness sets, as otherwise for any two small witness sets $\{u\}$ and $\{v\}$ such that u and v are non-adjacent, the graph $G - \{u, v\}$ would be disconnected, contradicting the assumption that G is 3-connected. Since all the big witness sets of $\mathcal{W}$ contain at most $2k$ vertices in total due to Observation 1, this implies that $|V(G)| \leq 2k+3$. This yields the desired contradiction to the assumption that $|V(G)| \geq 2k+4$.   □

Rule 2 *If G contains a block B on at least $k + 2$ vertices and $V(G) \setminus V(B) \neq \emptyset$, then return a trivial no-instance if $|V(G) \setminus V(B)| \geq k+1$, and return the instance $(G', k - |V(G) \setminus V(B)|)$ otherwise, where G' is the graph obtained from G by exhaustively contracting a vertex of $V(G) \setminus V(B)$ onto one of its neighbors.*

Lemma 5. *Rule 2 is safe.*

Proof. Suppose G contains a block B on at least $k+2$ vertices and $V(G) \setminus V(B) \neq \emptyset$. Then G is not 2-connected and contains at least two blocks.

Suppose (G, k) is a yes-instance, and let $\mathcal{W}$ be C-witness structure of G, where C is a cycle to which G is k-contractible. Since $|V(B)| \geq k + 2$, there must be at least two witness sets of $\mathcal{W}$ that contain vertices of B due to Observation 1. This implies that for every block $B' \neq B$ of G, all the vertices of B' must be contained in one witness set of $\mathcal{W}$, as otherwise there would be two vertex-disjoint paths in G between vertices of B and B'. For the same reason, every witness set of $\mathcal{W}$ contains at least one vertex of B. Consequently, no vertex of $|V(G) \setminus V(B)|$ appears in a small witness set of $\mathcal{W}$.

The above arguments, together with Observation 1, imply that (G, k) is a no-instance if $|V(G) \setminus V(B)| \geq k + 1$. It also implies that the instances (G, k) and $(G', k - |V(G) \setminus V(B)|)$ are equivalent otherwise.   □

Rule 3 *If G contains a block B on at most $k + 1$ vertices and $|V(G) \setminus V(B)| \geq k + 2$, then return the instance $(G', k - |V(B)|)$, where G' is the graph obtained from G by exhaustively contracting a vertex of $V(B)$ onto one of its neighbors.*

Lemma 6. *Rule 3 is safe.*

Proof. Let (G, k) be an instance of CYCLE CONTRACTION, and suppose G has a block B on at most $k + 1$ vertices such that $|V(G) \setminus V(B)| \geq k + 2$. Suppose (G, k) is a yes-instance, and let $\mathcal{W}$ be a C-witness structure of G for some cycle C to which G is k-contractible. Using arguments similar to the ones in the proof of Lemma 5, it can be seen that all the vertices of B must be contained in a single witness set of $\mathcal{W}$, and this witness set contains at least one vertex of $V(G) \setminus V(B)$. This shows that the instances $(G', k - |V(B)|)$ and (G, k) are equivalent.   □

Rule 4 *If G is 2-connected and G contains two vertices u and v such that $d_G(u) = d_G(v) = 2$, the two neighbors p and q of u both have degree 2 in G, and the graph $G - \{u, v\}$ has exactly two connected components that contain at least $k + 2$ vertices each, then return the instance (G', k), where G' is the graph obtained from G by contracting u onto p.*

Lemma 7. *Rule 4 is safe.*

Proof. Let (G, k) be an instance of CYCLE CONTRACTION on which Rule 4 can be applied. Suppose (G, k) is a yes-instance. Let C be an optimal cycle for G, and let $\mathcal{W}$ be a C-witness structure of G. Due to statement (iii) in Lemma 3, $\{u\}$ and $\{v\}$ are small witness sets of $\mathcal{W}$. Then $\mathcal{W}' = \mathcal{W} \setminus \{u\}$ is a C'-witness structure of G', where C' is a cycle containing one less vertex than C. Since the big witness sets of $\mathcal{W}'$ and $\mathcal{W}$ coincide, G' is k-contractible to C'. Hence (G', k) is a yes-instance of CYCLE CONTRACTION.

For the reverse direction, suppose (G', k) is a yes-instance. Let C' be an optimal cycle for G', and let $\mathcal{W}'$ be a C'-witness structure of G'. Consider the vertices p and v in G'. Note that $d_{G'}(p) = d_{G'}(v) = 2$, and that $G' - \{p, v\}$ has exactly two connected components G'_1 and G'_2 that contain at least $k + 1$ vertices each. Hence $\{p\}$ and $\{v\}$ are small witness sets of $\mathcal{W}'$ due to statement (iii) in Lemma 3. For similar reasons, considering the pair (q, v) instead of (p, v), we find that $\{q\}$ is a small witness set of $\mathcal{W}'$. In particular, p and q are in separate small witness sets of $\mathcal{W}'$. Now let $\mathcal{W}$ be the partition of $V(G)$ obtained from $\mathcal{W}'$ by adding the set $\{u\}$. Then $\mathcal{W}$ clearly is a C-witness structure of G, where C is a cycle that has one more vertex than C'. Since the big witness sets of $\mathcal{W}$ and $\mathcal{W}'$ coincide, we conclude that G is k-contractible to C, and hence (G, k) is a yes-instance of CYCLE CONTRACTION. □

Theorem 4. *The* CYCLE CONTRACTION *problem admits a kernel with at most* $6k + 6$ *vertices.*

Proof. We describe a kernelization algorithm for CYCLE CONTRACTION. Given an instance of CYCLE CONTRACTION, the algorithm starts by exhaustively applying the four reduction rules defined above. Let (G, k) be the obtained instance. If G is 3-connected, then $|V(G)| \leq 2k + 3$, as otherwise Rule 1 could be applied. Suppose G is not 2-connected. Since G is connected by assumption, G has at least two blocks. Let B be any block of G. Then $|V(B)| \leq k + 1$, as otherwise Rule 2 could be applied. Moreover, $|V(G) \setminus V(B)| \leq k + 1$ due to the assumption that Rule 3 cannot be applied. Hence $|V(G)| \leq 2k + 2$. Now suppose G is 2-connected. We then apply a final reduction rule: if $|V(G)| \geq 6k + 7$, then return a trivial no-instance. Before showing why this final reduction rule is safe, let us point out that after the application of this final reduction rule, we have obtained an instance (G', k') such that G' has at most $6k + 6$ vertices.

To see why the final reduction rule is safe, suppose, for contradiction, that (G, k) is a yes-instance of CYCLE CONTRACTION such that G is a 2-connected graph on at least $6k + 7$ vertices. Let C be an optimal cycle for G, and let $\mathcal{W}$ be a C-witness structure of G. By Observation 1, at most $2k$ vertices of G belong to big witness sets, which implies that at least $4k + 7$ vertices of G belong to small witness sets. Since $\mathcal{W}$ has at most k big witness sets by Observation 1, there are at most $2k$ vertices in small witness sets that have degree more than 2 in G, namely the ones adjacent to big witness sets. Consequently, there are at least $2k + 7$ vertices in small witness sets that have degree exactly 2, and there must be three small witness sets $\{p\}, \{u\}, \{q\}$ such that $d_G(p) = d_G(u) = d_G(q) = 2$ and p and q are the two neighbors of u. Let $\{v\}$ be a small witness set of $\mathcal{W}$ such that $v \notin \{p, u, q\}$ and $d_G(v) = 2$, and such that the two connected components G_1 and G_2 of the graph $G - \{u, v\}$ contain at least $k + 2$ small witness sets of $\mathcal{W}$ each. Since, apart from the vertices p, u, q and v, there are at least $2k + 3$ other vertices that have degree 2 in G and belong to small witness sets, such a set $\{v\}$ exists. This implies that Rule 4 could have been applied on (G, k), yielding the desired contradiction.

The correctness of our algorithm follows directly from Lemmas 4–7 and from the above proof that the final reduction rule is safe. It remains to argue that our kernelization algorithm runs in polynomial time. It is clear that every reduction rule can be applied in polynomial time. When applying any of the reduction rules, either the number of vertices in the graph or the parameter strictly decreases. This implies that we only apply the reduction rules a polynomial number of times, so the algorithm runs in polynomial time. □

4   Concluding Remarks

We showed that REGULAR CONTRACTION has a linear vertex kernel when $d = 2$. We expect that we can use similar arguments to obtain the same result for BOUNDED DEGREE CONTRACTION when $d = 2$. A more interesting question is whether both problems admit polynomial kernels when $d = 3$.

Acknowledgments. We would like to thank Marcin Kamiński and Dimitrios M. Thilikos for fruitful discussions on the topic. We also thank the three anonymous referees for insightful comments.

References

1. Asano, T., Hirata, T.: Edge-contraction problems. J. Comput. Syst. Sci. 26(2), 197–208 (1983)
2. Brouwer, A.E., Veldman, H.J.: Contractibility and NP-completeness. J. Graph Theory 11, 71–79 (1987)
3. Cai, L.: Parameterized complexity of cardinality constrained optimization problems. The Computer Journal 51(1), 102–121 (2008)
4. Diestel, R.: Graph Theory, Electronic Edition. Springer-Verlag (2005)
5. Downey, R.G., Fellows, M.R.: Parameterized Complexity. Springer (1999)
6. Fellows, M.R., Hermelin, D., Rosamond, F., Vialette, S.: On the parameterized complexity of multiple-interval problems. Theor. Comp. Sci. 410, 53–61 (2009)
7. Golovach, P.A., van 't Hof, P., Paulusma, D.: Obtaining planarity by contracting few edges. Theor. Comp. Sci. 476, 38–46 (2013)
8. Golovach, P.A., Kamiński, M., Paulusma, D., Thilikos, D.M.: Increasing the minimum degree of a graph by contractions. Theor. Comp. Sci. 481, 74–84 (2013)
9. Heggernes, P., van 't Hof, P., Lévêque, B., Lokshtanov, D., Paul, C.: Contracting graphs to paths and trees. Algorithmica (to appear) doi:10.1007/s00453-012-9670-2
10. Heggernes, P., van 't Hof, P., Lokshtanov, D., Paul, C.: Obtaining a bipartite graph by contracting few edges. In: FSTTCS 2011, LIPIcs, vol. 13, pp. 217–228 (2011)
11. Lewis, J.M., Yannakakis, M.: The node-deletion problem for hereditary properties is NP-complete. J. Comp. System Sci. 20, 219–230 (1980)
12. Marx, D., O'Sullivan, B., Razgon, I.: Finding small separators in linear time via treewidth reduction. In: ACM Trans. Algorithms (to appear), Manuscript available at http://www.cs.bme.hu/~{}dmarx/papers/marx-tw-reduction-talg.pdf
13. Mathieson, L., Szeider, S.: Editing graphs to satisfy degree constraints: A parameterized approach. J. Comput. Syst. Sci. 78, 179–191 (2012)
14. Moser, H., Thilikos, D.M.: Parameterized complexity of finding regular induced subgraphs. J. Discr. Algorithms 7, 181–190 (2009)
15. Yannakakis, M.: Edge-deletion problems. SIAM J. Comput. 10(2), 297–309 (1981)

Declarative Dynamic Programming
as an Alternative Realization of Courcelle's Theorem

Bernhard Bliem, Reinhard Pichler, and Stefan Woltran

Institute of Information Systems, Vienna University of Technology
{bliem,pichler,woltran}@dbai.tuwien.ac.at

Abstract. Many computationally hard problems become tractable if the graph structure underlying the problem instance exhibits small treewidth. A recent approach to put this idea into practice is based on a declarative interface to specify dynamic programming over tree decompositions, delegating the computation to dedicated solvers. In this paper, we prove that this method can be applied to any problem whose fixed-parameter tractability follows from Courcelle's Theorem.

1 Introduction

Many computationally hard problems become tractable if the graph structure underlying the problem instance at hand exhibits certain properties. An important structural parameter of this kind is treewidth. By using a seminal result due to Courcelle [1] several fixed-parameter tractability results have been proven in the last decade. To turn such theoretical tractability results into efficient computation in practice, two contrary approaches can be found in the literature (see also the excellent upcoming survey [2]). Either the user designs a suitable dynamic programming algorithm that works directly on tree decompositions of the instances (see, e.g., [3]), or a declarative description of the problem in terms of monadic second-order logic (MSO) is used with generic methods that automatically employ a fixed-parameter tractable algorithm where the concepts of tree decomposition and dynamic programming are used "inside", i.e., hidden from the user (see, e.g., [4,5] or the recent approach [6,7]). The obvious disadvantage of the first strategy is its purely procedural nature, thus a practical implementation requires considerable programming effort. The second approach lacks possibilities to incorporate domain-specific knowledge which is typically exploited in tailor-made dynamic programming solutions and thus crucial for efficient solutions.

In order to combine the best of the two worlds, a recent LISP-based approach called `Autograph` (see, e.g., [8]) allows to specify the problem at hand via combinations of (pre-defined) fly-automata; hereby, domain-specific knowledge is incorporated on the automata level. Another recent approach employs Answer Set Programming (ASP) [9] in combination with a system called D-FLAT[1] [10]. In this approach, it is possible to entirely describe the dynamic programming algorithm by declarative means. D-FLAT heuristically generates a tree decomposition of an input structure and provides the data structures that are propagated during dynamic programming. The task of solving each

---

[1] *Dynamic Programming Framework with Local Execution of ASP on Tree Decompositions. Available as free software at http://www.dbai.tuwien.ac.at/research/project/dynasp/dflat/.

G. Gutin and S. Szeider (Eds.): IPEC 2013, LNCS 8246, pp. 28–40, 2013.

subproblem is delegated to an efficient ASP system that executes a problem-specific encoding. Such specifications typically reflect the problem solving intuition due to the possibility of using a *Guess & Check* technique, and the rich ASP language (including, e.g., aggregates) allows for concise, easy-to-read encodings.

So far, D-FLAT has only been applied to some sample problems lying in NP [10]. It has been left open if this approach is more generally applicable. In this work, we present a slight extension of the D-FLAT approach and prove that this new method can indeed be used to solve *any* MSO-definable problem parameterized by the treewidth in fixed-parameter linear time. We introduce semantic trees as a tool for MSO model checking (MC). Semantic trees are closely related to the approaches from [6,11] but have properties that better suit our needs. Complementing the practically oriented exposition of D-FLAT in [10], the current work gives a theoretical result: We present an ASP-based description of a dynamic programming algorithm of the MSO MC problem via semantic trees and thus show the general applicability of the D-FLAT method.

2   Semantic Trees and Tree Decompositions

In this section we present our approach to MSO MC based on *semantic trees*, which are closely related to the game-theoretic techniques of [6] and the so-called *characteristic trees* of [11]. Below, we recall some basic notions and then highlight our method.

MSO Model Checking over Finite Structures. Let $\sigma = \{R_1, \ldots, R_K\}$ be a set of relation symbols. A *finite structure* $\mathcal{A}$ over σ (i.e., a "σ-structure", for short) is given by a finite domain $dom(\mathcal{A}) = A$ and relations $R_i^{\mathcal{A}} \subseteq A^\alpha$, where α denotes the arity of R_i.

We study the MSO model checking problem (i.e., the problem of evaluating an MSO sentence) over σ-structures. To simplify the presentation, we consider MSO sentences of the form $\phi = \exists Y_1 \exists z_1 \forall Y_2 \forall z_2 \ldots \exists Y_{n-1} \exists z_{n-1} \forall Y_n \forall z_n \psi$, s.t. n is even and ψ is a quantifier-free formula. Note that an *atom* in ϕ can either be of the form $R(z_{i_1}, \ldots, z_{i_\alpha})$ for some $R \in \sigma$ or of the form $Y_i(z_j)$. Let $At(\phi)$ denote the set of atoms occurring in ϕ.

An *interpretation* I of ψ over $\mathcal{A}$ is given by a tuple $(C_1, \ldots, C_n, d_1, \ldots, d_n)$, where $C_i \subseteq dom(\mathcal{A})$ is the interpretation of set-variable Y_i and $d_i \in dom(\mathcal{A})$ is the interpretation of the individual variable z_i. In a *partial interpretation*, we may assign the special value $undef$ to the individual variables z_i in ψ. The truth value $I(p)$ of an atom p in a partial interpretation I is defined in the obvious way: If at least one individual variable in the atom p is assigned the value $undef$ in I, then we also set $I(p) = undef$. Otherwise, $I(p)$ yields true or false exactly as for complete interpretations.

In order to systematically enumerate all possible interpretations for the quantifier-free part ψ of ϕ and to represent the truth value of ψ in each of these interpretations, we introduce the notion of *semantic trees*.

Definition 1. *For an MSO-formula ϕ and σ-structure $\mathcal{A}$, we define the* semantic tree *for ϕ and $\mathcal{A}$ as the following rooted, node-labeled tree with $2n + 2$ levels:*

- *Level 0 consists of the root.*
- *The nodes at levels 1 through $2n$ correspond to the variables $Y_1, z_1, Y_2, z_2, \ldots, Y_n, z_n$ in this order.*
- *Each of the nodes at level $2n + 1$ corresponds to the result of evaluating the atoms in ψ in one partial interpretation.*

The rank and label ℓ of each node N must satisfy the following conditions: the root has an empty label; every node at level $0, 2, 4, \ldots, 2n - 2$ has $|2^{dom(\mathcal{A})}|$ child nodes, s.t. each subset $B \subseteq dom(\mathcal{A})$ occurs as the label of one of these child nodes; every node at level $1, 3, 5, \ldots, 2n-1$ has $|dom(\mathcal{A})|+1$ child nodes, s.t. each $d \in dom(\mathcal{A}) \cup \{undef\}$ occurs as the label of one of these child nodes; for every node N at level $2n$, we define $I(N)$ as the partial assignment where the labels along the path from the root to N are assigned to the variables $Y_1, z_1, Y_2, z_2, \ldots, Y_n, z_n$. Then every such node N has exactly one child node, whose label is a pair (At^+, At^-), s.t. At^+ and At^- are the sets of atoms in ψ that evaluate to true or, respectively, false in $I(N)$.

For an MSO-formula ϕ and σ-structure $\mathcal{A}$, we can use the corresponding semantic tree $\mathcal{S}$ to get a naive MSO MC procedure: first delete the subtree rooted at every node N from $\mathcal{S}$ whenever $\ell(N) = undef$; then reduce the MSO MC problem to a Boolean circuit evaluation problem by replacing the nodes in $\mathcal{S}$ by $\vee$ or $\wedge$ depending on whether the corresponding quantifier in the quantifier prefix of ϕ is existential or universal. The leaf nodes of the Boolean circuit are labeled "true" or "false" depending on the truth value of ψ in the interpretation represented by this branch.

Compression of Semantic Trees for a Given Tree Decomposition. Of course, the MC procedure via semantic trees requires exponential time in the size of $\mathcal{A}$. We now show how semantic trees can be compressed in the presence of a tree decomposition of $\mathcal{A}$.

A *tree decomposition* of a structure $\mathcal{A}$ is a pair (T, χ) where $T = (V, E)$ is a (rooted) tree and $\chi : V \rightarrow 2^{dom(\mathcal{A})}$ maps nodes to so-called *bags* such that (1) for every $a \in dom(\mathcal{A})$, there is a $t \in V$ with $a \in \chi(t)$, (2) for every relation symbol R_i and every tuple $(a_1, \ldots, a_\alpha) \in R_i^{\mathcal{A}}$ there is a $t \in V$ with $\{a_1, \ldots, a_\alpha\} \subseteq \chi(t)$, and (3) for every $a \in dom(\mathcal{A})$, the set $\{t \in V \mid a \in \chi(t)\}$ induces a connected subtree of T. The latter is also known as the *connectedness condition*. The *width* of (T, χ) is defined as $\max_{t \in V}(|\chi(t)|) - 1$. The *treewidth* of $\mathcal{A}$ is the minimum width over all its tree decompositions. The notation $t \in \mathcal{T}$ expresses that t is a node of a tree decomposition $\mathcal{T}$. We write $\mathcal{T}_t$ and $\mathcal{A}_t$ to denote the subtree of $\mathcal{T}$ rooted at t, and the substructure of $\mathcal{A}$ induced by the domain elements occurring in the bags in $\mathcal{T}_t$, respectively.

By [12], we may assume that each node $t \in \mathcal{T}$ is of one of the following four types: It is either a *leaf node*, an *introduce node* (having one child t' with $\chi(t') \subseteq \chi(t)$ and $|\chi(t) \setminus \chi(t')| = 1$), a *forget node* (having one child t' with $\chi(t') \supseteq \chi(t)$ and $|\chi(t') \setminus \chi(t)| = 1$) or a *join node* (having two children t_1, t_2 with $\chi(t) = \chi(t_1) = \chi(t_2)$). Moreover, we may assume that the root of $\mathcal{T}$ has an empty bag.

The idea of our decision procedure for $\mathcal{A} \models \phi$ is to compute the semantic tree for every substructure $\mathcal{A}_t$ of $\mathcal{A}$. At the root node r of the tree decomposition, we thus get the semantic tree for the unrestricted structure $\mathcal{A}$, which we can then use for checking $\mathcal{A} \models \phi$ by a reduction to the Boolean circuit evaluation problem. We formally define this semantic tree for substructure $\mathcal{A}_t$ below.

Definition 2. *Consider an MSO-formula ϕ and σ-structure $\mathcal{A}$ with tree decomposition $\mathcal{T}$. For $t \in \mathcal{T}$, we say that $\mathcal{S}_t$ is the* local semantic tree at t *if $\mathcal{S}_t$ is the semantic tree of the MSO-formula ϕ and the induced substructure $\mathcal{A}_t$ of $\mathcal{A}$.*

To reach a fixed-parameter tractable algorithm w.r.t. treewidth, we introduce a compression of the semantic tree at each node t in the tree decomposition. The compression proceeds in two steps: First, we restrict the labels of the semantic trees to the domain elements present in $\chi(t)$. Second, if some node in a semantic tree has two child nodes with identical subtrees, then it suffices to retain only one of these subtrees.

Note that the concrete values of the labels at the internal nodes (i.e., the nodes corresponding to set variables or individual variables) in a semantic tree are irrelevant. Indeed, in the above reduction to Boolean circuits, only the tree structure of the semantic tree and the truth values (At^+, At^-) at the leaf nodes matter. As will be explained below, it is also convenient to slightly manipulate the truth values of some atoms which should be undefined according to the above definition of partial truth assignments. Since all subtrees with an undefined variable in one of the labels are ultimately removed from the semantic tree anyway, this has no effect on the evaluation of formula ϕ over structure $\mathcal{A}$. In summary, we get the following notion of *compressed, local semantic trees*.

Definition 3. *Consider an MSO-formula ϕ and σ-structure $\mathcal{A}$ with tree decomposition $\mathcal{T}$. For $t \in \mathcal{T}$, let $\mathcal{S}_t$ denote the local semantic tree at t. We call $\mathcal{C}_t$ a compressed, local semantic tree at t if $\mathcal{C}_t$ is obtained from $\mathcal{S}_t$ by applying rule L followed by rule A and then exhaustively applying rule R defined below:*

Rule L (changing Labels)

- *For every node corresponding to a set variable (i.e., levels $1, 3, \ldots, 2n - 1$), the label $B \subseteq dom(\mathcal{A})$ is replaced $B \cap \chi(t)$.*
- *For every node corresponding to an individual variable (i.e., levels $2, 4, \ldots, 2n$), the label d is replaced by a special symbol $\star$ if $d \in dom(\mathcal{A}) \setminus \chi(t)$, and left unchanged otherwise, i.e. if $d \in \chi(t) \cup \{undef, \star\}$.*

Rule A (modification of Atom set At^-). *For every node at level $2n + 1$, let I denote the interpretation along the path from the root to this node. In the label (At^+, At^-), replace At^- by $At^- \cup \{R(z_1, \ldots, z_\alpha) \in At(\phi) \mid \exists i, j \text{ s.t. } I(z_i) = undef \text{ and } I(z_j) = \star\}$.*

Rule R (eliminating Redundancy). *Let N be a node in $\mathcal{S}_t$ and let N_1, N_2 be two distinct child nodes of N. If the subtree rooted at N_1 and the subtree rooted at N_2 are identical, then we delete N_2 and the entire subtree rooted at N_2 from $\mathcal{S}_t$.*

The intuition of rule A is the following: Recall that the meaning of $I(z_j) = \star$ is that z_j is set to some value occurring in the subtree below node t in the tree decomposition but not in $\chi(t)$. The idea of letting $I(z_i) = undef$ is to set z_i to some value neither occurring $\chi(t)$ nor in the subtree below t. But then, by the connectedness condition of tree decompositions, we know that such atoms can never become true, no matter how the undefined variable will eventually be interpreted.

MSO Model Checking via Compressed, Local Semantic Trees. Given a finite structure $\mathcal{A}$ with a tree decomposition $\mathcal{T}$ and an MSO sentence ϕ, our MC procedure works in two steps: First, we compute a compressed, local semantic tree at every node t in $\mathcal{T}$ by a bottom-up traversal of $\mathcal{T}$. Then we evaluate ϕ over $\mathcal{A}$ by reducing the compressed, local semantic tree at the root node r of $\mathcal{T}$ to a Boolean circuit. Fixed-parameter linearity (w.r.t. the treewidth) of this algorithm is obtained as follows:

Theorem 1. *For the MSO model checking problem $\mathcal{A} \models \phi$, let $\mathcal{T}$ be a tree decomposition of $\mathcal{A}$. Then we can compute in time $O(f(\tau(\mathcal{T}), \phi) \cdot \|\mathcal{T}\|)$ a compressed, local semantic tree $\mathcal{C}_t$ at every node t in $\mathcal{T}$. Here, $\tau(\mathcal{T})$ denotes the width of $\mathcal{T}$ and f is a function not depending on $\mathcal{A}$.*

Proof (Sketch). The computation of a compressed, local semantic tree $\mathcal{C}_t$ for every node $t \in \mathcal{T}$ proceeds in a bottom-up manner from the leaf nodes of $\mathcal{T}$ to the root. For this computation, we distinguish the four possible types that a node t of $\mathcal{T}$ can have:

(1) If t is a *leaf node*, it can be shown that $\mathcal{C}_t$ simply coincides with the local semantic tree $\mathcal{S}_t$ at t, i.e., none of the rules L, A, and R is applicable.

(2) Let t be an *introduce node* with child node t', s.t. $\chi(t') = \chi(t) \setminus \{b\}$. Then $\mathcal{C}_t$ is obtained from $\mathcal{C}_{t'}$ by copying subtrees of $\mathcal{C}_{t'}$ and modifying the labels of the copies as follows. Every node N in $\mathcal{C}_{t'}$ with $\ell(N) \subseteq \chi(t')$ gives rise to two nodes in $\mathcal{C}_t$: one with unchanged label $\ell(N)$ and one with label $\ell(N) \cup \{b\}$. Similarly, every node N in $\mathcal{C}_{t'}$ with $\ell(N) = undef$ gives rise to two nodes in $\mathcal{C}_t$: one with unchanged label $undef$ and one with label b. Note that this corresponds to the intended meaning of the value $undef$, which is that a value shall be assigned to this individual variable "outside" the current subtree of the tree decomposition. For the adaptation of the truth values (At^+, At^-) at the leaf nodes of $\mathcal{C}_t$, the connectedness condition of tree decompositions is crucial. Finally, $\mathcal{C}_t$ is compressed via rule R.

(3) Let t be a *forget node* with child node t', s.t. $\chi(t) = \chi(t') \setminus \{b\}$. Then $\mathcal{C}_t$ is obtained from $\mathcal{C}_{t'}$ by first applying rule L. This means that we delete b from every set B in $\mathcal{C}_{t'}$. Moreover, if an individual variable is interpreted as b in $\mathcal{C}_{t'}$, we replace this interpretation by $\star$. For the truth values (At^+, At^-) at the leaf nodes of $\mathcal{C}_t$, it is now important to apply rule A from Definition 3. Finally, $\mathcal{C}_t$ is compressed via rule R.

(4) Finally, let t be a *join node* with child nodes t_1 and t_2. By definition of join nodes, we have $\chi(t) = \chi(t_1) = \chi(t_2)$. The nodes of $\mathcal{C}_t$ are obtained by combining "compatible" nodes of $\mathcal{C}_{t_1}$ and $\mathcal{C}_{t_2}$. For an odd level $i < 2n$ in $\mathcal{C}_t$ (i.e., the labels of these nodes provide the interpretation of a set variable in ϕ), a node N_1 in $\mathcal{C}_{t_1}$ and a node N_2 in $\mathcal{C}_{t_2}$ are compatible if $\ell(N_1) = \ell(N_2)$. Compatibility in case of an even level $0 < i < 2n$ in $\mathcal{C}_t$ (i.e., a node whose label interprets an individual variable) holds if either (a) $\ell(N_1) = \ell(N_2)$ and $\ell(N_i) \neq \star$ or (b) one of $\ell(N_1), \ell(N_2)$ is $undef$. In case (a), the node N in $\mathcal{C}_t$ resulting from combining N_1 and N_2 simply gets the label $\ell(N) = \ell(N_1) = \ell(N_2)$. In case (b), the label of the resulting node N is set to $\ell(N_i)$ with $\ell(N_i) \neq undef$. Note that in (a), it is important to exclude the combination of nodes N_1 and N_2 with $\ell(N_1) = \ell(N_2) = \star$. This is due to the intended meaning of $\star$, which stands for some domain element in the subtree below t in the tree decomposition s.t. this value no longer occurs in the bag of t. Hence, the two occurrences of $\star$ in $\mathcal{C}_{t_1}$ and $\mathcal{C}_{t_2}$ stand for different values. The label (At^+, At^-) at a leaf node of $\mathcal{C}_t$ is obtained from the labels of the corresponding nodes in $\mathcal{C}_{t_1}$ and $\mathcal{C}_{t_2}$ by taking the component-wise union. Finally, we compress $\mathcal{C}_t$ via rule R.   □

Our approach via (compressed) semantic trees has close links to the approaches based on extended MC games in [6] and on characteristic trees in [11]. The most significant difference is that we explicitly introduce a special symbol $\star$ for domain elements not present anymore in a given bag of a tree decomposition. This allows us to define our reduction of semantic trees by a simple equality test, while the reduce-operation in [6] is based on an isomorphism criterion (which would not allow for a simple ASP realization). The characteristic trees in [11] are used in the context of structures of bounded rank-width and are computed by a bottom-up traversal of a given t-labeled parse tree decomposition. The reduction of characteristic trees is also based on an equality criterion. However, in contrast to tree decompositions, the notions of a "bag" and of a special symbol $\star$ (for domain elements not present anymore in some bag) are not applicable.

3   ASP and D-FLAT

In this section, we give brief introductions to Answer Set Programming (ASP) [9] and the D-FLAT system [10]. We thus set the stage for presenting our main result, i.e., that D-FLAT possesses enough expressive power for solving any MSO-definable problem parameterized by the treewidth in fixed-parameter linear time.

ASP is a declarative language where a *program* Π is a set of *rules*

$$a_1 \vee \cdots \vee a_k \leftarrow b_1, \ldots, b_m, \text{not } b_{m+1}, \ldots, \text{not } b_n.$$

The constituents of a rule $r \in \Pi$ are $h(r) = \{a_1, \ldots, a_k\}$, $b^+(r) = \{b_1, \ldots, b_m\}$ and $b^-(r) = \{b_{m+1}, \ldots, b_n\}$. Intuitively, r states that if an answer set contains all of $b^+(r)$ and none of $b^-(r)$, then it contains some element of $h(r)$. A set of atoms I satisfies a rule r iff $I \cap h(r) \neq \emptyset$ or $b^-(r) \cap I \neq \emptyset$ or $b^+(r) \setminus I \neq \emptyset$. I is a *model* of a set of rules iff it satisfies each rule. I is an *answer set* of a program Π iff it is a subset-minimal model of the program $\Pi^I = \{h(r) \leftarrow b^+(r) \mid r \in \Pi, b^-(r) \cap I = \emptyset\}$ [13].

ASP programs can be viewed as succinctly representing problem solving specifications following the *Guess & Check* principle. A "guess" can, for example, be performed using disjunctive rules which non-deterministically open up the search space. Constraints (i.e., rules r with $h(r) = \emptyset$), on the other hand, amount to a "check" by imposing restrictions that solutions must obey.

In this paper, we use the language of the grounder *Gringo* [14,15] where programs may contain variables that are instantiated by all ground terms (elements of the Herbrand universe, i.e., constants and compound terms containing function symbols) before a solver computes answer sets according to the propositional semantics stated above.

Example 1. The following program solves the INDEPENDENT DOMINATING SET problem for graphs that are given as facts using the predicates vertex and edge.

```
{ in(X) : vertex(X) }.                                          1
← edge(X,Y), in(X;Y).                                          2
dominated(X) ← in(Y), edge(Y,X).                               3
← vertex(X), not in(X), not dominated(X).                      4
```

Let (V, E) denote the input graph and recall that a set $S \subseteq V$ is an independent dominating set of (V, E) iff $E \cap S^2 = \emptyset$ and for each $x \in V$ either $x \in S$ or there is some $y \in S$ with $(y, x) \in E$. Note that this program not only solves the decision variant of the problem, which is NP-complete, but also allows for solution enumeration.

Informally, the first rule (a so-called *choice rule* having an empty body) states that in is to be guessed to comprise any subset of V. The colon controls the instantiation of the

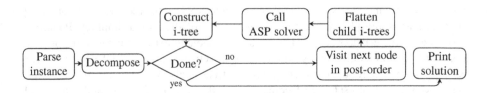

Fig. 1. Control flow in D-FLAT

variable X such that it is only instantiated with arguments of vertex from the input. The rule in line 2 – where in(X;Y) is shorthand for in(X), in(Y) – checks the independence property. Lines 3 and 4 finally ensure that each vertex not in the guessed set is dominated by this set.

In order to take advantage of this Guess & Check approach in a decomposed setting, we make use of the D-FLAT system [10]. To perform dynamic programming on tree decompositions, D-FLAT needs data structures to propagate the partial solutions. To this end, it equips each node t in a tree decomposition $\mathcal{T}$ of an input structure $\mathcal{A}$ with a so-called *i-tree*. By this we mean a tree where each node is associated with a set of ground terms called *items*. D-FLAT executes a user-supplied ASP program at each node $t \in \mathcal{T}$ (feeding it in particular the i-trees of children of t as input) and parses the answer sets to construct the i-tree of t. This procedure is depicted in Figure 1. To keep track of its origin, each i-tree node N is associated with a set of *extension pointers*, i.e., tuples referencing i-tree nodes from the child nodes of t that have given rise to N. For instance, if t has k children, the set of extension pointers of N consists of tuples $(N_1, \ldots, N_k)$, where each N_j is an i-tree node of the jth child of t. This allows us to obtain complete solutions by combining the item sets along a chain of extension pointers.

As input to the user's encoding, D-FLAT declares the fact final if the current node $t \in \mathcal{T}$ is the root; current(v) for any $v \in \chi(t)$; if t has a child t', introduced(v) or removed(v) for any $v \in \chi(t) \setminus \chi(t')$ or $v \in \chi(t') \setminus \chi(t)$, respectively; root($r$) if t has a child whose i-tree is rooted at r; sub(N, N') for any pair of nodes N, N' in a child's i-tree, if N' is a child of N; and childItem(N, i) if the item set of node N from a child's i-tree contains the element i. Finally, D-FLAT also provides the input structure as a collection of ground facts.

The answer sets specify the i-tree of the current tree decomposition node. Each answer set describes a branch in the i-tree. Atoms of the following form are relevant for this: length(l) declares that the branch consists of $l + 1$ nodes; extend(l, j) causes that j is added to the extension pointers of the node at depth l of the branch. item(l, i) states that the node at depth l of the branch contains i in its item set. All atoms using extend and item with the same depth argument constitute what we call a *node specification*.

To determine where branches diverge, D-FLAT uses the following recursive condition: Two node specifications coincide (i.e., describe the same i-tree node) iff *(1)* their depths, item sets and extension pointers are equal, and *(2)* both are at depth 0, or their parent node specifications coincide. In this way, an i-tree is obtained from the answer sets. It might however contain sibling subtrees that are equal w.r.t. item sets. If so, one of the subtrees is discarded and the extension pointers associated to its nodes are added to the extension pointers of the corresponding nodes in the remaining subtree. D-FLAT exhaustively performs this action to eliminate redundancies.

Example 2. Listing 1.1 shows a D-FLAT encoding for INDEPENDENT DOMINATING SET. All i-trees have height 1 (due to line 1); their roots are always empty and their leaves contain items involving the function symbols in and dominated. Note that lines 7–10 resemble the program from Example 1, while the rest of the program is required for appropriately extending and combining partial solutions from child nodes.

Suppose D-FLAT is currently processing a forget node. Then there is one child i-tree. For illustration, assume it consists of two branches whose respective leaf item sets are $\emptyset$ and $\{in(a), dominated(b)\}$. This i-tree is provided to the program in Listing 1.1 by means of the following input facts:

```
root(r). sub(r,s1). sub(r,s2).
childItem(s2,in(a)). childItem(s2,dominated(b)).
```

Each answer set of the program corresponds to a branch in the new i-tree, and each branch extends one branch from the child i-tree. The root of the new i-tree therefore always extends the root of the child i-tree (line 2). Which branch is extended is guessed in line 3. Lines 5 and 6 derive which vertices are "in" or "dominated" according to this guess, and line 10 enforces the dominance condition. Note that *it is not until a vertex is removed* that it can be established to violate this condition, since as long as a vertex is not removed potential neighbors dominating it could still be introduced. So, if the vertex c has been removed, then the constraint in line 10 would eliminate the answer set extending branch "s2", since c is neither "in" nor "dominated". Lines 11 and 12 fill the leaf item set with only those items that apply to vertices still in the current bag. (This ensures that the maximum size of an i-tree only depends on the decomposition width.) So if the branch with leaf "s2" is extended and vertex a is forgotten, these lines cause that the answer set specifies the item dominated(b), but not in(a).

In introduce nodes, line 7 guesses whether the introduced vertex is "in" the partial solution or not. Line 8 enforces the independence condition and line 9 determines dominated vertices. Line 4 ensures that in join nodes a pair of branches is only extended if these branches agree on which of the common vertices are "in".

4   MSO MC on Tree Decompositions with ASP

We now present an encoding for MSO MC in the style of the approach from Section 2 in order to show that ASP with D-FLAT can solve any MSO-definable problem in linear time for bounded treewidth. In the following, let $\mathcal{A}$ and $\mathcal{T}$ denote the input structure and one of its tree decompositions, respectively. For the sake of readability, we only consider the case where $\mathcal{A}$ is a graph, given by the predicates vertex and edge. As in Section 2, we assume the MSO formula ϕ for which $\mathcal{A} \models \phi$ is to be decided to be of the form $\exists Y_1 \exists z_1 \forall Y_2 \forall z_2 \ldots \exists Y_{n-1} \exists z_{n-1} \forall Y_n \forall z_n \psi$. Here we additionally assume that ψ is in CNF. Our encoding can, however, be easily generalized. In particular, the quantifier alternation is not required in principle but facilitates presentation. Much could be done to improve the MSO model checker that emerges from this work; but this is outside the scope of this paper whose focus is on the general applicability of D-FLAT.

```
length(1).                                                            1
extend(0,R) ← root(R).                                                2
1 { extend(1,S) : sub(R,S) } 1 ← extend(0,R).                         3
← extend(1,S;T), childItem(S,in(X)), not childItem(T,in(X)).          4
in(X) ← extend(1,S), childItem(S,in(X)).                              5
dominated(X) ← extend(1,S), childItem(S,dominated(X)).                6
{ in(X) : introduced(X) }.                                            7
← edge(X,Y), in(X;Y).                                                 8
dominated(X) ← in(Y), edge(Y,X).                                      9
← removed(X), not in(X), not dominated(X).                            10
item(1,in(X)) ← in(X), current(X).                                    11
item(1,dominated(X)) ← dominated(X), current(X).                      12
```

Listing 1.1. Computing independent dominating sets with D-FLAT

The formula ϕ is specified in ASP as follows. If the quantifier rank is i, then the fact length($i{+}1$) is declared. (This will cause each i-tree branch to have length $i{+}1$.) Each individual variable x or set variable X bound by the ith quantifier is declared by a fact of the form iVar(i, x) or sVar(i, X), respectively. The atoms $x \in X$ and membership in the edge relation are represented as in(x, X) and edge(x, y), respectively. Facts of the form pos(c, a) or neg(c, a) respectively denote that the atom a occurs positively or negatively in the clause c. For convenience, we supply a fact clause(c) for each clause c, and var(i, x) for each individual or set variable x bound by the ith quantifier.

Let t be the current node during a bottom-up traversal of a tree decomposition $\mathcal{T}$ of $\mathcal{A}$. The i-tree at t shall represent a compressed, local semantic tree. In particular, an item set of a (non-leaf) i-tree node shall encode the label of the respective semantic tree node. With each i-tree branch b we can thus associate a (partial) interpretation I_b of the variables in ϕ. I_b assigns $\star$ to variables with values not in $\chi(t)$, but we can extend it to all possible assignments I_b^+ without $\star$ values by following the extension pointers. As we assume ϕ to be in CNF, in the leaf of b we simply keep track of the clauses that have been satisfied by I_b^+ so far. We only use items of the following form: assign($x, \_nn$) denotes that $I_b(x) = \star$; assign(x, v) with $v \in \chi(t)$ denotes that $I_b(x) = v$; assign(X, v) denotes that $v \in I_b(X)$; true(c), which only occurs in leaf item sets, indicates that the clause c is true under I_b^+. For any individual variable x, the absence of any assign item whose first argument is x means that x is still undefined.

Listing 1.2 shows the ASP encoding that is to be executed at each node $t \in \mathcal{T}$ to construct an i-tree representing $\mathcal{C}_t$, the compressed, local semantic tree at t. As input, the encoding is provided with a set of facts describing ϕ as well as $\mathcal{T}$ together with the i-trees from the children of t (see Section 3). We say that D-FLAT accepts the input $\mathcal{A}$ if the program executed at the root node of $\mathcal{T}$ has at least one answer set.

Theorem 2. *An MSO MC instance $\mathcal{A} \models \phi$ is positive iff D-FLAT, when executed on Listing 1.2 together with ϕ (represented in ASP as a set of facts), accepts input $\mathcal{A}$.*

Proof (Sketch). Let $\mathcal{A}$ be the input graph with a tree decomposition $\mathcal{T}$, let $t \in \mathcal{T}$ be the node currently processed by D-FLAT during the bottom-up traversal, and let $\mathcal{C}_t$ denote the compressed, local semantic tree at t after executing the encoding at t. Again, $\mathcal{S}_t$ denotes the (non-compressed) local semantic tree at t, while $\mathcal{S}$ is the (complete) semantic tree for ϕ and $\mathcal{A}$. We first show that $\mathcal{C}_t$ is always constructed as desired according to the

```
assignedIn(X,S) ← childItem(S,assign(X,_)).                                       1
% Evaluation (only in the root)                                                   2
itemSet(0,R) ← final, root(R).                                                    3
itemSet(L+1,S) ← itemSet(L,R), sub(R,S).                                          4
exists(S) ← itemSet(L,S), L #mod 4 < 2, sub(S,_).                                 5
forall(S) ← itemSet(L,S), L #mod 4 > 1, sub(S,_).                                 6
invalid(S) ← iVar(L,X), itemSet(L,S), not assignedIn(X,S).                        7
bad(S) ← length(L), itemSet(L,S), clause(C),                                      8
    not childItem(S,true(C)).
bad(S) ← forall(S), not invalid(S), sub(S,T), bad(T).                             9
bad(S) ← exists(S), not invalid(S), not good(S).                                 10
good(S) ← exists(S), sub(S,T), not invalid(T), not bad(T).                       11
% Guess a branch for each child i-tree                                           12
extend(0,R) ← root(R).                                                           13
1 { extend(L+1,S) : sub(R,S) } 1 ← extend(L,R), sub(R,_).                        14
← extend(_,S), bad(S).                                                           15
← extend(_,S), invalid(S).                                                       16
% Preserve and extend assignment                                                17
{ assign(X,V) : var(_,X) } ← introduced(V).                                      18
assign(X,V) ← extend(_,S), childItem(S,assign(X,V)),                             19
    not removed(V).
assign(X,_nn) ← extend(L,S), childItem(S,assign(X,V)),                           20
    removed(V), iVar(L,X).
% Check: Only join compatible branches; the resulting assignment must be valid   21
← iVar(L,X), assign(X,V;W), V ≠ W.                                               22
← extend(L,S;T), S ≠ T, childItem(S;T,assign(X,_nn)).                            23
← extend(L,S;T), var(L,X), childItem(S,assign(X,V)),                             24
    not childItem(T,assign(X,V)), vertex(V).
% Determine clauses that have become true                                       25
assigned(X) ← iVar(L,X), extend(L,S), assignedIn(X,S).                           26
true(C) ← extend(_,S), childItem(S,true(C)).                                     27
true(C) ← pos(C,edge(X,Y)), assign(X,V), assign(Y,W),                            28
    edge(V,W).
true(C) ← neg(C,edge(X,Y)), assign(X,V), assign(Y,W),                            29
    vertex(V;W), not edge(V,W).
true(C) ← neg(C,edge(X,Y)), extend(_,S),                                         30
    childItem(S,assign(X,V)), removed(V), not assigned(Y).
true(C) ← neg(C,edge(X,Y)), extend(_,S),                                         31
    childItem(S,assign(Y,V)), removed(V), not assigned(X).
true(C) ← pos(C,in(X,Y)), assign(X,V), assign(Y,V).                              32
true(C) ← neg(C,in(X,Y)), assign(X,V), vertex(V),                               33
    not assign(Y,V).
% Declare resulting item sets                                                   34
item(L,assign(X,V)) ← var(L,X), assign(X,V).                                     35
item(L,true(C)) ← length(L), true(C).                                           36
```

Listing 1.2. MSO model checking with D-FLAT

proof of Theorem 1. Then we show that from C_t we can always construct S_t, and that this gives us S at the root of $\mathcal{T}$. The computation of C_t depends on the type of t.

(1) If t is a *leaf*, we guess a valid (partial) variable assignment without any $\star$ values (lines 18 and 22) and declare the appropriate item sets (line 35). Additionally, we add the clauses that are satisfied by the assignment (cf. rules deriving `true`) into the leaf item set (line 36). Eventually, D-FLAT's processing of the resulting answer sets (see Section 3) yields an i-tree representing S_t, which coincides with C_t.

(2) If t is an *introduce node* with child t', we guess a predecessor branch of the i-tree of t' (lines 13 and 14) whose assignment is preserved (line 19) and non-deterministically extended (lines 18 and 22). Already satisfied clauses remain so (line 27). Again, clauses that become satisfied are determined and the appropriate item sets are filled.

(3) If t is a *forget node*, we also guess a predecessor branch. We retain each `assign` item unless it involves the removed vertex (line 19), and we set the value of each individual variable that was assigned this vertex to $\star$ (line 20). Determining satisfied clauses and declaring item sets proceed as before. This yields an i-tree where the removed vertex is eliminated from the interpretation of each set variable, and individual variables previously set to that value are now assigned $\star$. Note that clauses might become satisfied due to the reasons for rule A from Section 2.

(4) If t is a *join node* with children t_1 and t_1, $\chi(t) = \chi(t_1) = \chi(t_2)$ holds. Here, we guess *a pair* of predecessor branches (lines 13 and 14). We generate C_t by combining "compatible" branches b_1 and b_2 from C_{t_1} and C_{t_2}, respectively. The notion of compatibility is the same as in the proof of Theorem 1, and enforced in lines 23 and 24. Thus the two assignments corresponding to b_1 and b_2 can simply be unified to yield the assignment of the new branch b (line 19). The set of clauses true under the assignment of b is now simply the union of the clauses true in b_1 and the clauses true in b_2 (line 27).

(5) If t is the *root node* of $\mathcal{T}$ (by assumption a forget node with an empty bag; see Section 2), the child i-tree nodes are organized with `exists`, `forall`, `invalid` and `bad`. Following the assumed form of the quantifier prefix of ϕ, non-leaf i-tree nodes at levels $4j$ and $4j+1$ (for $j \geq 0$) are marked with "exists", while those at levels $4j+2$ and $4j + 3$ are marked with "forall". A non-leaf node at level l is "invalid" if the lth quantifier binds an individual variable left uninterpreted by that node, and it is "bad" if the subformula of ϕ starting after the lth quantifier cannot be true. For this purpose, we start by labeling each leaf with "bad" if it does not report all clauses to be satisfied (line 8). By following extension pointers, it can be verified that none of the interpretations represented by the respective branch satisfies the matrix of ϕ due to our bookkeeping of satisfied clauses. All leaves that are neither "invalid" nor "bad" conversely correspond to interpretations satisfying the matrix of ϕ. We then propagate truth values toward the root (lines 9–11): A "forall" node is "bad" iff one of its children is "bad", and an "exists" node is "bad" iff it has only "bad" or "invalid" children. To ensure correctness and to only enumerate interpretations without undefined individual variables, the guessed predecessor branch must contain neither "bad" nor "invalid" nodes (lines 15 and 16).

Finally, we show that $\mathcal{A} \models \phi$ holds iff the root of the i-tree at the child of the root of $\mathcal{T}$ is not "bad". The i-tree of any $t \in \mathcal{T}$ below the root of $\mathcal{T}$ can be used to construct S_t by means of the extension pointers, as can be seen by induction. Furthermore, the clauses satisfied by the interpretation corresponding to a branch of S_t are exactly those in the respective leaf item set. If t is the child of the root node, we obtain S in this way. If t is the root of $\mathcal{T}$, the propagation of truth values in the child i-tree (lines 1–11) corresponds to the propagation of truth values in the Boolean circuit used for evaluation. If this propagation finally yields "false", line 15 ensures that no answer set exists because the i-tree root at the child of t is then "bad". Otherwise, there is a branch consisting only of nodes that are neither "bad" nor "invalid", and D-FLAT accepts the input.   □

Given an input structure $\mathcal{A}$ whose treewidth is below some fixed integer, one can construct a tree decomposition of $\mathcal{A}$ in linear time. The total runtime for deciding $\mathcal{A} \models \phi$ for fixed ϕ is then linear, since the tree decomposition has linear size and the search space in each ASP call is bounded by a constant. Note that Theorems 1 and 2 together thus amount to an alternative proof of Courcelle's Theorem.

5 Conclusion

There is vivid interest in turning theoretical tractability results obtained via Courcelle's Theorem into concrete computation which is feasible in practice [2]. In this paper, we have shown that the ASP-based D-FLAT approach is one candidate for reaching this goal, having provided a realization of a suitable dynamic programming algorithm for the MSO model checking problem. Since MSO model checking is often impractical despite bounded treewidth [16], it is advisable to implement problem-specific algorithms. Experiments reported in [10] suggest that D-FLAT is a promising means to do so. In contrast to recent MSO-based systems [6,7] where the problem is expressed in a monolithic way, D-FLAT allows to define the dynamic programming algorithm on a tree decomposition via ASP. Like in the Datalog approach [17], this admits a declarative specification while still being able to take advantage of domain knowledge. However, the approach in [17] aims at a single call to a Datalog engine, thus the very restrictive language of monadic Datalog is required to guarantee linear running times. Therefore, encoding the dynamic programming algorithm at hand is rather tedious (for instance, to handle set operations) making this approach less practical. In contrast, D-FLAT calls an ASP-solver in each node of the tree decomposition. This not only ensures the linear running times (assuming that D-FLAT encodings only use information from the current bag) but also allows one to take advantage of a richer modeling language, reducing the actual effort for the user. This leads to implementations of algorithms that leverage bounded treewidth in a natural way, as the examples in Section 3 and [10] show. In the current paper, we have shown that these were not just lucky coincidences – D-FLAT is indeed applicable to *any* MSO-definable problem. Future work in particular includes a comparison of the ASP-based D-FLAT approach with the LISP-based Autograph approach [8] regarding both the range of theoretical applicability and practical efficiency.

Acknowledgments. This work is supported by the Austrian Science Fund (FWF) projects P25518 and P25607. We also thank the anonymous referees for helpful comments.

References

1. Courcelle, B.: The monadic second-order logic of graphs. I. Recognizable sets of finite graphs. Inf. Comput. 85(1), 12–75 (1990)
2. Langer, A., Reidl, F., Rossmanith, P., Sikdar, S.: Practical algorithms for MSO model-checking on tree-decomposable graphs (2013),
   `http://tcs.rwth-aachen.de/~sikdar/index_files/article.pdf`
3. Niedermeier, R.: Invitation to Fixed-Parameter Algorithms. Oxford Lecture Series in Mathematics And Its Applications. Oxford University Press (2006)
4. Flum, J., Frick, M., Grohe, M.: Query evaluation via tree-decompositions. J. ACM 49(6), 716–752 (2002)
5. Klarlund, N., Møller, A., Schwartzbach, M.I.: MONA implementation secrets. Int. J. Found. Comput. Sci. 13(4), 571–586 (2002)
6. Kneis, J., Langer, A., Rossmanith, P.: Courcelle's theorem – a game-theoretic approach. Discrete Optimization 8(4), 568–594 (2011)
7. Langer, A., Reidl, F., Rossmanith, P., Sikdar, S.: Evaluation of an MSO-solver. In: Proc. ALENEX, pp. 55–63. SIAM / Omnipress (2012)
8. Courcelle, B., Durand, I.: Computations by fly-automata beyond monadic second-order logic. CoRR abs/1305.7120 (2013)
9. Brewka, G., Eiter, T., Truszczyński, M.: Answer set programming at a glance. Commun. ACM 54(12), 92–103 (2011)
10. Bliem, B., Morak, M., Woltran, S.: D-FLAT: Declarative problem solving using tree decompositions and answer-set programming. TPLP 12(4-5), 445–464 (2012)
11. Langer, A., Rossmanith, P., Sikdar, S.: Linear-time algorithms for graphs of bounded rankwidth: A fresh look using game theory - (extended abstract). In: Ogihara, M., Tarui, J. (eds.) TAMC 2011. LNCS, vol. 6648, pp. 505–516. Springer, Heidelberg (2011)
12. Kloks, T.: Treewidth. LNCS, vol. 842. Springer, Heidelberg (1994)
13. Gelfond, M., Lifschitz, V.: Classical negation in logic programs and disjunctive databases. New Generation Comput. 9(3/4), 365–386 (1991)
14. Gebser, M., Kaminski, R., Kaufmann, B., Ostrowski, M., Schaub, T., Thiele, S.: A user's guide to gringo, clasp, clingo, and iclingo. Preliminary Draft (2010),
    `http://potassco.sourceforge.net`
15. Gebser, M., Kaminski, R., Kaufmann, B., Schaub, T.: Answer Set Solving in Practice. Synthesis Lectures on Artificial Intelligence and Machine Learning. Morgan & Claypool Publishers (2012)
16. Frick, M., Grohe, M.: The complexity of first-order and monadic second-order logic revisited. Ann. Pure Appl. Logic 130(1-3), 3–31 (2004)
17. Gottlob, G., Pichler, R., Wei, F.: Monadic datalog over finite structures of bounded treewidth. ACM Trans. Comput. Log. 12(1) (2010)

The Fine Details of Fast Dynamic Programming over Tree Decompositions

Hans L. Bodlaender[1], Paul Bonsma[2,*], and Daniel Lokshtanov[3]

[1] Department of Computing Sciences, Utrecht University, P.O. Box 80.089,
3508 TB Utrecht, The Netherlands
[2] Faculty of EEMCS, University of Twente, The Netherlands, P.O. Box 217,
7500 AE Enschede, The Netherlands
p.s.bonsma@ewi.utwente.nl
[3] Department of Informatics, University of Bergen, P.O. Box 7803,
5020 Bergen, Norway

Abstract. We study implementation details for dynamic programming over tree decompositions. Firstly, a fact that is overlooked in many papers and books on this subject is that it is not clear how to test adjacency between two vertices in time bounded by a function of k, where k is the width of the given tree decomposition. This is necessary to obtain linear time dynamic programming algorithms. We address this by giving a simple $O(kn)$ time and space preprocessing procedure that enables adjacency testing in time $O(k)$, where n is the number of vertices of the graph.

Secondly, we show that a large class of NP-hard problems can be solved in time $O(q^{k+1}n)$, where q^{k+1} is the natural size of the dynamic programming tables. The key improvement is that we avoid a polynomial factor in k. This holds for all problems that can be formulated as a Min Weight Homomorphism problem: given a (di)graph G on n vertices and a (di)graph H on q vertices, with integer vertex and edge weights, is there a homomorphism from G to H with total (vertex and edge image) weight at most M? This result implies e.g. $O(2^k n)$ algorithms for Max Independent Set and Max Cut, and a $O(q^{k+1}n)$ algorithm for q-Colorability. The table building techniques we develop are also useful for many other problems.

1 Introduction

Dynamic programming over tree decompositions has become an important algorithmic technique, that is used as a central subroutine in many parameterized, exact and approximation algorithms for NP-hard problems. The key property that is often used is that for any constant k, many NP-hard graph problems can be solved in linear time, if a tree decomposition of the graph of width at most k is provided. Examples of early results of this kind are [3,4,7,9,19]. Good introductions can be found in [6,8,15]. Recent breakthrough results appear in [5,10,16].

---

* Corresponding author.

G. Gutin and S. Szeider (Eds.): IPEC 2013, LNCS 8246, pp. 41–53, 2013.
© Springer International Publishing Switzerland 2013

In this paper, we study implementation details of dynamic programming over tree decompositions. Throughout, denote by n the number of vertices of the input graph, and by k the width of the given tree decomposition. (k is viewed as a parameter, not a constant.) Firstly, a fact that has been overlooked in many papers and books on this subject, is that it is not clear how to test adjacency between two vertices in time bounded by any function of k. This is necessary to obtain linear time algorithms. Note that we cannot simply assume that an adjacency matrix is given: graphs of bounded treewidth are sparse (they have fewer than kn edges), and therefore an n^2 bit adjacency matrix cannot be constructed in linear time from a typical $O(kn \log n)$ bit input encoding, e.g. based on adjacency lists. (Note that for all cases where this method is relevant, k is much smaller than n.) We remark that if using quadratic space is allowed, then there is an easy linear time preprocessing procedure that enables constant time adjacency testing, using a lazy (i.e. uninitialized) adjacency matrix [1, Exercise 2.12]. However, there seems to be no straightforward way of obtaining linear time and space dynamic programming algorithm. In Section 3 we discuss these claims in more detail, and also present our first result: a simple $O(kn)$ time and space preprocessing procedure that, given a graph on n vertices and a tree decomposition on $O(n)$ nodes of width k, enables adjacency testing in time $O(k)$. This is not very deep, but also not obvious, and solves a gap in the existing literature.

There are more examples of seemingly trivial problems on sparse graphs where the fact that adjacency testing cannot be done in constant time is surprisingly problematic. For instance, it is well-known that a graph is a series parallel graph if and only if it can be reduced to a K_2 by iteratively suppressing vertices of degree two, and replacing multi-edges by single edges. Assuming that adjacency testing can be done in constant time, this characterization would easily yield a linear time algorithm for recognizing series parallel graphs. Nevertheless, without this assumption, a significantly more sophisticated algorithm is needed; see Valdes et al [18, Section 3.3].

Secondly, we consider the dependency of the complexity on the parameter k. For many NP-hard problems, $O(\text{poly}(k)c^k n)$ time algorithms are known, for some constant c (poly(k) denotes a polynomial factor in k). For problems where solutions can be characterized using local properties, this has been known for a long time, see e.g. [17]. In recent breakthrough results, such a complexity has also been obtained for problems with global connectivity constraints [5,10]. However, we study the simpler local problems here. For various of these problems, $O(\text{poly}(k)c^k n)$ algorithms are known, where c^k is the natural size of the dynamic programming tables. For various problems such as Max Independent Set, Max Cut and q-Colorability (problems without complex join operations), such a complexity is relatively easy to prove (see e.g. [6,15]). For other problems, such as in particular Min Dominating Set, this is significantly harder, but a $(3^k k^2 n)$ complexity has been achieved using the fast subset convolution technique [16], improving on the previous $O(4^k n)$ algorithm [2].

Assuming the Strong Exponential Time Hypothesis [12], it has been shown in [14] that the constant c in the exponential factor c^k cannot be improved for the aforementioned problems. Therefore, we study the question whether the $\text{poly}(k)$ factor can be removed. For problems addressed by the fast subset convolution technique, this seems impossible. However, for a large class of other problems this can be done: in Section 4 we consider the Minimum Weight Homomorphism (MWH) problem: given a graph G on n vertices, and a graph H on q vertices with integer vertex and edge weights, find a minimum weight homomorphism from G to H, or decide that none exists. This weight is the sum of the vertex and edge image weights. The graphs G and H may be directed and may have loops. MWH generalizes well-studied problems such as Max Independent Set, Max Cut and q-Colorability, and many others. We give an $O(q^{k+1}n)$ dynamic programming algorithm for MWH in Section 5.

Removing the $\text{poly}(k)$ factor requires precise treatment of various dynamic programming details that can usually be ignored, such as bag and table ordering, fast table building, and enabling constant time adjacency checking. One of the few papers that also discusses some of these in detail is the paper by Alber and Niedermeier [2], presenting an $O(4^k n)$ algorithm for Min Dominating Set (see also [15]). To be precise, only the join operation requires time $O(4^k)$ in [2]. For the case of path decompositions, where no join operation is required, they give an $O(k3^k n)$ algorithm. To illustrate that our techniques can be applied to a wider variety of problems, in the full version of this paper we will show that the $O(k)$ factor can be removed in this result. More precisely, using our techniques an $O(3^k n)$ algorithm for Min Dominating Set can be constructed, when a path decomposition of width k is given. In [6], another algorithm without $\text{poly}(k)$ factor is presented: a $O(2^k n)$ time algorithm for Max Independent Set is sketched, although various details are omitted. This inspired the current study.

We remark that from a purely theoretical asymptotic analysis viewpoint, removing $\text{poly}(k)$ factors seems irrelevant. Indeed, abusing the O-notation, for any $\epsilon > 0$ one may for instance write $O(k^2 3^k n) \subseteq O((3 + \epsilon)^k n)$. Nevertheless, we note that e.g. $k^2 3^k < 4^k$ only holds when $k \geq 22$. Since at this point, dynamic programming tables cannot be stored in a normal computer memory anymore, we conclude that the previous $O(4^k n)$ MDS algorithm by Alber and Niedermeier [2] is still the most efficient one in practice! (Compared to [16].) Similarly, the complexity improvements we present are important in practice. We start in Section 2 with basic definitions. Statements for which proofs will be given in the full version are marked with a star.

2 Preliminaries

For basic graph theory notations, see [11]. By uv and (u,v) we denote undirected and directed edges, respectively. By $N(u)$, $N^+(u)$ and $N^-(u)$ we denote the (undirected) neighborhood, out-neighborhood and in-neighborhood of u, respectively.

Definition 1. *A* tree decomposition *of a (di)graph G is a 2-tuple (T, X) where T is a tree and $X = \{X_v : v \in V(T)\}$ is a set of subsets X_v of $V(G)$ such that the following properties hold:*

1. *For every $xy \in E(G)$ (resp. $(x,y) \in E(G)$), there is a $v \in V(T)$ with $\{x,y\} \subseteq X_v$.*
2. *For every $x \in V(G)$, the subgraph of T induced by $X^{-1}(x) = \{v \in V(T) : x \in X_v\}$ is non-empty and connected.*

To distinguish between vertices of G and T, the latter are called *nodes*. For $v \in V(T)$, the set X_v is also called the *bag* of v. The *width* of a tree decomposition is $\max_{v \in V(T)} |X_v| - 1$. The *treewidth* $\text{tw}(G)$ of a graph G is the minimum width over all tree decompositions of G. A *rooted tree decomposition* $(T, X), r$ of G is obtained by additionally choosing a root $r \in V(T)$, which defines a child/parent relation between every pair of adjacent nodes, and ancestors/descendants in the usual way. A node without children is called a *leaf*.

Definition 2. *A rooted tree decomposition $(T, X), r$ of G is nice if every node $u \in V(T)$ is of one of the following types:*

Leaf: *u has no children.*
Forget: *u has one child v with $X_u \subset X_v$ and $|X_u| = |X_v| - 1$.*
Introduce: *u has one child v with $X_v \subset X_u$ and $|X_u| = |X_v| + 1$.*
Join: *u has two children v and w with $X_u = X_v = X_w$.*

The tree decomposition is called very nice *if in addition, for every leaf node u it holds that $|X_u| = 1$.*

If u is an introduce node with child v and $X_u \setminus X_v = \{x\}$, then we say x is *introduced* in u. The following fact is well-known and easy to prove: if G is a graph on n vertices with tree width k, then G has at most kn edges. For dynamic programming *(DP)* over tree decompositions, the following definitions are important. Let $(T, X), r$ be a rooted tree decomposition of G. For a node $u \in V(T)$, we denote $X(u) = \cup_v X_v$, where the union is taken over all descendants v of u, including u itself. The subgraph $G(u)$ is then defined as $G(u) = G[X(u)]$. DP algorithms rely on the following two key properties, which follow easily from Definition 1: firstly, $G(r) = G$. Secondly, for every $u \in V(T)$, the only vertices of $G(u)$ that (in G) may be incident with edges that are not in $G(u)$ are vertices in X_u.

Computation Model and Assumptions. We use the standard computation model (i.e. the RAM model, see e.g. [1]). The memory consists of an unbounded number of registers r_i, $i \in \mathbb{N}$, which can hold integer values. In constant time, we can read or write any r_i, execute an elementary program control instruction, or carry out a basic arithmetic operation. As basic arithmetic operations, we only require addition, subtraction, multiplication, and testing whether an integer is positive.

For convenience, we assume that for all graphs G we consider, $V(G) = \{1, \ldots, n\}$. The following standard (but usually implicit) assumption is important when discussing linear time (space) algorithms: if for an algorithmic result, no input encoding is specified, the result should hold for all "reasonable input encodings". For the case of a graph G with n vertices and m edges, and $\text{tw}(G) = k$, this implies in particular that for designing linear time algorithms, we cannot assume that an adjacency matrix is given. This is because an n^2 bit adjacency matrix cannot be constructed in linear time (which is $O(m) \subseteq O(kn)$) from adjacency lists, or edge lists. On the other hand, an adjacency list representation of G can be constructed in linear time from all other reasonable representations, so we may assume that an adjacency list encoding of G is given. Similarly, if a tree decomposition (T, X) of G is given, we should assume that the bags X_u are given as unordered lists of vertices of G, and that no additional information is available (such as a list of edges of $G[X_u]$).

3   Enabling Fast Adjacency Testing

We first discuss the useful and simple quadratic space solution using *lazy adjacency matrices*. Note that we may not assume that all registers are initialized to zero at the beginning of the computation, so using an n^2 bit adjacency matrix requires (non-linear) initialization time $\Theta(n^2)$. However this can be avoided using the following known trick (see e.g. [1, Exercise 2.12]): Store a list of edge objects in a consecutive memory block of size $O(m) \subseteq O(kn)$. Reserve n^2 registers for the adjacency matrix, called $a_{i,j}$ for $i, j \in V(G)$, but do not initialize these. Instead, for every edge $e = \{i, j\}$, initialize $a_{i,j}$ and $a_{j,i}$ with a pointer to the register containing the edge object e. Observe that this now allows constant time adjacency checking.

Nevertheless, if we insist on using linear space, there is no obvious reason why known dynamic programming algorithms can be implemented in linear time. For instance, graphs of treewidth k may contain vertices for which the degree is not bounded by any function of k. Furthermore, it may be that in any low width nice tree decomposition, such high degree vertices are introduced many times. So if for every introduce node, one considers the entire neighbor list of the introduced vertex, this does not yield a linear time algorithm.

Let T be a tree with root r. For any set $S \subseteq V(T)$, define a node $v \in S$ to be a *top node* if the unique (r, v)-path contains no vertices from S other than v. We will need the following basic properties of top nodes.

Proposition 3. *Let T, r be a rooted tree, and $S \subseteq V(T)$ with $T[S]$ connected.*

(i) S contains exactly one top node; denote this node by $\top(S)$.

(ii) Let $u \in S$ with parent v. Then $u = \top(S)$ if and only if $v \notin S$.

(iii) For any $S' \subseteq V(T)$ with $T[S']$ connected and $S \cap S' \neq \emptyset$: $\top(S') \in S$ or $\top(S) \in S'$.

Proof: (i): Suppose to the contrary that S contains two top nodes u and v. Let P_u and P_v be the unique (r, u)-path and (r, v)-path in T, respectively. Let w be the last vertex on P_u that is also in P_v (i.e. the lowest common ancestor of u and v). Combining the subpath of P_u from u to w and the subpath of P_v from w to v yields a path P_{uv} from u to v with an internal vertex w, with $w \notin S$. But then P_{uv} is the unique path from u to v in T, so deleting w separates u from v, contradicting that $T[S]$ is connected.

(ii): The forward direction is trivial. For the backward direction, suppose to the contrary that the unique (r, u)-path contains another vertex w with $w \in S$. Then the unique path from u to w contains $v \notin S$, contradicting that S is connected.

(iii): Consider a top node u of $S \cap S'$. Then the parent v of u is not in $S \cap S'$. If $v \notin S$, then by (ii), $u = \top(S)$. Analogously, if $v \notin S'$, then $v = \top(S')$.   □

Theorem 4. *Given a rooted tree decomposition of width k on $O(n)$ nodes, of a (di)graph G with $V(G) = \{1, \ldots, n\}$, there is an $O(kn)$ time preprocessing procedure that enables testing whether $x \in N(y)$ (resp. $x \in N^+(y)$ or $x \in N^-(y)$) in time $O(k)$, for all $x, y \in V(G)$.*

Proof: Let $(T, X), r$ be the given tree decomposition of width k, of a graph G on n vertices. For $x \in V(G)$, denote $\top(x) = \top(X^{-1}(x))$, i.e. the top node of all nodes that contain x in their bag. We now argue that in time $O(kn)$, $\top(x)$ can be computed for every $x \in V(G)$. For every node $u \in V(T)$: if u has a parent v, then mark all $x \in X_v$. Next, set $\top(x) = u$ for all unmarked $x \in X_u$. This is correct by Proposition 3(ii). Finally, reset the markings for all $x \in X_v$. For a single node, this procedure takes time $O(k)$. During this process, we can also compute for every node $u \in V(T)$ a list $L(u)$ of vertices x with $\top(x) = u$.

For $x \in V(G)$, define $N_{\text{top}}(x) = N(x) \cap X_{\top(x)}$. We argue that in time $O(kn)$, $N_{\text{top}}(x)$ can be computed for every $x \in V(G)$. For every node u, first mark all vertices in X_u. Next, for every vertex $x \in L(u)$, construct $N_{\text{top}}(x)$ by considering all neighbors and adding those that are marked to $N_{\text{top}}(x)$. Finally, reset the markings for all $x \in X_u$. For a single node u this procedure takes time $O(k) + O(\sum_{x \in L(u)} d(x))$. Since G has at most $O(kn)$ edges, this yields a complexity of $O(kn)$.

We argue that for any two vertices $x, y \in V(G)$, $xy \in E(G)$ if and only if $x \in N_{\text{top}}(y)$ or $y \in N_{\text{top}}(x)$. Suppose $xy \in E(G)$. By Definition 1, both $X^{-1}(x)$ and $X^{-1}(y)$ are connected, and $X^{-1}(x) \cap X^{-1}(y) \neq \emptyset$. So by Proposition 3(iii), $\top(x) \in X^{-1}(y)$ or $\top(y) \in X^{-1}(x)$ holds. Thus $y \in N_{\text{top}}(x)$ or $x \in N_{\text{top}}(y)$.

Finally, since every bag contains at most $k + 1$ vertices, $|N_{\text{top}}(x)| \leq k + 1$ holds for all $x \in V(G)$. So testing whether $x \in N_{\text{top}}(y)$ or $y \in N_{\text{top}}(x)$ can be done in time $O(k)$.

A simple modification of the above proof yields the statement for the case of digraphs. (The key statement then is that $(x, y) \in E(G)$ if and only if $x \in N_{\text{top}}^-(y)$ or $y \in N_{\text{top}}^+(x)$.)   □

4   Dynamic Programming Rules for Min-Weight Homomorphism

Let G and H be digraphs, possibly with loops. A *homomorphism* from G to H is a function $f : V(G) \to V(H)$ such that for all $(u, v) \in E(G)$, $(f(u), f(v)) \in E(H)$. Let $a : V(H) \to \mathbb{N}$ and $b : E(H) \to \mathbb{N}$ be weight functions. The *weight* of a homomorphism f from G to H is then

$$w(f) = \sum_{v \in V(G)} a(f(v)) + \sum_{(u,v) \in E(G)} b(f(u), f(v)).$$

The problem *Min-Weight Homomorphism (MWH)* is defined as follows: given digraphs G and H, with vertex and edge weights a and b, decide whether a homomorphism from G to H exists, and if so, compute one of minimum weight. Note that the analog problem where both G and H are undirected is a special case, since undirected edges can be replaced by a pair of oppositely directed edges. This problem generalizes many well-studied problems, such as:

- Max Independent Set (Min Vertex Cover): choose H to be an undirected graph on two vertices u and v, with an edge between them and a loop on u. The two edges and v have weight zero, u has weight one.
- q-Colorability: choose H to be a complete undirected graph on q vertices. The weights are irrelevant, since this is a decision problem.
- Max-Cut (Min Edge Bipartization): choose H to be an undirected graph on two vertices, with an edge between them and loops on both. The loops both receive weight one, and the other edge and both vertices receive weight zero.

For notational convenience, we modify H as follows: for every $u, v \in V(H)$ (including $u = v$), if $(u,v) \notin E(H)$ then add an edge (u,v) with weight ∞. So the original graph admits a homomorphism if and only if the new graph admits a homomorphism of finite weight. From now on, assume that $(u,v) \in E(H)$ for all u, v, possibly with infinite weight.

Let $(T, X), r$ be a rooted tree decomposition of G. For a node $u \in V(T)$, our DP computes values $val_u(f)$ for every $f : X_u \to V(H)$, defined as follows:

$$val_u(f) = \min\{w(h) \mid h : X(u) \to V(H) \text{ s.t. } h|_{X_u} = f\}.$$

So $val(f)$ is the minimum weight of a homomorphism from $G(u)$ to H that coincides with f. Then since $G(r) = G$, the minimum weight of a homomorphism from G to H is computed by taking the minimum value of $val_r(f)$ over all $f : X_r \to V(H)$. The values $val_u(f)$ can be computed as follows, in case $(T, X), r$ is a nice tree decomposition:

Lemma 5 (*). *Let $(T, X), r$ be a nice tree decomposition, and let $u \in V(T)$.*

Leaf: *If u is a leaf node, then $val_u(f) = w(f)$.*
Forget: *If u is a forget node with child v, then $val_u(f) = \min\{val_v(h) \mid h : X_v \to V(H) \text{ s.t. } h|_{X_u} = f\}$.*
Introduce: *If u is an introduce node with child v and $X_u \setminus X_v = \{x\}$, then*

$$val_u(f) = val_v(f|_{X_v}) + a(f(x)) +$$

$$\sum_{y \in N^+(x) \cap X_u} b(f(x), f(y)) + \sum_{y \in N^-(x) \cap X_v} b(f(y), f(x)).$$

Join: *If u is a join node with children v and x, then $val_u(f) = val_v(f) + val_x(f) - w(f)$.*

(For the introduce case, summing over $y \in N^+(x) \cap X_u$ and $y \in N^-(x) \cap X_v$ is done to guarantee that a possible loop on x is only considered once.)

5   Implementation

We now show how the above rules can be implemented to yield a total DP complexity of $O(q^{k+1}n)$. First note that we use nice tree decompositions, and not *very* nice tree decompositions, i.e. we do not require for leaf nodes u that $|X_u| = 1$. This has the downside that the computation for leaf nodes becomes more complicated. However, the problem with very nice tree decompositions is that they may have more than $O(n)$ nodes (recall that we do not view k as a constant!): Consider the graph G_n with vertex set $\{x_1, \ldots, x_n\} \cup \{y_1, \ldots, y_n\}$, and edges $x_i x_j$ for all i, j, and $x_i y_j$ for all $i \neq j$. This graph has treewidth $n - 1$: it has a tree decomposition (T, X) where T is a $K_{1,n}$, where the central node u has $X_u = \{x_1, \ldots, x_n\}$, and every y_i is contained in the bag for exactly one leaf. All bags have size n. It can be verified that any very nice tree decomposition of this graph, of width $k = \text{tw}(G_n) = n - 1$, has $\Omega(kn)$ nodes. This makes it hard to prove the desired complexity, so we use nice tree decompositions instead, for which it is well-known that they have at most $4n$ nodes (see e.g. [13, Lemma 13.1.2]). Algorithmically, using the fact that for any $uv \in E(T)$, $X_u \setminus X_v$ and $X_v \setminus X_u$ can be computed in time $O(k)$ (see Section 3), one can prove the following:

Lemma 6 (*). *Let G be a graph on n vertices. Given a tree decomposition (T, X) of G of width k, on $O(n)$ nodes, in time $O(kn)$, a nice tree decomposition (T', X') of G of width k can be constructed, with at most $4n$ nodes.*

(We remark that in the end, this distinction is not so important; our computation method for leaf nodes can also be viewed as a way of analyzing very nice tree decompositions that have a specific form, which can always be guaranteed.) So now it suffices to prove that for any single node u, all values $\text{val}_u(f)$ can be computed in time $O(q^{k+1})$. Since there may be q^{k+1} functions $f : X_u \to V(H)$, this requires that every value can be computed in (amortized) constant time. In the case of a join node (which is the most challenging to implement), the value $\text{val}_v(f) + \text{val}_w(f) - w(f|_{X_u})$ should be be computed in constant time. This provides two challenges: computing $w(f|_{X_u})$ in constant time for every f, and looking up the matching values $\text{val}_v(f)$ and $\text{val}_w(f)$ in constant time. The latter challenge is addressed by storing the values $\text{val}_u(f)$ and $\text{val}_v(f)$ for all f in two tables (for u and for v), which have to be ordered the same way. We first discuss details related to this ordering.

Ordered Bags and Tables. W.l.o.g. we assume throughout that $V(H) = Q = \{0, \ldots, q - 1\}$. Let $u \in V(H)$ and $X_u = \{x_1, \ldots, x_p\}$. Functions $f : X_u \to Q$ will be represented by a vector $(c_1, \ldots, c_p)$ where $c_i = f(x_i)$. This requires a complete order on the bag vertices. We assume that the vertices of X_u are ordered $(x_1, \ldots, x_p)$ such that for all $i < j$, $x_i < x_j$ holds. (Recall that $V(G) = \{1, \ldots, n\}$.) Ordered bags are denoted as tuples instead of sets. The values $\text{val}_u(f)$ are then stored in a table (array) T_u of length q^p, according to the order given by $\text{index}(f) = \sum_{i=1}^{p} q^{i-1} c_i$. So $T_u[\text{index}(f)] = \text{val}_u(f)$. (Note that index($f$) is a bijection from all possible functions f to $\{0, \ldots, q^p - 1\}$.) If this ordering method is used, we say that T_u *is a table representing all functions* $f : X_u \to Q$, *ordered according to* $(x_1, \ldots, x_p)$. For a join node u with children v and w, we now have for every i that $T_u[i] = T_v[i] + T_w[i] - w(f|_{X_u})$ (where $i = \text{index}(f)$), so we can easily find matching values in the tables of v and w. The desired order on the vertices of each bag can be guaranteed using a simple preprocessing step: Using any $O(k \log k)$ time sorting algorithm, this can be done within time $O(nk \log k)$ for all nodes.

However, this introduces a small problem for forget and introduce nodes: e.g. for an introduce node u with child v, it is convenient to assume that $X_u = (x_1, \ldots, x_{p+1})$ and $X_v = (x_1, \ldots, x_p)$, i.e. the newly introduced vertex is the last one. This means that the tables may need to be reordered. Since this reordering corresponds to 'swapping only one coordinate', this can be done in time $O(q^p)$, as shown in the next lemma.

Lemma 7. *Let* $X = \{x_1, \ldots, x_p\}$, *and* $Q = \{0, \ldots, q - 1\}$. *Let* T *and* T' *be tables representing all functions* $f : X \to Q$, *ordered according to* $(x_1, \ldots, x_p)$ *and* $(x_1, \ldots, x_{i-1}, x_{i+1}, \ldots, x_p, x_i)$, *respectively. Then* T *and* T' *can be computed from each other in time* $O(q^p)$.

Proof: For $f : X \to Q$ with $f(x_i) = c_i$ for all i, recall that $\text{index}(f) = \sum_{j=1}^{p} q^{j-1} c_j$. Define $\text{index}'(f) = \sum_{j=1}^{i-1} q^{j-1} c_j + \sum_{j=i+1}^{p} q^{j-2} c_j + q^{p-1} c_i$. This way, for every f it holds that $T[\text{index}(f)] = T'[\text{index}'(f)]$.

Define $x(f) = \sum_{j=1}^{i-1} q^{j-1} c_j$, $y(f) = \sum_{j=1}^{p-i} q^{j-1} c_{j+i}$ and $z(f) = c_i$. Then we can alternatively write $\text{index}(f) = x(f) + q^i y(f) + q^{i-1} z(f)$ and $\text{index}'(f) = x(f) + q^{i-1} y(f) + q^{p-1} z(f)$. In addition, observe that for every combination of values $x \in \{0, \ldots, q^{i-1} - 1\}$, $y \in \{0, \ldots, q^{p-i} - 1\}$ and $z \in \{0, \ldots, q-1\}$ there is a function f with $x(f) = x$, $y(f) = y$ and $z(f) = z$. Therefore, the tables T and T' can be computed from each other by looping over all possible combinations of values x, y and z, and using the equality $T[x + q^i y + q^{i-1} z] = T'[x + q^{i-1} y + q^{p-1} z]$. The values q^i, q^{i-1} and q^{p-1} can be computed beforehand, which ensures that the computation for a single combination of x, y and z requires only a constant number of elementary arithmetic operations (addition and multiplication). Since there are exactly q^p combinations of x, y and z, this proves the statement. □

Table Computation We now show for every type of node u how the table T_u can be computed efficiently.

Lemma 8. *Let* u *be a forget node with child* v, *for which the table* T_v *is known. Let* $p = |X_v|$. *Then in time* $O(q^p)$, *the table* T_u *can be computed.*

Proof: First suppose that $X_u = (x_1, \ldots, x_{p-1})$ and $X_v = (x_1, \ldots, x_p)$, i.e. the last vertex in the (ordered) bag X_v is forgotten. Then we compute the values $T_u[i]$ as follows. First, initialize $T_u[i] = \infty$ for all $i \in \{0, \ldots, q^{p-1} - 1\}$. Next, for all combinations of $i \in \{0, \ldots, q - 1\}$ and $j \in \{0, \ldots, q^{p-1} - 1\}$, reassign $T_u[j] := T_v[j + iq^{p-1}]$, if the latter value is smaller than the current value of $T_u[j]$. Because of the way the tables are ordered, and because $X_v \setminus X_u = \{x_p\}$, this correctly computes $T_u[\text{index}(f)] = \min_h T_v[\text{index}(h)]$ over all $h : X_v \to Q$ with $h|_{X_u} = f$. This computation takes constant time for one entry of T_v (using the precomputed value q^{p-1}), so time $O(q^p)$ in total.

In the case that $X_v = (x_1, \ldots, x_p)$ and $X_u = (x_1, \ldots, x_{i-1}, x_{i+1}, \ldots, x_p)$ for $i < p$, we can first translate the table T_v into a table T' ordered according to $(x_1, \ldots, x_{i-1}, x_{i+1}, \ldots, x_p, x_i)$, in time $O(q^p)$ (Lemma 7), and then apply the above procedure.   □

Next, we show how to efficiently compute the table T_u for an introduce node u. Recall that if u is an introduce node with child v and $X_u \setminus X_v = \{x\}$, then $\text{val}_u(f)$ can be computed by adding to $\text{val}_v(f|_{X_v})$ a *correction term* of $a(f(x)) + \sum_{y \in N^+(x) \cap X_u} b(f(x), f(y)) + \sum_{y \in N^-(x) \cap X_v} b(f(y), f(x))$. We will show how to compute this correction term in (amortized) constant time. This involves looping over all $y \in X_u$, and deciding whether $y \in N^+(x)$ or $y \in N^-(x)$, respectively. This needs to be done in constant time, instead of time $O(k)$, as given by the method from Section 3. To this end, we use an initial preprocessing step that computes a *local adjacency matrix* A^u for every node u. To define A^u, we use the bag order introduced above again. Let $X_u = (x_1, \ldots, x_p)$. Then for $i, j \in \{1, \ldots, p\}$, $A^u_{i,j} = 1$ if and only if $(x_i, x_j) \in E(G)$, and $A^u_{i,j} = 0$ otherwise. Using the techniques from Section 3, one can easily prove the next statement.

Proposition 9 (*). *Let (T, X) be a tree decomposition on $O(n)$ nodes of width k, of a graph G on n vertices, with ordered bags. In time $O(k^2 n)$, local adjacency matrices can be computed for every node.*

Furthermore, for H we store a vertex weight vector and edge weight matrix in the memory, to ensure that for any $x, y \in V(H)$, the values $a(x)$ and $b(x, y)$ can be retrieved in constant time. (Recall that $V(H) = \{0, \ldots, q - 1\}$). This introduces at most a negligible term $O(q^2)$ to the complexity.

Lemma 10. *Let u be an introduce node with child v, for which the table T_v is known. Let $p = |X_u|$. Then in time $O(q^p)$, the table T_u can be computed.*

Proof: We assume that $X_u = (x_1, \ldots, x_p)$ and $X_v = (x_1, \ldots, x_{p-1})$, so x_p is the vertex that is introduced to the bag. This assumption is justified after using an $O(q^{p-1})$ preprocessing step, similar to the case of forget nodes (see above).

For $\alpha \in Q$ and $j \in \{0, \ldots, p-1\}$, define $X_u^j = \{x_1, \ldots, x_j, x_p\}$ (in particular $X_u^0 = \{x_p\}$) and

$$\text{cor}_j^\alpha(f) = a(\alpha) + \sum_{y \in N^+(x_p) \cap X_u^j} b(\alpha, f(y)) + \sum_{y \in N^-(x_p) \cap X_u^j \setminus \{x_p\}} b(f(y), \alpha),$$

for any function $f : X_u^j \to Q$ with $f(x_p) = \alpha$. Define C_j^α to be a table containing these values, ordered according to $(x_1, \ldots, x_j)$. (More precisely: $C_j^\alpha[\text{index}(f)] = \text{cor}_j^\alpha(f)$, where $\text{index}(f) = \sum_{i=1}^j q^{i-1} f(x_i)$. The table C_j^α has length q^j.)

The table T_u can then be computed using the equality $T_u[i + q^{p-1}\alpha] = T_v[i] + C_{p-1}^\alpha[i]$, for all $i \in \{0, \ldots, q^{p-1} - 1\}$ and $\alpha \in \{0, \ldots, q - 1\}$. So it now suffices to show how, for every $\alpha \in Q$, the table C_{p-1}^α can be computed in time $O(q^{p-1})$. These tables can be computed as follows: C_0^α has a single entry, with value $a(\alpha)$ if there is no loop on x_p, and value $a(\alpha) + b(\alpha, \alpha)$ otherwise. For every $j \geq 1$, C_j^α can be computed from C_{j-1}^α as follows. Loop over all values $\beta := f(x_j) \in \{0, \ldots, q-1\}$. For every such β, the next segment of C_j^α (of length q^{j-1}) consists of a copy of the table C_{j-1}^α, with a term $\mathrm{val}^- + \mathrm{val}^+$ added to each entry, where

- $\mathrm{val}^- = b(\beta, \alpha)$ if $x_j \in N^-(x_p)$, and $\mathrm{val}^- = 0$ otherwise, and
- $\mathrm{val}^+ = b(\alpha, \beta)$ if $x_j \in N^+(x_p)$, and $\mathrm{val}^+ = 0$ otherwise.

We use the entries $A_{j,p}^u$ and $A_{p,j}^u$ of the local adjacency matrix A^u to decide in constant time whether $x_j \in N^-(x_p)$ or $x_j \in N^+(x_p)$. Recall that using a precomputed edge weight matrix, the value $b(\beta, \alpha)$ can be retrieved in constant time. Since every entry computation now takes constant time, C_j^α can be computed from C_{j-1}^α this way in time $O(q^j)$, for every $j \in \{0, \ldots, p-1\}$. In total, this gives a complexity of $q + q^2 + \ldots + q^{p-1} = \frac{q^p - 1}{q-1} - 1 \in O(q^{p-1})$. $\qquad\square$

Lemma 11. *Let u be a leaf, with $p = |X_u|$. Then in time $O(q^p)$, the table T_u can be computed.*

Proof: Let $X_u = (x_1, \ldots, x_p)$. Essentially, computing T_u is done by turning the nice tree decomposition into a very nice tree decomposition: the leaf u is replaced by a path $u_p, u_{p-1}, \ldots, u_1$ (rooted at u_p), with $X_{u_i} = (x_1, \ldots, x_i)$. In particular, $X_{u_p} = X_u$.

The table for node u_1 can be trivially computed in time $O(q)$. For $i \geq 2$, u_i is an introduce node, and by Lemma 10, the table T_{u_i} can be computed in time $O(q^i)$. Since $q + q^2 + \ldots + q^p = \frac{q^{p+1} - 1}{q-1} - 1 \in O(q^p)$, this shows that computing the table $T_{u_p} = T_u$ can be done in time $O(q^p)$. $\qquad\square$

Lemma 12. *Let u be a join node, with $p = |X_u|$, and children v and x for which the tables T_v and T_x are known. Then in time $O(q^p)$, the table T_u can be computed.*

Proof: Recall that $\mathrm{val}_u(f) = \mathrm{val}_v(f) + \mathrm{val}_x(f) - w(f)$ holds for every $f : X_u \to Q$. Because the tables for T_v and T_x are ordered the same way, the only remaining challenge lies in computing the correction terms $w(f)$. Essentially, this is done by introducing a new leaf child ℓ for u, with $X_\ell = X_u$ (ordered the same way as X_u). Then we can write $\mathrm{val}_u(f) = \mathrm{val}_v(f) + \mathrm{val}_x(f) - \mathrm{val}_\ell(f)$ for every f. In other words, $T_u[i] = T_v[i] + T_x[i] - T_\ell[i]$ for every index i. By Lemma 11, the table T_ℓ can be computed in time $O(q^p)$, which concludes the proof. $\qquad\square$

Now we can prove our main result.

Theorem 13. *Let G be a digraph on n vertices, for which a tree decomposition (T, X) on $O(n)$ nodes is given, of width k. Let H be a digraph with $|V(H)| = q \geq 2$, and nonnegative integer vertex and edge weights. Then in time $O(nq^{k+1})$, it can be decided whether a homomorphism $f : V(G) \to V(H)$ exists, and if so, one of minimum weight can be computed.*

Proof: The algorithm can be summarized as follows. First, in time $O(kn)$ we transform the given tree decomposition into a nice tree decomposition of width k, on $O(n)$ nodes (Lemma 6). Next, we order the bags for every node, in time $O(nk \log k)$. Then we use the $O(kn)$ time preprocessing procedure from Theorem 4 to ensure that (directed) adjacency testing can be done in time $O(k)$. Finally, we use this to compute local adjacency matrices for every node, as defined above, in total time $O(k^2 n)$ (Proposition 9). This preprocessing phase has a total complexity of $O(k^2 n)$. For H, we precompute a vertex weight vector, edge weight matrix and power vector, to ensure that for every $x, y \in V(H)$, the values $a(x)$ and $b(x, y)$ can be retrieved in constant time, and for every $i \in \{0, \dots, k\}$, the value q^i can be retrieved in constant time. This adds only a negligible term to the total complexity. Define $b(x, y) = \infty$ if $(x, y) \notin E(H)$.

At this point, the tree decomposition is in the desired form, and auxiliary data structures have been built, so that Lemmas 8–12 can be applied. Since bags contain at most $k+1$ vertices, and there are at most $O(n)$ nodes, this shows that computing all tables can be done in time $O(q^{k+1}n)$. Let r be the root node of the nice tree decomposition. By definition of the tables, and because $G(r) = G$, there exists a homomorphism f from G to H of weight at most m if and only if the table T_r contains a value of at most m. This gives the algorithm for the decision problem, with a total complexity of $O(k^2 n) + O(q^{k+1} n) \subseteq O(q^{k+1} n)$. Constructing a minimum weight homomorphism f can subsequently be done in a standard way: mark a minimum entry in the table for T_r, and use this to mark the corresponding entries in all other nodes, in a top down way (for forget nodes, see the proof of Lemma 8). Then for every vertex $x \in V(G)$, $f(x)$ can be computed by considering the marked entry in $\top(x)$. This can be implemented such that the complexity does not increase by more than a constant factor. □

6   Discussion

The fast DP table building techniques we use here can be used for many other DP problems. For instance, in the full version of this paper we will use them to show that Min Dominating Set can be solved in time $O(3^k n)$, if a *path* decomposition of width k is given. (For join nodes, no $O(3^k)$ implementation seems possible.) This is somewhat surprising since values in introduce nodes tables are the minimum of multiple values in the child table. These can nevertheless be computed in constant time on average, by computing a table of correction *pointers*, similar to the table of correction terms in Lemma 10.

References

1. Aho, A.V., Hopcroft, J.E., Ullman, J.D.: The Design and Analysis of Computer Algorithms. Addison-Wesley, Reading (1974)
2. Alber, J., Niedermeier, R.: Improved tree decomposition based algorithms for domination-like problems. In: Rajsbaum, S. (ed.) LATIN 2002. LNCS, vol. 2286, pp. 613–627. Springer, Heidelberg (2002)

3. Arnborg, S., Proskurowski, A.: Linear time algorithms for NP-hard problems restricted to partial k-trees. Discrete Applied Mathematics 23(1), 11–24 (1989)
4. Bodlaender, H.L.: Dynamic programming on graphs with bounded treewidth. In: Lepistö, T., Salomaa, A. (eds.) ICALP 1988. LNCS, vol. 317, pp. 105–118. Springer, Heidelberg (1988)
5. Bodlaender, H.L., Cygan, M., Kratsch, S., Nederlof, J.: Deterministic single exponential time algorithms for connectivity problems parameterized by treewidth. In: Fomin, F.V., Freivalds, R., Kwiatkowska, M., Peleg, D. (eds.) ICALP 2013, Part I. LNCS, vol. 7965, pp. 196–207. Springer, Heidelberg (2013)
6. Bodlaender, H.L., Koster, A.M.C.A.: Combinatorial optimization on graphs of bounded treewidth. The Computer Journal 51(3), 255–269 (2008)
7. Borie, R.B.: Recursively Constructed Graph Families. PhD thesis, School of Information and Computer Science, Georgia Institute of Technology (1988)
8. Borie, R.B., Gary Parker, R., Tovey, C.A.: Solving problems on recursively constructed graphs. ACM Computing Surveys 41(4) (2008)
9. Courcelle, B.: The monadic second-order logic of graphs. I. Recognizable sets of finite graphs. Information and Computation 85(1), 12–75 (1990)
10. Cygan, M., Nederlof, J., Pilipczuk, M., van Rooij, J.M.M., Wojtaszczyk, J.O.: Solving connectivity problems parameterized by treewidth in single exponential time. In: FOCS 2011, pp. 150–159. IEEE (2011)
11. Diestel, R.: Graph Theory, 4th edn. Springer (2010)
12. Impagliazzo, R., Paturi, R.: On the complexity of k-SAT. Journal of Computer and System Sciences 62(2), 367–375 (2001)
13. Kloks, T.: Treewidth. LNCS, vol. 842. Springer, Heidelberg (1994)
14. Lokshtanov, D., Marx, D., Saurabh, S.: Known algorithms on graphs of bounded treewidth are probably optimal. In: SODA 2011, pp. 777–789. SIAM (2011)
15. Niedermeier, R.: Invitation to Fixed-Parameter Algorithms. Oxford Lecture Series in Mathematics and its Applications, vol. 31. Oxford University Press, Oxford (2006)
16. van Rooij, J.M.M., Bodlaender, H.L., Rossmanith, P.: Dynamic programming on tree decompositions using generalised fast subset convolution. In: Fiat, A., Sanders, P. (eds.) ESA 2009. LNCS, vol. 5757, pp. 566–577. Springer, Heidelberg (2009)
17. Telle, J.A., Proskurowski, A.: Algorithms for vertex partitioning problems on partial k-trees. SIAM Journal on Discrete Mathematics 10(4), 529–550 (1997)
18. Valdes, J., Tarjan, R.E., Lawler, E.L.: The recognition of series parallel digraphs. In: STOC 1979, pp. 1–12. ACM (1979)
19. Wimer, T.V.: Linear Algorithms on k-Terminal Graphs. PhD thesis, Dept. of Computer Science, Clemson University (1987)

On Subexponential and FPT-Time Inapproximability*

Edouard Bonnet, Bruno Escoffier, Eun Jung Kim, and Vangelis Th. Paschos**

PSL Research University, Université Paris-Dauphine, LAMSADE, CNRS UMR 7243
{escoffier,eun-jung.kim,paschos}@lamsade.dauphine.fr

Abstract. Fixed-parameter algorithms, approximation algorithms and moderately exponential algorithms are three major approaches to algorithms design. While each of them being very active in its own, there is an increasing attention to the connection between these different frameworks. In particular, whether INDEPENDENT SET would be better approximable once endowed with subexponential-time or FPT-time is a central question. In this article, we provide new insights to this question using two complementary approaches; the former makes a strong link between the linear PCP conjecture and inapproximability; the latter builds a class of equivalent problems under approximation in subexponential time.

1 Introduction

Fixed-parameter algorithms, approximation algorithms and moderately exponential/subexponential algorithms are major approaches for efficiently solving NP-hard problems. These three areas, each of them being very active in its own, have been considered as foreign to each other until recently. Polynomial-time approximation algorithm produces a solution whose quality is guaranteed to lie within a certain range from the optimum. One illustrative problem indicating the development of this area is INDEPENDENT SET. The approximability of INDEPENDENT SET within constant ratios has remained as the most important open problems for a long time in the field. It was only after the novel characterization of **NP** by PCP theorem [2] that such inapproximability was proven assuming **P** $\neq$ **NP**. Subsequent improvements of the original PCP theorem led to much stronger result for INDEPENDENT SET: it is inapproximable within ratios $\Omega(n^{\varepsilon-1})$ for any $\varepsilon > 0$, unless **P** = **NP** [3].

Moderately exponential (subexponential, respectively) computation allows exponential (subexponential, respectively) running time for the sake of optimality. In this case, the endeavor lies in limiting the growth of the running time function as slow as possible. Parameterized complexity provides an alternative framework to analyze the running time in a more refined way [4]. Given an instance with

---

* Research supported by the French Agency for Research under the DEFIS program TODO, ANR-09-EMER-010.
** Also, Institut Universitaire de France.

G. Gutin and S. Szeider (Eds.): IPEC 2013, LNCS 8246, pp. 54–65, 2013.

a parameter k, the aim is to get an $O(f(k) \cdot n^c)$-time (or equivalently, FPT-time) algorithm for some constant c, where the constant c is independent of k. As these two research programs offer a generous running time when compared to that of classic approximation algorithms, a growing amount of attention is paid to them as a way to cope with hardness in approximability. The first one yields *moderately exponential approximation*. In moderately exponential approximation, the core question is whether a problem is approximable in moderately exponential time while such approximation is impossible in polynomial time. Suppose a problem is solvable in time $O^*(\gamma^n)$, but it is NP-hard to approximate within ratio r. Then, we seek for r-approximation algorithms of running time significantly faster than $O^*(\gamma^n)$. This issue has been considered for several problems [5,6,7,13,17].

The second research program handles approximation by fixed parameter algorithms. We say that a minimization (maximization, respecitvely) problem Π, together with a parameter k, is *parameterized r-approximable* if there exists an FPT-time algorithm which computes a solution of size at most (at least, respectively) rk whenever the input instance has a solution of size at most (at least, respectively) k. This line of research was initiated by three independent works [15,9,11]. For an excellent overview, see [22]. In what follows, parameterization means "standard parameterization", i.e., where problems are parameterized by the cost of the optimal solution.

Several natural questions can be asked dealing with these two programs. In particular, the following ones have been asked several times [22,15,17,7].

Q1: can a problem, which is highly inapproximable in polynomial time, be well-approximated in subexponential time?

Q2: does a problem, which is highly inapproximable in polynomial time, become well-approximable in FPT-time?

Few answers have been obtained until now. Regarding **Q1**, negative results can be directly obtained by gap-reductions for certain problems. For instance, COLORING is not approximable within ratio $4/3 - \epsilon$ since this would allow to determine whether a graph is 3-colorable or not in subexponential time. This contradicts a widely-acknowledged computational assumption [19]:

Exponential Time Hypothesis (ETH): There exists an $\epsilon > 0$ such that no algorithm solves 3SAT in time $2^{\epsilon n}$, where n is the number of variables.

Regarding **Q2**, [15] shows that assuming FPT $\neq$ W[2], for any r the INDEPENDENT DOMINATING SET problem is not r-approximable[1] in FPT-time.

Among interesting problems for which **Q1** and **Q2** are worth being asked are INDEPENDENT SET, COLORING and DOMINATING SET. They fit in the frame of both **Q1** and **Q2** above: they are hard to approximate in polynomial time while their approximability in subexponential or in parameterized time is still open.

In this paper, we study parameterized and subexponential (in)approximability of natural optimization problems. In particular, we follow two guidelines:

---

[1] Actually, the result is even stronger: it is impossible to obtain a ratio $r = g(k)$ for any function g.

(i) getting inapproximability results under some conjecture and (ii) establishing classes of uniformly inapproximable problems under approximability preserving reductions.

Following the first direction, we establish a link between a major conjecture in PCP theorem and inapproximability in subexponential-time and in FPT-time, assuming ETH. Just below, we state this conjecture while the definition of PCP is deferred to the next section.

> *Linear PCP Conjecture* (LPC): $3\text{SAT} \in \text{PCP}_{1,\beta}[\log |\phi| + D, E]$ for some $\beta \in (0, 1)$, where $|\phi|$ is the size of the 3SAT instance (sum of lengths of clauses), D and E are constant.

Unlike ETH which is arguably recognized as a valid statement, LPC is a wide open question. In Lemma 1 stated in Section 2, we claim that if LPC turns out to hold, it implies that one of the most interesting questions in subexponential and parameterized approximation is answered in the negative. In particular, the followings hold for INDEPENDENT SET on n vertices, for *any* constant $0 < r < 1$ assuming ETH:

(i) There is no r-approximation algorithm in time $O(2^{n^{1-\delta}})$ for any $\delta > 0$.
(ii) There is no r-approximation algorithm in time $O(2^{o(n)})$, if LPC holds.
(iii) There is no r-approximation algorithm in time $O(f(k)n^{O(1)})$, if LPC holds.

Remark that (i) is not conditional upon LPC. In fact, this is an immediate consequence of near-linear PCP construction achieved in [14]. Note that similar inapproximability results under ETH for MAX-3SAT and MAX-3LIN for some subexponential running time have been obtained in [24].

Following the second guideline, we show that a number of problems are equivalent with respect to approximability in subexponential time. Designing a family of equivalent problems is a common way to provide an evidence in favor of hardness of these problems. One prominent example is the family of problems complete under SERF-reducibility [19] which leads to equivalent formulations of ETH. More precisely, for a given problem Π, let us formulate the following hypothesis, which can be seen as the approximate counterpart of ETH.

Hypothesis 1 (APETH(Π)). *There exist two constants $\epsilon > 0$ and r ($r < 1$ if Π is a maximization problem, $r > 1$, otherwise), such that Π is not r-approximable in time $2^{\epsilon n}$.*

We prove that several well-known problems are equivalent with respect to the APETH (APETH-equivalent). To this end, a notion called the *approximation preserving sparsification* is proposed. A recipe to prove that two problems A and B are APETH-equivalent consists of two steps. The first is to reduce an instance of A into a family of instances in "bounded" version (bounded degree for graph problems, bounded occurrence for satisfiability problems), which are equivalent with respect to approximability. This step is where the proposed notion comes into play. The second is to use standard approximability preserving reductions to derive equivalences between bounded versions of A and B. In this paper, we consider L-reductions [25] for this purpose. Furthermore, we show that if

APETH fails for one of these problems, then *any* problem in MaxSNP would be approximable for *any* constant ratio in subexponential FPT-time $2^{o(k)}$, which is also an evidence toward the validity of APETH. This result can be viewed as an extension of [10], which states that none of MaxSNP hard problems allows $2^{o(k)}$-time algorithm under ETH.

Some preliminaries and notation are given in Section 2. Results derived from PCP and LPC are given in Section 3. The second direction on equivalences between problems is described in Section 4.

2 Preliminaries and Notation

We denote by $\text{PCP}_{\alpha,\beta}[q,p]$ (see for instance [2] for more on PCP systems) the set of problems for which there exists a PCP verifier which uses q random bits, reads at most p bits in the proof and is such that: (1) if the instance is positive, then there exists a proof such that V(erifier) accepts with probability at least α; (2) if the instance is negative, then for any proof V accepts with probability at most β. The following theorem is proved in [14] (see also Theorem 7 in [24]), presenting a further refinement of the characterization of NP.

Theorem 1. *[14] For every $\epsilon > 0$,*

$$3\text{SAT} \in \text{PCP}_{1,\epsilon}[(1+o(1))\log n + O(\log(1/\epsilon)), O(\log(1/\epsilon))]$$

A recent improvement [24] of Theorem 1 (a PCP Theorem with two-query projection tests, sub-constant error and almost-linear size) has some important corollaries in polynomial approximation. In particular:

Corollary 1. *[24] Under ETH, for every $\epsilon > 0$, and $\delta > 0$, it is impossible to distinguish between instances of* MAX-3SAT *with m clauses where at least $(1-\epsilon)m$ are satisfiable from instances where at most $(7/8+\epsilon)m$ are satisfiable, in time $O(2^{m^{1-\delta}})$.*

Under LPC, a stronger version of this result follows from standard argument[2].

Lemma 1. *If LPC[3] and ETH hold, then there exists $r < 1$ such that for every $\epsilon > 0$ it is impossible to distinguish between instances of* MAX-3SAT *with m clauses where at least $(1-\epsilon)m$ are satisfiable from instances where at most $(r+\epsilon)m$ are satisfiable, in time $2^{o(m)}$.*

This (conditional) hardness result of approximating MAX-3SAT will be the basis of the negative results of parameterized approximation in Section 3.1.

Let us now present two useful gap amplification results for INDEPENDENT SET. First, as noted in [16], the so-called self-improvement property [18] can be proven for INDEPENDENT SET also in the case of parameterized approximation.

---

[2] All missing proofs can be found in the extended version of the paper [1].

[3] Note that LPC as expressed in this article implies the result even with replacing $(1-\epsilon)m$ by m. However, we stick with this lighter statement $(1-\epsilon)m$ in order, in particular, to emphasize the fact that perfect completeness is not required in the LPC conjecture.

Lemma 2. *[16] If there exists a parameterized r-approximation algorithm for some $r \in (0,1)$ for* INDEPENDENT SET, *then this is true for any $r \in (0,1)$.*

It is also well known that the very powerful tool of expander graphs allows to derive the following gap amplification for INDEPENDENT SET (see [1]).

Theorem 2. *Let G be a graph on n vertices (for a sufficiently large n) and $a > b$ be two positive real numbers. Then for any real $r > 0$ one can build in polynomial time a graph G_r and specify constants a_r and b_r such that: (i) G_r has $N \leqslant Cn$ vertices, where C is some constant independent of G (but may depend on r); (ii) if $\omega(G) \leqslant bn$ then $\omega(G_r) \leqslant b_r N$; (iii) if $\omega(G) \geqslant an$ then $\omega(G_r) \geqslant a_r N$; (iv) $b_r/a_r \leqslant r$.*

Finally, we will use in the sequel the well known sparsification lemma [19]. Intuitively, this lemma allows to work with 3-SAT formula with linear lengths (the sum of the lengths of clauses is linearly bounded in the number of variables).

Lemma 3. *[19] For all $\epsilon > 0$, a 3-SAT formula ϕ on n variables can be written as the disjunction of at most $2^{\epsilon n}$ 3-SAT formula ϕ_i on (at most) n variables such that ϕ_i contains each variable in at most c_ϵ clauses for some constant c_ϵ. Moreover, this reduction takes at most $p(n)2^{\epsilon n}$ time.*

3   Some Consequences of (Almost-)Linear Size PCP System

3.1   Parameterized Inapproximability Bounds

It is shown in [12] that, under ETH, for any function f no algorithm running in time $f(k)n^{o(k)}$ can determine whether there exists an independent set of size k, or not (in a graph with n vertices). A challenging question is to obtain a similar result for approximation algorithms for INDEPENDENT SET. In the sequel, we propose a reduction from MAX-3SAT to INDEPENDENT SET that, based upon the negative result of Corollary 1, only gives a negative result for *some* function f (because Corollary 1 only avoids *some* subexponential running times). However, this reduction gives the inapproximability result sought, if the consequence of LPC given in Lemma 1 (which strengthens Corollary 1 and seems to be a much weaker assumption than LPC) is used instead. We emphasize the fact that the results in this section are valid as soon as a hardness result for MAX-3SAT as that in Lemma 1 holds.

The proof of the following theorem essentially combines the parameterized reduction in [12] and a classic gap-preserving reduction.

Theorem 3. *Under LPC and ETH, there exists $r < 1$ such that, no parameterized approximation algorithm for* INDEPENDENT SET *running in time $f(k)n^{o(k)}$ can achieve approximation ratio r in graphs of order n.*

The following result follows from Lemma 2 and Theorem 3.

Corollary 2. *Under LPC and* ETH, *for any* $r \in (0, 1)$ *there is no r-approxi-mation parameterized algorithm for* INDEPENDENT SET *(i.e., an algorithm that runs in time* $f(k)p(n)$ *for some function f and some polynomial p).*

Let us now consider DOMINATING SET which is known to be W[2]-hard [4]. The existence of parameterized approximation algorithms for this problem is open [15]. Here, we present an approximation preserving reduction (fitting the parameterized framework) which, given a graph $G(V, E)$ on n vertices where V is a set of K cliques $C_1, \cdots, C_K$, builds a graph $G'(V', E')$ such that G has an independent set of size α if and only if G' has a dominating set of size $2K - \alpha$. Using the fact that the graphs produced in the proof of Theorem 3 are of this form (vertex set partitioned into cliques), this reduction will allow us to obtain a lower bound (based on the same hypothesis) for the approximation of MIN DOMINATING SET from Theorem 3.

The graph G' is built as follows. For each clique C_i in G, add a clique C'_i of the same size in G'. Add also: an independent set S_i of size $3K$, each vertex in S_i being adjacent to all vertices in C'_i and a special vertex t_i adjacent to all the vertices in C'_i. For each edge $e = (u, v)$ with u and v *not* in the same clique in G, add an independent set W_e of size $3K$. Suppose that $u \in C_i$ and $v \in C_j$. Then, each vertex in W_e is linked to t_i and to all vertices in C'_i but u, and t_j and all vertices in C'_j but v.

Informally, the reduction works as follows. The set S_i ensures that we have to take at least one vertex in each C'_i, the fact that $|W_e| = 3K$ ensures that it is never interesting to take a vertex in W_e. If we take t_i in a dominating set, this will mean that we do not take any vertex in the set C_i in the corresponding independent set in G. If we take one vertex in C'_i (but not t_i), this vertex will be in the independent set in G. Let us state this property in the following lemma.

Lemma 4. *G has an independent set of size* α *if and only if* G' *has a dominating set of size* $2K - \alpha$.

Theorem 4. *Under LPC and* ETH, *there exists an* $r > 1$ *such that there is no r-approximation algorithm for* DOMINATING SET *running in time* $f(k)n^{o(k)}$ *where n is the order of the graph.*

Such a lower bound immediately transfers to SET COVER since a graph on n vertices for DOMINATING SET can be easily transformed into an equivalent instance of SET COVER with ground set and set system both of size n.

Corollary 3. *Under LPC and* ETH, *there exists* $r > 1$ *such that there is no r-approximation algorithm for* SET COVER *running in time* $f(k)m^{o(k)}$ *in instances with m sets.*

3.2 On the Approximability of INDEPENDENT SET and Related Problems in Subexponential Time

As mentioned in Section 2, an almost-linear size PCP construction [24] for 3SAT allows to get the negative result stated in Corollary 1. In this section, we present further consequences of Theorem 1, based upon a combination of known reductions with (almost) linear size amplifications of the instance.

First, Theorem 1 combined with the reduction in [2] showing inapproximability results for INDEPENDENT SET in polynomial time and the gap amplification of Theorem 2, leads to the following result.

Theorem 5. *Under* ETH, *for any* $r > 0$ *and any* $\delta > 0$, *there is no r-approximation algorithm for* INDEPENDENT SET *running in time* $O(2^{n^{1-\delta}})$, *where n is the order of the input graph.*

Since (for $k \leqslant n$), $n^{k^{1-\delta}} = O(2^{n^{1-\delta'}})$, for some $\delta' < \delta$, the following holds.

Corollary 4. *Under* ETH, *for any* $r > 0$ *and any* $\delta > 0$, *there is no r-approximation algorithm for* INDEPENDENT SET *(parameterized by k) running in time* $O(n^{k^{1-\delta}})$, *where n is the order of the input graph.*

The results of Theorem 5 and Corollary 4 can be immediately extended to problems that are linked to INDEPENDENT SET by approximability preserving reductions (that preserve at least constant ratios) that have linear amplifications of the sizes of the instances, as in the following proposition.

Proposition 1. *Under* ETH, *for any* $r > 0$ *and any* $\delta > 0$, *there is no r-approximation algorithm for either* SET PACKING *or* BIPARTITE SUBGRAPH *running in time* $O(2^{n^{1-\delta}})$ *in a graph of order n.*

Dealing with minimization problems, Theorem 5 and Corollary 4 can be extended to COLORING, using the reduction given in [21]. Note that this reduction uses the particular structure of graphs produced in the inapproximability result in [2] (as in Theorem 5). Hence, the following result can be derived.

Proposition 2. *Under* ETH, *for any* $r > 1$ *and any* $\delta > 0$, *there is no r-approximation algorithm for* COLORING *running in time* $O(2^{n^{1-\delta}})$ *in a graph of order n.*

Concerning the approximability of VERTEX COVER and MIN-SAT in subexponential time, the following holds.

Proposition 3. *Under* ETH, *for any* $\epsilon > 0$ *and any* $\delta > 0$, *there is no* $(7/6 - \epsilon)$-approximation algorithm for VERTEX COVER *running in time* $O(2^{n^{1-\delta}})$ *in graphs of order n, nor for* MIN-SAT *running in time* $2^{m^{1-\delta}}$ *in CNF formulæ with m clauses.*

All the results given in this section are valid under ETH and rule out some ratios in subexponential time of the form $2^{n^{1-\delta}}$. It is worth noticing that if LPC holds, then all these results would hold for *any* subexponential time. Note that this is in some sense optimal since it is easy to see that, for any increasing and unbounded function $r(n)$, INDEPENDENT SET is approximable within ratio $1/r(n)$ in subexponential time (simply consider all the subsets of V of size at most $n/r(n)$ and return the largest independent set among these sets).

Corollary 5. *If* LPC *holds, under* ETH *the negative results of Theorem 5 and Propositions 1, 2 and 3 hold for any time complexity $2^{o(n)}$.*

4 Subexponential Approximation Preserving Reducibility

In this section, we study subexponential approximation preserving reducibility. Recall that APETH(Π) (Hypothesis 1) states that it is hard to approximate in subexponential time problem Π, within some constant ratio r. We exhibit that a set of problems are APETH-equivalent using the notion of *approximation preserving sparsification*. We then link APETH with approximation in subexponential FPT-time.

4.1 Approximation Preserving Sparsification and APETH Equivalences

Recall that the sparsification lemma for 3SAT reduces a formula ϕ to a set of formulae ϕ_i with bounded occurrences of variables such that solving the instances ϕ_i would allow to solve ϕ. We attempt to build an analogous construction for subexponential approximation using the notion of *approximation preserving sparsification*. Given an optimization problem Π and some parameter of the instance, Π-B denotes the problem restricted to instances where the parameter is at most B. For example, we can prescribe the maximum degree of a graph or the maximum number of literal occurrences as the parameter. Then Π-B would be the problems restricted to instances with the parameter bounded by B.

Definition 1. *An approximation preserving sparsification from a problem Π to a bounded parameter version Π-B of Π is a pair (f, g) of functions such that, given any $\epsilon > 0$ and any instance I of Π:*

1. *f maps I into a set $f(I, \epsilon) = (I_1, I_2, \ldots, I_t)$ of instances of Π, where $t \leqslant 2^{\epsilon n}$ and $n_i = |I_i| \leqslant n$; moreover, there exists a constant B_ϵ (independent on I) such that any I_i has parameter at most B_ϵ;*
2. *for any $i \leqslant t$, g maps a solution S_i of an instance I_i (in $f(I, \epsilon)$) into a solution S of I;*
3. *there exists an index $i \leqslant t$ such that if a solution S_i is an r-approximation in I_i, then $S = g(I, \epsilon, I_i, S_i)$ is an r-approximation in I;*
4. *f is computable in time $O^*(2^{\epsilon n})$, and g is polynomial with respect to $|I|$.*

With a slight abuse of notation, let APETH(Π-B) denote the hypothesis: $\exists B$ such that APETH(Π-B), meaning that Π is hard to approximate in subexponential time even for some bounded parameter family of instances. Then the following holds[4].

Theorem 6. *If there exists an approximation preserving sparsification from Π to Π-B, then APETH(Π) if and only if APETH(Π-B).*

---

[4] Note that we could consider a more general definition, leading to the same theorem, by allowing (1) a slight amplification of the size of I_i ($n_i \leqslant \alpha n$ for some fixed α in item 1), (2) an expansion of the ratio in item 3 (if S_i is r-approximate S is $h(r)$ approximate where $h(r)$ goes to one when r goes to one) and (3) a computation time $O^*(2^{\epsilon n})$ for g in item 4.

We now illustrate this technique on some problems. It is worth noticing that the sparsification lemma for 3SAT in [19] is *not* approximation preserving[5]; one cannot use it to argue that approximating MAX-3SAT (in subexponential time) is equivalent to approximating MAX-3SAT with bounded occurrences.

Proposition 4. *There exists an approximation preserving sparsification from* INDEPENDENT SET *to* INDEPENDENT SET-B *and one from* VERTEX COVER *to* VERTEX COVER-B.

Proof. Let $\epsilon > 0$. It is well known that the positive root of $1 = x^{-1} + x^{-1-B}$ goes to one when B goes to infinity. Then, consider a B_ϵ such that this root is at most 2^ϵ. Our sparsification is obtained via a branching tree: the leaves of this tree will be the set of instances I_i; f consists of building this tree; a solution of an instance in the leaf corresponds, via the branching path leading to this leaf, to a solution of the root instance, and that is what g makes.

More precisely, for INDEPENDENT SET, consider the following usual branching tree, starting from the initial graph G: as long as the maximum degree is at least B_ϵ, consider a vertex v of degree at least B_ϵ, and branch on it: either take v in the independent set (and remove $N[v]$), or do not take it. The branching stops when the maximum degree of the graph induced by the unfixed vertices is at most $B_\epsilon - 1$. When branching, at least $B_\epsilon + 1$ vertices are removed when taking v, and one when not taking v; thus the number of leaves is $t \leqslant 2^{\epsilon n}$ (by the choice of B_ϵ). Then, f and g satisfy items 1 and 2 of the definition. For item 3, it is sufficient to note that g maps S_i in S by adding adequate vertices. Then, if we consider the path in the tree corresponding to an optimal solution S^*, leading to a particular leaf G_i, we have that $|S^*| = |S^* \cap G_i| + k$ for some $k \geqslant 0$, and the solution S computed by g is of size $|S| = |S_i| + k$. So, $\frac{|S|}{|S^*|} \geqslant \frac{|S_i|}{|S^* \cap G_i|} \geqslant r$ if S_i is an r-approximation for G_i. The same argument holds also for VERTEX COVER.   □

Analogous arguments apply more generally to any problem where we have a "sufficiently good" branching rule when the parameter is large. Indeed, suppose we can ensure the decrease in instance size by $g(B)$ for nondecreasing and unbounded function g in all (possibly except for one) branches. Then such a branching rule can be utilized to yield an approximation preserving sparsification as in Proposition 4.

We give another approximation preserving sparsification, where there is no direct branching rule allowing to remove a sufficiently large number of vertices. Let GENERALIZED DOMINATING SET be defined as follows: given a graph $G = (V, E)$ where V is partitioned into V_1, V_2, V_3, we ask for a minimum size set of vertices $V' \subseteq V_1 \cup V_2$ which dominates all vertices in $V_2 \cup V_3$. Of course, the case $V_2 = V$ corresponds to the usual DOMINATING SET problem. Note that GENERALIZED DOMINATING SET is also a generalization of SET COVER, with $V_2 = \emptyset$, V_3 being the ground set and V_1 being the set system.

---

[5] One of the reasons is that when a clause C is contained in a clause C', a reduction rule removes C', that is safe for the satisfiability of the formula, but not when considering approximation.

Proposition 5. *There exists an approximation preserving sparsification from* GENERALIZED DOMINATING SET *to* GENERALIZED DOMINATING SET-*B*.

Combining Proposition 5 with some reductions, the following can be shown.

Lemma 5. APETH*(*DOMINATING SET*) implies* APETH*(*INDEPENDENT SET-*B).*

Note that similarly, APETH(SET COVER) implies APETH(INDEPENDENT SET-*B*), when the complexity of SET COVER is measured by $n + m$.

Then, we have the following set of equivalent problems.

Theorem 7. SET COVER, INDEPENDENT SET, INDEPENDENT SET-*B*, VERTEX COVER, VERTEX COVER-*B*, DOMINATING SET, DOMINATING SET-*B*, MAX CUT-*B*, 3SAT-*B*, MAX-*k*SAT-*B* *(for any $k \geqslant 2$) are* APETH-*equivalent.*

Proof. The equivalences between VERTEX COVER-*B*, INDEPENDENT SET-*B*, MAX CUT-*B*, 3SAT-*B*, MAX-2SAT-*B*, DOMINATING SET-*B* follow immediately from [25]. Indeed, for these problems [25] provides *L*-reductions with linear size amplification. The equivalence between MAX-*k*SAT-*B* problems is also well known (just replace a clause of size k by $k - 1$ clauses of size 3).

The equivalence between INDEPENDENT SET and INDEPENDENT SET-*B*, VERTEX COVER and VERTEX COVER-*B* follows from Proposition 4. Finally, Lemma 5 allows us to conclude for DOMINATING SET.   □

4.2   APETH and Parameterized Approximation

The equivalence drawn in Theorem 7 gives a first intuition that the corresponding problems should be hard to approximate in subexponential time for some ratio. In this section we show another argument towards this hypothesis: if it fails, then *any* MaxSNP problem admits for *any r* < 1 a parameterized *r*-approximation algorithm in subexponential time $2^{o(k)}$, which would be quite surprising. The following theorem can be construed as an extension of [10].

Theorem 8. *The following statements are equivalent:*

(i) APETH(Π) holds for one (equivalently all) problem(s) in Theorem 7;

(ii)   there exist a MaxSNP-complete problem Π, some ratio r < 1 and a constant $\epsilon > 0$ such that there is no parameterized r-approximation algorithm for Π with running time $O(2^{\epsilon k} poly(|I|))$;

(iii)   for any MaxSNP-complete problem Π, there exist a ratio r < 1 and an $\epsilon > 0$ such that there is no parameterized r-approximation algorithm for Π with running time $O(2^{\epsilon k} poly(|I|))$.

As an interesting complement of the above theorem, we show that trade-offs between (exponential) running time and approximation ratio do exist for any MaxSNP problem. In [8], it is shown that every MaxSNP problem Π is fixed-parameter tractable in time $2^{O(k)}$ for the standard parameterization, while in [25] it is shown that Π is approximable in polynomial time within a constant ratio

ρ_{Π}. We prove here that there exists a family of parameterized approximation algorithms achieving ratio $\rho_{\Pi} + \epsilon$, for any $\epsilon > 0$, and running in time $2^{O(\epsilon k)}$. This is obtained as a consequence of a result in [20].

Proposition 6. *Let Π be a standard parameterization of a MaxSNP-complete problem. For any $\epsilon > 0$, there exists a parameterized $(\rho_{\Pi} + \epsilon)$-approximation algorithm for Π running in time $\gamma^{\epsilon k} \cdot poly(|I|)$ for some constant γ.*

5   Conclusion

More interesting questions remain untouched in the junction of approximation and (sub)exponential-time/FPT-time computations. This paper is only a first step in this direction and we wish to motivate further research. Among a range of problems to be tackled, we propose the followings.

- Our inapproximability results are conditional upon Linear PCP Conjecture. Is it possible to relax the condition to a more plausible one?
- Or, we dare ask whether (certain) inapproximability results in FPT-time imply strong improvement in PCP theorem. For example, would the converse of Lemma 1 hold?
- Can we design approximation preserving sparsifications for problems like MAX CUT or MAX-3SAT? It seems to be difficult to design a sparsifier based on branching rules, so a novel idea is needed.

Note that we have considered in this article constant approximation ratios. As noted earlier, ratio $1/r(n)$ is achievable in subexponential time for any increasing and unbounded function r. However, dealing with parameterized approximation algorithms, achieving a non-constant ratio is also an open question. More precisely, finding in FPT-time an independent set of size $g(k)$ when there exists an independent set of size k is not known for *any* unbounded and increasing function g.

Finally, let us note that, in the same vein of the first part of our work, Mathieson [23] recently studied a proof checking view of parameterized complexity. Possible links between these two approaches are worth being investigated in future works.

References

1. Bonnet, É., Escoffier, B., Kim, E.J., Paschos, V.T.: On Subexponential and FPT-time Inapproximability. CoRR abs/1211.6656 (2012)
2. Arora, S., Lund, C., Motwani, R., Sudan, M., Szegedy, M.: Proof verification and intractability of approximation problems. J. ACM 45, 501–555 (1998)
3. Zuckerman, D.: Linear degree extractors and the inapproximability of max clique and chromatic number. In: Proc. STOC 2006, pp. 681–690 (2006)
4. Downey, R.G., Fellows, M.R.: Parameterized complexity. Monographs in Computer Science. Springer, New York (1999)

5. Björklund, A., Husfeldt, T., Koivisto, M.: Set partitioning via inclusion-exclusion. SIAM J. Comput. 39, 546–563 (2009)
6. Bourgeois, N., Escoffier, B., Paschos, V.T.: Efficient approximation of min coloring by moderately exponential algorithms. Inform. Process. Lett. 109, 950–954 (2009)
7. Bourgeois, N., Escoffier, B., Paschos, V.T.: Approximation of max independent set, min vertex cover and related problems by moderately exponential algorithms. Discrete Appl. Math. 159, 1954–1970 (2011)
8. Cai, L., Chen, J.: On fixed-parameter tractability and approximability of NP optimization problems. J. Comput. System Sci. 54, 465–474 (1997)
9. Cai, L., Huang, X.: Fixed-parameter approximation: Conceptual framework and approximability results. In: Bodlaender, H.L., Langston, M.A. (eds.) IWPEC 2006. LNCS, vol. 4169, pp. 96–108. Springer, Heidelberg (2006)
10. Cai, L., Juedes, D.W.: On the existence of subexponential parameterized algorithms. J. Comput. Syst. Sci. 67(4), 789–807 (2003)
11. Chen, Y.-J., Grohe, M., Grüber, M.: On parameterized approximability. In: Bodlaender, H.L., Langston, M.A. (eds.) IWPEC 2006. LNCS, vol. 4169, pp. 109–120. Springer, Heidelberg (2006)
12. Chen, J., Huang, X., Kanj, I.A., Xia, G.: Strong computational lower bounds via parameterized complexity. J. Comput. System Sci. 72, 1346–1367 (2006)
13. Cygan, M., Pilipczuk, M.: Exact and approximate bandwidth. Theoret. Comput. Sci. 411, 3701–3713 (2010)
14. Dinur, I.: The PCP theorem by gap amplification. J. ACM 54 (2007)
15. Downey, R.G., Fellows, M.R., McCartin, C.: Parameterized approximation problems. In: Bodlaender, H.L., Langston, M.A. (eds.) IWPEC 2006. LNCS, vol. 4169, pp. 121–129. Springer, Heidelberg (2006)
16. Escoffier, B., Paschos, V.T., Tourniaire, E.: Moderately exponential and parameterized approximation: some structural results (unpublished manuscript)
17. Fürer, M., Gaspers, S., Kasiviswanathan, S.P.: An exponential time 2-approximation algorithm for bandwidth. In: Chen, J., Fomin, F.V. (eds.) IWPEC 2009. LNCS, vol. 5917, pp. 173–184. Springer, Heidelberg (2009)
18. Garey, M.R., Johnson, D.S.: Computers and intractability. In: A Guide to the Theory of NP-completeness, W. H. Freeman, San Francisco (1979)
19. Impagliazzo, R., Paturi, R., Zane, F.: Which problems have strongly exponential complexity? J. Comput. System Sci. 63, 512–530 (2001)
20. Kim, E.J., Williams, R.: Improved parameterized algorithms for above average constraint satisfaction. In: Marx, D., Rossmanith, P. (eds.) IPEC 2011. LNCS, vol. 7112, pp. 118–131. Springer, Heidelberg (2012)
21. Lund, C., Yannakakis, M.: On the hardness of approximating minimization problems. J. Assoc. Comput. Mach. 41, 960–981 (1994)
22. Marx, D.: Parameterized complexity and approximation algorithms. The Computer Journal 51, 60–78 (2008)
23. Mathieson, L.: A proof checking view of parameterized complexity. CoRR abs/1206.2436 (2012)
24. Moshkovitz, D., Raz, R.: Two query PCP with sub-constant error. In: Proc. FOCS 2008, pp. 314–323 (2008)
25. Papadimitriou, C.H., Yannakakis, M.: Optimization, approximation and complexity classes. J. Comput. System Sci. 43, 425–440 (1991)
26. Vazirani, V.: Approximation algorithms. Springer, Berlin (2001)

Multi-parameter Complexity Analysis for Constrained Size Graph Problems: Using Greediness for Parameterization*

Édouard Bonnet, Bruno Escoffier,
Vangelis Th. Paschos**, and Émeric Tourniaire

PSL Research University, Université Paris-Dauphine, LAMSADE, CNRS UMR 7243
{escoffier,edouard.bonnet,paschos,emeric.tourniaire}@lamsade.dauphine.fr

Abstract. We study the parameterized complexity of a broad class of problems called "local graph partitioning problems" that includes the classical fixed cardinality problems as MAX k-VERTEX COVER, k-DENSEST SUBGRAPH, etc. By developing a technique that we call "greediness-for-parameterization", we obtain fixed parameter algorithms with respect to a pair of parameters k, the size of the solution (but *not* its value) and Δ, the maximum degree of the input graph. In particular, greediness-for-parameterization improves asymptotic running times for these problems upon random separation (that is a special case of color coding) and is more intuitive and simple. Then, we show how these results can be easily extended for getting standard-parameterization results (i.e., with parameter the value of the optimal solution) for a well known local graph partitioning problem.

1 Introduction

A local graph partitioning problem is a problem defined on some graph $G = (V, E)$ with two integers k and p. Feasible solutions are subsets $V' \subseteq V$ of size exactly k. The value of their solutions is a linear combination of sizes of edge-subsets and the objective is to determine whether there exists a solution of value at least or at most p. Problems as MAX k-VERTEX COVER, k-DENSEST SUBGRAPH, k-LIGHTEST SUBGRAPH, MAX $(k, n-k)$-CUT and MIN $(k, n-k)$-CUT, also known as fixed cardinality problems, are local graph partitioning problems. When dealing with graph problems, several natural parameters, other than the size p of the optimum, can be of interest, for instance, the maximum degree Δ of the input graph, its treewidth, etc. To these parameters, common for any graph problem, in the case of local graph partitioning problem handled here, one more natural parameter of very great interest can be additionally considered, the size k of V'. For instance, most of these problems have mainly been studied in [1,2],

---

* Research supported by the French Agency for Research under the program TODO, ANR-09-EMER-010.
** Also, Institut Universitaire de France.

G. Gutin and S. Szeider (Eds.): IPEC 2013, LNCS 8246, pp. 66–77, 2013.
© Springer International Publishing Switzerland 2013

from a parameterized point of view, with respect to parameter k, and have been proven W[1]-hard. Dealing with standard parameterization, the only problems that, to the best of our knowledge, have not been studied yet, are the MAX $(k, n - k)$-CUT and the MIN $(k, n - k)$-CUT problems.

In this paper we develop a technique for obtaining multi-parameterized results for local graph partitioning problems. Informally, the basic idea behind it is the following. Perform a branching with respect to a vertex chosen upon some greedy criterion. For instance, this criterion could be to consider some vertex v that maximizes the number of edges added to the solution under construction. Without branching, such a greedy criterion is not optimal. However, if at each step either the greedily chosen vertex v, or some of its neighbors (more precisely, a vertex at bounded distance from v) are a good choice (they are in an optimal solution), then a branching rule on neighbors of v leads to a branching tree whose size is bounded by a function of k and Δ, and at least one leaf of which is an optimal solution. This method, called "greediness-for-parameterization", is presented in Section 2 together with interesting corollaries about particular local graph partitioning problems.

The results of Section 2 can sometimes be easily extended to standard parameterization results. In Section 3 we study standard parameterization of the two still unstudied fixed cardinality problems MAX and MIN $(k, n-k)$-CUT. We prove that the former is fixed parameter tractable (FPT), while, unfortunately, the status of the latter one remains still unclear. In order to handle MAX $(k, n-k)$-CUT we first show that when $p \leqslant k$ or $p \leqslant \Delta$, the problem can be solved in polynomial time. So, the only "non-trivial" case occurs when $p > k$ and $p > \Delta$. That case is handled by greediness-for-parameterization. Unfortunately, this method concludes inclusion of MIN $(k, n-k)$-CUT in FPT only for some particular cases. Note that in a very recent technical report by [3], Fomin et al., the following problem is considered: given a graph G and two integers k, p, determine whether there exists a set $V' \subset V$ of size *at most* k such that at most p edges have exactly one endpoint in V'. They prove that this problem is FPT with respect to p. Let us underline the fact that looking for a set of size at most k seems to be radically different from looking for a set of size exactly k (as in MIN $(k, n - k)$-CUT). For instance, in the case $k = n/2$, the former becomes the MIN CUT problem that is in P, while the latter becomes the MIN BISECTION problem that is NP-hard.

In Section 4.1, we mainly revisit the parameterization by k but we handle it from an approximation point of view. Given a problem Π parameterized by parameter ℓ and an instance I of Π, a parameterized approximation algorithm with ratio $g(.)$ for Π is an algorithm running in time $f(\ell)|I|^{O(1)}$ that either finds an approximate solution of value at least/at most $g(\ell)\ell$, or reports that there is no solution of value at least/at most ℓ. We prove that, although W[1]-hard for the exact computation, MAX $(k, n - k)$-CUT has a parameterized approximation schema with respect to k and MIN $(k, n - k)$-CUT a randomized parameterized approximation schema. These results exhibit two problems which are hard with respect to a given parameter but which become easier when we relax exact computation requirements and seek only (good) approximations. To our knowledge, the only other problem having similar behaviour is another fixed cardinality problem, the MAX k-VERTEX COVER problem, where one has to find the subset of k vertices which cover the greatest number of edges [4]. Note that the existence of problems having this behaviour but with respect to the standard parameter is an open (presumably very difficult to answer) question in [4]. Let us note that polynomial approximation of MIN $(k, n - k)$-CUT has been studied in [5] where it is proven that, if $k = O(\log n)$, then the problem admits a randomized polynomial time approximation schema, while, if $k = \Omega(\log n)$, then it admits an approximation ratio $(1 + \frac{\varepsilon k}{\log n})$, for any $\varepsilon > 0$. Approximation of MAX $(k, n - k)$-CUT has been studied in several papers and a ratio $1/2$ is achieved in [6] (slightly improved with a randomized algorithm in [7]), for all k.

Finally, in Section 4.2, we handle parameterization of local graph partitioning problems by the treewidth tw of the input graph and show, using a standard dynamic programming technique, that they admit an $O^*(2^{\text{tw}})$-time FPT algorithm, where the $O^*(\cdot)$ notation ignores polynomial factors. Let us note that the interest of this result, except its structural aspect (many problems for the price of a single algorithm), lies also in the fact that some local partitioning problems (this is the case, for instance, of MAX and MIN $(k, n - k)$-CUT) do not fit Courcelle's Theorem [8]. Indeed, MAX and MIN BISECTION are not expressible in MSO since the equality of the cardinality of two sets is not MSO-definable. In fact, if one could express that two sets have the same cardinality in MSO, one would be able to express in MSO the fact that a word has the same number of a's and b's, on a two-letter alphabet, which would make that the set $E = \{w : |w|_a = |w|_b\}$ is MSO-definable. But we know that, on words, MSO-definability is equivalent to recognizability; we also know by the standard pumping lemma (see, for instance, [9]) that E is not recognizable [10], a contradiction. Hence, MAX and MIN $(k, n - k)$-CUT are not expressible in MSO; consequently, the fact that those two problems, parameterized by treewidth (tw) are FPT cannot be obtained by Courcelle's Theorem. Furthermore, even several known extended variants of MSO which capture more problems [11], do not seem to be able to express the equality of two sets either.

For reasons of limits to the paper's size, some of the results of the paper are given without proofs that can be found in [12].

2   Greediness-for-Parameterization

We first formally define the class of local graph paritioning problems.

Definition 1. *A* local graph partitioning problem *is a problem having as input a graph $G = (V, E)$ and two integers k and p. Feasible solutions are subsets $V' \subseteq V$ of size exactly k. The value of a solution, denoted by $\text{val}(V')$, is a linear combination $\alpha_1 m_1 + \alpha_2 m_2$ where $m_1 = |E(V')|$, $m_2 = |E(V', V \setminus V')|$ and $\alpha_1, \alpha_2 \in \mathbb{R}$, where $E(X)$ is the set of edges in the subgraph $G[X]$ induced by X, and $E(X, Y)$ is the set of edges having one endpoint in X and one endpoint in Y. The goal is to determine whether there exists a solution of value at least p (for a maximization problem) or at most p (for a minimization problem).*

Note that $\alpha_1 = 1$, $\alpha_2 = 0$ corresponds to k-DENSEST SUBGRAPH and k-SPARSEST SUBGRAPH, while $\alpha_1 = 0$, $\alpha_2 = 1$ corresponds to $(k, n-k)$-CUT, and $\alpha_1 = \alpha_2 = 1$ gives k-COVERAGE. As a local graph partitioning problem is entirely defined by α_1, α_2 and goal $\in \{\min, \max\}$ we will unambiguously denote by $\mathcal{L}(\text{goal}, \alpha_1, \alpha_2)$ the corresponding problem. For conciseness and when no confusion is possible, we will use *local problem* instead. In the sequel, k always denotes the size of feasible subset of vertices and p the standard parameter, i.e., the solution-size. Moreover, as a partition into k and $n - k$ vertices, respectively, is completely defined by the subset V' of size k, we will consider it to be the solution. A *partial solution* T is a subset of V' with less than k vertices. Similarly to the value of a solution, we define the value of a partial solution, and denote it by $\text{val}(T)$.

Informally, we devise algorithms for local problems that add vertices to an initially empty set T (for "taken" vertices) and stop when T becomes of size k, i.e., when T itself becomes a feasible solution. A vertex introduced in T is irrevocably introduced there and will be not removed later.

Definition 2. *Given a local graph partitioning problem $\mathcal{L}(\text{goal}, \alpha_1, \alpha_2)$, the contribution of a vertex v within a partial solution T (such that $v \in T$) is defined by $\delta(v, T) = \frac{1}{2}\alpha_1 |E(\{v\}, T)| + \alpha_2 |E(\{v\}, V \setminus T)|$.*

Note that the value of any (partial) solution T satifies $\text{val}(T) = \Sigma_{v \in T}\delta(v, T)$. One can also remark that $\delta(v, T) = \delta(v, T \cap N(v))$, where $N(v)$ denotes the (open) neighbourhood of the vertex v. Function δ is called the *contribution function* or simply the *contribution* of the corresponding local problem.

Definition 3. *Given a local graph partitioning problem $\mathcal{L}(\text{goal}, \alpha_1, \alpha_2)$, a contribution function is said to be degrading if for every v, T and T' such that $v \in T \subseteq T'$, $\delta(v, T) \leqslant \delta(v, T')$ for goal $= \min$ (resp., $\delta(v, T) \geqslant \delta(v, T')$ for goal $= \max$).*

Note that it can be easily shown that for a maximization problem, a contribution function is degrading if and only if $\alpha_2 \geqslant \alpha_1/2$ ($\alpha_2 \leqslant \alpha_1/2$ for a minimization problem). So in particular MAX k-VERTEX COVER, k-SPARSEST SUBGRAPH and MAX $(k, n-k)$-CUT have a degrading contribution function.

Theorem 1. *Every local partitioning problem having a degrading contribution function can be solved in $O^*(\Delta^k)$.*

Proof. With no loss of generality, we carry out the proof for a minimization local problem $\mathcal{L}(\min, \alpha_1, \alpha_2)$. We recall that T will be a partial solution and eventually a feasible solution. Consider the following algorithm ALG1(T,k) which branches upon the closed neighborhood $N[v]$ of a vertex v minimizing the greedy criterion $\delta(v, T \cup \{v\})$:

- set $T = \emptyset$;
  - if $k > 0$ then:
    * pick the vertex $v \in V \setminus T$ minimizing $\delta(v, T \cup \{v\})$;
    * for each vertex $w \in N[v] \setminus T$ run ALG1$(T \cup \{w\}, k - 1)$;
  - else ($k = 0$), store the feasible solution T;
  - output the best among the solutions stored.

The branching tree of `ALG1` has depth k, since we add one vertex at each recursive call, and arity at most $\max_{v \in V} |N[v]| = \Delta + 1$, where $N[v]$ denotes the closed neighbourhood of v. Thus, the algorithm runs in $O^*(\Delta^k)$.

For the optimality proof, we use a classical hybridation technique between some optimal solution and the one solution computed by `ALG1`.

Consider an optimal solution V'_{opt} different from the solution V' computed by `ALG1`. A node s of the branching tree has two characteristics: the partial solution $T(s)$ at this node (denoted simply T if no ambiguity occurs) and the vertex chosen by the greedy criterion $v(s)$ (or simply v). We say that a node s of the branching tree *conforms* with the optimal solution V'_{opt} if $T(s) \subseteq V'_{opt}$. A node s *deviates* from the optimal solution V'_{opt} if none of its sons conforms with V'_{opt}.

We start from the root of the branching tree and, while possible, we move to a conform son of the current node. At some point we reach a node s which deviates from V'_{opt}. We set $T = T(s)$ and $v = v(s)$. Intuitively, T corresponds to the shared choices between the optimal solution and `ALG1` made along the branch from the root to the node s of the branching tree. Setting $V_n = V'_{opt} \setminus T$, V_n does not intersect $N[v]$, otherwise s would not be deviating.

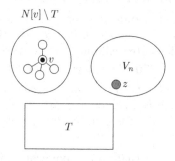

Fig. 1. Situation of the input graph at a deviating node of the branching tree. The vertex v can substitute z since, by the hypothesis, $N[v] \setminus T$ and V_n are disjoint and the contribution of a vertex can only decrease when we later add some of its neighbors in the solution.

Choose any $z \in V'_{opt} \setminus T$ and consider the solution induced by the set $V_e = V'_{opt} \cup \{v\} \setminus \{z\}$. We show that this solution is also optimal. Let $V_c = V'_{opt} \setminus \{z\}$. We have $\mathrm{val}(V_e) = \Sigma_{w \in V_c} \delta(w, V_e) + \delta(v, V_e)$. Besides, $\delta(v, V_e) = \delta(v, V_e \cap N(v)) = \delta(v, T \cup \{v\})$ since $V_e \setminus (T \cup \{v\}) = V_n$ and according to the last remark of the previous paragraph, $N(v) \cap V_n = \emptyset$. By the choice of v, $\delta(v, T \cup \{v\}) \leqslant \delta(z, T \cup \{z\})$, and, since δ is a degrading contribution, $\delta(z, T \cup \{z\}) \leqslant \delta(z, V'_{opt})$. Summing up, we get $\delta(v, V_e) \leqslant \delta(z, V'_{opt})$ and $\mathrm{val}(V_e) \leqslant \Sigma_{w \in V_c} \delta(w, V_e) + \delta(z, V'_{opt})$. Since v is not in the neighborhood of $V'_{opt} \setminus T = V_n$ only z can degrade the contribution of those vertices, so $\Sigma_{w \in V_c} \delta(w, V_e) \leqslant \Sigma_{w \in V_c} \delta(w, V'_{opt})$, and $\mathrm{val}(V_e) \leqslant \Sigma_{w \in V_c} \delta(w, V'_{opt}) + \delta(z, V'_{opt}) = \mathrm{val}(V'_{opt})$.

Thus, by repeating this argument at most k times, we can conclude that the solution computed by `ALG1` is as good as V'_{opt}.

Corollary 1. MAX k-VERTEX COVER, k-SPARSEST SUBGRAPH *and* MAX $(k, n - k)$-CUT *can be solved in* $O^*(\Delta^k)$.

As mentioned before, the local problems mentioned in Corollary 1 have a degrading contribution.

Theorem 2. *Every local partitioning problem can be solved in* $O^*((\Delta k)^{2k})$.

Proof (Sketch of proof). Once again, with no loss of generality, we prove the theorem in the case of minimization, i.e., $\mathcal{L}(\min, \alpha_1, \alpha_2)$. The proof of Theorem 2 involves an algorithm fairly similar to ALG1 but instead of branching on a vertex chosen greedily and its neighborhood, we will branch on sets of vertices inducing connected components (also chosen greedily) and the neighborhood of those sets.

Let us first state the following straightforward lemma that bounds the number of induced connected components and the running time to enumerate them.

Lemma 1. *One can enumerate the connected induced subgraphs of size up to k in time $O^*(\Delta^{2k})$.*

Consider now the following algorithm ALG2(T,k):

- set $T = \emptyset$;
  - if $k > 0$ then, for each i from 1 to k:
    * find $S_i \in V \setminus T$ minimizing $val(T \cup S_i)$ with S_i inducing a connected component of size i;
    * for each i, for each $v \in S_i$, run ALG2$(T \cup \{v\}, k - 1)$;
  - else ($k = 0$), stock the feasible solution T;
  output the stored feasible solution T minimizing $val(T)$.

The branching tree of ALG2 has size $O(k^{2k})$. Computing the S_i in each node takes time $O^*(\Delta^{2k})$ according to Lemma 1. Thus, the algorithm runs in $O^*((\Delta k)^{2k})$.

For the optimality of ALG2, we use the following lemma.

Lemma 2. *Let A,B,X,Y be pairwise disjoint sets of vertices such that* val $(A \cup X) \leqslant$ val $(B \cup X)$, $N[A] \cap Y = \emptyset$ *and* $N[B] \cap Y = \emptyset$. *Then,* val $(A \cup X \cup Y) \leqslant$ val $(B \cup X \cup Y)$.

We now show that ALG2 is sound, using again hybridation between an optimal solution V'_{opt} and the one solution found by ALG2. We keep the same notation as in the proof of the soundness of ALG1. Node s is a node of the branching tree which deviates from V'_{opt}, all nodes in the branch between the root and s conform with V'_{opt}, the shared choices constitute the set of vertices $T = T(s)$ and, for each i, set $S_i = S_i(s)$ (analogously to $v(s)$ in the previous proof, s is now linked to the subsets S_i computed at this node). Set $V_n = V'_{opt} \setminus T$. Take a maximal connected (non empty) subset H of V_n. Set $S = S_{|H|}$ and consider $V_e = V'_{opt} \setminus H \cup S = (T \cup V_n) \setminus H \cup S = T \cup S \cup (V_n \setminus H)$. Note that, by hypothesis, $N[S] \cap V_n = \emptyset$ since s is a deviating node. By the choice of S at the node s, val $(T \cup S) \leqslant$ val $(T \cup H)$. So, val $(V_e) =$ val $(T \cup S \cup (V_n \setminus H)) =$ val $(T \cup H \cup (V_n \setminus H)) =$ val $(T \cup V_n) =$ val (V'_{opt}) according to Lemma 2, since by construction neither $N[H]$ nor $N[S]$, do intersect $V_n \setminus H$. Iterating the argument at most k times we get to a leaf of the branching tree of ALG2 which yields a solution as good as V'_{opt}. The proof of the theorem is now completed.

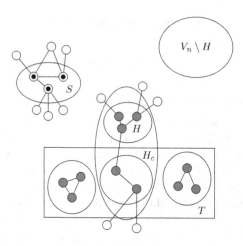

Fig. 2. Illustration of the proof, with filled vertices representing the optimal solution V'_{opt} and dotted vertices representing the set $S = S_{|H|}$ computed by ALG2 which can substitute H, since V_n does not interact with H_c nor with S

Corollary 2. k-DENSEST SUBGRAPH *and* MIN $(k, n - k)$-CUT *can be solved in* $O^*((\Delta k)^{2k})$.

Here also, simply observe that the problems mentioned in Corollary 2 are local graph partitioning problems.

Theorem 1 improves the $O^*(2^{(\Delta+1)k} ((\Delta + 1)k)^{\log((\Delta+1)k)})$ time complexity for the corresponding problems given in [13] obtained there by the *random separation* technique, and Theorem 2 improves it whenever $k = 2^{o(\Delta)}$. Recall that random separation consists of randomly guessing if a vertex is in an optimal subset V' of size k (white vertices) or if it is in $N(V') \setminus V'$ (black vertices). For all other vertices the guess has no importance. As a right guess concerns at most only $k + k\Delta$ vertices, it is done with high probability if we repeat random guesses $f(k, \Delta)$ times with a suitable function f. Given a random guess, i.e., a random function $g : V \to \{\text{white,black}\}$, a solution can be computed in polynomial time by dynamic programming. Although random separation (and a fortiori *color coding* [14]) have also been applied to other problems than local graph partitioning ones, greediness-for-parameterization seems to be quite general and improves both running time and easiness of implementation since our algorithms do not need complex derandomizations.

Let us note that the greediness-for-parameterization technique can be even more general, by enhancing the scope of Definition 1 and can be applied to problems where the objective function takes into account not only edges but also vertices. The value of a solution could be defined as a function val : $\mathcal{P}(V) \to \mathbb{R}$ such that val $(\emptyset) = 0$, the contribution of a vertex v in a partial solution T is $\delta(v, T) =$ val $(T \cup v) -$ val (T). Thus, for any subset T, val $(T) =$ val $(T \setminus \{v_k\}) + \delta(v_k, T \setminus \{v_k\})$ where k is the size of T and v_k is the last vertex added to the solution. Hence, val $(T) = \Sigma_{1 \leqslant i \leqslant k} \delta(v_i, \{v_1, \ldots, v_{i-1}\}) +$ val $(\emptyset) = \Sigma_{1 \leqslant i \leqslant k} \delta(v_i, \{v_1, \ldots, v_{i-1}\})$.

Now, the only hypothesis we need to show Theorem 2 is the following: for each T' such that $(N(T') \setminus T) \cap (N(v) \setminus T) = \emptyset$, $\delta(v, T \cup T') = \delta(v, T)$.

Notice also that, that under such modification, MAX k-DOMINATING SET, asking for a set V' of k vertices that dominate the highest number of vertices in $V \setminus V'$ fulfills the enhancement just discussed. We therefore derive the following.

Corollary 3. MAX k-DOMINATING SET *can be solved in* $O^*((\Delta k)^{2k})$.

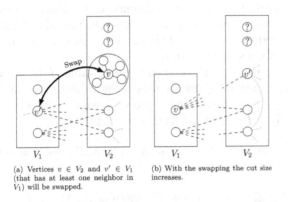

(a) Vertices $v \in V_2$ and $v' \in V_1$ (that has at least one neighbor in V_1) will be swapped.

(b) With the swapping the cut size increases.

Fig. 3. Illustration of a swapping

3 Standard Parameterization for Max and Min (k, n − k)-Cut

3.1 Max (k, n − k)-Cut

In the sequel, we use the standard notation $G[U]$ for any $U \subseteq V$ to denote the subgraph induced by the vertices of U. In this section, we show that MAX $(k, n - k)$-CUT parameterized by the standard parameter, i.e., by the value p of the solution, is FPT. Using an idea of bounding above the value of an optimal solution by a swapping process (see Figure 3), we show that the non-trivial case satisfies $p > k$. We also show that $p > \Delta$ holds for non trivial instances and get the situation illustrated in Figure 4. The rest of the proof is an immediate application of Corollary 1.

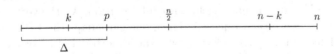

Fig. 4. Location of parameter p, relatively to k and Δ

Lemma 3. *In a graph with minimum degree r, the optimal value* opt *of a* MAX *$(k,n\text{-}k)$-cut satisfies* opt $\geqslant \min\{n - k, rk\}$.

Proof. We divide arbitrarily the vertices of a graph $G = (V, E)$ into two subsets V_1 and V_2 of size k and $n - k$, respectively. Then, for every vertex $v \in V_2$, we check if v has a neighbor in V_1. If not, we try to swap v and a vertex $v' \in V_1$ which has strictly less than r neighbors in V_2 (see Figure 3). If there is no such vertex, then every vertex in V_1 has at least r neighbors in V_2, so determining a cut of value at least rk. When swapping is possible, as the minimum degree is r and the neighborhood of v is entirely contained in V_2, moving v from V_2 to V_1 will increase the value of the cut by at least r. On the other hand, moving v' from V_1 to V_2 will reduce the value of the cut by at most $r - 1$. In this way, the value of the cut increases by at least 1.

Finally, either the process has reached a cut of value rk (if no more swap is possible), or every vertex in V_2 has increased the value of the cut by at least 1 (either immediately, or after a swapping process), which results in a cut of value at least $n - k$, and the proof of the lemma is completed.

Corollary 4. *In a graph with no isolated vertices, the optimal value for* MAX *$(k, n - k)$-CUT is at least $\min\{n - k, k\}$.*

Then, Corollary 1 suffices to conclude the proof of the the following theorem.

Theorem 3. *The* MAX *$(k, n - k)$-CUT problem parameterized by the standard parameter p is FPT.*

3.2   Min (k, n − k)-Cut

Unfortunately, unlike what have been done for MAX $(k, n - k)$-CUT, we have not been able to show until now that the case $p < k$ is "trivial". So, Algorithm ALG2 in Section 2 cannot be transformed into a standard FPT algorithm for this problem.

However, we can prove that if $p \geqslant k$, then MIN $(k, n - k)$-CUT parameterized by the value p of the solution is FPT. This is an immediate corollary of the following proposition.

Proposition 1. MIN *$(k, n - k)$-CUT parameterized by $p + k$ is FPT.*

Proof. Each vertex v such that $|N(v)| \geqslant k + p$ has to be in $V \setminus V'$ (of size $n - k$). Indeed, if one puts v in V' (of size k), among its $k + p$ incident edges, at least $p + 1$ leave from V'; so, it cannot yield a feasible solution. All the vertices v such that $|N(v)| \geqslant k + p$ are then rejected. Thus, one can adapt the FPT algorithm in $k + \Delta$ of Theorem 2 by considering the k-neighborhood of a vertex v not in the whole graph G, but in $G[T \cup U]$. One can easily check that the algorithm still works and since in those subgraphs the degree is bounded by $p + k$ we get an FPT algorithm in $p + k$.

In [5], it is shown that, for any $\varepsilon > 0$, there exists a randomized $(1 + \frac{\varepsilon k}{\log n})$-approximation for MIN $(k, n - k)$-CUT. From this result, we can easily derive that when $p < \frac{\log n}{k}$ then the problem is solvable in polynomial time (by a randomized algorithm). Indeed, fixing $\varepsilon = 1$, the algorithm in [5] is a $(1 + \frac{k}{\log(n)})$-approximation. This approximation ratio is strictly better than $1 + \frac{1}{p}$. This means that the algorithm outputs a solution of value lower than $p + 1$, hence at most p, if there exists a solution of value at most p.

We now conclude this section by claiming that, when $p \leqslant k$, MIN $(k, n-k)$-CUT can be solved in time $O^*(n^p)$.

Proposition 2. *If $p \leqslant k$, then* MIN $(k, n-k)$-CUT *can be solved in time $O^*(n^p)$.*

4 Other Parameterizations

4.1 Parameterization by k and Approximation of Max and Min (k, n − k)-Cut

Recall that both MAX and MIN $(k, n - k)$-CUT parameterized by k are W[1]-hard [2,1]. In this section, we give some approximation algorithms working in FPT time with respect to parameter k.

Proposition 3. MAX $(k, n - k)$-CUT, *parameterized by k has a fixed-parameter approximation schema. On the other hand,* MIN $(k, n - k)$-CUT *parameterized by k has a randomized fixed-parameter approximation schema.*

Finding approximation algorithms that work in FPT time with respect to parameter p is an interesting question. Combining the result of [5] and an $O(\log^{1.5}(n))$-approximation algorithm in [7] we can show that the problem is $O(k^{3/5})$ approximable in polynomial time by a randomized algorithm. But, is it possible to improve this ratio when allowing FPT time (with respect to p)?

4.2 Parameterization by the Treewidth

When dealing with parameterization of graph problems, some classical parameters arise naturally. One of them, very frequently used in the fixed parameter literature is the treewidth of the graph.

It has already been proven that MIN and MAX $(k, n - k)$-CUT, as well as k-DENSEST SUBGRAPH can be solved in $O^*(2^{tw})$ [15,16]. We show here that the algorithm in [15] can be adapted to handle the whole class of local problems, deriving so the following result.

Proposition 4. *Any local graph partitioning problem can be solved in time $O^*(2^{tw})$.*

Corollary 5. *Restricted to trees, any local graph partitioning problem can be solved in polynomial time.*

Corollary 6. MIN BISECTION *parameterized by the treewidth of the input graph is FPT.*

It is worth noticing that the result easily extends to the weighted case (where edges are weighted) and to the case of partitioning V into a constant number of classes (with a higher running time).

PERSPECTIVES. Of course, the main remaining open question is the parameterized complexity of MIN $(k, n-k)$-CUT with respect to the value of the solution p. Another problem of interest is to look for a better algorithm for local graph partitioning problem in general. For instance, we can not rule out a time-complexity in $O^*((a\Delta)^{bk})$, with a and b two constants, which would be really closer to the $O^*(\Delta^k)$ complexity of the degrading contribution case.

References

1. Cai, L.: Parameter complexity of cardinality constrained optimization problems. The Computer Journal 51, 102–121 (2008)
2. Downey, R.G., Estivill-Castro, V., Fellows, M.R., Prieto, E., Rosamond, F.A.: Cutting up is hard to do: the parameterized complexity of k-cut and related problems. Electronic Notes in Theoretical Computer Science, vol. 78, pp. 205–218. Elsevier (2003)
3. Fomin, F.V., Golovach, P.A., Korhonen, J.H.: On the parameterized complexity of cutting a few vertices from a graph. Technical report, CoRR, abs/1304.6189 (2013)
4. Marx, D.: Parameterized complexity and approximation algorithms. The Computer Journal 51, 60–78 (2008)
5. Feige, U., Krauthgamer, R., Nissim, K.: On cutting a few vertices from a graph. Discrete Appl. Math. 127, 643–649 (2003)
6. Ageev, A.A., Sviridenko, M.I.: Approximation algorithms for maximum coverage and max cut with given sizes of parts. In: Cornuéjols, G., Burkard, R.E., Woeginger, G.J. (eds.) IPCO 1999. LNCS, vol. 1610, pp. 17–30. Springer, Heidelberg (1999)
7. Feige, U., Langberg, M.: Approximation algorithms for maximization problems arising in graph partitioning. J. Algorithms 41, 174–211 (2001)
8. Courcelle, B.: The monadic second-order logic of graphs. i. recognizable sets of finite graphs. Information and Computation 85, 12–75 (1990)
9. Lewis, H.R., Papadimitriou, C.H.: Elements of the theory of computation. Prentice-Hall (1981)
10. Maneth, S.: Logic and automata. Lecture 3: Expressiveness of MSO graph properties. Logic Summer School (2006)
11. Szeider, S.: Monadic second order logic on graphs with local cardinality constraints. In: Ochmański, E., Tyszkiewicz, J. (eds.) MFCS 2008. LNCS, vol. 5162, pp. 601–612. Springer, Heidelberg (2008)
12. Bonnet, E., Escoffier, B., Paschos, V.T., Tourniaire, E.: Multi-parameter complexity analysis for constrained size graph problems: using greediness for parameterization. CoRR abs/1306.2217 (2013)
13. Cai, L., Chan, S.M., Chan, S.O.: Random separation: a new method for solving fixed-cardinality optimization problems. In: Bodlaender, H.L., Langston, M.A. (eds.) IWPEC 2006. LNCS, vol. 4169, pp. 239–250. Springer, Heidelberg (2006)

14. Alon, N., Yuster, R., Zwick, U.: Color-coding. J. Assoc. Comput. Mach. 42, 844–856 (1995)
15. Bourgeois, N., Giannakos, A., Lucarelli, G., Milis, I., Paschos, V.T.: Exact and approximation algorithms for DENSEST k-SUBGRAPH. In: Ghosh, S.K., Tokuyama, T. (eds.) WALCOM 2013. LNCS, vol. 7748, pp. 114–125. Springer, Heidelberg (2013)
16. Kloks, T.: Treewidth. LNCS, vol. 842. Springer, Heidelberg (1994)

Chain Minors Are FPT

Jarosław Błasiok[1] and M. Kamiński[2]

[1] Instytut Informatyki
Uniwersytet Warszawski
jb291202@students.mimuw.edu.pl
[2] Département d'Informatique
Université libre de Bruxelles
and
Instytut Informatyki
Uniwersytet Warszawski
mjk@mimuw.edu.pl

Abstract. Given two finite posets P and Q, P is a *chain minor* of Q if there exists a partial function f from the elements of Q to the elements of P such that for every chain in P there is a chain C_Q in Q with the property that f restricted to C_Q is an isomorphism of chains.

We give an algorithm to decide whether a poset P is a chain minor of a poset Q that runs in time $\mathcal{O}(|Q| \log |Q|)$ for every fixed poset P. This solves an open problem from the monograph by Downey and Fellows [*Parameterized Complexity*, 1999] who asked whether the problem was fixed parameter tractable.

Keywords: partially ordered sets, parameterized complexity, data structures and algorithms.

1 Introduction

It is widely believed that NP-hard problems do not admit polynomial-time deterministic algorithms. Nevertheless, such problems tend to appear in practical applications and it is necessary to deal with them anyway. Among many approaches to NP-hard problems *parameterized complexity* has recently received a lot of attention. It was first studied systematically by Downey and Fellows in [2]. The main idea of parameterized complexity is to equip the instance of a problem with a parameter and confine the superpolynomial behaviour of the algorithm to the parameter. Here we can efficiently solve large instances of the problem as long as the parameter is small.

Parameterized complexity. More formally, an instance of a parameterized problem is a pair (I, k) where $k \in \mathbb{N}$. XP is the class of parameterized problems such that for every k there is an algorithm that solves that problem in time $\mathcal{O}(|I|^{f(k)})$, for some function f (that does not depend on I). One example is the CLIQUE problem parameterized by the size of the clique defined as follows: given (G, k) where G is graph and k is a natural number, is there a clique of size k in G?

G. Gutin and S. Szeider (Eds.): IPEC 2013, LNCS 8246, pp. 78–83, 2013.
© Springer International Publishing Switzerland 2013

One can simply enumerate all k-subsets of vertices to solve the problem in time $\mathcal{O}(n^{k+2})$, hence, in time polynomial for every fixed k.

Much more desirable parameterized complexity is FPT. A parameterized problem is called *fixed parameter tractable* (FPT) if there is an algorithm that for every instance (I, k) solves the problem in time $\mathcal{O}(f(k)|I|^c)$ for some function f (that does not depend on $|I|$). That is, for a fixed parameter k, the problem is solvable in polynomial time and the degree of the polynomial does not depend on k. SATISFABILITY of boolean formulae parameterized by number of variables is FPT; it can be solved by a brute force algorithm in time $\mathcal{O}(2^k m)$ where m is size of the instance.

In their monograph [2], Downey and Fellows included a list of open problems, asking whether they admit an FPT solution ("FPT suspects") or are hard by means of parameterized complexity ("tough customers"). Recently, Fomin and Marx have revised this list of problems [3]. Many of the problems from the original list have been solved since the publication of [2], yet CHAIN MINOR remains open. It was listed as a "tough customer" – suspecting it is not fixed parameter tractable. However, we prove otherwise.

Chain minors. Chain minors were introduced by Möring and Müller in [6] in the context of scheduling stochastic project networks and first studied systematically by Gustedt in [4] and in his PhD thesis [5]. Gustedt proved that finite posets are *well quasi ordered* by chain minors, that is, in any infinite sequence of posets there is a pair of posets such that one is a chain minor of the other. A consequence of this fact is that any class of posets closed under taking chain minors can be characterized by a finite family of minimal forbidden posets.

The CHAIN MINOR problem is to decide, given two posets P and Q, whether P is a chain minor of Q. The parameterized approach to CHAIN MINOR is justified as Gustedt showed in [4] that CHAIN MINOR is NP-hard (giving a reduction from PRECEDENCE CONSTRAINED SCHEDULING). Note that it is not known whether CHAIN MINOR is NP-complete. There is no obvious nondeterministic polynomial-time algorithm for that problem, except for a very simple case — Gustedt in his PhD thesis has proved that CHAIN MINOR is NP-complete when restricted to posets of height at most 3.

Our results. Gustedt also gave an XP algorithm for the CHAIN MINOR problem [5]. More specifically, he gave an algorithm that checks whether P is a chain minor of Q in time $\mathcal{O}(|P|^2|Q|^{|P|} + f(|P|))$. We improve his result, giving two fixed parameter tractable algorithms (parameterized by $|P|$) — randomized and deterministic — where the former one runs in $\mathcal{O}(f(|P|)|Q|)$ time and the latter in $\mathcal{O}(f(|P|)|Q| \log |Q|)$ time. Both algorithms need linear memory.

The technique that we use to design the FPT algorithm is called *color coding* and was originally developed by Alon, Yster, and Zwick in [1] to give the first FPT algorithm for the K-PATH problem (= finding a path of size k in a given graph). Since then, this technique has been successfully applied many times, yet in most of those examples colors where introduced artificially (as in K-PATH). In our case, they are naturally derived from the problem definition.

2   Definitions and Basic Facts

A finite partially ordered set (*poset*) is a pair $(V, <)$ where V is a finite set and $<$ is a binary relation on V that is transitive, irreflexive, and antisymmetric. A *chain* in a poset is a sequence of elements $(v_1, v_2, \ldots v_n)$, $v_i \in V$ such that $v_i < v_j$, for all $1 \leq i < j \leq n$.

Given two finite posets $P = (V_P, <_P)$ and $Q = (V_Q, <_Q)$, we say that P is a *chain minor* of Q ($P \preceq Q$) if and only if there exists a partial function $f : V_Q \longrightarrow V_P$ with a property that for every chain $(c_1, c_2, \ldots, c_n)$ in P there is a chain $(c_1', c_2', \ldots c_n')$ in Q such that $f(c_i') = c_i$. In this case, we call f a *witness* for $P \preceq Q$ and we write $P \preceq^f Q$. It is easy to check that $\preceq$ is a quasi-order (transitive and antisymmetric). One can easily check that if $V_P \subseteq V_Q$ and $<_P$ is induced by $<_Q$ (that is, P is subposet of Q), then P is also a chain minor of Q.

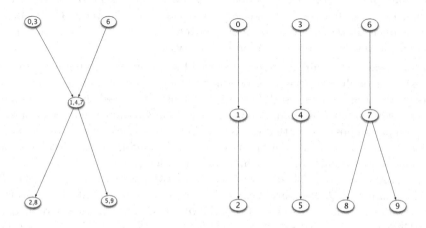

Fig. 1. The left poset P is a chain minor of the right poset Q as certified by the witness function from the elements of the Q to the elements of P

3   Algorithm

Our goal is to present a deterministic FPT algorithm. We will start with a randomized algorithm and use a standard technique (of splitters) to derandomize it at the price of slightly worse time complexity. However, we need some auxiliary lemmas first.

Lemma 1. *There is a deterministic algorithm which given two posets P and Q and a partial function $f : V_Q \longrightarrow V_P$ determines whether $P \preceq^f Q$ in time $\mathcal{O}(2^{|P|}|Q|)$.*

Proof. For $q \in Q$, let $\text{pred}(q)$ be the set of elements less or equal to q in Q, that is, $\text{pred}(q) = \{q' \in Q : q' \leq q\}$. It is enough to iterate over all chains of P, and for every chain $(c_1, c_2, \ldots, c_p)$ consider only those vertices of V_Q which are mapped by f to any of c_i — let us call them Q'. Let us now consider vertices from Q' in topological order. For every vertex q, let us compute the maximum j such that one can find a chain $c'_1, c'_2, \ldots, c'_j$ in set $\mathrm{pred}(q)$ (we require $f(c'_i) = c_i$). Let us call that value $\mathrm{maxc}(q)$.

To calculate $\mathrm{maxc}(q)$ knowing maxc of every predecessor, we just take

$$\mathrm{maxc}(q) = \begin{cases} j & \text{if } \max_{v < q} \mathrm{maxc}(v) = j - 1 \wedge f(q) = c_j \\ \max_{v < q} \mathrm{maxc}(v) & \text{otherwise} \end{cases}$$

The solution can be read off from maxc values. $\qquad\qquad\qquad\qquad\Box$

Lemma 2. *Let P and Q be finite posets and $k = |P|$. If $P \preceq^f Q$, then there is a subposet Q^0 of Q of size at most $2^k k$ such that if f' is equal to f on Q^0, then $P \preceq^{f'} Q$.*

Proof. Let $P \preceq^f Q$ and let $\mathcal{C}$ be the set of all chains in P. $\mathcal{C}$ has at most 2^k elements (as any subset of the elements from V_P forms at most one chain). For every chain $c = (c_1, \ldots, c_{n_c}) \in \mathcal{C}$, take an arbitrary chain $(c'_1, \ldots, c'_{n_c})$ in Q such that $f(c'_i) = c_i$, for $i = 1, \ldots, n_c$.

Now let V_{Q^0} be $\bigcup_c \{c'_1, c'_2, \ldots, c'_{n_c}\}$. Notice that $|V_{Q^0}| \leq \sum_{c \in \mathcal{C}} n_c \leq \sum_{c \in \mathcal{C}} k \leq 2^k k$. If f' is equal to f on Q^0, we have to check that given a chain $c_1, c_2, \ldots c_{n_c}$ in P one can find preimages with respect to f' of the elements of that chain such that the preimages form a chain in Q. It suffices to take the elements c'_i from above; they belong to Q^0 by definition, thus $f'(c'_i) = f(c'_i) = c_i$ for $i = 1, \ldots, n_c$ and the elements $c'_1, \ldots, c'_n$ were chosen to be a chain. $\qquad\Box$

3.1 Randomized Algorithm

Now we will state and prove a key lemma for Theorem 1.

Lemma 3. *If $P \preceq Q$, then a function $g : V_Q \longrightarrow V_P$ taken uniformly at random from the set of all such functions is a witness for $P \preceq Q$ with probability at least $k^{-2^k k}$, where $k = |P|$.*

Proof. Let f be a witness for $P \preceq Q$. Now take Q^0 as in Lemma 2. It follows from Lemma 2 that it is sufficient to show that a function g taken uniformly at random is equal to f on Q^0 with high probability, as the probability of g being a witness for $P \preceq Q$ is at least as large. Now the lemma follows from the following simple calculation.

$$\mathbb{P}(P \preceq^g Q) \geq \mathbb{P}(g|Q^0 = f|Q^0)$$
$$= \prod_{v \in Q^0} \mathbb{P}(g(v) = f(v))$$
$$= \prod_{v \in Q^0} \frac{1}{k}$$
$$= k^{-2^k k}.$$

$\qquad\qquad\qquad\qquad\qquad\qquad\qquad\qquad\qquad\qquad\qquad\qquad\qquad\Box$

Now we are ready to prove Theorem 1.

Theorem 1. *There is a randomized algorithm for* CHAIN MINOR *with time complexity* $\mathcal{O}(|P|^{2^{|P|}}|P|2^{|P|}|Q|)$ *and linear space complexity.*

Proof. Let $k = |P|$. It is enough to repeat the following procedure $k^{2^k}k$ times: take a random function g and check whether it is a witness for $P \preceq Q$. If any of those function is a witness, return YES; otherwise, return NO. The desired time and space complexity follow from Lemma 1. Lemma 3 bounds the probability of an error by a constant. Indeed, if $P \preceq Q$, then the probability that the algorithm answers NO is not greater then $(1 - 1/p_k)^{p_k}$, where $p_k = k^{2^k}k$, which is bounded by $(1 - 1/2)^2$, for $k \geq 2$ (and tends to $1/e$ as k tends to infinity).   □

3.2 Deterministic Algorithm

We will derandomize the algorithm from Theorem 1 using a well-known derandomization technique of splitters. A (n, k, l)-splitter is a family of functions $\mathcal{F}$, $\mathcal{F} \ni f : \{1, \ldots n\} \longrightarrow \{1, \ldots, l\}$, such that for every $W \subseteq \{1 \ldots n\}$ of size at most k there is some function $f \in \mathcal{F}$ which is injective on W. We will need the following theorem by Naor, Schulman, and Srinivasan from [7].

Theorem 2. *([7]) There exists a (n, k, k)-splitter that can be constructed in time* $\mathcal{O}(e^k k^{\mathcal{O}(\log k)} n \log n)$.

Theorem 3. *There is a deterministic algorithm for* CHAIN MINOR *with time complexity* $\mathcal{O}(|P|^{2^{|P|}}|P|2^{|P|+\mathcal{O}(\log^2 |P|)}|Q| \log |Q|)$ *and linear space complexity.*

Proof. Given P and Q, let us take $k = |P|$, $n = |Q|$. Fix a bijection between $\{1 \ldots n\}$ and V_Q. Let's fix a $(n, 2^k k, 2^k k)$ splitter, and iterate through every function from that splitter and every function from the set $\{1, \ldots, 2^k k\}$ to P. Then check whether the composition of these two functions is a witness for $P \preceq Q$.

To prove correctness of the algorithm, let us consider $P \preceq^f Q$ and take Q^0 as in Lemma 2. It follows from the definition of splitters that there exists a function f, such that f is injective on Q^0. Then, just because we iterate over all functions from the set $\{1, \ldots k2^k\}$ to P at some point we take one, such that the composition equals f when restricted to Q^0. This pair yields a witness for $P \preceq Q$.   □

4 Conclusions

1. It is easy to prove that every class of posets closed under taking of chain minors can be characterized by a set of minimal forbidden chain minors. Gustedt proved in [5] that posets are well quasi ordered. Consequently, each such set of forbidden chain minors is finite. Gustedt also gave an XP algorithm to decide whether a poset H is a chain minor of a poset Q when parameterized by the number of elements of H. These two results show that for every class of posets $\mathcal{P}$ closed under taking chain minors there exists a polynomial-time algorithm deciding whether the input poset Q is in $\mathcal{P}$. (The exponent of the polynomial depends on the class.)

We give an FPT algorithm to test whether a poset H is a chain minor of a poset Q when parameterized by the number of elements of H. A consequence of our result is that for every class of posets $\mathcal{P}$ closed under taking chain minors there exists a $\mathcal{O}(|Q|\log|Q|)$ algorithm deciding whether the input poset Q is in $\mathcal{P}$.

2. The project of Graph Minors of Robertson and Seymour is arguably one of the most significant achievements in modern graph theory. Robertson and Seymour proved that graphs are well quasi ordered under graph minors and gave an FPT algorithm to decide whether a graph H is a minor of a graph G when parameterized by H. They were also able to describe the structure of graphs that do not contain a fixed graph as a minor.

Is there a parallel theory possible for chain minors in posets? Gustedt proved in [5] that chain minors are well quasi ordered and this work gives an FPT algorithm for the CHAIN MINOR problem. However, neither of the two elucidates the structure of posets with a forbidden chain minor. Is a structural characterization possible?

In particular, it looks that characterizing posets without pC_q as a chain minor is already the first challenge. (pC_q is a poset consisting of p disjoint chains each on q vertices.) Note that any poset of size p and height q is a chain minor of 2^pC_q. It is also quite straightforward that posets without C_q chain minor are just posets of height less then q but even a characterization of posets without $2C_q$ as a chain minor seems elusive.

3. Let us recall that Gustedt showed in [5] that the CHAIN MINOR problem is NP-hard but it is not known whether the problem is NP-complete. This is an interesting question. In particular, given two posets P, Q and a function $w : Q \longrightarrow P$, is there a polynomial-time deterministic algorithm deciding whether w is a witness for $P \preceq Q$? Such algorithm would naturally give rise to an NP algorithm for CHAIN MINOR.

4. Finally, both our algorithms are double exponential in the parameter. Could this be improved to get a single exponential dependence?

References

1. Alon, N., Yuster, R., Zwick, U.: Color-coding. Journal of the ACM 42(4), 844–856 (1995)
2. Downey, R.G., Fellows, M.R.: Parameterized Complexity. Springer, NewYork (1999)
3. Fomin, F.V., Marx, D.: FPT suspects and tough customers: Open problems of Downey and Fellows (submitted)
4. Gustedt, J.: Algorithmic Aspects of Ordered Structures. PhD thesis, Berlin (1992)
5. Gustedt, J.: Well Quasi Ordering Finite Posets and Formal Languages. Journal of Combinatorial Theory, Series B 65(1), 111–124 (1995)
6. Möhring, R.H., Müller, R.: A combinatorial approach to obtain bounds for stochastic project networks. Tech. report, Technische Universität Berlin (1992)
7. Naor, M., Schulman, L.J., Srinivasan, A.: Splitters and near-optimal derandomization. In: Proceedings of the 36th Annual Symposium on Foundations of Computer Science FOCS (1995)

Incompressibility of H-Free Edge Modification*

Leizhen Cai** and Yufei Cai***

Department of Computer Science and Engineering,
The Chinese University of Hong Kong, Shatin, Hong Kong SAR, China
`lcai@cse.cuhk.edu.hk`, `cai@mathematik.uni-marburg.de`

Abstract. Given a fixed graph H, the H-FREE EDGE DELETION (resp., COMPLETION, EDITING) problems ask whether it is possible to delete from (resp., add to, delete from or add to) the input graph at most k edges so that the resulting graph is H-free, i.e., contains no induced subgraph isomorphic to H. These H-free edge modification problems are well known to be FPT for every fixed H. In this paper, we study the nonexistence of polynomial kernels for them in terms of the structure of H, and completely characterize their nonexistence for H being paths, cycles or 3-connected graphs. As a very effective tool, we have introduced a constrained satisfiability problem PROPAGATIONAL SATISFIABILITY to cope with the propagation of edge additions/deletions, and we expect the problem to be useful in studying the nonexistence of polynomial kernels.

1 Introduction

Edge modification problems are concerned with adding edges to or deleting edges from input graphs to obtain graphs with desired properties, and have been studied extensively under frameworks of both traditional complexity and parameterized complexity. In this paper, we focus on edge modification problems concerning the property of being H-*free* for a fixed graph H, i.e., our desired graph contains no induced subgraph isomorphic to H. Such problems are fundamental as any hereditary property is H-free for every graph H in a set of forbidden induced subgraphs. We consider the following H-free edge modification problems.

H-FREE EDGE DELETION
Instance: Graph G, and *parameter k*.
Question: Can we delete from G at most k edges to make it H-free?

H-FREE EDGE COMPLETION and H-FREE EDGE EDITING are defined similarly by replacing "delete from" with "add to" and "delete from or add to" respectively.

---

* Partially based on the MPhil Thesis of the 2nd author under the supervision of the 1st author.

** Partially supported by GRF grant CUHK410409 of the Research Grants Council of Hong Kong.

*** Current address: Philipps-Universitaet, Marburg Mehrzweckgebaeude 05D06, Hans-Meerwein Straße, 35032 Marburg, Gemany.

G. Gutin and S. Szeider (Eds.): IPEC 2013, LNCS 8246, pp. 84–96, 2013.

The above H-free edge modification problems are FPT for every fixed H following a general result of the first author [6]. In IWPEC'06 [2], the same author raised the issue of determining the existence of polynomial kernels for H-FREE EDGE DELETION in terms of the structure of H. Kratsch and Wahlström [11] constructed the first H for which neither H-FREE EDGE DELETION nor H-FREE EDGE EDITING admits polynomial kernels, and Guillemot et al. [10] established the nonexistence of polynomial kernels for H-FREE EDGE DELETION when H is a path P_l with $l \geq 13$ or a cycle C_l with $l \geq 12$, provided that coNP $\not\subseteq$ NP/poly. On the other hand, Gramm et al. [9] obtained polynomial kernels for P_3-FREE EDGE DELETION, COMPLETION and EDITING, and Guillemot et al. [10] presented polynomial kernels for P_4-FREE EDGE DELETION, COMPLETION and EDITING. Other than the above results, very little was known regarding polynomial kernels of H-free edge modification problems.

In this paper, we study the nonexistence of polynomial kernels for H-free modification problems in terms of the structure of H. We fully characterize 3-connected H for which H-free edge modification problems admit no polynomial kernel, and determine exactly when P_l- or C_l-free edge modification problems admit no polynomial kernel, assuming coNP $\not\subseteq$ NP/poly.

- For 3-connected H, H-FREE EDGE DELETION and EDITING admit no polynomial kernel iff H is not a complete graph.
- For 3-connected H, H-FREE EDGE COMPLETION admits no polynomial kernel iff H misses at least two edges.
- For H being a path or cycle, H-FREE EDGE DELETION, COMPLETION and EDITING admit no polynomial kernel iff H has at least 4 edges.

We assume that the reader is familiar with the general framework for kernelization lower bounds [1, 3–5, 8]. In the paper, our kernels refer to *generalized kernels* [4] (called *bikernels* by Alon et al. [1]).

Definition 1. [1, 4] *A generalized kernelization from a parameterized problem Π into another parameterized problem Π' is an algorithm that takes any instance $(I, k) \in \Pi$ as input, runs in time polynomial in $|I| + k$, and outputs an instance $(I', k') \in \Pi'$ such that*

(a) *(I, k) is a yes-instance of Π iff (I', k') is a yes-instance of Π', and*
(b) *both $|I'|$ and k' are bounded by a function $g(k)$ on k alone.*

The output (I', k') is called a generalized kernel, and it is a polynomial kernel if $g(k)$ is a polynomial.

A *polynomial parameter transformation* (Bodlaender et al. [5]) from a parameterized problem Π into another parameterized problem Π' is the same as a generalized kernelization with condition (b) changed to "*the value of parameter k' is bounded by a polynomial of k*". For simplicity, we call a parameterized problem *incompressible* if it has no polynomial kernel unless coNP $\not\subseteq$ NP/poly.

To obtain our results, first we introduce a constrained satisfiability problem PROPAGATIONAL SATISFIABILITY and prove its incompressibility (Section 2).

Then we use it as our seed problem for polynomial parameter transformations to establish the incompressibility of some "quarantined" H-free edge modification problems where we have a restriction on edges that can be added/deleted (Section 3). Finally we lift the quarantine by using "enforcers" (Section 4), and discuss some open problems (Section 5).

Our results significantly improve our knowledge on the incompressibility of H-free edge modification problems, and our PROPAGATIONAL SATISFIABILITY problem is very useful in coping with the propagation of edge deletions/additions and thus the incompressibility of edge modification problems. We hope that our ideas will be useful in the discovery of a dichotomy theorem on the incompressibility of H-free edge modification problems, and we also expect PROPAGATIONAL SATISFIABILITY to be useful in studying the nonexistence of polynomial kernels in general.

2 Satisfiability of Propagational Formulas

One main complication of H-free edge modification problems lies in the possibility of introducing new induced copies of H when we add/delete edges, which causes a propagation of edge additions/deletions. To cope with this, we introduce in this section a constrained satisfaction problem PROPAGATIONAL SATISFIABILITY and establish its incompressibility, and we will use the problem extensively to show the incompressibility of our edge modification problems.

Definition 2. *A ternary Boolean function $f(x, y, z)$, where x, y and z are either Boolean variables or constants 0 or 1, is* propagational *if it satisfies $f(1, 0, 0) = 0$ and $f(0, 0, 0) = f(1, 0, 1) = f(1, 1, 0) = f(1, 1, 1) = 1$.*

In other words, $f(x, y, z)$ is propagational if it is true when either $x = y = z = 0$ or "$x = 1$ implies $y = 1$ or $z = 1$". There are eight different propagational functions f in total due to the freedom of defining the value of f for the other three assignments of variables.

Example 3. *The following three functions are propagational:*

$$f_1(x, y, z) = \overline{x} \vee y \vee z,$$
$$f_2(x, y, z) = x \, \mathsf{XOR} \, (y \, \mathsf{NOR} \, z),$$
$$\mathsf{Not\text{-}1\text{-}in\text{-}3}(x, y, z) = (\overline{x} \vee y \vee z) \wedge (x \vee \overline{y} \vee z) \wedge (x \vee y \vee \overline{z}).$$

Propagational functions $f(x, y, z)$ generalize function $\mathsf{Not\text{-}1\text{-}in\text{-}3}$ of Kratsch and Wahlström [11], and capture the relation that "whatever happens to x must happen to either y or z", which is of great use when we deal with edge modification problems because of propagations of edge deletions/additions. The following example of C_4-FREE EDGE DELETION explains such a connection. Suppose that we want to delete some light edges from the graph in Fig. 1 to obtain a C_4-free graph. When we delete edge x, we create a new induced C_4 in the graph, and we must delete either edge y or edge z or both in order to destroy the new C_4.

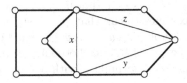

Fig. 1. Realization of a propagational function $f(x, y, z)$ by C_4-free edge deletion

Therefore we can use the graph to realize a propagational function $f(x, y, z)$, which also represents the propagation of edge deletions from x to y or z.

For a Boolean function $f(x, y, z)$, a *conjunctive formula* φ is of the form

$$f(x_1, y_1, z_1) \wedge f(x_2, y_2, z_2) \wedge \cdots \wedge f(x_m, y_m, z_m).$$

Each $f(x_i, y_i, z_i)$ is a *clause* of φ, and the *Hamming weight* of an assignment of 0's and 1's to variables is the number of 1's in the assignment. For φ, the *degree of a variable* is its number of occurrences in φ, and the *degree of φ* is the maximum degree of its variables. We say that φ is *t-regular* if all its variables have degree t.

PROPAGATIONAL SATISFIABILITY
Instance: Conjunctive formula φ of a propagational function f, *parameter* k.
Question: Does φ have a satisfying assignment of Hamming weight $\leq k$?

We will establish the incompressibility of the above problem in two steps: first prove its NP-completeness, and then show that it is OR-compositional.

Lemma 4. *For any propagational function $f(x, y, z)$,* PROPAGATIONAL SATIS-
FIABILITY *is NP-complete on degree-6 conjunctive formulas with one occurrence of constant 1.*

Proof. The problem is clearly in NP, and we give a polynomial reduction from the classical VERTEX COVER problem. For an arbitrary instance (G, k) of VERTEX COVER, we first construct a conjunctive formula φ': each vertex of G is a variable and each edge uv of G corresponds to a clause $f(1, u, v)$, which forces us to choose either u or v in order to satisfy $f(1, u, v)$. Clearly G has a vertex cover of size $\leq k$ iff φ' can be satisfied with $\leq k$ true variables.

Next we convert φ' into a degree-6 conjunctive formula φ with one occurrence of constant 1. Note that two clauses $f(x, y, 0)$ and $f(y, x, 0)$ ensure that x and y have the same value, and we write $(x = y)$ as a short hand for $f(x, y, 0) \wedge f(y, x, 0)$. Given any $p = 2^q$, we can make variables $w_1, \ldots, w_{2p-1}$ take the same value by adding the following set $F(w)$ of clauses

$(w_1 = w_2) \wedge (w_1 = w_3) \wedge$
$(w_2 = w_4) \wedge (w_2 = w_5) \wedge (w_3 = w_6) \wedge (w_3 = w_7) \wedge$

$$\vdots \qquad\qquad \vdots$$

$(w_{p/2} = w_p) \wedge (w_{p/2} = w_{p+1}) \wedge \cdots \wedge (w_{p-1} = w_{2p-2}) \wedge (w_{p-1} = w_{2p-1}).$

Among these clauses, w_1 appears four times, $w_p, \ldots, w_{2p-1}$ appear twice, and other variables appear six times (recall that $(w_i = w_j)$ means two appearances of w_i and w_j each). The variables $w_1, w_p, \ldots, w_{2p-1}$ can be used in other clauses.

Let m be the number of edges of G, and choose $p = 2^q$ between $3m$ and $6m-1$. We construct from φ' a degree-6 formula φ with one occurrence of constant 1.

1. Add a variable w_1 not occurring in φ' and the clause $f(1, w_1, 0)$, which forces w_1 to take value 1.
2. Add variables $w_2, \ldots, w_{2p-1}$ not occurring in φ' to represent occurrences of 1 in φ', and add clauses $F(w)$ to force all $w_2, \ldots, w_{2p-1}$ to take value 1.
3. For every variable v of φ', add variables $v_1, \ldots, v_{2p-1}$ and clauses $F(v)$ to force all $v_1, \ldots, v_{2p-1}$ to have the same value.
4. For the i-th clause $f(1, u, v)$ of φ', add clause $f(w_{p+3i-3}, u_{p+3i-2}, v_{p+3i-1})$. Since $i \leq m$ and $3m \leq p$, we will never run out of variables.

If φ' is satisfiable with $\leq k$ true variables, we can satisfy φ with $(k+1)(2p-1) \leq 12(k+1)m$ true variables, consisting of $w_1, \ldots, w_{2p-1}$ and those $v_1, \ldots, v_{2p-1}$ for $v = 1$ in the satisfying assignment to φ'. The converse is also true. $\square$

Lemma 5. *For any propagational function $f(x, y, z)$,* Propagational Satisfiability *is OR-compositional on degree-6 conjunctive formulas with one occurrence of constant 1.*

Proof. We describe a composition algorithm very similar to the one in Lemma 2 of Kratsch and Wahlström [11]. Let $(\varphi_1, k), \ldots, (\varphi_t, k)$ be t instances of the problem such that each φ_i has degree 6 and one occurrence of 1. Note that each φ_i can be solved in $O(3^k |\varphi_i|)$ time by bounded search tree. If $t > 2^k$, we have enough time to solve each φ_i and output a dummy yes- or no-instance accordingly.

Therefore we assume $t \leq 2^k$, and let $p = 2^q$ be the power of two between t and $2t - 1$. Construct a conjunctive formula φ' as follows.

1. Rename variables of $\varphi_1, \ldots, \varphi_t$ so that they are all distinct.
2. Add all clauses of $\varphi_1, \ldots, \varphi_t$ to φ'.
3. For each φ_i, replace the occurrence of 1 with a distinct variable w_{i+p-1}.
4. Add the following clauses so that $\geq q$ variables from $w_2, \ldots, w_{2p-1}$, including one of $w_p, \ldots, w_{2p-1}$, are forced to take value 1:

$$f(1, w_2, w_3) \wedge$$
$$f(w_2, w_4, w_5) \wedge f(w_3, w_6, w_7) \wedge$$
$$\vdots \qquad\qquad \vdots$$
$$f(w_{p/2}, w_p, w_{p+1}) \wedge \cdots \wedge f(w_{p-1}, w_{2p-2}, w_{2p-1}).$$

Set $k' = k + q \leq 2k$. If some (φ_i, k) is a yes-instance, then we can satisfy φ' with q true variables from $w_2, \ldots, w_{2p-1}$ including w_{i+p-1} and $\leq k$ additional true variables satisfying clauses of φ_i, for a total of k' true variables. The clauses of

φ_j with $j \neq i$ are satisfied with 0 in all 3 positions. Conversely, if (φ', k') is a yes-instance, then $\geq q$ variables from $w_2, \ldots, w_{2p-1}$ are forced to be true, including some w_{i+p-1}. Then clauses of φ_i are satisfied with the remaining quota of $\leq k$ true variables. $\square$

Theorem 6. *For any propagational function* $f(x, y, z)$, Propagational Satisfiability *on 6-regular conjunctive formulas is incompressible.*

Proof. By Lemma 4 and Lemma 5 and the work in [4], Propagational Satisfiability is incompressible on degree-6 conjunctive formulas with one occurrence of 1. We can easily modify a degree-6 conjunctive formula into an equivalent 6-regular conjunctive formula: For each variable x of degree d, add $(6 - d)$ clauses of the form $f(1, 1, x)$. $\square$

3   Incompressibility: Quarantined Edge Modification

To ease the complication in tackling H-free modification problems, we first add a restriction to edges that can be added or deleted, which forms "quarantined" edge modification problems. We then use our incompressible propagational satisfiability problems to show that "quarantined" edge modification problems are incompressible for H being a 4-cycle, 5 cycle, or 3-connected graph, which forms the base for our main results.

Quarantined H-Free Edge Deletion
Instance: Graph G, forbidden set $F \subseteq E(G)$, and *parameter k.*
Question: Can we delete at most k edges from $E(G) - F$ to make G H-free?

Edges in F are *forbidden edges*, edges in $E(G) - F$ are *allowed edges*, and allowed edges form the *allowed subgraph*.

Quarantined H-Free Edge Completion.
Instance: Graph G, forbidden set $F \subseteq E(\overline{G})$, and *parameter k.*
Question: Can we add at most k edges from $E(\overline{G}) - F$ to make G H-free?

Note that $\overline{G}$ is the complement of G. Edges in F are *forbidden nonedges*, edges in $E(\overline{G}) - F$ are *allowed nonedges*, and allowed nonedges form the *allowed complement* of G.

Theorem 7. Quarantined C_4-Free Edge Deletion *is incompressible on graphs whose allowed subgraphs contain no C_4 as a partial subgraph.*

Proof. We give a polynomial parameter transformation from Propagational Satisfiability on 6-regular conjunctive formulas of propagational function Not-1-in-3. We need the three components in Fig. 2, where an edge marked with a letter, say x, will be referred to as an x-edge.

For an arbitrary instance (φ, k) of our Propagational Satisfiability, we construct an instance (G, F, k') of Quarantined C_4-Free Edge Deletion as follows (see Fig. 3 for an example).

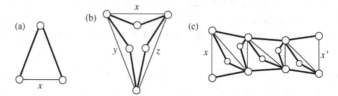

Fig. 2. Components for Quarantined C_4-Free Edge Deletion with thick edges indicating forbidden edges in F: (a) truth-setting component T(x), (b) satisfaction-testing component $S(x, y, z)$, and (c) communication component $C(x)$. Note that $S(x, y, z)$ realizes Not-1-in-3(x, y, z) as $S(x, y, z)$ itself is C_4-free and we need to delete at least two edges from $\{x, y, z\}$ to ensure that $S(x, y, z)$ stays C_4-free.

1. Create a truth-setting component $T(x)$ for each variable x of φ, and a satisfaction-testing component $S(x, y, z)$ for each clause $f(x, y, z)$ of φ.
2. For each clause $f(x, y, z)$, consider each $v \in \{x, y, z\}$. If $v \in \{0, 1\}$, then the v-edge in $S(x, y, z)$ is deleted if $v = 1$ and marked as forbidden if $v = 0$. Otherwise v is a variable, and we add a communication component $C(v)$, identify the v-edge of $T(v)$ with the v-edge of $C(v)$ and identify the v'-edge of $C(v)$ with the v-edge of $S(x, y, z)$.
3. Let G be the resultant graph, F the set of forbidden edges in all components, and set $k' = 37k$.

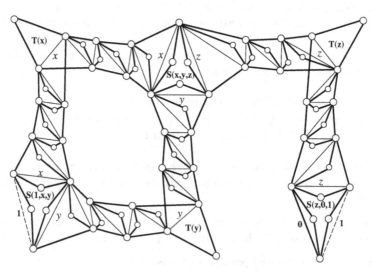

Fig. 3. Graph G for $\varphi = f(1, x, y) \wedge f(x, y, z) \wedge f(z, 0, 1)$ using components in Fig. 2. For clarity of illustration, φ is not 6-regular here. Thick edges are forbidden edges F, and dashed lines indicate deleted edges of components $S(x, y, z)$.

It is easy to see that the allowed subgraph of G contains no C_4 as a partial subgraph, and the transformation is a polynomial parameter transformation. We show that φ is satisfiable with $\leq k$ true variables iff G can be made C_4-free by deleting $\leq k'$ allowed edges.

($\Rightarrow$) Consider a satisfying assignment of φ with $\leq k$ true variables, and let E' be allowed edges of all copies of communication components in G for all true variables. Each variable x has 6 communication components with one shared x-edge and contributes 37 edges to E', implying $|E'| \leq 37k = k'$. It is easily checked that $G - E'$ is C_4-free.

($\Leftarrow$) Let E' be a set of $\leq k'$ allowed edges in G whose deletion results in a C_4-free graph. Observe that for a communication component $C(x)$ in G, as far as C_4-freeness is concerned, deleting its x-edge will force the deletion of all its allowed edges, including x'-edge. It follows that for every truth-setting component $T(x)$, the 37 allowed edges of communication components attached to $T(x)$ are either all deleted or none deleted. Assign $x = 1$ if the x-edge of $T(x)$ is in E' and assign $x = 0$ otherwise, and we have set $\leq k$ variables true. For each clause $f(x, y, z)$, it is ensured by the C_4-freeness of its satisfaction-testing component $S(x, y, z)$ after deleting E' that $f(x, y, z)$ satisfies Not-1-in-3 and thus is true. $\qquad\square$

With the above theorem, we can easily give a polynomial parameter transformation from QUARANTINED C_4-FREE EDGE DELETION to $\overline{P_5}$-FREE EDGE DELETION, where $\overline{P_5}$ is the same as the house graph $C_5 + e$ [7].

Corollary 8. $\overline{P_5}$-FREE EDGE DELETION *is incompressible.*

The construction and proof in Theorem 7 highlight the basic ideas in establishing the incompressibility of QUARANTINED H-FREE EDGE DELETION and COMPLETION:

1. Use $T(x)$ to decide whether to assign 0 or 1 to x.
2. Use $S(x, y, z)$ to realize a propagational function f.
3. Use $C(x)$ to represent the propagation of edge deletions/additions from x-edges to x'-edges, and connect $T(x)$ with satisfaction-testing components.

Indeed, we can establish the incompressibility of QUARANTINED C_4-FREE EDGE COMPLETION in a way almost identical to the proof of Theorem 7: use the components in Fig. 4, instead of those in Fig. 2. We also use a different propagational function $f(x, y, z) = x\, \text{XOR}\, (y\, \text{NOR}\, z)$, instead of Not-1-in-3.

Theorem 9. QUARANTINED C_4-FREE EDGE COMPLETION *is incompressible on graphs whose allowed complements have girth greater than 4.*

As in the case of $\overline{P_5}$-free edge deletion, we can use Theorem 9 to construct an easy polynomial parameter transformation for the incompressibility of $\overline{P_5}$-free edge completion [7].

Corollary 10. QUARANTINED $\overline{P_5}$-FREE EDGE COMPLETION *is incompressible.*

Very similar constructions also work for C_5-FREE EDGE DELETION and COMPLETION. Here we only give the key components, satisfaction-testing components

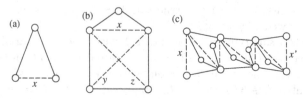

Fig. 4. Components for QUARANTINED C_4-FREE EDGE COMPLETION (where allowed nonedges are denoted by dashed lines and forbidden nonedges are invisible): (a) truth-setting component T(x), (b) satisfaction-testing component $S(x, y, z)$, and (c) communication component $C(x)$. Note that $S(x, y, z)$ realizes $f(x, y, z) = x\,\mathsf{XOR}\,(y\,\mathsf{NOR}\,z)$: when we add some edges in $\{x, y, z\}$, the resulting graph is C_4-free iff the string xyz is 000, 101, 110 or 111.

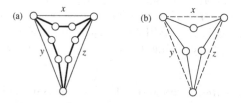

Fig. 5. Satisfaction-testing components $S(x, y, z)$ for (a) C_5-FREE EDGE DELETION and (b) C_5-FREE EDGE COMPLETION. Both realize propagational function Not-1-in-3.

$S(x, y, z)$, in Fig. 5, and full proofs are available in [7] where it describes a general scheme for this type of constructions.

We now turn to 3-connected H for which similar constructions also work. Here, again we only give the construction of satisfaction-testing components $S(x, y, z)$, and full proofs are available in [7] and will be given in the full paper.

QUARANTINED H-FREE EDGE DELETION *for H being 3-connected but not complete.* H contains an induced $P_3 = a, b, c$. Let x be the nonedge ac, y edge ab and z edge bc; and we set $S(x, y, z)$ to $H + x$. Regard edges x, y and z as Boolean variables. For an edge $e \in \{x, y, z\}$, assign to it value 1 iff it is deleted from $S(x, y, z)$, and define Boolean function $f(x, y, z) = 1$ iff the graph obtained from $S(x, y, z)$ is H-free when we delete from $S(x, y, z)$ edges in $\{x, y, z\}$ with value 1. It is easily checked that $f(0, 0, 0) = f(1, 0, 1) = f(1, 1, 0) = f(1, 1, 1) = 1$ but $f(1, 0, 0) = 0$, implying that f is propagational.

QUARANTINED H-FREE EDGE COMPLETION *for H being 3-connected with at least 2 nonedges.* Let x be an arbitrary edge and y, z two nonedges of H, and we delete edge x from H to form $S(x, y, z)$. For an edge $e \in \{x, y, z\}$, assign to it value 1 iff it is added to $S(x, y, z)$, and define Boolean function $f(x, y, z) = 1$ iff the graph obtained from $S(x, y, z)$ is H-free when we add to $S(x, y, z)$ edges in $\{x, y, z\}$ with value 1. Again, $f(0, 0, 0) = f(1, 0, 1) = f(1, 1, 0) = f(1, 1, 1) = 1$ but $f(1, 0, 0) = 0$, and we get a propagational f.

We now summarize our results of the section in the following theorem, which will be used in the next section to obtain our main results.

Theorem 11. QUARANTINED H-FREE EDGE DELETION *is incompressible if H is C_4, C_5, $\overline{P_5}$, or a 3-connected graph with at least one nonedge.* QUARANTINED H-FREE EDGE COMPLETION *is incompressible if H is C_4, C_5, $\overline{P_5}$, or a 3-connected graph with at least two nonedges.*

4   Lifting the Quarantine

In the previous section, we have shown the incompressibility of quarantined H-free edge deletion/completion problems. We now discuss how to lift the quarantine so that our results extends to our original unquarantined H-free edge deletion/completion problems. Furthermore, our tools will also allow us to easily extend incompressibility of edge deletion/completion to edge editing.

We need a way to prevent an edge from being deleted for edge deletion and a nonedge from being added for edge completion. This is in fact pretty straightforward: for each forbidden edge $e \in F$ attach $k+1$ vertex-disjoint copies of $H + e'$ (where e' is any nonedge of H) by identifying e' with e, and for each forbidden nonedge $e \in F$ attach $k+1$ vertex-disjoint copies of $H - e'$ (where e' is any edge of H) by identifying e' with e. *The trick is to prevent the introduction of new induced H in the process.*

Definition 12. *An H-free deletion enforcer is an H-free graph H' with a distinguished edge e' such that (a) $H' - e'$ has an induced H, and (b) the identification of e' with any edge e of a vertex-disjoint H-free graph produces an H-free graph.*

Definition 13. *An H-free completion enforcer is an H-free graph H' with a distinguished nonedge e' such that (a) $H' + e'$ has an induced H, and (b) the identification of e' with any nonedge e of a vertex-disjoint H-free graph produces an H-free graph.*

It is clear that the above notion of enforcers will enable us to lift quarantine by attaching $k+1$ enforcers to prevent an edge from being deleted or a nonedge from being added, without introducing unwanted copies of induced H. As a bonus, completion (resp., deletion) enforcers establish incompressibility of edge editing problems directly from that of edge deletion (resp., completion) problems: *by forbidding all nonedge with completion enforcers, editing is forced to be deletion; and by forbidding all edges with deletion enforcers, editing is forced to be completion.*

Lemma 14. H-FREE EDGE DELETION *(resp.* H-FREE EDGE COMPLETION*) is incompressible if* QUARANTINED H-FREE EDGE DELETION *(resp.* QUARANTINED H-FREE EDGE COMPLETION*) is incompressible and there exists an H-free deletion (resp. completion) enforcer.*

Lemma 15. *H*-FREE EDGE EDITING *is incompressible if either H*-FREE EDGE
DELETION *is incompressible and there exists an H-free completion enforcer or*
H-FREE EDGE COMPLETION *is incompressible and there exists an H-free dele-*
tion enforcer.

The following claims can be easily verified.

1. For any $t \geq 4$, adding any chord e' to C_t yields a C_t-free deletion enforcer,
   and deleting any edge e' from C_t yields a C_t-free completion enforcer.
2. If H is 3-connected and has a nonedge e', then $H + e'$ is an H-free deletion
   enforcer.
3. If H is 3-connected then for any edge e', $H - e'$ is an H-free completion
   enforcer.

We also need the following easy to see but very useful facts.

Lemma 16. *H*-FREE EDGE DELETION *is equivalent to* $\overline{H}$-FREE EDGE COM-
PLETION, *and H*-FREE EDGE EDITING *is equivalent to* $\overline{H}$-FREE EDGE EDITING.

Now we are ready to state our characterization for 3-connected H.

Theorem 17. *Let H be 3-connected and assume* coNP $\not\subseteq$ NP/poly.

1. *H*-FREE EDGE COMPLETION *admits no polynomial kernel iff H has* ≥ 2
   nonedges.
2. *H*-FREE EDGE DELETION *and H*-FREE EDGE EDITING *admit no polyno-*
   mial kernel iff H is not a complete graph.

Proof. Incompressibility follows from Theorem 11, the existence of enforcers,
Lemma 14, and Lemma 15.

For the cases that admit polynomial kernels, *H*-FREE EDGE COMPLETION is
trivial for H being a complete graph, and easily solved in $O(kn^t)$ for H being
$K_t - e$ for some constant t (for each H found in G just add the missing edge):
both solvable in polynomial time and thus have trivial kernels.

When H is a complete graph K_t for some constant t, K_t-FREE EDGE EDITING
is equivalent to K_t-FREE EDGE DELETION. The latter admits a polynomial
kernel by reducing it to HITTING SET where each subset has size $\binom{t}{2}$, and we
can make the kernel an instance of K_t-FREE EDGE DELETION if one insists [7].
                                                                                  □

Theorem 18. *Let H be a path or cycle and assume* coNP $\not\subseteq$ NP/poly. *H*-FREE
EDGE DELETION, COMPLETION *and* EDITING *have no polynomial kernel iff H*
has at least 4 edges.

Proof. For $H = P_t$ with $t \leq 4$, polynomial kernels for these problems are found
by Gramm et al. [9] and Guillemot et al. [10]. Since $C_3 = K_3$, polynomial kernels
exist for these problems when $H = C_3$ as discussed in the proof of Theorem 17.

For the incompressibility part, if $H = C_t$ or P_t with $t \geq 6$ then $\overline{H}$ is 3-
connected with at least two nonedges, and the incompressibility of these problems follow from Theorem 17, Lemma 14 and Lemma 16. For the remaining
three cases $H = C_4, C_5$ or P_5, their incompressibility follow from Theorem 11,
Lemma 14, Lemma 15, and Lemma 16.   □

5   Conclusion: Towards a Dichotomy Theorem

Our incompressibility for 3-connected H is actually much more powerful than it looks in Theorem 17. Because of Lemma 16, Theorem 17 implies the following result which covers a very extensive range of H.

Corollary 19. *For any fixed H, H-FREE EDGE DELETION (resp. COMPLETION, and EDITING) is incompressible whenever H or $\overline{H}$ is 3-connected with at least two nonedges.*

From this we can deduce that for most trees H and for most disconnected H, H-free edge modification problems are incompressible as $\overline{H}$ is 3-connected for most such H. In fact for trees H, we know that H-free edge modification problems are incompressible for all but a small number of trees [7]. In this regards, $H = K_{1,3}$ (the claw graph) is a very challenging case.

Problem 20. *Determine whether claw-free edge modification problems admit polynomial kernels.*

For general H, we pretty much know how blocks and connected components in H affect the incompressibility of H-free modification problems [7]. This leaves 2-connected H a very important case. Note that DIAMOND-FREE EDGE DELETION admits a polynomial kernel [7].

Conjecture 21. *For any fixed 2-connected H, H-FREE EDGE DELETION and EDITING are incompressible unless H is complete or the diamond graph $K_4 - e$, and H-FREE EDGE COMPLETION is incompressible unless H misses at most one edge.*

Since most hereditary families of graphs are characterized by several forbidden subgraphs, it is also meaningful and important to study the incompressibility of their corresponding edge modification problems.

Problem 22. *Let $\mathcal{F}$ be a family of graphs. What is the relation between the incompressibility of $\mathcal{F}$-free edge modification problems and that of H-free edge modification problems for every $H \in \mathcal{F}$? In particular, does the incompressibility of H_1- and H_2-free edge modification problems imply that of $\{H_1, H_2\}$-free edge modification problem?*

We hope that our work in the paper will be useful towards a dichotomy theorem on incompressibility of H-free edge modification problems, or perhaps even a dichotomy theorem for the general $\mathcal{F}$-free edge modification problems.

References

1. Alon, N., Gutin, G., Kim, E.J., Szeider, S., Yeo, A.: Solving Max-r-SAT above a tight lower bound. Algorithmica 61, 638–655 (2011)
2. Bodlaender, H.L., Cai, L., Chen, J., Fellows, M.R., Telle, J.A., Marx, D.: Open problems in parameterized and exact computation — IWPEC 2006. Utrecht University Technical Report UU-CS-2006-052 (2006)

3. Bodlaender, H.L., Downey, R.G., Fellows, M.R., Hermelin, D.: On problems without polynomial kernels. Journal of Computer and System Sciences 75(8), 423–434 (2009)
4. Bodlaender, H.L., Jansen, B.M.P., Kratsch, S.: Cross-composition: a new technique for kernelization lower bounds. arXiv:1011.4224v2 (2012), http://arxiv.org/abs/1011.4224
5. Bodlaender, H.L., Thomassé, S., Yeo, A.: Analysis of data reduction: transformations give evidence for non-existence of polynomial kernels. Utrecht University Technical Report UU-CS-2008-030 (2008)
6. Cai, L.: Fixed-parameter tractability of graph modification problems for hereditary properties. Information Processing Letters 58(4), 157–206 (1996)
7. Cai, Y.: Polynomial Kernelisation of H-free Edge Modification Problems. MPhil Thesis, Department of Computer Science and Engineering, The Chinese University of Hong Kong, Hong Kong SAR, China (2012), http://www.uni-marburg.de/fb12/ps/team/cai-masterarbeit.pdf
8. Fortnow, L., Santhanam, R.: Infeasibility of instance compression and succinct PCPs for NP. Journal of Computer and System Sciences 77(1), 91–106 (2011)
9. Gramm, J., Guo, J., Hüffner, F., Niedermeier, R.: Graph-modeled data clustering: fixed parameter algorithms for clique generation. Theory of Computing Systems 38(4), 373–392 (2005)
10. Guillemot, S., Paul, C., Perez, A.: On the (non-)existence of polynomial kernels for P_l-free edge modification problems. In: Raman, V., Saurabh, S. (eds.) IPEC 2010. LNCS, vol. 6478, pp. 147–157. Springer, Heidelberg (2010)
11. Kratsch, S., Wahlström, M.: Two edge modification problems without polynomial kernels. In: Chen, J., Fomin, F.V. (eds.) IWPEC 2009. LNCS, vol. 5917, pp. 264–275. Springer, Heidelberg (2009)

Contracting Few Edges
to Remove Forbidden Induced Subgraphs

Leizhen Cai* and Chengwei Guo

Department of Computer Science and Engineering,
The Chinese University of Hong Kong, Hong Kong S.A.R., China
{lcai,cwguo}@cse.cuhk.edu.hk

Abstract. For a given graph property Π (i.e., a collection Π of graphs), the Π-CONTRACTION problem is to determine whether the input graph G can be transformed into a graph satisfying property Π by contracting at most k edges, where k is a parameter. In this paper, we mainly focus on the parameterized complexity of Π-CONTRACTION problems for Π being H-free (i.e., containing no induced subgraph isomorphic to H) for various fixed graphs H.

We show that CLIQUE CONTRACTION (equivalently, P_3-FREE CONTRACTION for connected graphs) is FPT (fixed-parameter tractable) but admits no polynomial kernel unless $NP \subseteq coNP/poly$, and prove that CHORDAL CONTRACTION (equivalently, $\{C_l : l \geq 4\}$-FREE CONTRACTION) is W[2]-hard. We completely characterize the parameterized complexity of H-FREE CONTRACTION for all fixed 3-connected graphs H: FPT but no polynomial kernel unless $NP \subseteq coNP/poly$ if H is a complete graph, and W[2]-hard otherwise. We also show that H-FREE CONTRACTION is W[2]-hard whenever H is a fixed cycle C_l for some $l \geq 4$ or a fixed path P_l for some odd $l \geq 5$.

1 Introduction

Edge contraction is a fundamental operation in graph theory, and plays a crucial role in the celebrated graph minor theory. An *edge contraction* in a graph identifies two endpoints of an edge, and eliminates loop and multiple edges in the resulting graph. For a given graph property Π (i.e., a collection Π of graphs), the Π-CONTRACTION problem asks whether the input graph can be modified into a Π-*graph*, i.e. a graph satisfying property Π, by at most k edge contractions.

The complexity of edge contraction problems has been studied in the literature, but does not receive as much attention as graph modification problems in terms of vertex and edge addition/deletion. Watanabe *et al.* [15] and Asano and Hirata [1,2] proved that Π-CONTRACTION is NP-complete if Π is finitely characterized by 3-connected forbidden subgraphs, or Π is hereditary on contractions and is determined by biconnected components.

* Partially supported by GRF grant CUHK410409 of the Research Grants Council of Hong Kong.

G. Gutin and S. Szeider (Eds.): IPEC 2013, LNCS 8246, pp. 97–109, 2013.
© Springer International Publishing Switzerland 2013

Recently, researchers have studied edge contraction problems from the perspective of parameterized complexity. Heggernes *et al.* [11] have obtained an FPT algorithm for BIPARTITE CONTRACTION that asks whether a graph can be modified into a bipartite graph by at most k edge contractions. Later Heggernes *et al.* [10] presented a $4.98^k n^{O(1)}$ time algorithm for TREE CONTRACTION and a $2^{k+o(k)} + n^{O(1)}$ time algorithm for PATH CONTRACTION. Golovach *et al.* [7] considered Π-CONTRACTION for Π being the class of graphs of minimum degree at least d and showed that the problem is FPT when both d and k are parameters, but W[1]-hard when only k is the parameter and NP-complete when $d = 14$. Furthermore, Golovach et al. [8] showed that PLANAR CONTRACTION is FPT.

In this paper, we focus on the parameterized complexity of the following H-FREE CONTRACTION problems, where a graph is H-*free* if it contains no induced copy of H, i.e., an induced subgraph isomorphic to H. We note that several important graph classes (e.g., cographs, triangle-free graphs, and claw-free graphs) are characterized by H-freeness.

H-FREE CONTRACTION
Instance: Graph G, positive integer k as parameter.
Question: Can we obtain an H-free graph from G by at most k edge contractions?

It is easy to see that whenever H is a fixed complete graph K_t, H-FREE CONTRACTION is FPT as the only way to destroy a copy of K_t is to contract some edges in the copy, which implies an FPT algorithm by the bounded search tree method. However, the situation for H other than complete graphs is very complicated as contractions can occur for edges not involved in any induced copies of H. In this paper, we try to determine the parameterized complexity of H-FREE CONTRACTION in terms of the structure of H, and we have made important progress towards this goal by the following results:

- CLIQUE CONTRACTION (equivalently, P_3-FREE CONTRACTION for connected graphs) is FPT but admits no polynomial kernel unless $NP \subseteq coNP/poly$, and P_l-FREE CONTRACTION is W[2]-hard for every fixed path P_l with odd $l \geq 5$.
- C_3-FREE CONTRACTION is FPT but admits no polynomial kernel unless $NP \subseteq coNP/poly$, and C_l-FREE CONTRACTION is W[2]-hard for every fixed cycle C_l with $l \geq 4$.
- CHORDAL CONTRACTION is W[2]-hard, which is in contrast to that both CHORDAL COMPLETION and CHORDAL DELETION are FPT [3,12,13].
- For every fixed 3-connected graph H, H-FREE CONTRACTION is W[2]-hard whenever H is not a complete graph. Otherwise, it is FPT but admits no polynomial kernel unless $NP \subseteq coNP/poly$.

Our FPT algorithm for CLIQUE CONTRACTION first finds a large "seed clique" in the input graph, and then uses a branch-and-search algorithm to contract other edges into the clique. This idea is useful for other edge contraction problems such as SPLIT CONTRACTION, which will appear in our future paper. For the W[2]-hardness proofs, all FPT reductions in this paper are from the classical DOMINATING SET problem that takes an integer k as parameter, and asks whether an input graph G contains a *dominating k-set*, i.e., at most k vertices V' s.t. every vertex in $V(G) - V'$ is adjacent to some vertex in V'.

All graphs in the paper are simple, finite, and undirected. For a graph G, we denote its vertex set and edge set by $V(G)$ and $E(G)$ respectively. A graph is *chordal* if it has no induced cycle of size greater than 3. For an integer t, K_t is a complete graph on t vertices, C_t is a cycle on t vertices, and P_t is a path on t vertices. The *contraction* of edge uv in G removes u and v from G, and replaces them by a new vertex adjacent to precisely those vertices that were adjacent to at least one of u or v. For a set of edges $F \subseteq E(G)$, we use G/F to denote the graph obtained from G by sequentially contracting all edges in F. If a graph H with vertex set $\{h_1, \cdots, h_l\}$ can be obtained from graph G by a sequence of edge contractions, then G is *contractible* to H. In this case, G has a *H-witness structure*: a partition of $V(G)$ into l sets $W(h_1), \cdots, W(h_l)$, called *witness sets*, such that each $W(h_i)$ induces a connected subgraph of G and for any two $W(h_i)$ and $W(h_j)$, there is an edge between $W(h_i)$ and $W(h_j)$ in G iff $h_i h_j \in E(H)$. We obtain H from G by contracting vertices in each $W(h_i)$ into a single vertex.

2 Path-Free Contraction

We start with P_l-FREE CONTRACTION problems for fixed $l \geq 3$. Since edge contractions preserve the connectedness of a graph and a graph is a complete graph iff it is P_3-free and connected, P_3-FREE CONTRACTION for connected graphs is equivalent to CLIQUE CONTRACTION that asks whether we can transform the input graph into a clique (i.e., complete graph) by contracting at most k edges.

We note that transforming a n-vertex graph G into a clique by contracting k edges is equivalent to finding a $(n-k)$-clique minor of G as an edge contraction reduces the number of vertices by one. Thus CLIQUE CONTRACTION is a parametric dual of MAXIMUM CLIQUE MINOR that takes as input a graph G and an integer h, and asks whether G contains a clique K_h as a minor. MAXIMUM CLIQUE MINOR is NP-complete as shown by Eppstein [6], and FPT when parameterized by h following a celebrated result on graph minors by Robertson and Seymour [14]. The NP-completeness of CLIQUE CONTRACTION directly follows from that of MAXIMUM CLIQUE MINOR. Here we present an FPT algorithm for CLIQUE CONTRACTION, which combines bounded search tree with a kernelization of the problem from the second author's PhD dissertation [9].

Theorem 1. CLIQUE CONTRACTION *can be solved in* $O(2^{7k} k^{2k+5} + m)$ *time, but admits no polynomial kernel unless* $NP \subseteq coNP/poly$.

Proof. For a vertex set A, we denote by $E[A]$ the set of edges whose both endpoints are in A. For any two disjoint vertex sets B and C, we use $E[B, C]$ to denote the set of edges whose one endpoint is in B and the other is in C.

Since each edge contraction affects only two vertices, a n-vertex graph G must contain a clique of $(n - 2k)$ vertices V_c in order for G to be contractible

to a clique by at most k contractions. We start by using an FPT algorithm for
VERTEX COVER to find such vertex set V_c. Next, we construct a bounded search
tree and consider all possible edges in the solution set. In the search tree, we
branch out by contracting edges of $E(G - V_c)$, edges of $E[V_c, V(G) - V_c]$, and
edges of $E[V_c]$ in sequence. See Fig. 1 for an illustration. Note that the number
of edges in $E[V_c, V(G) - V_c]$ or $E[V_c]$ might be very large. However, we do not
need to consider all edges. The trick is to compress the possible choices into a
special set of edges whose size is bounded by a function of k. Our algorithm
consists of the following steps:

1. Determine whether there is a set V_c of $n - 2k$ vertices that induces a clique
   in G. If yes, find V_c and let $V_k = V(G) - V_c$; otherwise, return "NO".
2. We construct a search tree and label the root by the input instance (G, k).
   We branch out at the root by contracting every possible set of at most k
   edges in $E[V_k]$ and label the new node of the tree by the resulting instance
   (G', k'), where k' is the number of remaining edge contractions.
3. For each node (G', k') obtained in Step 2, we assume that vertices of V_k are
   contracted into vertices V_k' in G'. We branch out by every possible partition
   $V_k' = (V_p, R)$ (V_p corresponds to the subset of V_k' consisting of vertices not
   involved in edge contractions). Let $T = \{v \in V_c \mid \exists w \in V_p, \ wv \notin E(G')\}$. If
   $|R| > k'$ or $|T| > 2k'$, discard this node.

   We continue to branch by contracting every possible set of $|R|$ edges in
   $E[R, T]$ that covers all vertices in R and label the new node of the tree by
   the resulting instance (G'', k''). Here $k'' = k' - |R|$, and vertices in R are
   merged into the large clique $G''[V_c]$.
4. For each node (G'', k'') obtained in Step 3, we arbitrarily choose a vertex
   $u \in V_c - T$. We branch by contracting every k''-subset of $E[T \cup \{u\}]$.
5. If there exists a leaf in this search tree labelled with a clique (in Step 4),
   then return "YES"; otherwise, return "NO".

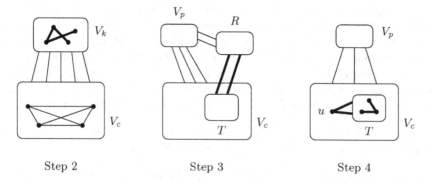

Step 2   Step 3   Step 4

Fig. 1. Edges being considered for contractions (thick edges) in Step 2-4

In Step 1, finding a $(n - 2k)$-clique is equivalent to finding a $2k$-vertex cover in the complement graph of G, which costs $O(1.2738^{2k} + kn)$ time following a known algorithm by Chen $et.$ $al.$ [4]. In Step 2, the root has at most $\sum_{k'} \binom{\binom{|V_k|}{2}}{k-k'} \leq \sum_{k'} (2k^2)^{k-k'}$ children. In Step 3, the total number of different partitions $V'_k = (V_p, R)$ is $2^{|V'_k|} \leq 2^{2k}$, and for each partition we branch into at most $|R||T| \leq 2k'^2$ nodes. In Step 4, for each node we branch into at most $(\binom{|T|+1}{2})^{k''} \leq (2k^2)^{k'}$ leaves. Therefore the size of this search tree is bounded by $\sum_{k'} (2k^2)^{k-k'} 2^{2k} 2k'^2 (2k^2)^{k'} = O(2^{3k} k^{2k+3})$, and each node of the tree takes $O(m)$ time to generate. Thus, the total running time of our branching algorithm is $O(1.2738^{2k} + kn) + O(2^{3k} k^{2k+3}) O(m) = O(2^{3k} k^{2k+3} m)$. Following a general result in the second author's PhD dissertation [9], CLIQUE CONTRACTION has a kernel of $O(2^{2k} k)$ vertices, which can be constructed in linear time. Combining this exponential kernel with our branching algorithm, we obtain an FPT algorithm running in time $O(2^{7k} k^{2k+5} + m)$.

For the correctness of the algorithm, it is easy to see that (G, k) has a solution when our algorithm outputs "YES". On the other hand, suppose that G contains a solution set S of size k. Our branching algorithm indeed simulates the procedure of contracting S in G. First after contracting edges $S \cap E[V_k]$, vertex set V_k is modified into a set $V'_k = V_p \cup R$ where V_p consists of vertices that are not involved in any edges of $S \setminus E[V_k]$. Note that T is the set of vertices in V_c that are not adjacent to at least one vertex of V_p. To make G into a clique, every vertex in T must be incident with some edge in $S \setminus E[V_k]$, implying that $T \leq 2k'$ where $k' = |S \setminus E[V_k]|$. For an arbitrary vertex $u \in V_c - T$, we construct an edge set S^* from S by removing edges $\{xy \in S : x, y \in V_c - T\}$ and replacing every xy in S with $x \in T$ and $y \in V_c - T$ by xu. It can be shown that S^* is also a solution of (G, k), and by Step 4 there always be a leaf in the search tree labelled with G/S^*, implying that the algorithm outputs "YES". The complete proof will be given in the full paper.

We now turn to the non-existence of polynomial kernels for CLIQUE CONTRACTION. Due to space limit, we will only sketch the main idea here and give the complete proof in the full paper. First we show that the following ONE-SIDED DOMINATING SET problem admits no polynomial kernel unless $NP \subseteq coNP/poly$: Given a bipartite graph $G = (X, Y; E)$ and an integer t with $|X|$ being the parameter, does X have a subset of at most t vertices that dominates Y? The NP-completeness of the unparameterized version of the problem easily follows from that of DOMINATING SET, and we can show that ONE-SIDED DOMINATING SET is OR-compositional, implying that it admits no polynomial kernel unless $NP \subseteq coNP/poly$. Note that this problem is different from RED-BLUE DOMINATING SET (defined by Dom et $al.$ [5]) whose solution set is in Y instead of X.

Next we give a *polynomial parameter transformation* from ONE-SIDED DOMINATING SET to CLIQUE CONTRACTION. The main idea of the transformation is as follows: First we construct a bipartite graph $G' = (X', Y'; E')$ from $G = (X, Y; E)$ by adding $|X| - t$ new vertices Z to X and make them adjacent to every vertex of Y, and adding a new vertex w to Y and make it adjacent to every vertex of X (see Fig. 2). Note that each vertex in Z must combine with some vertices in X to form a dominating set for $Y' = Y \cup \{w\}$. It is easy to see that X has a dominating t-set for Y iff X' can be partitioned into $|X| - t + 1$ disjoint dominating sets for Y'.

Then we replace Y' by $2(|X'| - t') + 1 = 2|X| - 1$ copies $Y_1, \cdots, Y_{2|X|-1}$ of Y where $t' = |X| - t + 1$, connect every $a \in Y_i$ to $b \in X'$ iff $ab \in E'$ for $i = 1, \cdots, 2|X| - 1$, and make X' and $Y_1 \cup \cdots \cup Y_{2|X|-1}$, respectively, into two cliques to form graph G'' (see Fig. 2). If X' can be partitioned into t' disjoint sets $S_1, \cdots, S_{t'}$ each of which dominates Y', then we can contract vertices in each S_i into a single vertex to make G'' into a clique. The total number of edge contractions we use is $\Sigma_i(|S_i| - 1) = (\Sigma_i|S_i|) - t' = |X'| - t'$. Conversely if G'' contains $|X'| - t'$ edges whose contractions yield a clique, then obviously there exists some Y_j whose vertices are not involved in edge contractions. It is easy to see that vertices in each witness set of X' form a dominating set for Y_j and the number of different witness sets in X' is at least t', implying that X' can be partitioned into t' disjoint dominating sets for Y'.

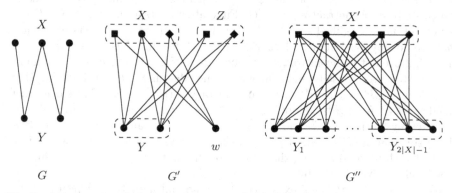

Fig. 2. An example of the transformation from ONE-SIDED DOMINATING SET to CLIQUE CONTRACTION with $t = 1$

Since X has a dominating t-set for Y in G iff G'' can be modified into a clique by using $|X'| - t' = |X| - 1$ edge contractions, CLIQUE CONTRACTION admits no polynomial kernel unless $NP \subseteq coNP/poly$. □

Because P_3-FREE CONTRACTION on connected graphs is equivalent to CLIQUE CONTRACTION, we immediately have the following result.

Corollary 2. P_3-FREE CONTRACTION *is FPT but admits no polynomial kernel unless* $NP \subseteq coNP/poly$.

On the other hand, P_l-FREE CONTRACTION is hard for every odd $l \geq 5$.

Theorem 3. *For every fixed odd* $l \geq 5$, P_l-FREE CONTRACTION *is W[2]-hard.*

Proof. First we note the following easy FPT reduction from P_l-FREE CONTRACTION to P_{l+2}-FREE CONTRACTION for every $l \geq 3$: For any graph G and positive integer k, we construct a graph G' by attaching $k+1$ leaves to each vertex v of G, i.e., adding $k+1$ new vertices and connecting them to v with new edges. It is easy to see that (G, k) is a yes-instance of P_l-FREE CONTRACTION iff (G', k) is a yes-instance of P_{l+2}-FREE CONTRACTION.

Therefore we need only prove the theorem for the base case $l = 5$. For this purpose, we give an FPT reduction from DOMINATING SET to P_5-FREE CONTRACTION.

Given an instance (G, k) with $V(G) = \{v_1, \cdots, v_n\}$, we construct in polynomial time a graph G' as follows (see Fig. 3 for an illustration):

- Create an independent set $\{x_1, \cdots, x_n\}$ and a clique $\{y_1, \cdots, y_n\}$.
- Make x_i adjacent to y_j iff $i = j$ or $v_i v_j \in E(G)$.
- Create a new vertex u and make it adjacent to every vertex of $\{y_1, \cdots, y_n\}$.
- Create a $(k+1)$-clique $\{z_1, \cdots, z_{k+1}\}$, where each z_i is made adjacent to u and has a new vertex w_i attaching to it.

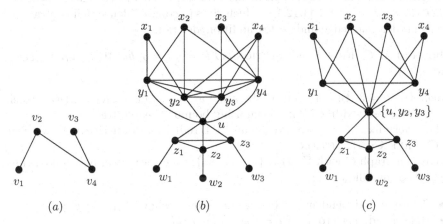

Fig. 3. (a) Graph G with dominating set $\{v_2, v_3\}$; (b) Graph G' obtained from G; (c) P_5-free graph G^* obtained from G' by contracting $\{uy_2, uy_3\}$

We claim that G has a dominating k-set iff G' can be made into P_5-free by contracting at most k edges.

Suppose that T is a dominating k-set in G. We contract k edges $\{uy_i : v_i \in T\}$ in G' to obtain a graph G^*. Note that u is made adjacent to every vertex of $\{x_1, \cdots, x_n\}$ in G^*. It is easy to see that G^* contains no induced 5-path.

Conversely, suppose that G' contains at most k edges F whose contraction results in a P_5-free graph. We show that there exists a dominating k-set in G. We may assume $k < n$, otherwise G always has a dominating k-set. Observe that at least one induced path (u, z_r, w_r) for some $1 \le r \le k+1$ survives after contracting F, which implies that each induced 3-path from some x_i to u must be destroyed to make G' P_5-free. Thus distance $d_{G'/F}(u, x_i) \le 1$ for $i = 1, \cdots, n$. We now use F to obtain a dominating set of G. Let $R = \{x_i : \exists v, vx_i \in F\}$ and $R^* = \{y_i : x_i \in R\}$. Let S be a set of vertices in $\{y_1, \cdots, y_n\}$ that are finally in the same witness set with u in the graph G'/F. It is easy to see that $|R^*| + |S| = |R| + |S| \leq |F| \leq k$. Since for each $1 \leq i \leq n$, $d_{G'/F}(u, x_i) \leq 1$, vertex x_i is either contained in R or adjacent to some vertex of S in G'. This implies that $R^* \cup S$ dominates $\{x_1, \cdots, x_n\}$, and thus G has a dominating set of at most k vertices.   □

The reduction in Theorem 3 does not work for even number l, and new ideas are needed to deal with even l.

3   Cycle-Free Contraction

In this section, we consider contraction problems concerning cycles. We show that C_l-FREE CONTRACTION is W[2]-hard for every fixed $l \geq 4$, and the reduction in our proof also implies that CHORDAL CONTRACTION, which is the same as $\{C_l : l \geq 4\}$-FREE CONTRACTION, is also W[2]-hard. It is worth noting that two related graph modification problems CHORDAL COMPLETION and CHORDAL DELETION are both FPT [3,12,13], which gives us some evidence that contraction seems harder than edge and vertex addition/deletion.

Theorem 4. C_l-FREE CONTRACTION *is FPT for $l = 3$, but W[2]-hard for every fixed $l \geq 4$.*

Proof. For $l = 3$, the problem is the same as K_3-FREE CONTRACTION which can be easily solved in $O(3^k n^3)$ time using bounded search tree.

For every fixed $l \geq 4$, we provide an FPT reduction from DOMINATING SET to C_l-FREE CONTRACTION.

Given a graph G with $V(G) = \{v_1, \cdots, v_n\}$, we construct in polynomial time a graph G' as follows:

- Create an independent set $\{x_1, \cdots, x_n\}$ and a clique $\{y_1, \cdots, y_n\}$.
- Make x_i adjacent to y_j iff $i = j$ or $v_i v_j \in E(G)$.
- Create a new vertex u and make it adjacent to every vertex of $\{y_1, \cdots, y_n\}$.
- For each x_i, create a length-2 path and a length-$(l-2)$ path whose two ends are identified with u and x_i, these two paths form an induced l-cycle H_i.

For convenience, we refer to these n induced l-cycles $H_1, \cdots, H_n$ as u-*cycles*. We claim that G has a dominating k-set iff G' can be made into a C_l-free graph by contracting at most k edges.

Suppose that T is a dominating k-set in G, we contract k edges $\{uy_i : v_i \in T\}$ in G'. In the resulting graph G^*, u is made adjacent to all vertices of $\{x_1, \cdots, x_n\}$. Therefore all u-cycles are destroyed and the size of the largest induced cycle in G^* is $l - 1$, implying that G' is C_l-free.

Conversely, suppose that G' contains at most k edges F whose contraction results in a C_l-free graph. In particular, all u-cycles are destroyed in G'/F. We may assume $k < n$, otherwise G always has a dominating k-set.

We consider the intersection between F and u-cycles. Let $F_i = F \cap E(H_i)$ for $i = 1, \cdots, n$. Observe that the only u-cycle destroyed by contraction of F_i is

H_i, which can also be destroyed by contracting x_iy_i. Thus for every F_i that is non-empty, we replace F_i by a single edge $\{x_iy_i\}$, and then obtain a set F^* from F whose contraction destroys all u-cycles in G'. Since none edge of F^* lies in any u-cycle, then for each $1 \leq i \leq n$, u is either made adjacent to x_i or identified with x_i by contracting F^*, i.e., $d_{G'/F^*}(u, x_i) \leq 1$. Using the same argument in Theorem 3, we can use F^* to obtain a dominating set of G containing at most $|F^*| \leq k$ vertices.   □

Our proof in Theorem 4 actually shows that Π-CONTRACTION is W[2]-hard for Π being the class of graphs without induced cycles of length $\geq l$ for any fixed $l \geq 4$. We note that for $l = 3$, Π coincides with forests, and the problem becomes FPT as shown by Heggernes *et al.* [10]. For $l = 4$, Π is exactly the class of chordal graphs, and thus we have the following theorem for CHORDAL CONTRACTION.

Theorem 5. CHORDAL CONTRACTION *is W[2]-hard.*

4   *H*-Free for 3-Connected *H*

Asano and Hirata [1] showed that Π-CONTRACTION is NP-complete whenever Π is characterized by a finite forbidden set of 3-connected graphs. However, their reduction is not an FPT reduction and not useful in dealing with the parameterized complexity of H-FREE CONTRACTION. In this section, we fully characterize the parameterized complexity of H-FREE CONTRACTION for 3-connected H.

Theorem 6. *Let H be a fixed 3-connected graph. If H is a complete graph, then H-FREE CONTRACTION is FPT but admits no polynomial kernel unless $NP \subseteq coNP/poly$. Otherwise H-FREE CONTRACTION is W[2]-hard.*

Proof. If H is a complete graph K_t with $t \geq 3$, we can easily obtain an FPT algorithm running in $O(\binom{t}{2}^k n^t)$ time by bounded search tree as the only way to destroy a copy of K_t is to contract some edges in the copy. To show that the problem has no polynomial kernel, we introduce a constrained satisfiability problem RESTRICTED-1s-IN-4 SAT, prove that it is NP-complete and OR-compositional and thus admits no polynomial kernel unless $NP \subseteq coNP/poly$, and then give a polynomial parameter transformation from it to our problem K_t-FREE CONTRACTION. Due to space limit, we omit the lengthy proofs here, which are available from the PhD dissertation (§5.2) of the second author [9].

For the W[2]-hardness part of the theorem, we consider two cases in terms of the structure of H.

Case 1. H is not chordal. We give an FPT reduction from DOMINATING SET to H-FREE CONTRACTION. For a graph G with $V(G) = \{v_1, \cdots, v_n\}$, we construct a graph G' as follows (see Fig. 4 for an illustration):

- Create an independent set $\{x_1, \cdots, x_n\}$ and a clique $\{y_1, \cdots, y_n\}$.
- Make x_i adjacent to y_j iff $i = j$ or $v_iv_j \in E(G)$.

- Create a new vertex u and make it adjacent to every vertex of $\{y_1, \cdots, y_n\}$.
- Replicate n copies $H_1, \cdots, H_n$ of H. For each H_i, arbitrarily choose two non-adjacent vertices s and t in its largest induced cycle, and identify u with s, and x_i with t.

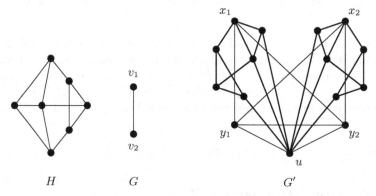

Fig. 4. An example of the reduction from DOMINATING SET to H-FREE CONTRACTION when H is 3-connected and non-chordal

We claim that G has a dominating k-set iff G' can be made into an H-free graph by contracting at most k edges. Suppose that T is a dominating k-set in G. We contract k edges $\{uy_i : v_i \in T\}$ in G' to obtain a graph G^*, where u is made adjacent to all vertices of $\{x_1, \cdots, x_n\}$. We show that G^* is H-free.

Let $l\ (\geq 4)$ be the size of the largest induced cycle in H, and $t\ (\geq 1)$ be the number of different induced l-cycles in H. If G^* contains an induced subgraph H^* that is isomorphic to H, then H^* also has t different induced C_l. Observe that $G^* - (H_1 \cup \cdots \cup H_n)$ is chordal, and each H_i in G^* contains at most $t - 1$ induced C_l because u and x_i is adjacent now. Thus there exists $p \neq q$ such that $V(H^*) \cap V(H_p) \neq \emptyset$ and $V(H^*) \cap V(H_q) \neq \emptyset$, which implies that $x_p, x_q, u \in V(H^*)$. However, removal of u, x_p will disconnect H^*, contradicting to the fact that H^* is 3-connected. Therefore G^* is an H-free graph.

Conversely, suppose that G' contains at most k edges F whose contraction results in a H-free graph. Similar to the proof in Theorem 4, there exists a set F^* of at most k edges such that contraction of F^* destroys all induced copies $H_1, \cdots, H_n$ of H, and none edge of F^* lies in these copies. We can use F^* to obtain a dominating set of G containing at most $|F^*| \leq k$ vertices.

Case 2. H is chordal. The reduction for Case 1 does not work for 3-connected chordal H, because the constructed graph G' is a chordal graph. We will modify the reduction by subdividing the clique $\{y_1, \cdots, y_n\} \cup \{u\}$ and forcing contractions to occur in a specified set of edges.

Given an arbitrary instance (G, k) of DOMINATING SET, we construct an instance $(G', 2k)$ of H-FREE CONTRACTION in FPT time. Let $V(G) = \{v_1, \cdots, v_n\}$ and we construct graph G' as follows:

- Create two independent sets: $\{x_1, \cdots, x_n\}$ and $\{y_1, \cdots, y_n\}$.
- Make x_i adjacent to y_j and mark this edge iff $i = j$ or $v_i v_j \in E(G)$.
- Create a new vertex u.
- For every pair of vertices $\{a, b\}$ in $\{y_1, \cdots, y_n\} \cup \{u\}$, create a degree-2 vertex $w_{a,b}$ which is made adjacent to a and b. All these vertices constitute the subdivision of a $(n + 1)$-vertex clique.
- Replicate n copies $H_1, \cdots, H_n$ of H. For each H_i, arbitrarily choose two non-adjacent vertices s and t, and identify u with s, and x_i with t. Mark all edges in H_i.

For every marked edge e, we will prevent it from being contracted. For this purpose, we need the following operation of *attaching an expanded-H* to an edge e of a graph: subdivide an edge uv of a copy of H by a vertex w, and identify edge uw with edge e. Note that after this operation, the contraction of e will generate a copy of H. We attach $2k + 1$ vertex-disjoint expanded-H's to e to prevent e from being contracted since contracting e will generate $2k + 1$ induced copies of H that cannot be destroyed by $2k$ edge contractions. Thus we can only contract edges in $\{aw_{a,b}, bw_{a,b} : a, b \in \{y_1, \cdots, y_n\} \cup \{u\}\}$.

We claim that G has a dominating k-set iff G' can be made into an H-free graph by at most $2k$ edge contractions. Suppose that T is a dominating k-set in G. We contract $2k$ edges $\{uw_{u,y_i}, y_i w_{u,y_i} : v_i \in T\}$ of G to obtain a graph G^*. Note that u is identified with $\{y_i : v_i \in T\}$ in G^* and therefore is adjacent to every vertex of $\{x_1, \cdots, x_n\}$, implying that $H_1, \cdots, H_n$ are destroyed. We show that G^* is H-free.

Assume that G^* contains an induced subgraph H^* isomorphic to H. By the 3-connectivity of H^*, it is clear that H^* is entirely inside the part of G' before attaching expanded-H's. Since the subgraph of G^* induced by $\{x_1, \cdots, x_n\} \cup \{y_1, \cdots, y_n\} \cup \{u\} \cup \{w_{a,b} : a, b \in \{y_1, \cdots, y_n\} \cup \{u\}\}$ is triangle-free, H^* intersects H_i for some $1 \le i \le n$. If H^* contains a vertex outside H_i, then deleting u and x_i will disconnect this 3-connected graph H^*, implying a contradiction. Thus, the vertex set $V(H^*)$ is exactly the set $V(H_i)$. However, the subgraph induced by $V(H_i)$ in G^* has one more edge ux_i than H, contradicting to the fact that H^* is isomorphic to H. Therefore G^* is an H-free graph.

Conversely, suppose that G' contains at most $2k$ edges F whose contraction results in an H-free graph. Note that subgraphs $H_1, \cdots, H_n$ in G' are destroyed by contracting F, which implies that u is made adjacent to each x_i in G'/F. Let S be a set of vertices in $\{y_1, \cdots, y_n\}$ that are finally in the same witness set with u in G'/F. We have $2|S| \le |F|$ and S dominates $\{x_1, \cdots, x_n\}$, implying that G has a dominating set of at most $|S| \le k$ vertices. $\qquad \square$

5   Concluding Remarks

We have studied H-FREE CONTRACTION problems in an attempt to obtain a dichotomy theorem for their parameterized complexity in terms of the structure of H, and we believe that techniques in the paper will be useful for further study of Π-Contraction problems. There are many natural and interesting problems about H-FREE CONTRACTION and Π-CONTRACTION in general, and we will now discuss some open problems and propose some conjectures.

Unlike edge and vertex addition/deletions, edge contraction changes the structure of a graph less locally, and we feel that this nature makes edge contraction problems much more harder than edge and vertex modification problems. In general, we believe that H-FREE CONTRACTION is fixed-parameter intractable unless H has a very special structure which limits the change.

Conjecture 7. *For any fixed connected graph H, H-FREE CONTRACTION is W[2]-hard unless H is a complete graph or some graph with at most 5 vertices.*

In light of the above conjecture, it will be important to determine whether H-FREE CONTRACTION is FPT for small graphs, in particular for H being P_4 and $K_{1,3}$. Note that $K_{1,t}$-FREE CONTRACTION is W[2]-hard for every fixed $t \geq 4$ [9].

Problem 8. *Determine whether P_4-FREE CONTRACTION (or COGRAPH CONTRACTION) and CLAW-FREE CONTRACTION are FPT.*

In connection with Conjecture 7, a confirmation of the following conjecture will be useful.

Conjecture 9. *Let H' be an induced subgraph of H. Then H-FREE CONTRACTION is W[2]-hard whenever H'-FREE CONTRACTION is.*

For P_l-FREE CONTRACTION with even $l \geq 6$, we feel that it is useful to investigate how to prevent an edge from being contracted in order to settle the following conjecture.

Conjecture 10. *For every fixed $l \geq 6$, P_l-FREE CONTRACTION is W[2]-hard.*

Let $\mathcal{F}$ be a family of forbidden graphs. The $\mathcal{F}$-FREE CONTRACTION problem asks whether we can contract at most k edges in G to obtain a graph that is H-free for all $H \in \mathcal{F}$. Our work on H-FREE CONTRACTION may shed light on this general problem, and the following problem may serve as a good starting point.

Problem 11. *Is it true that $\{H_1, H_2\}$-FREE CONTRACTION is W[2]-hard when both H_1-FREE CONTRACTION and H_2-FREE CONTRACTION are W[2]-hard, and FPT when both are FPT?*

Finally, we believe that FPT algorithms for CLIQUE CONTRACTION and K_t-FREE CONTRACTION can be improved.

Problem 12. *Design faster FPT algorithms for CLIQUE CONTRACTION and K_t-FREE CONTRACTION.*

References

1. Asano, T., Hirata, T.: Edge-deletion and edge-contraction problems. In: Proceedings of STOC 1982, pp. 245–254 (1982)
2. Asano, T., Hirata, T.: Edge-contraction problems. Journal of Computer and System Sciences 26(2), 197–208 (1983)
3. Cai, L.: Fixed-parameter tractability of graph modification problems for hereditary properties. Information Processing Letters 58(4), 171–176 (1996)
4. Chen, J., Kanj, I.A., Xia, G.: Improved upper bounds for vertex cover. Theoretical Computer Science 411(40-42), 3736–3756 (2010)
5. Dom, M., Lokshtanov, D., Saurabh, S.: Incompressibility through colors and IDs. In: Albers, S., Marchetti-Spaccamela, A., Matias, Y., Nikoletseas, S., Thomas, W. (eds.) ICALP 2009, Part I. LNCS, vol. 5555, pp. 378–389. Springer, Heidelberg (2009)
6. Eppstein, D.: Finding large clique minors is hard. Journal of Graph Algorithms and Applications 13(2), 197–204 (2009)
7. Golovach, P.A., Kamiński, M., Paulusma, D., Thilikos, D.M.: Increasing the minimum degree of a graph by contractions. Theoretical Computer Science 481, 74–84 (2013)
8. Golovach, P.A., van't Hof, P., Paulusma, D.: Obtaining planarity by contracting few edges. Theoretical Computer Science 476, 38–46 (2013)
9. Guo, C.: Parameterized Complexity of Graph Contraction Problems. PhD Thesis, The Chinese University of Hong Kong, Hong Kong S.A.R, China (2013), http://www.cse.cuhk.edu.hk/~cwguo/PhdThesis.pdf
10. Heggernes, P., van 't Hof, P., Lévêque, B., Lokshtanov, D., Paul, C.: Contracting graphs to paths and trees. In: Marx, D., Rossmanith, P. (eds.) IPEC 2011. LNCS, vol. 7112, pp. 55–66. Springer, Heidelberg (2012)
11. Heggernes, P., van't Hof, P., Lokshtanov, D., Paul, C.: Obtaining a bipartite graph by contracting few edges. In: Supratik, C., Amit, K. (eds.) FSTTCS 2011. LIPIcs, vol. 13, pp. 217–228. Leibniz-Zentrum für Informatik, Schloss Dagstuhl (2011)
12. Kaplan, H., Shamir, R., Tarjan, R.: Tractability of parameterized completion problems on chordal, strongly chordal, and proper interval graphs. SIAM Journal on Computing 28(5), 1906–1922 (1999)
13. Marx, D.: Chordal deletion is fixed-parameter tractable. Algorithmica 57(4), 747–768 (2010)
14. Robertson, N., Seymour, P.D.: Graph minors. XIII. The disjoint paths problem. Journal of Combinatorial Theory, Series B 63(1), 65–110 (1995)
15. Watanabe, T., Ae, T., Nakamura, A.: On the removal of forbidden graphs by edge-deletion or by edge-contraction. Discrete Applied Mathematics 3(2), 151–153 (1981)

Fixed-Parameter and Approximation Algorithms:
A New Look[*]

Rajesh Chitnis[1,**], MohammadTaghi Hajiaghayi[1,**], and Guy Kortsarz[2,***]

[1] Department of Computer Science, University of Maryland at College Park, USA
{rchitnis,hajiagha}@cs.umd.edu
[2] Rutgers University, Camden, NJ
guyk@camden.rutgers.edu

Abstract. A Fixed-Parameter Tractable (FPT) ρ-approximation algorithm for a minimization (resp. maximization) parameterized problem P is an FPT algorithm that, given an instance $(x, k) \in P$ computes a solution of cost at most $k \cdot \rho(k)$ (resp. $k/\rho(k)$) if a solution of cost at most (resp. at least) k exists; otherwise the output can be arbitrary. For well-known intractable problems such as the W[1]-hard Clique and W[2]-hard Set Cover problems, the natural question is whether we can get any FPT-approximation. It is widely believed that both Clique and Set-Cover admit no FPT ρ-approximation algorithm, for any increasing function ρ. However, to the best of our knowledge, there has been no progress towards proving this conjecture. Assuming standard conjectures such as the Exponential Time Hypothesis (ETH) [11] and the Projection Games Conjecture (PGC) [18], we make the first progress towards proving this conjecture by showing that

- Under the ETH and PGC, there exist constants $F_1, F_2 > 0$ such that the Set Cover problem does not admit a FPT approximation algorithm with ratio k^{F_1} in $2^{k^{F_2}} \cdot \text{poly}(N, M)$ time, where N is the size of the universe and M is the number of sets.
- Unless NP $\subseteq$ SUBEXP, for every $1 > \delta > 0$ there exists a constant $F(\delta) > 0$ such that Clique has no FPT cost approximation with ratio $k^{1-\delta}$ in $2^{k^F} \cdot \text{poly}(n)$ time, where n is the number of vertices in the graph.

In the second part of the paper we consider various W[1]-hard problems such as Directed Steiner Tree, Directed Steiner Forest, Directed Steiner Network and Minimum Size Edge Cover. For all these problem we give polynomial time $f(\text{OPT})$-approximation algorithms for some small function f (the largest approximation ratio we give is OPT^2). Our results indicate a potential separation between the classes W[1] and W[2]; since no W[2]-hard problem is known to have a polynomial time $f(\text{OPT})$-approximation for any function f. Finally, we answer a question by Marx [14] by showing the well-studied Strongly Connected Steiner Subgraph problem (which is W[1]-hard and does not have any polynomial time constant factor approximation) has a constant factor FPT-approximation.

[*] A full version of the paper is available at http://arxiv.org/abs/1308.3520

[**] Supported in part by NSF CAREER award 1053605, NSF grant CCF-1161626, ONR YIP award N000141110662, DARPA/AFOSR grant FA9550-12-1-0423, and a University of Maryland Research and Scholarship Award (RASA). The first author is also supported by a Simons Award for Graduate Students in Theoretical Computer Science. The second author is also with AT&T Labs.

[***] Supported in part by NSF grant number 434923.

G. Gutin and S. Szeider (Eds.): IPEC 2013, LNCS 8246, pp. 110–122, 2013.
© Springer International Publishing Switzerland 2013

1 Introduction

Parameterized Complexity is a two-dimensional generalization of "P vs. NP" where in addition to the overall input size n, one studies the effects on the computational complexity of a secondary measurement that captures additional relevant information. This additional information can be, for example, a structural restriction on the input distribution considered, such as a bound on the treewidth of an input graph or the size of a solution. For general background on the theory see [4]. For decision problems with input size n, and a parameter k, the two dimensional analogue (or generalization) of P, is solvability within a time bound of $O(f(k)n^{O(1)})$, where f is a function of k alone. Problems having such an algorithm are said to be *fixed parameter tractable* (FPT). The W-hierarchy is a collection of computational complexity classes: we omit the technical definitions here. The following relation is known amongst the classes in the W-hierarchy: FPT $= W[0] \subseteq W[1] \subseteq W[2] \subseteq \ldots$. It is widely believed that FPT $\neq W[1]$, and hence if a problem is hard for the class $W[i]$ (for any $i \geq 1$) then it is considered to be fixed-parameter intractable. We say that a problem is W-hard if it is hard for the class W[i] for some $i \geq 1$. When the parameter is the size of the solution then the most famous examples of W[1]-hard and W[2]-hard problems are Clique and Set Cover respectively. We define these two problems below:

Clique
Input : An undirected graph $G = (V, E)$, and an integer k
Problem: Does G have a clique of size at least k?
Parameter: k

Set Cover
Input: Universe $U = \{u_1, u_2, \ldots, u_n\}$ and a collection $\mathcal{S} = \{S_1, S_2, \ldots, S_m\}$ of subsets of U such that $\bigcup_{j=1}^{m} S_j = U$.
Problem: Is there a subcollection $\mathcal{S}' \subseteq \mathcal{S}$ such that $\mathcal{S}' \leq k$ and $\bigcup_{S_i \in \mathcal{S}'} S_i = U$?
Parameter: k

The next natural question is whether these fixed-parameter intractable problems at least admit parameterized approximation algorithms.

1.1 Parameterized Approximation Algorithms

We follow the notation from Marx [15]. Any NP-optimization problem can be described as $O = (I, \text{sol}, \text{cost}, \text{goal})$, where I is the set of instances, $\text{sol}(x)$ is the set of feasible solutions for instance x, the positive integer $\text{cost}(x; y)$ is the cost of solution y for instance x, and goal is either min or max. We assume that $\text{cost}(x, y)$ can be computed in polynomial time, $y \in \text{sol}(x)$ can be decided in polynomial time, and $|y| = |x|^{O(1)}$ holds for every such y.

Definition 1. *Let $\rho : \mathbb{N} \to \mathbb{R}_{\geq 1}$ be a computable function such that $\rho(k) \geq 1$ for every $k \geq 1$; if goal=min then $k \cdot \rho(k)$ is nondecreasing and if the goal=max then $k/\rho(k)$ is unbounded and nondecreasing. An **FPT approximation algorithm** with approximation* *ratio ρ for O is an algorithm $\mathbb{A}$ that, given an input $(x, k) \in \Sigma^* \times \mathbb{N}$ satisfying* $\text{sol}(x) \neq \emptyset$ *and*

$$\begin{cases} opt(x) \leq k & \text{if goal} = \min \\ opt(x) \geq k & \text{if goal} = \max \end{cases} \tag{*}$$

computes $y \in \text{sol}(x)$ such that

$$\begin{cases} \text{cost}(x, y) \leq k \cdot \rho(k) & \text{if goal} = \min \\ \text{cost}(x, y) \geq k/\rho(k) & \text{if goal} = \max \end{cases} \tag{**}$$

We require that on input (x, k) the algorithm $\mathbb{A}$ runs in $f(k) \cdot |x|^{O(1)}$ time for some computable function f.

Note that if the input does not satisfy (*), then the output can be arbitrary.

Remark 1. Given an output $y \in sol(x)$ we can check in FPT time if it satisfies *(**)*. Hence we can assume that an FPT approximation algorithm always[1] either outputs a $y \in sol(x)$ that satisfies *(**)* or outputs a default value `reject`. We call such an FPT approximation algorithm that has this property as **normalised**.

Classic polynomial-time approximation algorithms determine the performance ratio by comparing the output with the optimum. In FPT approximation algorithms there is a subtle difference: we compare the output to the parameter to determine the approximation ratio. Fellows [6] asked about finding an FPT approximation algorithm for W[2]-hard Dominating Set (which is a special case of Set Cover), or ruling out such a possibility. The following conjecture due to Marx (personal communication) is widely believed in the FPT community:

Conjecture 1. Both Set Cover and Clique do not admit an FPT algorithm with approximation ratio ρ, for any function ρ.

However to the best of our knowledge there has been no progress towards proving this conjecture, even under assumptions from complexity theory. In this paper we take a first step towards proving Conjecture 1, under well-known and reasonable[2] assumptions from complexity theory like the Exponential Time Hypothesis (ETH) of Impagliazzo et al. [11] and the Projection Games Conjecture (PGC) of Moshkovitz [18].

For both minimization and maximization problems, the most interesting and practical case is the input (x, k) when $k = OPT(x)$. This motivates the definition of the following variant of FPT approximation algorithms:

Definition 2. *Let $\rho : \mathbb{N} \to \mathbb{R}_{\geq 1}$ be a computable function such that $\rho(k) \geq 1$ for every $k \geq 1$; if goal=min then $k \cdot \rho(k)$ is nondecreasing and if goal=max then $k/\rho(k)$ is unbounded and nondecreasing. An **FPT optimum approximation algorithm** for O*

---

[1] Even if the input does not satisfy (*).

[2] It is very important to only work under well-believed assumptions, since otherwise we will be able to prove pretty much what we want, but it is of no value.

with approximation ratio ρ is an algorithm $\mathbb{A}'$ that, given an input $x \in \Sigma^*$ satisfying $\text{sol}(x) \neq \emptyset$ *outputs a* $y \in \text{sol}(x)$ *such that*

$$\begin{cases} \text{cost}(x,y) \leq OPT(x) \cdot \rho(OPT(x)) & \textit{if } \text{goal} = \min \\ \text{cost}(x,y) \geq OPT(x)/\rho(OPT(x)) & \textit{if } \text{goal} = \max \end{cases} \tag{1}$$

We require that on input x *the algorithm* $\mathbb{A}$ *runs in* $f(OPT(x)) \cdot |x|^{O(1)}$ *time for some computable function* f.

In Section 2.2, we show the following theorem:

Theorem 1. *Let* O *be a minimization problem in NP, and* $\mathbb{A}$ *be an FPT approximation algorithm for* O *with ratio* ρ. *On input* (x, k) *let the running time of* $\mathbb{A}$ *be* $f(k) \cdot |x|^{O(1)}$ *for some non-decreasing computable function* f. *Then* O *also has an FPT optimum approximation algorithm* $\mathbb{A}'$ *with approximation ratio* ρ, *and whose running time on input* x *is also* $f(OPT(x)) \cdot |x|^{O(1)}$

Hence for minimization problems, it is enough to prove hardness results only for the notion of FPT optimum approximation algorithms (see Definition 2). We do not know any relation between the two definitions for maximization problems, and hence we prove hardness results for both Definition 1 and Definition 2.

2   Our Results

We make the first progress towards proving Conjecture 1, under standard assumptions from complexity theory. In particular for Set Cover we assume the Exponential Time Hypothesis (ETH) [11] and the Projection Games Conjecture (PGC) [18][3]. The PGC gives a reduction from SAT to Projection Games. Composing this with the standard reduction from Projection Games to Set Cover gives a reduction from SAT to Set Cover. Since the ETH gives a lower bound on the running time of SAT, we are able to show the following inapproximability result in Section 3:

Theorem 2. *Under the ETH and PGC,*

1. *There exist constants* $F_1, F_2 > 0$ *such that the Set Cover problem does not admit an FPT **optimum** approximation algorithm with ratio* $\rho(OPT) = OPT^{F_1}$ *in* $2^{OPT^{F_2}} \cdot \text{poly}(N, M)$ *time, where* N *is the size of the universe and* M *is the number of sets.*
2. *There exist constants* $F_3, F_4 > 0$ *such that the Set Cover problem does not admit an FPT approximation algorithm with ratio* $\rho(k) = k^{F_3}$ *in* $2^{k^{F_4}} \cdot \text{poly}(N, M)$ *time, where* N *is the size of the universe and* M *is the number of sets.*

In Section 4, we consider the Clique problem. We use the result of Zuckerman [21] which states that it is NP-hard to get an $O(n^{1-\epsilon})$-approximation for Clique. Given any problem $X \in$ NP, by using the Zuckerman reduction from X to Clique allows us to show the following result.

---

[3] The PGC is stated in Section 3.1. The ETH is the hypothesis that 3-SAT cannot be solved in $2^{o(n)}$ time where n is the number of variables. [11]

Theorem 3. *Unless NP ⊆ SUBEXP, for every $1 > \delta > 0$*

1. *There exists a constant $F(\delta) > 0$ such that Clique has no FPT **optimum** approximation with ratio $\rho(OPT) = OPT^{1-\delta}$ in $2^{OPT^F} \cdot poly(n)$ time, where n is the number of vertices in the graph.*
2. *There exists a constant $F'(\delta) > 0$ such that Clique has no FPT approximation with ratio $\rho(k) = k^{1-\delta}$ in $2^{k^{F'}} \cdot poly(n)$ time, where n is the number of vertices in the graph.*

2.1 Polytime $f(OPT)$-Approximation for W-Hard Problems

We also deal with the following question: given that a problem is W-hard, can we maybe get a good polynomial-time approximation for the problem? Any problem in NP can be solved in $2^{n^{O(1)}}$ time by simply enumerating all candidates for the witness. If the parameter k is at least $\log n$, then we immediately have $2^k \geq n$ and the problem can be solved in $2^{n^{O(1)}} \leq 2^{2^{k^{O(1)}}}$ time which is FPT time in k. So for large values of the parameter the brute force algorithm itself becomes an FPT algorithm. Hence the intrinsic hardness to obtain FPT algorithms for intractable problems is when the parameter k is small (say at most $\log n$). In this case, we show how to replace the impossible FPT solution by a good approximation, namely $f(OPT)$ approximation for some small function f. We systematically design polynomial-time $f(OPT)$ approximation algorithms for a number of W[1]-hard minimization problems such as Minimum Size Edge Cover, Strongly Connected Steiner Subgraph, Directed Steiner Forest and Directed Steiner Network. Each of the aforementioned problems is known to have strong inapproximability (in terms of input size). Since we can assume OPT is small, this implies $f(OPT)$ is small as well. Therefore for these W[1]-hard problems, if the parameter is large then we can get an FPT algorithm, otherwise if the parameter is small (then OPT is small as well, otherwise we can reject for these minimization problems)and we obtain a reasonable approximation in polynomial time. These results point towards a separation between the classes W[1] and W[2] since we do not know any W[2]-hard problem which has a polynomial-time $f(OPT)$-approximation, for any function f. In fact, Marx (personal communication) conjectured that the W[2]-hard Set Cover problem does not have a polynomial-time $f(OPT)$-approximation for any function f.

Finally in Section 6 we show that the well-studied W[1]-hard Strongly Connected Steiner Subgraph problem has an FPT 2-approximation algorithm. This answers a question by Marx [14] regarding finding a problem which is fixed-parameter intractable, does not have a constant factor approximation in polynomial time but admits a constant factor FPT approximation. To the best of our knowledge no such W[2]-hard problem (parameterized by solution size) is known, and this indicates another potential difference between W[1] and W[2].

2.2 Proof of Theorem 1

Let $x \in \Sigma^*$ be the input for $\mathbb{A}'$. The algorithm $\mathbb{A}'$ runs the algorithm $\mathbb{A}$ on the instances $(x, 1), (x, 2), \ldots$ until the first k such that the output of $\mathbb{A}$ on (x, k) is a solution of cost at most $k \cdot \rho(k)$. Then $\mathbb{A}'$ outputs $\mathbb{A}(x, k)$. By Definition 1, we know that $k \leq OPT(x)$. Hence $k \cdot \rho(k) \leq OPT(x) \cdot \rho(OPT(x))$. It remains to analyze the running time of $\mathbb{A}'$.

Since $k \leq OPT(x)$, the running time of $\mathbb{A}'$ is upper bounded by $\sum_{i=1}^{k} f(i) \cdot |x|^{O(1)} \leq \sum_{i=1}^{OPT(x)} f(i) \cdot |x|^{O(1)} = \left(\sum_{i=1}^{OPT(x)} f(i) \right) \cdot |x|^{O(1)} \leq OPT(x) \cdot f(OPT(x)) \cdot |x|^{O(1)} = f(OPT(x)) \cdot |x|^{O(1)}$, since f is non-decreasing and $OPT(x) \leq |x|$.

3 An FPT Inapproximability Result for Set Cover

The goal of this section is to prove Theorem 2.

3.1 The Projection Games Conjecture

First we define a *projection game*. Note that with a loss of factor two we can assume that the alphabet is the same for both sides. The input to a projection game consists of:

- A bipartite graph $G = (V_1, V_2, E)$
- A finite alphabets Σ
- Constraints (also called projections) given by $\pi_e : \Sigma \to \Sigma$ for every $e \in E$.

The goal is to find an assignment $\phi : V_1 \cup V_2 \to \Sigma$ that *satisfies* as many of the edges as possible. We say that an edge $e = \{a, b\} \in E$ is satisfied, if the projection constraint holds, i.e., $\pi_e(\phi(a)) = \phi(b)$. We denote the size of a projection game by $n = |V_1| + |V_2| + |E|$.

Conjecture 2. (Projection Games Conjecture [18]) There exists $c > 0$ such that for every $\epsilon > 0$, there is a polynomial reduction RED-1 from SAT[4] to Projection Games which maps an instance I of SAT to an instance I_1 of Projection Games such that:

1. If a YES instance I of SAT satisfies $|I|^c \geq \frac{1}{\epsilon}$, then all edges of I_1 can be satisfied.
2. If a NO instance I of SAT satisfies $|I|^c \geq \frac{1}{\epsilon}$, then at most ϵ-fraction of the edges of I_1 can be satisfied.
3. The size of I_1 is almost-linear in the size of I, and is given by $|I_1| = n = |I|^{1+o(1)} \cdot poly(\frac{1}{\epsilon})$.
4. The alphabet Σ for I_1 has size $poly(\frac{1}{\epsilon})$.

A weaker version of the conjecture is known, but the difference is that the alphabet in [19] has size $\exp(\frac{1}{\epsilon})$. As pointed out in [18], the Projection Games Conjecture is an instantiation of the Sliding Scale Conjecture of Bellare et al. [2] from 1993. Thus, in fact this conjecture is actually 20 years old. But we have reached a state of knowledge now that it seems likely that the Projection Games Conjecture will be proved not long from now (see Section 1.2 of [18]). Thus it seems that posing this conjecture is quite reasonable. In contrast to this is the Unique Games Conjecture [12]. On the positive side, it seems that the Unique Games Conjecture is much more influential than the Projection Games Conjecture. But it seems unlikely (to us) that the Unique Games Conjecture will be resolved in the near future.

---

[4] SAT is the standard Boolean satisfiability problem.

3.2 Reduction from Projection Games to Set Cover

The following reduction from Projection Games to Set Cover is known, see [1,13]. For completeness, we give a proof in the full version of the paper [5].

Theorem 4. *There is a reduction* RED-2 *from Projection Games to Set Cover which maps an instance* $I_1 = (G = (V_1, V_2, E), \Sigma, \pi)$ *of Projection Games to an instance* I_2 *of Set Cover such that:*

1. *If all edges of* I_1 *can be satisfied, then* I_2 *has a set cover of size* $|V_1| + |V_2|$.
2. *If at most* ϵ-*fraction of edges of* I_1 *can be satisfied, then the size of a minimum set cover for* I_2 *is at least* $\frac{|V_1| + |V_2|}{\sqrt{32\epsilon}}$
3. *The instance* I_2 *has* $|\Sigma| \times (|V_1| + |V_2|)$ *sets and the size of the universe is* $2^{O(\frac{1}{\sqrt{\epsilon}})} \times |\Sigma|^2 \times |E|$
4. *The time taken for the reduction is upper bounded by* $2^{O(\frac{1}{\sqrt{\epsilon}})} \times \text{poly}(|\Sigma|) \times \text{poly}(|E| + |V_1| + |V_2|)$

3.3 Composing the Two Reductions

Composing the reductions from Conjecture 2 and Theorem 4 we get:

Theorem 5. *There exists* $c > 0$, *such that for every* $\epsilon > 0$ *there is a reduction* RED-3 *from SAT to Set Cover which maps an instance* I *of SAT to an instance* I_2 *of Set Cover such that*

1. *If a YES instance* I *of SAT satisfies* $|I|^c \geq \frac{1}{\epsilon}$, *then* I_2 *has a set cover of size* β.
2. *If a NO instance* I *of SAT satisfies* $|I|^c \geq \frac{1}{\epsilon}$, *then* I_2 *does not have a set cover of size less than* $\frac{\beta}{\sqrt{32\epsilon}}$.
3. *The size* N *of the universe for the instance* I' *is* $2^{O(\frac{1}{\sqrt{\epsilon}})} \times \text{poly}(\frac{1}{\epsilon}) \times \text{poly}(|I|)$.
4. *The number* M *of sets for the set cover instance* I' *is* $\text{poly}(\frac{1}{\epsilon}) \times \text{poly}(|I|)$.
5. *The total time required for* RED-3 *is* $emph(|I|) + 2^{O(\frac{1}{\sqrt{\epsilon}})} \times \text{poly}(\frac{1}{\epsilon}) \times \text{poly}(|I|)$.

where $\beta \leq |I_1| = |I|^{1+o(1)} \cdot \text{poly}(\frac{1}{\epsilon})$. *Note that the number of elements is very large compared to the number of sets.*

Proof. We apply the reduction from Theorem 4 with $|\Sigma| = \text{poly}(\frac{1}{\epsilon})$ and $|V_1| + |V_2| + |E| = n = |I|^{1+o(1)} \cdot \text{poly}(\frac{1}{\epsilon})$. Substituting these values in Conjecture 2 and Theorem 4, we get the parameters as described in the given theorem. We work out each of the values below:

1. If I is a YES instance of SAT satisfying $\epsilon \geq \frac{1}{|I|^c}$, then RED-1 maps it to an instance $I_1 = (G = (V_1, V_2, E), \Sigma, \pi)$ of Projection Games such that all edges of I_1 can be satisfied. Then RED-2 maps I_1 to an instance I_2 of Set Cover such that I_2 has a set cover of size $\beta = |V_1| + |V_2| \leq |V_1| + |V_2| + |E| = |I_1| = |I|^{1+o(1)} \cdot \text{poly}(\frac{1}{\epsilon})$.

---

[5] A full version of the paper is available at http://arxiv.org/abs/1308.3520

2. If I is a NO instance of SAT satisfying $\epsilon \geq \frac{1}{|I|^c}$, then RED-1 maps it to an instance $I_1 = (G = (V_1, V_2, E), \Sigma, \pi)$ of Projection Games such that at most ϵ-fraction of the edges of I_1 can be satisfied. Then RED-2 maps I_1 to an instance I_2 of Set Cover such that I_2 does not have a set cover of size $\frac{\beta}{\sqrt{32\epsilon}}$, where β is as calculated above.

3. By Theorem 4(3), the size of the universe is $2^{O(\frac{1}{\sqrt{\epsilon}})} \times |\Sigma|^2 \times |E|$. Observing that $|\Sigma| = \text{poly}(\frac{1}{\epsilon})$ and $|E| \leq |I_1| = |I|^{1+o(1)} \cdot \text{poly}(\frac{1}{\epsilon})$, it follows that the size of the universe is $2^{O(\frac{1}{\sqrt{\epsilon}})} \times \text{poly}(\frac{1}{\epsilon}) \times \text{poly}(|I|)$.

4. By Theorem 4(3), the number of sets is $|\Sigma| \times (|V_1| + |V_2|)$. Observing that $|\Sigma| = \text{poly}(\frac{1}{\epsilon})$ and $|V_1| + |V_2| \leq |I_1| = |I|^{1+o(1)} \cdot \text{poly}(\frac{1}{\epsilon})$, it follows that the number of sets is $\text{poly}(\frac{1}{\epsilon}) \times \text{poly}(|I|)$.

5. Since RED-3 is the composition of RED-1 and RED-2, the time required for RED-3 is the summation of the times required for RED-1 and RED-2. By Conjecture 2, the time required for RED-1 is $\text{poly}(|I|)$. By Theorem 4(4), the time required for RED-2 is at most $2^{O(\frac{1}{\sqrt{\epsilon}})} \times \text{poly}(|\Sigma|) \times \text{poly}(|E| + |V_1| + |V_2|)$. Observing that $|\Sigma| = \text{poly}(\frac{1}{\epsilon})$ and $|V_1| + |V_2| + |E| = |I_1| = |I|^{1+o(1)} \cdot \text{poly}(\frac{1}{\epsilon})$, it follows that the time required for RED-2 is at most $2^{O(\frac{1}{\sqrt{\epsilon}})} \times \text{poly}(\frac{1}{\epsilon}) \times \text{poly}(|I|)$. Adding up the two, the time required for RED-3 is at most $\text{poly}(|I|) + 2^{O(\frac{1}{\sqrt{\epsilon}})} \times \text{poly}(\frac{1}{\epsilon}) \times \text{poly}(|I|)$.

Finally we are ready to prove Theorem 2.

3.4   Proof of Theorem 2(1)

The roadmap of the proof is as follows: suppose to the contrary there exists an FPT **optimum** approximation algorithm, say $\mathbb{A}$, for Set Cover with ratio $\rho(OPT) = OPT^{F_1}$ in $2^{OPT^{F_2}} \cdot \text{poly}(N, M)$ time, where N is the size of the universe and M is the number of sets (recall Definition 2). We will choose the constant F_1 such that using RED-3 from Theorem 5 (which assumes PGC), the algorithm $\mathbb{A}$ applied to the instance I_2 will be able to decide the instance I_1 of SAT. Then to violate ETH we will choose the constant F_2 such that the running time of $\mathbb{A}$ summed up with the time required for RED-3 is subexponential in $|I|$.

Let $c > 0$ be the constant from Conjecture 2. Fix some constant $1 > \delta > 0$ and let $c^* = \min\{c, 2 - 2\delta\}$. Note that $\frac{c^*}{2} \leq 1 - \delta$. Choosing $\epsilon = \frac{1}{|I|^{c^*}}$ implies $\epsilon \geq \frac{1}{|I|^c}$, since $c \geq c^*$. We carry out the reduction RED-3 given by Theorem 5. From Conjecture 2(3), we know that $|I_1| = |I|^{1+o(1)} \cdot \text{poly}(\frac{1}{\epsilon})$. Let $\lambda > 0$ be a constant such that the $\text{poly}(\frac{1}{\epsilon})$ is upper bounded by $(\frac{1}{\epsilon})^\lambda$. Then Theorem 5 implies $\beta \leq |I|^{1+o(1)} \cdot (\frac{1}{\epsilon})^\lambda$. However we have chosen $\epsilon = \frac{1}{|I|^{c^*}}$, and hence asymptotically we get

$$\beta \leq |I|^2 \cdot |I|^{\lambda c^*} = |I|^{2+\lambda c^*} \tag{2}$$

Choose the constant F_1 such that $\frac{c^*}{4(2 + \lambda c^*)} \geq F_1$. Suppose Set Cover has an FPT optimum approximation algorithm $\mathbb{A}$ with ratio $\rho(OPT) = OPT^{F_1}$ (recall Definition 2).

We show that this algorithm $\mathbb{A}$ can decide the SAT problem. Consider an instance I of SAT, and let $I_2 =$ RED-3(I) be the corresponding instance of Set Cover. Run the FPT approximation algorithm on I_G, and let $\mathbb{A}(I_2)$ denote the output of ALG. We have the following two cases:

- $\frac{\beta}{\sqrt{32\epsilon}} \leq \mathbb{A}(I_2)$: Then we claim that I is a NO instance of SAT. Suppose to the contrary that I is a YES instance of SAT. Then Theorem 5(1) implies $\beta \geq OPT(I_2)$. Hence $\frac{\beta}{\sqrt{32\epsilon}} \leq \mathbb{A}(I_2) \leq OPT \cdot \rho(OPT) = OPT^{1+F_1} = \beta^{1+F_1} \Rightarrow \frac{1}{\sqrt{32\epsilon}} \leq \beta^{F_1}$.

  However, asymptotically we have $\frac{1}{\sqrt{32\epsilon}} = \frac{|I|^{\frac{c^*}{2}}}{\sqrt{32}} > |I|^{\frac{c^*}{4}} \geq (|I|^{2+\lambda c^*})^{F_1} = \beta^{F_1}$, where the last two inequalities follows from the choice of F_1 and Equation 2 respectively. This leads to a contradiction, and therefore I is a NO instance of SAT.

- $\frac{\beta}{\sqrt{32\epsilon}} > \mathbb{A}(I_2)$: Then we claim that I is a YES instance of SAT. Suppose to the contrary that I is a NO instance of SAT. Then Theorem 5(2) implies $OPT(I_2) \geq \frac{\beta}{\sqrt{32\epsilon}}$. Therefore we have $\frac{\beta}{\sqrt{32\epsilon}} > \mathbb{A}(I_2) \geq OPT(I_2) \geq \frac{\beta}{\sqrt{32\epsilon}}$.

Therefore we run the algorithm $\mathbb{A}$ on the instance I_2 and compare the output $\frac{\beta}{\sqrt{32\epsilon}}$ with n^ϵ. As seen above, this comparison allows us to decide the SAT problem. We now choose the constant F_2 such that the running time of $\mathbb{A}$ summed up with the time required for RED-3 is subexponential in $|I|$. By Theorem 5(5), the total time taken by RED-3 is poly$(|I|) + 2^{O(\frac{c^*}{\sqrt{\epsilon}})} \times$ poly$(\frac{1}{\epsilon}) \times$ poly$(|I|) =$ poly$(|I|) + 2^{O(|I|^{\frac{c^*}{2}})} \times$ poly$(|I|^{\frac{c^*}{2}}) \times$ poly$(|I|) =$ poly$(|I|) + 2^{o(I)} \cdot$ poly$(|I|)$ since $\frac{c^*}{2} \leq 1 - \delta$. Hence total time taken by RED-3 is subexponential in I. We now show that there exists a constant F_2 such that the claimed running time of $2^{OPT^{F_2}} \cdot$ poly(N, M) for the algorithm $\mathbb{A}$ is subexponential in $|I|$, thus contradicting ETH. We do not have to worry about the poly(N, M) factor: the reduction time is subexponential in $|I|$, and hence $\max\{N, M\}$ is also upper bounded by a subexponential function of $|I|$. Hence, we essentially want to choose a constant $F_2 > 0$ such that $OPT^{F_2} \leq M^{F_2} = o(|I|)$. From Theorem 5(4), we know that $M \leq |\Sigma| \times |V_1 + V_2|$. Since $|\Sigma| =$ poly$(\frac{1}{\epsilon})$, let $\alpha > 0$ be a constant such that the $|\Sigma| \leq (\frac{1}{\epsilon})^\alpha$. We have seen earlier in the proof that $|V_1 + V_2| \leq |I_1| \leq |I|^2 \cdot (\frac{1}{\epsilon})^\lambda = |I|^{2+c^*\lambda}$. Therefore $M^{F_2} \leq (|I|^{2+c^*\lambda+c^*\alpha})^{F_2}$. Choosing $F_2 < \frac{1}{2+\lambda c^*+2\alpha c^*}$ gives OPT$^{F_2} = o(|I|)$, which is what we wanted to show. $\square$

3.5   Proof of Theorem 2(2)

Observe that due to Theorem 1, Theorem 2(1) implies Theorem 2(2).

4   An FPT Inapproximability Result for Clique

We use the following theorem due to Zuckerman [21], which in turn is a derandomization of a result of Håstad [10] .

Theorem 6. [10,21] *Let X be any problem in NP. For any constant $\epsilon > 0$ there exists a polynomial time reduction from X to Clique so that the gap between the clique sizes corresponding to the YES and NO instances of X is at least $n^{1-\epsilon}$, where n is the number of vertices of the Clique instances.*

4.1 Proof of Theorem 3(1)

Fix a constant $1 > \delta > 0$. Set $0 < \epsilon = \frac{\delta}{\delta+2}$, or equivalently $\delta = \frac{2\epsilon}{1-\epsilon}$. Let X be any problem in NP. Let the Hastad-Zuckerman reduction from X to Clique [10,21] which creates a gap of at least $n^{1-\epsilon}$ map an instance I of X to the corresponding instance I_G of Clique. Since the reduction is polynomial, we know that $n = |I_G| = |I|^D$ for some constant $D(\epsilon) > 0$. Note that D depends on ϵ, which in turn depends on δ. Hence, ultimately D depends on δ. If I is a YES instance of X, then I_G contains a clique of size at least $n^{1-\epsilon}$ since each graph has a trivial clique of size one and the gap between YES and NO instances of Clique is at least $n^{1-\epsilon}$. Similarly, observe that a graph on n vertices can have a clique of size at most n. To maintain the gap of at least $n^{1-\epsilon}$, it follows if I is a NO instance of X then the maximum size of a clique in I_G is at most n^ϵ. To summarize, we have

- If I is a YES instance, then $OPT(I_G) \geq n^{1-\epsilon}$
- If I is a NO instance, then $OPT(I_G) \leq n^\epsilon$

Suppose Clique has an FPT **optimum** approximation algorithm $\mathbb{A}$ with ratio $\rho(OPT) = OPT^{1-\delta}$ (recall Definition 2). We show that this algorithm $\mathbb{A}$ can decide the problem X. Consider an instance I of X, and let I_G be the corresponding instance of Clique. Run the FPT approximation algorithm on I_G, and let $\mathbb{A}(I_G)$ denote the output of $\mathbb{A}$. We have the following two cases:

- $n^\epsilon \geq \mathbb{A}(I_G)$: Then we claim that I is a NO instance of X. Suppose to the contrary that I is a YES instance of X, then we have $n^\epsilon \geq \mathbb{A}(I_G) \geq \frac{OPT_{I_G}}{\rho(OPT(I_G))} = (OPT(I_G))^\delta \geq (n^{1-\epsilon})^\delta = n^{2\epsilon}$, which is a contradiction.
- $n^\epsilon < \mathbb{A}(I_G)$: Then we claim that I is a YES instance of X. Suppose to the contrary that I is a NO instance of X, then we have $n^\epsilon < \mathbb{A}(I_G) \leq OPT(I_G) \leq n^\epsilon$, which is a contradiction.

We run the algorithm $\mathbb{A}$ on the instance I_G and compare the output $\mathbb{A}(I_G)$ with n^ϵ. As seen above, this comparison allows us to decide the problem X. We now show how to choose the constant F such that the running $2^{OPT^F} \cdot \text{poly}(n)$ is subexponential in $|I|$. We claim that $F = \frac{1}{D+1}$ works. Note that $OPT(I_G) \leq n$ always. Hence $2^{OPT^F} \cdot$
$\text{poly}(n) \leq 2^{n^F} \cdot \text{poly}(n) = 2^{(|I|^D)^F} \cdot \text{poly}(|I|^D) = 2^{|I|^{DF}} \cdot \text{poly}(|I|) = 2^{|I|^{\frac{D}{D+1}}} \cdot$
$\text{poly}(|I|) = 2^{o(I)} \cdot \text{poly}(|I|)$. This implies we can could solve X in subexponential time using $\mathbb{A}$. However X was any problem chosen from the class NP, and hence NP $\subseteq$ SUBEXP.   □

4.2 Proof of Theorem 3(2)

Fix a constant $1 > \delta > 0$. Set $0 < \epsilon = \frac{\delta}{\delta+1}$, or equivalently $\delta = \frac{\epsilon}{1-\epsilon}$. Let X be any problem in NP. Let the Hastad-Zuckerman reduction from X to Clique [10,21] which creates a gap of at least $n^{1-\epsilon}$ map an instance I of X to the corresponding instance I_G of Clique. Since the reduction is polynomial, we know that $n = |I_G| = |I|^D$ for some constant $D(\epsilon) > 0$. Note that D depends on ϵ, which in turn depends on δ. Hence, ultimately D depends on δ. If I is a YES instance of X, then I_G contains a clique of size at least $n^{1-\epsilon}$ since each graph has a trivial clique of size one and the gap between YES and NO instances of Clique is at least $n^{1-\epsilon}$. Similarly, observe that a graph on n vertices can have a clique of size at most n. To maintain the gap of at least $n^{1-\epsilon}$, it follows if I is a NO instance of X then the maximum size of a clique in I_G is at most n^{ϵ}.

Suppose Clique has an FPT approximation algorithm ALG with ratio $\rho(k) = k^{1-\delta}$ (recall Definition 1). We show that this algorithm ALG can decide the problem X. Set $k = n^{\epsilon}$. On the input (I_G, n^{ϵ}) to ALG, there are two possible outputs:

– ALG outputs `reject` $\Rightarrow OPT(I_G) < n^{\epsilon} \Rightarrow I$ is a NO instance of X
– ALG outputs a clique of size $\geq \frac{k}{\rho(k)} \Rightarrow OPT(I_G) \geq \frac{k}{\rho(k)} = \frac{k}{k^{1-\delta}} = k^{\delta} = (n^{\epsilon})^{\delta} = n^{1-\epsilon}$
  $\Rightarrow I$ is a YES instance of X

Therefore the FPT approximation algorithm ALG can decide the problem $X \in$ NP. We now show how to choose the constant F' such that the running $\exp(k^{F'}) \cdot \text{poly}(n)$ is subexponential in $|I|$. We claim that $F' = \frac{1}{\epsilon \cdot D + 1}$ works. This is because $2^{k^{F'}} \cdot \text{poly}(n) = 2^{n^{\epsilon F'}} \cdot \text{poly}(n) = 2^{|I|^{\epsilon D F'}} \cdot \text{poly}(|I|^D) = 2^{|I|^{\frac{\epsilon D}{\epsilon D + 1}}} \cdot \text{poly}(|I|) = 2^{o(I)} \cdot \text{poly}(|I|)$. This implies we can could solve X in subexponential time using ALG. However X was any problem chosen from the class NP, and hence NP $\subseteq$ SUBEXP.   □

5   Polytime $f(OPT)$-Approximation for W[1]-Hard Problems

In Section 2.1 we have seen the motivation for designing polynomial time $f(OPT)$-approximation algorithms for W[1]-hard problems such as Minimum Size Edge Cover, Strongly Connected Steiner Subgraph, Directed Steiner Forest and Directed Steiner Network. Our results are summarized in Figure 1 (all the proofs are deferred to the full version):

	W[1]-hardness	Polytime Approx. Ratio
Strongly Connected Steiner Forest	Guo et al. [7]	OPT^{ϵ}
Directed Steiner Forest	Follows from [7]	$OPT^{1+\epsilon}$
Directed Steiner Network	Follows from [7]	OPT^2
Minimum Edge Cover	Cai [3]	$OPT - 1$
Directed Multicut	Marx and Razgon [17]	$3 \cdot OPT$ [8]

Fig. 1. Polytime $f(OPT)$-approximation for W[1]-hard problems

6   Constant Factor FPT Approximation for SCSS

In this section we show that SCSS has an FPT 2-approximation. We define the problem formally:

Strongly Connected Steiner Subgraph (SCSS)

Input : An directed graph $G = (V, E)$, a set of terminals $T = \{t_1, t_2, \ldots, t_\ell\}$ and an integer p

Problem: Does there exists a set $E' \subseteq E$ such that $|E'| \leq p$ and the graph $G' = (V, E')$ has a $t_i \to t_j$ path for every $i \neq j$

Parameter: p

Lemma 1. *Strongly Connected Steiner Subgraph has an FPT 2-approximation.*

Proof. Let G_{rev} denote the reverse graph obtained from G, i.e., reverse the orientation of each edge. Any solution of SCSS instance must contain a path from t_1 to each terminal in $T \setminus t_1$ and vice versa. Consider the following two instances of the Directed Steiner Tree problem: $I_1 = (G, t_1, T \setminus t_1)$ and $I_2 = (G_{rev}, t_1, T \setminus t_1)$, and let their optimum be OPT_1, OPT_2 respectively. Let OPT be the optimum of given SCSS instance and k be the parameter. If $OPT > k$ then we output anything (see Definition 1). Otherwise we have $k \geq OPT \geq \max\{OPT_1, OPT_2\}$. We know that the Directed Steiner Tree problem is FPT parameterized by the size of the solution [5]. Hence we find the values OPT_1, OPT_2 in time which is FPT in k. Clearly the union of solutions for I_1 and I_2 os a solution for instance I of SCSS. The final observation is $OPT_1 + OPT_2 \leq OPT + OPT = 2 \cdot OPT$. □

Guo et al. [7] show that SCSS is W[1]-hard parameterized by solution size plus number of terminals. It is known that SCSS has no $\log^{2-\epsilon} n$-approximation in polynomial time for any fixed $\epsilon > 0$, unless NP has quasi-polynomial Las Vegas algorithms [9]. Combining these facts with Lemma 1 implies that SCSS is a W[1]-hard problem that is not known to admit a constant factor approximation in polynomial time but has a constant factor FPT approximation. This answers a question by Marx [14]. Previously the only such problem known was a variant of the Almost-2-SAT problem [20] called 2-ASAT-BFL, due to Marx and Razgon [16].

References

1. Arora, S., Lund, C.: Approximation algorithms for NP-hard problems. In: Hochbaum, D. (ed.) PWS Publishing Co., Boston (1997)
2. Bellare, M., Goldwasser, S., Lund, C., Russeli, A.: Efficient probabilistically checkable proofs and applications to approximations. In: STOC (1993)
3. Cai, L.: Parameterized complexity of cardinality constrained optimization problems. Comput. J. 51(1), 102–121 (2008)
4. Downey, R., Fellows, M.: Parameterized Complexity. Springer (1999)
5. Dreyfus, S.E., Wagner, R.A.: The steiner problem in graphs. Networks 1(3) (1971)
6. Fellows, M.R., Guo, J., Marx, D., Saurabh, S.: Data Reduction and Problem Kernels (Dagstuhl Seminar 12241). Dagstuhl Reports 2(6), 26–50 (2012)
7. Guo, J., Niedermeier, R., Suchý, O.: Parameterized complexity of arc-weighted directed steiner problems. SIAM J. Discrete Math. 25(2), 583–599 (2011)
8. Gupta, A.: Improved results for directed multicut. In: SODA, pp. 454–455 (2003)
9. Halperin, E., Krauthgamer, R.: Polylogarithmic inapproximability. In: STOC 2003: Proceedings of the Thirty-Fifth Annual ACM Symposium on Theory of Computing, pp. 585–594. ACM Press, New York (2003)

10. Håstad, J.: Clique is hard to approximate within $n^{1-\epsilon}$. In: FOCS (1996)
11. Impagliazzo, R., Paturi, R.: On the complexity of k-sat. J. Comput. Syst. Sci. 62(2) (2001)
12. Khot, S.: On the Unique Games Conjecture. In: FOCS, p. 3 (2005)
13. Lund, C., Yannakakis, M.: On the hardness of approximating minimization problems. J. ACM 41(5), 960–981 (1994)
14. Marx, D.: Parameterized complexity and approximation algorithms. Comput. J. 51(1) (2008)
15. Marx, D.: Completely inapproximable monotone and antimonotone parameterized problems. Journal of Computer and System Sciences 79(1) (2013)
16. Marx, D., Razgon, I.: Constant ratio fixed-parameter approximation of the edge multicut problem. Inf. Process. Lett. 109(20), 1161–1166 (2009)
17. Marx, D., Razgon, I.: Fixed-parameter tractability of multicut parameterized by the size of the cutset. In: STOC, pp. 469–478 (2011)
18. Moshkovitz, D.: The Projection Games Conjecture and The NP-Hardness of ln n-Approximating Set-Cover. In: Gupta, A., Jansen, K., Rolim, J., Servedio, R. (eds.) APPROX/RANDOM 2012. LNCS, vol. 7408, pp. 276–287. Springer, Heidelberg (2012)
19. Moshkovitz, D., Raz, R.: Two-query PCP with subconstant error. J. ACM 57(5) (2010)
20. Razgon, I., O'Sullivan, B.: Almost 2-sat is fixed-parameter tractable. J. Comput. Syst. Sci. 75(8), 435–450 (2009)
21. Zuckerman, D.: Linear degree extractors and the inapproximability of max clique and chromatic number. In: STOC, pp. 681–690 (2006)

Subgraphs Satisfying MSO Properties on z-Topologically Orderable Digraphs

Mateus de Oliveira Oliveira

School of Computer Science and Communication,
KTH Royal Institute of Technology, 100-44 Stockholm, Sweden
mdeoliv@kth.se

Abstract. We introduce the notion of z-topological orderings for digraphs. We prove that given a digraph G on n vertices admitting a z-topological ordering, together with such an ordering, one may count the number of subgraphs of G that at the same time satisfy a monadic second order formula φ and are the union of k **directed** paths, in time $f(\varphi, k, z) \cdot n^{O(k \cdot z)}$. Our result implies the polynomial time solvability of many natural counting problems on digraphs admitting z-topological orderings for constant values of z and k. Concerning the relationship between z-topological orderability and other digraph width measures, we observe that any digraph of **directed** path-width d has a z-topological ordering for $z \leq 2d+1$. On the other hand, there are digraphs on n vertices admitting a z-topological order for $z = 2$, but whose directed path-width is $\Theta(\log n)$. Since graphs of bounded **directed** path-width can have both arbitrarily large **undirected** tree-width and arbitrarily large clique width, our result provides for the first time a suitable way of partially transposing metatheorems developed in the context of the monadic second order logic of graphs of constant **undirected** tree-width and constant clique width to the realm of digraph width measures that are closed under taking subgraphs and whose constant levels incorporate families of graphs of arbitrarily large undirected tree-width and arbitrarily large clique width.

Keywords: Slice Theory, Digraph Width Measures, Monadic Second Order Logic of Graphs, Algorithmic Meta-theorems.

1 Introduction

Two cornerstones of parametrized complexity theory are Courcelle's theorem [13] stating that monadic second order logic properties may be model checked in linear time in graphs of constant undirected tree-width, and its subsequent generalization to counting given by Arnborg, Lagergren and Seese [2]. The importance of such metatheorems stem from the fact that several NP-complete problems such as Hamiltonicity, colorability, and their respective #P-hard counting counterparts, can be modeled in terms of MSO_2 sentences and thus can be efficiently solved in graphs of constant **undirected** tree-width.

In this work we introduce the notion of z-topological orderings for digraphs and provide a suitable way of partially transposing the metatheorems in [13,2] to digraphs admitting z-*topological orderings* for constant values of z. In order to state our main

G. Gutin and S. Szeider (Eds.): IPEC 2013, LNCS 8246, pp. 123–136, 2013.

result we will first give a couple of easy definitions: Let $G = (V, E)$ be a directed graph. For subsets of vertices $V_1, V_2 \subseteq V$ we let $E(V_1, V_2)$ denote the set of edges with one endpoint in V_1 and another endpoint in V_2. We say that a linear ordering $\omega = (v_1, v_2, ..., v_n)$ of the vertices of V is a z-topological ordering of G if for every **directed** simple path $p = (V_p, E_p)$ in G and every i with $1 \leq i \leq n$, we have that $|E_p \cap E(\{v_1 ..., v_i\}, \{v_{i+1}, ..., v_n\})| \leq z$. In other words, ω is a z-topological ordering if every **directed** simple path of G bounces back and forth at most z times along ω. The terminology z-topological ordering is justified by the fact that any topological ordering of a DAG G according to the usual definition, is a 1-topological ordering according to our definition. Conversely if a digraph admits a 1-topological ordering, then it is a DAG. We denote by MSO_2 the monadic second order logic of graphs with edge set quantification. An edge-weighting function for a digraph $G = (V, E)$ is a function $w : E \to \Omega$ where Ω is a finite commutative semigroup of size polynomial in $|V|$ whose elements are totally ordered. The weight of a subgraph $H = (V', E')$ of G is defined as $w(H) = \sum_{e \in E'} w(e)$. A maximal-weight subgraph of G satisfying a given property φ is a subgraph $H = (V', E')$ such that $H \models \varphi$ and such that for any other subgraph $H' = (V'', E'')$ of G for which $H' \models \varphi$ we have $w(H) \geq w(H')$. Now we are in a position to state our main theorem:

Theorem 1.1 (Main Theorem). *For each MSO_2 formula φ and each positive integers $k, z \in N$ there exists a computable function $f(\varphi, z, k)$ such that: Given a digraph $G = (V, E)$ on n vertices, a z-topological ordering ω of G, an edge-weighting function $w : E \to \Omega$, and a positive integer $l \leq n$, we can count in time $f(\varphi, z, k) \cdot n^{O(z \cdot k)}$ the number of subgraphs H of G simultaneously satisfying the following four properties:*

(i) $H \models \varphi$
(ii) H is the union of k directed paths[1]
(iii) H has l vertices
(iv) H has maximal weight

Our result implies the polynomial time solvability of many natural counting problems on digraphs admitting z-topological orderings for constant values of z and k. We observe that graphs admitting z-topological orderings for constant values of z can already have simultaneously unbounded tree-width and unbounded clique-width, and therefore the problems that we deal with here cannot be tackled by the approaches in [13,2,15]. For instance any DAG is 1-topologically orderable. In particular, the $n \times n$ directed grid in which all horizontal edges are directed to the left and all vertical edges oriented up is 1-topologically orderable, while it has both undirected tree-width $\Omega(n)$ and clique-width $\Omega(n)$.

2 Applications

To illustrate the applicability of Theorem 1.1 with a simple example, suppose we wish to count the number of Hamiltonian cycles on G. Then our formula φ will express that

---

[1] A digraph H is the union of k directed paths if $H = \cup_{i=1}^{k} p_i$ for not necessarily vertex-disjoint nor edge-disjoint directed paths $p_1, ..., p_k$.

the graphs we are aiming to count are cycles, namely, connected graphs in which each vertex has degree precisely two. Such a formula can be easily specified in MSO_2. Since any cycle is the union of two directed paths, we have $k = 2$. Since we want all vertices to be visited our $l = n$. Finally, the weights in this case are not relevant, so it is enough to set the semigroup Ω to be the one element semigroup $\{1\}$, and the weights of all edges to be 1. In particular the total weight of any subgraph of G according to this semigroup will be 1. By Theorem 1.1 we can count the number of Hamiltonian cycles in G in time $f(\varphi, k, z) \cdot n^{2z}$. We observe that Hamiltonicity can be solved within the same time bounds for other directed width measures, such as directed tree-width [33].

Interestingly, Theorem 1.1 allow us to count structures that are much more complex than cycles. And in our opinion it is rather surprising that counting such complex structures can be done in XP. For instance, we could choose to count the number of maximal Hamiltonian subgraphs of G which can be written as the union of k directed paths. We can repeat this trick with virtually any natural property that is expressible in MSO_2. For instance we can count the number of maximal weight 3-colorable subgraphs of G that are the union of k-paths. Or the number of subgraphs of G that are the union of k directed paths and have di-cuts of size $k/10$. Observe that, as it should be expected, our framework does not allow one to find in polynomial time a maximal di-cut of the whole graph G nor to determine in polynomial time whether the whole graph G is 3-colorable since these problems are already NP-complete for DAGs, i.e., for $z = 1$.

If $H = (V, E)$ is a digraph, then the underlying undirected graph of H is obtained from H by forgetting the direction of its edges. A very interesting application of Theorem 1.1 consists in counting the number of maximal-weight subgraphs of G which are the union of k paths and whose underlying undirected graph satisfy some structural property, such as, connectedness, planarity, bounded genus, bipartiteness, etc.

Corollary 2.1. *Let $G = (V, E)$ be a digraph on n vertices and $w : E \to \Omega$ be an edge weighting function. Then given a z-topological ordering ω of G one may count in time $O(n^{k \cdot z})$ the number of maximal-weight subgraphs of G that are the union of k directed paths and whose underlying undirected graph satisfy any combination of the following properties: 1) Connectedness, 2) Being a forest, 3) Bipartiteness, 4) Planarity, 5) Constant Genus g, 6) Outerplanarity, 7) Being Series Parallel, 8) Having Constant Treewidth t 9) Having Constant Branchwidth b, 10) Satisfy any minor closed property.*

Another family of applications for Theorem 1.1 arises from the fact that the monadic second order alternation hierarchy is infinite [37]. Additionally, each level r of the polynomial hierarchy has a very natural complete problem, the r-round-3-coloring problem, that also belongs to the r-th level of the monadic second order hierarchy (Theorem 11.4 of [1]). Thus by Theorem 1.1 we may count the number of r-round-3-colorable subgraphs of G that are the union of k directed paths in time $f(\varphi_r, z, k) \cdot n^{O(z \cdot k)}$.

We observe that the condition that the subgraphs we consider are the union of k directed paths is not as restrictive as it might appear at a first glance. For instance one can show that for any $a, b \in N$ the $a \times b$ undirected grid is the union of 4 directed paths. Additionally these grids have zig-zag number $O(\min\{a, b\})$. Therefore counting the number of maximal grids of height $O(z)$ on a digraph of zig-zag number z is a neat example of problem which can be tackled by our techniques but which cannot be formulated as a linkage problem, namely, the most successful class of problems that has been shown to be solvable in polynomial time for constant values of several digraph width measures [33].

3   Overview of the Proof of Theorem 1.1

We will prove Theorem 1.1 within the framework of regular slice languages, which was originally developed by the author to tackle several problems within the partial order theory of concurrency [17,18]. The main steps of the proof of Theorem 1.1 are as follows. To each regular slice language $\mathcal{L}$ we associate a possibly infinite set of digraphs $\mathcal{L}_\mathcal{G}$. In Section 6 we will define the notion of z-dilated-saturated regular slice language and show that given any digraph G together with a z-topological ordering $\omega = (v_1, v_2, ..., v_n)$ of G, and any z-dilated-saturated slice language $\mathcal{L}$, one may efficiently count the number of subgraphs of G that are isomorphic to some digraph in $\mathcal{L}_\mathcal{G}$ (Theorem 6.2). Then in Section 7 we will show that given any monadic second order formula φ and any natural numbers z, k one can construct a z-dilated-saturated regular slice language $\mathcal{L}(\varphi, z, k)$ representing the set of all digraphs that at the same time satisfy φ and are the union of k directed paths (Theorem 7.1). The construction of $\mathcal{L}(\varphi, z, k)$ is done once and for all for each φ, k and z, and is completely independent from the digraph G. Finally, the proof of Theorem 1.1 will follow by plugging Theorem 7.1 into Theorem 6.2. Proofs of intermediate results omitted for a matter of clarity or due to lack of space can be found in the full version of this work [19].

4   Comparison with Existing Work

Since the last decade, the possibility of lifting the metatheorems in [13,2] to the directed setting has been an active line of research. Indeed, following an approach delineated by Reeds [39] and Johnson, Robertson, Seymour and Thomas [33], several digraph width measures have been defined in terms of the number of cops needed to capture a robber in a certain evasion game on digraphs. From these variations we can cite for example, directed tree-width [39,33], DAG width [6], D-width [40,29], directed path-width [4], entanglement [7,8], Kelly width [32] and Cycle Rank [21,31]. All these width measures have in common the fact that DAGs have the lowest possible constant width (0 or 1 depending on the measure). Other width measures in which DAGs do not have necessarily constant width include DAG-depth [24], and Kenny-width [24].

The introduction of the digraph width measures listed above was often accompanied by algorithmic implications. For instance, certain linkage problems that are NP-complete for general graphs, e.g. Hamiltonicity, can be solved efficiently in graphs of constant directed tree-width [33]. The winner of certain parity games of relevance to the theory of μ-calculus can be determined efficiently in digraphs of constant DAG width [6], while it is not known if the same can be done for general digraphs. Computing disjoint paths of minimal weight, a problem which is NP-complete in general digraphs, can be solved efficiently in graphs of bounded Kelly width. However, except for such sporadic successful algorithmic implications, researchers have failed to come up with an analog of Courcelle's theorem for graph classes of constant width for any of the digraph width measures described above. It turns out that there is a natural barrier against this goal: It can be shown that unless all the problems in the polynomial hierarchy have sub-exponential algorithms, which is a highly unlikely assumption, MSO_2 model checking is intractable in any class of graphs that is closed under taking subgraphs and whose undirected tree-width is poly-logarithmic unbounded [34,35]. An analogous result can be proved with respect to model checking of MSO_1 properties if we assume a non-uniform version of the extended exponential time hypothesis [26,25]. All classes of digraphs of constant width with respect to the directed measures described above are closed under subgraphs and have poly-logarithmically unbounded tree-width, and thus fall into the impossibility theorem of [34,35]. It is worth noting that Courcelle, Makowsky and Rotics have shown that MSO_1 model checking is tractable in classes of graphs of constant clique-width [15,16], and that these classes are poly-logarithmic unbounded, but they are not closed under taking subgraphs.

We define the *zig-zag number* of a digraph G to be the minimum z for which G has a z-topological ordering, and denote it by $zn(G)$. The zig-zag number is a digraph width measure that is closed under taking subgraphs, poly-logarithmically unbounded and that has interesting connections with some of the width measures described above. In particular we can prove the following theorem stating that families of graphs of constant zig-zag number are strictly richer than families of graphs of constant directed path-width.

Theorem 4.1. *Let G be a digraph of directed path-width d. Then G has zig-zag number $z \leq 2d + 1$. Furthermore, given a directed path decomposition of G one can efficiently derive a z-topological ordering of G. On the other hand, there are digraphs on n vertices whose zig-zag number is 2 but whose directed path-width is $\Theta(\log n)$.*

Theorem 4.1 legitimizes the algorithmic relevance of Theorem 1.1 since path decompositions of graphs of constant directed path-width can be computed in polynomial time [42]. The same holds with respect to the cycle rank of a graph since constant cycle-rank decompositions[2] can be converted into constant directed-path decompositions in polynomial time [30]. Therefore all the problems described in Section 1 can be solved efficiently in graphs of constant directed path-width and in graphs of constant cycle rank. We should notice that our main theorem circumvents the impossibility results of [34,35,26,25] by confining the monadic second order logic properties to subgraphs that are the union of k *directed* paths.

A pertinent question consists in determining whether we can eliminate either z or k from the exponent of the running time $f(\varphi, k, z) \cdot n^{O(k \cdot z)}$ stated in Theorem 1.1. The following two theorems say that under strongly plausible parameterized complexity assumptions [20], namely that $W[2] \neq FPT$ and $W[1] \neq FTP$, the dependence of both k and z in the exponent of the running time is unavoidable.

Theorem 4.2 (Lampis-Kaouri-Mitsou[36]). *Determining whether a digraph G of cycle rank z has a Hamiltonian circuit is $W[2]$ hard with respect to z.*

Since by Theorem 4.1 constant cycle rank is less expressive than constant zig-zag number, the hardness result stated in Theorem 4.2 also works for zig-zag number. Given a sequence of $2k$ not necessarily distinct vertices $\sigma = (s_1, t_1, s_2, t_2, ..., s_k, t_k)$, a σ-linkage

---

[2] By cycle-rank decomposition we mean a direct elimination forest[30].

is a set of internally disjoint directed paths $p_1, p_2, ..., p_k$ where each p_i connects s_i to t_i.

Theorem 4.3 (Slivkins[41]). *Given a DAG G, determining whether G has a σ-linkage $\sigma = (s_1, t_1, s_2, t_2, ..., s_k, t_k)$ is hard for $W[1]$.*

It is not hard to see that σ-linkages are expressible in MSO_2. Additionally, since a σ-linkage is clearly union of k-paths, Theorem 4.3 implies that the dependence on k in the exponent of the running time in Theorem 1.1 is necessary even if z is fixed to be 1.

Below we compare the zig-zag number with several other digraph width measures. If G is a digraph, we write $dtw(G)$ for its directed tree-width [33], $Dw(G)$ for its D-width [30], $dagw(G)$ for its DAG-width [6], $dpw(G)$ for its directed path-width [4], $kellyw(G)$ for its Kelly-width [24], $ddp(G)$ for its DAG-depth [24], $Kw(G)$ for its K-width [24], $s(G)$ for its weak separator number [30] and $r(G)$ for its cycle rank [30]. We write $A \precsim B$ to indicate that there are graphs of constant width with respect to the measure A but unbounded width with respect to the measure B. We write $A \preceq B$ to express that A is not asymptotically greater than B.

$$zn(G) \underset{\sim}{\precsim} dpw(G) \overset{[30]}{\underset{\sim}{\precsim}} cr(G) \overset{[25]}{\underset{\sim}{\precsim}} \begin{cases} Kw(G) \\ ddp(G) \end{cases} \qquad \frac{cr(G)}{\log n} \overset{[30]}{\preceq} s(G) \qquad (1)$$

$$\frac{zn(G)}{\log n} \preceq s(G) \overset{[30]}{\preceq} Dw(G) \overset{[30]}{\preceq} dagw(G) \overset{[6]}{\preceq} dpw(G) \qquad (2)$$

$$\sqrt{\frac{zn(G)}{\log n}} \preceq dtw(G) \overset{[32]}{\preceq} kellyw(G) \qquad \sqrt{Dw(G)} \overset{[23]}{\preceq} dtw(G) \overset{[23]}{\preceq} Dw(G) \qquad (3)$$

The numbers above $\precsim$ and $\preceq$ point to the references in which these relations where established. The only new relations are $zn(G) \precsim dpw(G)$, $zn(G)/\log n \preceq s(G)$ and $\sqrt{zn(G)/\log n} \preceq dtw(G)$ which follow from Theorem 4.1 together with the already known relations between directed path-width and the other digraph width measures.

5   Regular Slice Languages

A slice $\mathbf{S} = (V, E, l, s, t)$ is a digraph comprising a set of vertices V, a set of edges E, a vertex labeling function $l : V \to \Gamma$ for some set of symbols Γ, and functions $s, t : E \to V$ which respectively associate to each edge $e \in E$, a source vertex e^s and a target vertex e^t. We notice that an edge might possibly have the same source and target ($e^s = e^t$). The vertex set V is partitioned into three disjoint subsets: an in-frontier $I \subseteq V$ a center $C \subseteq V$ and an out-frontier $O \subseteq V$. Additionally, we require that each frontier-vertex in $I \cup O$ is the endpoint of exactly one edge in E and that no edge in E has both endpoints in the same frontier. The frontier vertices in $I \cup O$ are labeled by l with numbers from the set $\{1, ..., q\}$ for some natural number $q \geq \max\{|I|, |O|\}$ in such a way that no two vertices in the same frontier receive the same number. Vertices belonging to different frontiers may on the other hand be labeled with the same number. The center vertices in C are labeled by l with elements from $\Gamma \setminus \{1, ..., q\}$. We say that a slice $\mathbf{S}$ is normalized if $l(I) = \{1, ..., |I|\}$ and $l(O) = \{1, ..., |O|\}$. Non-normalized slices will play an important role in Section 6 when we define the notion of sub-slice. Since we will deal with weighted graphs, we will also allow the edges of a slice to be weighted by a function $w : E \to \Omega$ where Ω is a finite commutative semigroup.

A slice $\mathbf{S}_1$ with frontiers (I_1, O_1) can be glued to a slice $\mathbf{S}_2$ with frontiers (I_2, O_2) provided $l_1(O_1) = l_2(I_2)$ and that for each $i \in l(O_1)$ there exist edges $e_1 \in \mathbf{S}_1$ and $e_2 \in \mathbf{S}_2$ such that either $e_1^t \in O_1$, $e_2^s \in I_2$ and $l_1(e_1^t) = l_2(e_2^s)$ or $e_1^s \in O_1$, $e_2^t \in I_2$ and $l_1(e_1^s) = l_2(e_2^t)$. In this case the glueing gives rise to the slice $\mathbf{S}_1 \circ \mathbf{S}_2$ with frontiers (I_1, O_2) which is obtained by fusing each such pair of edges e_1, e_2. The fusion of e_1 with e_2 proceeds as follows. First we create an edge e_{12}. If $e_1^t \in O_1$, $e_2^s \in I_2$, we set $e_{12}^s = e_1^s$, $e_{12}^t = e_2^t$ and delete e_1, e_2, e_1^t, e_2^s. Otherwise, if $e_1^s \in O_1$, $e_2^t \in I_2$, we set $e_{12}^s = e_2^s$, $e_{12}^t = e_1^t$ and delete e_1, e_2, e_1^s, e_2^t. Thus in the process of gluing two slices, the vertices in the glued frontiers disappear. If $\mathbf{S}_1$ and $\mathbf{S}_2$ are weighted by functions w_1 and w_2, then we add the requirement that the glueing of $\mathbf{S}_1$ with $\mathbf{S}_2$ can be performed if the weights of the edges touching the out-frontier of $\mathbf{S}_1$ agree with the weights of their corresponding edges touching the in-frontier of $\mathbf{S}_2$. In the opposite direction, any slice can be decomposed into a sequence of atomic parts which we call *unit slices*, namely, slices with at most one vertex on its center. Thus slices may be regarded as a graph theoretic analog of the knot theoretic braids [3], in which twists are replaced by vertices. Within automata theory, slices may be related to several formalisms such as graph automata [43,11], graph rewriting systems [12,5,22], and others [28,27,10,9]. In particular, slices may be regarded as a specialized version of the multi-pointed graphs defined in [22] but subject to a slightly different composition operation.

The width of a slice $\mathbf{S}$ with frontiers (I, O) is defined as $w(\mathbf{S}) = \max\{|I|, |O|\}$. In the same way that letters from an alphabet may be concatenated by automata to form infinite languages of strings, we may use automata or regular expressions over alphabets of slices of a bounded width to define infinite families of digraphs. Let $\Sigma_{\mathbb{S}}^{c,q}$ denote the set of all unit slices of width at most c and whose frontier vertices are numbered with numbers from $\{1, ..., q\}$ for $q \geq c$. We say that a slice is *initial* if its in-frontier is empty and *final* if its out-frontier is empty. A slice with empty center is called a *permutation slice*. Due to the restriction that each frontier vertex of a slice must be connected to precisely one edge, we have that each vertex in the in-frontier of a permutation slice is necessarily connected to a unique vertex in its out-frontier. The empty slice, denoted by ε, is the slice with empty center and empty frontiers. We regard the empty slice as a permutation slice. A subset $\mathcal{L}$ of the free monoid $(\Sigma_{\mathbb{S}}^{c,q})^*$ generated by $\Sigma_{\mathbb{S}}^{c,q}$ is a slice language if for every sequence of slices $\mathbf{S}_1 \mathbf{S}_2 ... \mathbf{S}_n \in \mathcal{L}$ we have that $\mathbf{S}_1$ is an initial slice, $\mathbf{S}_n$ a final slice and $\mathbf{S}_i$ can be glued to $\mathbf{S}_{i+1}$ for each $i \in \{1, ..., n-1\}$. We should notice that at this point the operation of the monoid in consideration is just the concatenation $\mathbf{S}_1 \mathbf{S}_2$ of slice symbols $\mathbf{S}_1$ and $\mathbf{S}_2$ and should not be confused with the composition $\mathbf{S}_1 \circ \mathbf{S}_2$ of slices. The unit of the monoid is just the empty symbol λ and not the empty slice, thus the elements of $\mathcal{L}$ are simply sequences of slices, regarded as dumb letters. To each slice language $\mathcal{L}$ over $\Sigma_{\mathbb{S}}^{c,q}$ we associate a graph language $\mathcal{L}_{\mathcal{G}}$ consisting of all digraphs obtained by composing the slices in each string in $\mathcal{L}$.

$$\mathcal{L}_{\mathcal{G}} = \{\mathbf{S}_1 \circ \mathbf{S}_2 ... \circ \mathbf{S}_n | \mathbf{S}_1 \mathbf{S}_2 ... \mathbf{S}_n \in \mathcal{L}\} \tag{4}$$

However we observe that a set $\mathcal{L}_G$ of digraphs may be represented by several different slice languages, since a digraph in $\mathcal{L}_G$ may be decomposed in several ways as a string of unit slices. We will use the term *unit decomposition* of a digraph H to denote any sequence of unit slices $\mathbf{U} = \mathbf{S}_1\mathbf{S}_2...\mathbf{S}_n$ whose composition $\mathbf{S}_1 \circ \mathbf{S}_2 \circ ... \circ \mathbf{S}_n$ yields H. We say that the unit decomposition $\mathbf{U}$ is *dilated* if it contains permutation slices, including possibly the empty slice. The slice-width of $\mathbf{U}$ is the minimal c for which $\mathbf{U} \in (\Sigma_{\mathbb{S}}^{c,q})^*$ for some q. In other words, the slice width of a unit decomposition is the width of the widest slice appearing in it.

A slice language is regular if it is generated by a finite automaton or regular expressions over slices. We notice that since any slice language is a subset of the free monoid generated by a slice alphabet $\Sigma_{\mathbb{S}}^{c,q}$, we do not need to make a distinction between regular and rational slice languages. Therefore, by Kleene's theorem, every slice language generated by a regular expression can be also generated by a finite automaton. Equivalently, a slice language is regular iff it can be generated by the slice graphs defined below [17]:

Definition 5.1 (Slice Graph). *A slice graph over a slice alphabet $\Sigma_{\mathbb{S}}^{c,q}$ is a labeled directed graph $\mathcal{SG} = (\mathcal{V}, \mathcal{E}, \mathcal{S}, \mathcal{I}, \mathcal{T})$ possibly containing loops but without multiple edges where $\mathcal{I} \subseteq \mathcal{V}$ is a set of initial vertices, $\mathcal{T} \subseteq \mathcal{V}$ a set of final vertices and the function $\mathcal{S} : \mathcal{V} \to \Sigma_{\mathbb{S}}^{c,q}$ satisfies the following conditions:*

- *$\mathcal{S}(\mathfrak{v})$ is a initial slice for every vertex $\mathfrak{v}$ in $\mathcal{I}$,*
- *$\mathcal{S}(\mathfrak{v})$ is final slice for every vertex $\mathfrak{v}$ in $\mathcal{T}$ and,*
- *$(\mathfrak{v}_1, \mathfrak{v}_2) \in \mathcal{E}$ implies that $\mathcal{S}(\mathfrak{v}_1)$ can be glued to $\mathcal{S}(\mathfrak{v}_2)$.*

We say that a slice graph is *deterministic* if none of its vertices has two out-neighbors labeled with the same slice and if there is no two initial vertices labeled with the same slice. In other words, in a deterministic slice graph no two distinct walks are labeled with the same sequence of slices. We denote by $\mathcal{L}(\mathcal{SG})$ the slice language generated by $\mathcal{SG}$, which we define as the set of all sequences slices $\mathcal{S}(\mathfrak{v}_1)\mathcal{S}(\mathfrak{v}_2)\cdots\mathcal{S}(\mathfrak{v}_n)$ where $\mathfrak{v}_1\mathfrak{v}_2\cdots\mathfrak{v}_n$ is a walk on $\mathcal{SG}$ from an initial vertex to a final vertex. We write $\mathcal{L}_G(\mathcal{SG})$ for the language of digraphs derived from $\mathcal{L}(\mathcal{SG})$.

6   Counting Subgraphs Specified by a Slice Language

A sub-slice of a slice $\mathbf{S}$ is a subgraph of $\mathbf{S}$ that is itself a slice. If $\mathbf{S}'$ is a sub-slice of $\mathbf{S}$ then we consider that the numbering in the frontiers of $\mathbf{S}'$ are inherited from the numbering of the frontiers of $\mathbf{S}$. Therefore, even if $\mathbf{S}$ is normalized, its sub-slices might not be. If $\mathbf{U} = \mathbf{S}_1\mathbf{S}_2...\mathbf{S}_n$ is a unit decomposition of a digraph G, then a sub-unit-decomposition of $\mathbf{U}$ is a unit decomposition $\mathbf{U}' = \mathbf{S}'_1\mathbf{S}'_2...\mathbf{S}'_n$ of a subgraph H of G such that $\mathbf{S}'_i$ is a sub-slice of $\mathbf{S}_i$ for $1 \le i \le n$. We observe that sub-unit-decompositions may be padded with empty slices. A unit decomposition $\mathbf{U} = \mathbf{S}_1\mathbf{S}_2...\mathbf{S}_n$ may have exponentially many sub-unit-decompositions of a given slice-width c. However, as we will state in Lemma 6.1 the set of all such sub-unit decompositions of $\mathbf{U}$ may be represented by a slice graph of size polynomial in n. A normalized unit decomposition is a unit decomposition $\mathbf{U} = \mathbf{S}_1\mathbf{S}_2...\mathbf{S}_n$ such that $\mathbf{S}_i$ is a normalized slice for each $i \in \{1, ..., n\}$.

A slice language is normalized if all unit decompositions in it are normalized. A slice-graph is normalized if all slices labeling its vertices are normalized. We notice that a regular slice language is normalized if and only if it is generated by a normalized slice graph.

Lemma 6.1. *Let G be a digraph with n vertices, $\mathbf{U} = \mathbf{S}_1\mathbf{S}_2...\mathbf{S}_n$ be a normalized unit decomposition of G of slice-width q, and let $c \in \mathbf{N}$ be such that $c \leq q$. Then one can construct in time $n \cdot q^{O(c)}$ an acyclic and deterministic slice graph $\mathcal{SUB}^c(\mathbf{U})$ on $n \cdot q^{O(c)}$ vertices whose slice language $\mathcal{L}(\mathcal{SUB}^c(\mathbf{U}))$ consists of all sub-unit-decompositions of $\mathbf{U}$ of slice-width at most c.*

Let $\omega = (v_1, v_2, ..., v_n)$ be a linear ordering of the vertices of a digraph H. We say that a dilated unit decomposition $\mathbf{U} = \mathbf{S}_1\mathbf{S}_2...\mathbf{S}_m$ of H is compatible with ω if v_i is the center vertex of $\mathbf{S}_{j_i}$ for each $i \in \{1, ..., n\}$ and if $j_i > j_{i-1}$ for each $i \in \{1, ..., n-1\}$ (observe that we need to use the subindex j_i instead of simply i because $\mathbf{U}$ is dilated and therefore some slices in $\mathbf{U}$ have no center vertex). Notice that for each ordering ω there might exist several unit decompositions of H that are compatible to ω. If ω is a z-topological ordering of a digraph G and if $\mathbf{U}$ is a dilated unit decomposition of G that is compatible with ω, then we say that $\mathbf{U}$ has zig-zag number z. The zig-zag number of a slice language $\mathcal{L}$ is the maximal zig-zag number of a unit decomposition in $\mathcal{L}$. If a dilated unit decomposition $\mathbf{U}$ has zig-zag number z then any of its sub-unit-decompositions has zig-zag number at most z.

Proposition 6.1. *Let $\mathbf{U}$ be a unit decomposition of zig-zag number z. Then any sub-unit-decomposition in $\mathcal{L}(\mathcal{SUB}^c(\mathbf{U}))$ has zig-zag number at most z.*

A slice language $\mathcal{L}$ is z-dilated-saturated, if $\mathcal{L}$ has zig-zag number at most z and if for every digraph $H \in \mathcal{L}_\mathcal{G}$, every z-topological ordering ω of H and every dilated unit decomposition $\mathbf{U}$ of H that is compatible with ω we have that $\mathbf{U} \in \mathcal{L}$. We should **emphasize** that the intersection of the graph languages generated by two slice graphs is not in general reflected by the intersection of their slice languages. Indeed, it is easy to define slice languages $\mathcal{L}, \mathcal{L}'$ for which $\mathcal{L}_\mathcal{G} = \mathcal{L}'_\mathcal{G}$ but for which $\mathcal{L} \cap \mathcal{L}' = \emptyset$! Additionally, a reduction from the Post correspondence problem [38] established by us in [17] implies that even determining whether the intersection of the graph languages generated by slice languages is empty, is undecidable. However this is not an issue if at least one of the intersecting languages is z-dilated-saturated, as stated in the next proposition.

Proposition 6.2. *Let $\mathcal{L}$ and $\mathcal{L}'$ be two slice languages over $\Sigma_\mathbb{S}^{c,q}$, such that $\mathcal{L}$ has zig-zag number z and such that $\mathcal{L}'$ is z-saturated. If we let $\mathcal{L}^\cap = \mathcal{L} \cap \mathcal{L}'$, then $\mathcal{L}_\mathcal{G}^\cap = \mathcal{L}_\mathcal{G} \cap \mathcal{L}'_\mathcal{G}$.*

If $\mathbf{S}$ is a normalized slice in $\Sigma_\mathbb{S}^{c,q}$ with in-frontier I and out-frontier O then a q-numbering of $\mathbf{S}$ is a pair of functions $in : I \to \{1, ..., q\}$, $out : O \to \{1, ..., q\}$ such that for each two vertices $v, v' \in I$, $l(v) < l(v')$ implies that $in(l(v)) < in(l(v'))$ and, for each two vertices $v, v' \in O$, $l(v) < l(v')$ implies that $out(l(v)) < out(l(v'))$. We let $(\mathbf{S}, in, out)$ denote the slice obtained from $\mathbf{S}$ by renumbering each frontier vertex $v \in I$ with $in(l(v))$ and each out frontier vertex $v \in O$ with the $out(l(v))$. The q-numbering-expansion of a normalized slice $\mathbf{S}$ is the set $\mathcal{N}(\mathbf{S})$ of all q-numberings of $\mathbf{S}$.

Let $\mathcal{SG} = (\mathcal{V}, \mathcal{E}, \mathcal{S}, \mathcal{I}, \mathcal{T})$ be a slice graph over $\Sigma_\mathbb{S}^{c,q}$. Then the q-numbering expansion of $\mathcal{SG}$ is the slice graph $\mathcal{N}^q(\mathcal{SG}) = (\mathcal{V}', \mathcal{E}', \mathcal{S}', \mathcal{I}', \mathcal{T}')$ defined as follows. For each vertex $\mathfrak{v} \in \mathcal{V}$ and each slice $(\mathbf{S}(\mathfrak{v}), in, out) \in \mathcal{N}^q(\mathcal{S}(\mathfrak{v}))$ we create a vertex $\mathfrak{v}_{in,out}$ in $\mathcal{V}'$ and label it with $(\mathbf{S}, in, out)$. Subsequently we connect $\mathfrak{v}_{in,out}$ to $\mathfrak{v}'_{in',out'}$ if there was an edge $(\mathfrak{v}, \mathfrak{v}') \in \mathcal{E}$ and if $(\mathbf{S}, in, out)$ can be glued to $(\mathbf{S}', in', out')$.

Theorem 6.1. *Let G be digraph, $\mathbf{U} = \mathbf{S}_1 \mathbf{S}_2 ... \mathbf{S}_n$ be a normalized unit decomposition of G of slice-width q and zig-zag number z, $\mathcal{SG}$ be a normalized z-dilated-saturated slice graph over $\Sigma_{\mathbb{S}}^{c,q}$ and $\mathcal{N}^q(\mathcal{SG})$ be the q-numbering expansion of $\mathcal{SG}$. Then the set of all sub-unit-decompositions of $\mathbf{U}$ of slice-width at most c whose composition yields a graph isomorphic to some graph in $\mathcal{L}_{\mathcal{G}}(\mathcal{SG})$ is represented by the regular slice language $\mathcal{L}(\mathcal{SUB}^c(\mathbf{U})) \cap \mathcal{L}(\mathcal{N}^q(\mathcal{SG}))$.*

Let $\mathcal{SG} = (\mathcal{V}, \mathcal{E}, \mathcal{S}, \mathcal{I}, \mathcal{T})$ be a slice graph and $(\Omega, +)$ be a finite commutative semi-group with an identity element 0. Then the Ω-*weight expansion* of $\mathcal{SG}$ is the slice graph $\mathcal{W}^\Omega(\mathcal{SG}) = (\mathcal{V}', \mathcal{E}', \mathcal{S}', \mathcal{I}', \mathcal{T}')$ defined as follows: For each vertex $\mathfrak{v} \in \mathcal{V}$ labeled with the slice $\mathcal{S}(\mathfrak{v}) = (V, E, l)$, we add the set of vertices $\{\mathfrak{v}_{w,tot}\}_w$ to $\mathcal{V}'$ where w ranges over all weighting functions $w : E \to \Omega$ and tot ranges over Ω. We label each $\mathfrak{v}_{w,tot}$ with the tuple $(\mathcal{S}(\mathfrak{v}), w, tot)$. Then we add an edge $(\mathfrak{v}_{w,tot}, \mathfrak{v}'_{w',tot'})$ to $\mathcal{E}'$ if and only if $(\mathfrak{v}, \mathfrak{v}') \in \mathcal{E}$, if the slice $(\mathcal{S}(\mathfrak{v}), w)$ can be glued to the slice $(\mathcal{S}(\mathfrak{v}'), w')$ and if $tot' = tot + \sum_{e \in E'out} w(e)$. The set of final vertices $\mathcal{T}'$ consists of all vertices in $\mathcal{V}'$ which are labeled with a triple $(\mathbf{S}, w, tot)$ where $\mathbf{S}$ is a final slice. The set of initial vertices $\mathcal{I}'$ consists of all vertices in $\mathcal{V}'$ which are labeled with a triple $(\mathbf{S}, w, 0)$ where $\mathbf{S}$ is an initial slice. Intuitively if $\mathcal{SG}$ generates a language of graphs $\mathcal{L}_G$, then $\mathcal{W}^\Omega(\mathcal{SG})$ generates the language $\mathcal{L}'_G$ of all possible weighted versions of graphs in $\mathcal{L}_g(\mathcal{SG})$. In Theorem 6.2 below q is the cut-width of G and therefore it can be as large as $O(n^2)$. The parameter c on the other hand is the slice-width of the subgraphs that are being counted.

Theorem 6.2 (Subgraphs in a Saturated Slice Language). *Let $G = (V, E)$ be a digraph of cut-width q with respect to a z-topological ordering $\omega = (v_1, v_2, ..., v_n)$ of its vertices, and let $\mathcal{SG}$ be a deterministic normalized z-dilated-saturated slice graph over $\Sigma_{\mathbb{S}}^{c,q}$ on r vertices. Let $w : E \to \Omega$ be an weighting function on E and $l \le n$ be a positive integer. Then we may count in time $r^{O(1)} \cdot n^{O(c)} \cdot q^{O(c)}$ the number of subgraphs of G of size l, that are isomorphic to some subgraph in $\mathcal{L}_{\mathcal{G}}(\mathcal{SG})$ and have maximal weight.*

Proof. Let $\mathbf{U} = \mathbf{S}_1 \mathbf{S}_2 ... \mathbf{S}_n$ be any normalized unit decomposition that is compatible with ω, i.e., such that v_i is the center vertex of $\mathbf{S}_i$ for $i = 1, ..., n$. Clearly such a unit decomposition can be constructed in polynomial time in n. Since $\mathcal{SG}$ is dilated saturated, by Theorem 6.1 the set of all subgraphs of G that are isomorphic to some digraph in $\mathcal{L}_{\mathcal{G}}(\mathcal{SG})$ is represented by the regular slice language $\mathcal{L}(\mathcal{SUB}^c(U)) \cap \mathcal{L}(\mathcal{W}^\Omega(\mathcal{N}^q(\mathcal{SG})))$. By Lemma 6.1 $\mathcal{SUB}^c(U)$ has $n \cdot q^{O(c)}$ vertices and can be constructed within the same time bounds. The numbering expansion $\mathcal{N}^q(\mathcal{SG})$ of $\mathcal{SG}$ has $\binom{q}{O(c)} \cdot r = r \cdot q^{O(c)}$ vertices and can be constructed within the same time bounds. The Ω-expansion $\mathcal{W}^\Omega(\mathcal{N}^q(\mathcal{SG}))$ of $\mathcal{N}^q(\mathcal{SG})$ has $|\Omega|^{O(c)} \cdot r \cdot q^{O(c)} = n^{O(c)} \cdot r \cdot q^{O(c)}$ vertices and can be constructed within the same time bounds. Let $\mathcal{SG}^\cap = \mathcal{W}^\Omega(\mathcal{N}^q(\mathcal{SG})) \cap \mathcal{SUB}^c(\mathbf{U})$. Since $\mathcal{SG}^\cap$ can be obtained by a product construction, it has $r \cdot n^{O(c)} \cdot q^{O(c)}$ vertices. Since $\mathcal{SUB}^c(\mathbf{U})$ is acyclic, $\mathcal{SG}^\cap$ is also acyclic. Therefore counting the subgraphs in G isomorphic to some graph in $\mathcal{L}_{\mathcal{G}}(\mathcal{SG})$ amounts to counting the number of simple directed paths from an initial to a final vertex in $\mathcal{SG}^{\cap}$. Since we are only interested in counting subgraphs with l vertices, we can intersect this acyclic slice graph with the slice graph $\mathcal{SG}^l$ generating all unit decomposition over $\Sigma_{\mathbf{S}}^{c,q}$ containing precisely l unit slices that are not permutation slices. Again the slice graph $\mathcal{SG}^{\cap} \cap \mathcal{SG}^l$ will be acyclic. Finally since we are only interested in counting maximal-weight subgraphs, we delete from $\mathcal{T}'$ those vertices labeled with triples $(\mathbf{S}, w, tot)$ in which tot is not maximal. The label of each path from an initial to a final vertex in this last slice graph identifies unequivocally a subgraph of G of size l and maximal weight. By standard dynamic programming we can count the number of paths in a DAG from a set of initial vertices to a set of final vertices in time polynomial on the number of vertices of the DAG. Thus we can determine the number of l-vertex maximal-weight subgraphs of G which are isomorphic to some digraph in $\mathcal{L}(\mathcal{SG})$ in time $r^{O(1)} n^{O(c)} q^{O(c)}$. $\qquad\square$

7   Subgraphs Satisfying a Given MSO Property

In this section we will only give the necessary definitions to state Lemma 7.1 and Theorem 7.1, which are crucial steps towards the proof of Theorem 1.1. For an extensive account on the monadic second order logic of graphs we refer the reader to the treatise [14] (in special Chapters 5 and 6). As it is customary, we will represent a digraph G by a relational structure $G = (V, E, s, t, l_V, l_E)$ where V is a set of vertices, E a set of edges, $s, t \subseteq E \times V$ are respectively the source and target relations, $l_V \subseteq V \times \Sigma_V$ and $l_E \subseteq V \times \Sigma_E$ are respectively the vertex-labeling and edge-labeling relations. We give the following semantics to these relations: $s(e, v)$ and $t(e, v')$ are true if v and v' are respectively the source and the target of the edge e; $l_V(v, a)$ is true if v is labeled with the symbol $a \in \Sigma_V$ while $l_E(e, b)$ is true if e is labeled with the symbol $b \in \Sigma_E$. We always assume that e is oriented from its source to its target. Let $\{x, y, z, z_1, y_1, z_1, ...\}$ be an infinite set of first order variables and $\{X, Y, Z, X_1, Y_1, Z_1, ...\}$ be an infinite set of second order variables. Then the set of MSO_2 formulas is the smallest set of formulas containing:

- the atomic formulas $x \in X$, $V(x)$, $E(x)$, $s(x, y)$, $t(x, y)$, $l_V(x, a)$ for each $a \in \Sigma_V$, $l_E(x, b)$ for each $b \in \Sigma_E$,
- the formulas $\varphi \wedge \psi$, $\varphi \vee \psi$, $\neg\varphi$, $\exists x.\varphi(x)$ and $\exists X.\varphi(X)$, where φ and ψ are MSO_2 formulas.

If $\mathcal{X}$ is a set of second order variables, and $G = (V, E)$ is a graph, then an interpretation of $\mathcal{X}$ over G is a function $M : \mathcal{X} \to 2^V$ that assigns to each variable in $\mathcal{X}$ a subset of vertices of V. The semantics of a formula $\varphi(\mathcal{X})$ over free variables $\mathcal{X}$ being true on a graph G under interpretation M is the usual one. A sentence is a formula φ without free variables. For a sentence φ and a graph G, if it is the case that φ is true in G, then we say that G satisfies φ and denote this by $G \models \varphi$. Now we are in a position to state a crucial Lemma towards the proof of Theorem 1.1. Intuitively it states that for any MSO_2 formula φ the set of all unit decompositions of a fixed width of digraphs satisfying φ forms a regular set.

Lemma 7.1. *For any MSO_2 sentence φ over digraphs and any $c \in N$, the set $\mathcal{L}(\varphi, \Sigma_{\mathbb{S}}^c)$ of all slice strings $\mathbf{S}_1\mathbf{S}_2...\mathbf{S}_k$ over $\Sigma_{\mathbb{S}}^c$ such that $\mathbf{S}_1 \circ \mathbf{S}_2 \circ ... \circ \mathbf{S}_k = G$ and $G \models \varphi$ is a regular subset of $(\Sigma_{\mathbb{S}}^c)^*$.*

Lemma 7.1 gives a slice theoretic analog of Courcelle's model checking theorem: In order to verify whether a digraph G of existential slice-width at most c satisfies a given MSO property φ, one just needs to find a slice decomposition $\mathbf{U} = \mathbf{S}_1\mathbf{S}_2...\mathbf{S}_n$ of G and subsequently verify whether the deterministic finite automaton (or slice graph) accepting $\mathcal{L}(\varphi, \Sigma_{\mathbb{S}}^c)$ accepts $\mathbf{U}$. However the goal of the present work is to make a rather different use of Lemma 7.1. Namely, next in Theorem 7.1 we will restrict Lemma 7.1 in such a way that it concerns only z-dilated-saturated regular slice languages, so that it can be coupled to Theorem 6.2, yielding in this way a proof of Theorem 1.1.

Theorem 7.1. *For any MSO_2 formula φ and any $k, z \in N$, one may effectively construct a z-dilated-saturated slice graph $\mathcal{SG}(\varphi, k, z)$ over the slice alphabet $\Sigma_{\mathbb{S}}^{k \cdot z}$ whose graph language $\mathcal{L}_{\mathcal{G}}(\mathcal{SG}(\varphi, k, z))$ consists precisely of the digraphs of zig-zag number at most z that satisfy φ and that are the union of k directed paths.*

Finally we are in a position to prove Theorem 1.1. The proof will follow from a combination of Theorems 7.1 and 6.2.

Proof of Theorem 1.1 Given a monadic second order formula φ, and positive integers k and z, first we construct the dilated-saturated slice graph $\mathcal{SG}(\varphi, z, k)$ over $\Sigma_{\mathbb{S}}^{k \cdot z}$ as in Theorem 7.1. Since the slice-width of a digraph is at most $O(n^2)$ if we plug $q = O(n^2)$, $r = |\mathcal{SG}(\varphi, z, k)|$ and $\mathcal{SG}(\varphi, z, k)$ into Theorem 6.2, and if we let $f(\varphi, z, k) = r^{O(1)}$, then we get an overall upper bound of $f(\varphi, z, k) \cdot n^{O(k \cdot z)}$ for computing the number of subgraphs of G that satisfy φ and that are the union of k-directed paths. $\square$

8   Final Comments

In this work we have employed slice theoretic techniques to obtain the polynomial time solvability of many natural combinatorial questions on digraphs of constant directed path-width, cycle rank, K-width and DAG-depth. We have done so by using the zig-zag number of a digraph as a point of connection between these directed width measures, regular slice languages and the monadic second order logic of graphs. Thus our results shed new light into a field that has resisted algorithmic metatheorems for more than a decade. More precisely, we showed that despite the severe restrictions imposed by the impossibility results in [34,35,26,25], it is still possible to develop logic-based algorithmic metatheorems for digraph width measures that are able to solve in polynomial time a considerable variety of interesting problems.

Acknowledgements. The author would like to thank Stefan Arnborg for interesting discussions about the monadic second order logic of graphs and for providing valuable comments and suggestions on this work.

References

1. Ajtai, M., Fagin, R., Stockmeyer, L.J.: The closure of monadic NP. J. Comput. Syst. Sci. 60(3), 660–716 (2000)
2. Arnborg, S., Lagergren, J., Seese, D.: Easy problems for tree-decomposable graphs. J. Algorithms 12(2), 308–340 (1991)
3. Artin, E.: The theory of braids. Annals of Mathematics 48(1), 101–126 (1947)
4. Barát, J.: Directed path-width and monotonicity in digraph searching. Graphs and Combinatorics 22(2), 161–172 (2006)
5. Bauderon, M., Courcelle, B.: Graph expressions and graph rewritings. Mathematical Systems Theory 20(2-3), 83–127 (1987)
6. Berwanger, D., Dawar, A., Hunter, P., Kreutzer, S., Obdržálek, J.: The DAG-width of directed graphs. J. Comb. Theory, Ser. B 102(4), 900–923 (2012)
7. Berwanger, D., Grädel, E.: Entanglement - A measure for the complexity of directed graphs with applications to logic and games. In: Baader, F., Voronkov, A. (eds.) LPAR 2004. LNCS (LNAI), vol. 3452, pp. 209–223. Springer, Heidelberg (2005)
8. Berwanger, D., Grädel, E., Kaiser, L., Rabinovich, R.: Entanglement and the complexity of directed graphs. Theor. Comput. Sci. 463, 2–25 (2012)
9. Borie, R.B., Parker, R.G., Tovey, C.A.: Deterministic decomposition of recursive graph classes. SIAM J. Discrete Math. 4(4), 481–501 (1991)
10. Bozapalidis, S., Kalampakas, A.: Recognizability of graph and pattern languages. Acta Inf. 42(8-9), 553–581 (2006)
11. Brandenburg, F.-J., Skodinis, K.: Finite graph automata for linear and boundary graph languages. Theoretical Computer Science 332(1-3), 199–232 (2005)
12. Courcelle, B.: Graph expressions and graph rewritings. Math. Syst. Theory 20, 83–127 (1987)
13. Courcelle, B.: Graph rewriting: An algebraic and logic approach. In: Handbook of Theoretical Computer Science, pp. 194–242 (1990)
14. Courcelle, B., Engelfriet, J.: Graph Structure and Monadic Second-Order Logic: A Language-Theoretic Approach, vol. 138. Cambridge University Press (2012)
15. Courcelle, B., Makowsky, J.A., Rotics, U.: Linear time solvable optimization problems on graphs of bounded clique-width. Th. of Comp. Syst. 33(2), 125–150 (2000)
16. Courcelle, B., Makowsky, J.A., Rotics, U.: On the fixed parameter complexity of graph enumeration problems definable in monadic second-order logic. Discrete Applied Mathematics 108(1-2), 23–52 (2001)
17. de Oliveira Oliveira, M.: Hasse diagram generators and Petri nets. Fundam. Inform. 105(3), 263–289 (2010)
18. de Oliveira Oliveira, M.: Canonizable partial order generators. In: Dediu, A.-H., Martín-Vide, C. (eds.) LATA 2012. LNCS, vol. 7183, pp. 445–457. Springer, Heidelberg (2012)
19. de Oliveira Oliveira, M.: Subgraphs satisfying MSO properties on z-topologically orderable digraphs. Preprint (full version of this paper) arXiv:1303.4443 (2013)
20. Downey, R.G., Fellows, M.R.: Fixed parameter tractability and completeness. In: Complexity Theory: Current Research, pp. 191–225 (1992)
21. Eggan, L.C.: Transition graphs and the star height of regular events. Michigan Mathematical Journal 10(4), 385–397 (1963)
22. Engelfriet, J., Vereijken, J.J.: Context-free graph grammars and concatenation of graphs. Acta Informatica 34(10), 773–803 (1997)
23. Evans, W., Hunter, P., Safari, M.: D-width and cops and robbers. Manuscript (2007)
24. Ganian, R., Hliněný, P., Kneis, J., Langer, A., Obdržálek, J., Rossmanith, P.: On digraph width measures in parameterized algorithmics. In: Chen, J., Fomin, F.V. (eds.) IWPEC 2009. LNCS, vol. 5917, pp. 185–197. Springer, Heidelberg (2009)

25. Ganian, R., Hliněný, P., Kneis, J., Meister, D., Obdržálek, J., Rossmanith, P., Sikdar, S.: Are there any good digraph width measures? In: Raman, V., Saurabh, S. (eds.) IPEC 2010. LNCS, vol. 6478, pp. 135–146. Springer, Heidelberg (2010)
26. Ganian, R., Hlinený, P., Langer, A., Obdrzálek, J., Rossmanith, P., Sikdar, S.: Lower bounds on the complexity of MSO1 model-checking. In: STACS 2012, vol. 14, pp. 326–337 (2012)
27. Giammarresi, D., Restivo, A.: Recognizable picture languages. International Journal Pattern Recognition and Artificial Intelligence 6(2-3), 241–256 (1992)
28. Giammarresi, D., Restivo, A.: Two-dimensional finite state recognizability. Fundam. Inform. 25(3), 399–422 (1996)
29. Gruber, H.: On the d-width of directed graphs. Manuscript (2007)
30. Gruber, H.: Digraph complexity measures and applications in formal language theory. Discrete Math. & Theor. Computer Science 14(2), 189–204 (2012)
31. Gruber, H., Holzer, M.: Finite automata, digraph connectivity, and regular expression size. In: Aceto, L., Damgård, I., Goldberg, L.A., Halldórsson, M.M., Ingólfsdóttir, A., Walukiewicz, I. (eds.) ICALP 2008, Part II. LNCS, vol. 5126, pp. 39–50. Springer, Heidelberg (2008)
32. Hunter, P., Kreutzer, S.: Digraph measures: Kelly decompositions, games, and orderings. Theor. Comput. Sci. 399(3), 206–219 (2008)
33. Johnson, T., Robertson, N., Seymour, P.D., Thomas, R.: Directed tree-width. J. Comb. Theory, Ser. B 82(1), 138–154 (2001)
34. Kreutzer, S.: On the parameterized intractability of monadic second-order logic. Logical Methods in Computer Science 8(1) (2012)
35. Kreutzer, S., Tazari, S.: Lower bounds for the complexity of monadic second-order logic. In: LICS, pp. 189–198 (2010)
36. Lampis, M., Kaouri, G., Mitsou, V.: On the algorithmic effectiveness of digraph decompositions and complexity measures. Discrete Optimization 8(1), 129–138 (2011)
37. Matz, O., Thomas, W.: The monadic quantifier alternation hierarchy over graphs is infinite. In: LICS, pp. 236–244 (1997)
38. Post, E.L.: A variant of a recursively unsolvable problem. Bulletion of the American Mathematical Society 52, 264–268 (1946)
39. Reed, B.A.: Introducing directed tree width. Electronic Notes in Discrete Mathematics 3, 222–229 (1999)
40. Safari, M.A.: D-width: A more natural measure for directed tree width. In: Jedrzejowicz, J., Szepietowski, A. (eds.) MFCS 2005. LNCS, vol. 3618, pp. 745–756. Springer, Heidelberg (2005)
41. Slivkins, A.: Parameterized tractability of edge-disjoint paths on directed acyclic graphs. In: Di Battista, G., Zwick, U. (eds.) ESA 2003. LNCS, vol. 2832, pp. 482–493. Springer, Heidelberg (2003)
42. Tamaki, H.: A polynomial time algorithm for bounded directed pathwidth. In: Kolman, P., Kratochvíl, J. (eds.) WG 2011. LNCS, vol. 6986, pp. 331–342. Springer, Heidelberg (2011)
43. Thomas, W.: Finite-state recognizability of graph properties. Theorie des Automates et Applications 172, 147–159 (1992)

Computing Tree-Depth Faster Than 2^n

Fedor V. Fomin*, Archontia C. Giannopoulou*, and Michał Pilipczuk*

Department of Informatics, University of Bergen,
P.O. Box 7803, 5020 Bergen, Norway
{fomin,archontia.giannopoulou,michal.pilipczuk}@ii.uib.no

Abstract. A connected graph has tree-depth at most k if it is a subgraph of the closure of a rooted tree whose height is at most k. We give an algorithm which for a given n-vertex graph G, in time $\mathcal{O}^*(1.9602^n)$ computes the tree-depth of G. Our algorithm is based on combinatorial results revealing the structure of minimal rooted trees whose closures contain G.

1 Introduction

The tree-depth of a graph G, denoted $\mathbf{td}(G)$, is the minimum number k such that there is a rooted forest F, not necessarily a subgraph of G, with the following properties.

- $V(G) = V(F)$,
- Every tree in F is of height at most k, i.e. the longest path between the root of the tree and any of its leaves contains at most k vertices,
- G is a subgraph of the closure of F, which is the graph obtained from F by adding all edges between every vertex of F and the vertices contained in the path from this vertex to the root of the tree that it belongs to.

This parameter has increasingly been receiving attention since it was defined by Nešetřil and Ossona de Mendez in [13] and played a fundamental role in the theory of classes of bounded expansion [14–17]. Tree-depth is a very natural graph parameter, and due to different applications, was rediscovered several times under different names as the vertex ranking number [2], the ordered coloring [10], and the minimum height of an elimination tree of a graph [20].

From the algorithmic perspective, it has been known that the problem of computing tree-depth is NP-hard even when restricted to bipartite graphs [2, 13]. However, it also admits polynomial time algorithms for specific graph classes [6, 12]. For example, when the input graph is a tree its tree-depth can be computed in linear time [20]. Moreover, as tree-depth is closed under minors, the results of Robertson and Seymour [18, 19] imply that the problem is in FPT when parameterized by the solution size. In [2], Bodlaender et al. showed that the computation of tree-depth is also in XP when parameterized by treewidth. From the point of view of

* Supported by European Research Council (ERC) Grant "Rigorous Theory of Preprocessing", reference 267959.

G. Gutin and S. Szeider (Eds.): IPEC 2013, LNCS 8246, pp. 137–149, 2013.

approximation, tree-depth can be approximated in polynomial time within a factor of $\mathcal{O}(\log^2 n)$ [4], where n is the number of vertices of the input graph. Moreover, there is a simple approximation algorithm that, given a graph G, returns a forest F such that G is contained in the closure of F and the height of F is at most $2^{\mathsf{td}(G)}$ [17]. Finally, it is easy to see (and will be described in Section 3) that there exists an exact algorithm for the computation of tree-depth running in $\mathcal{O}^*(2^n)$ time[1].

We are interested in tree-depth from the perspective of exact exponential time algorithms. Tree-depth is intimately related to another two well studied parameters, treewidth and pathwidth. The treewidth of a graph can be defined as the minimum taken over all possible completions into a chordal graph of the maximum clique size minus one. Similarly, path-width can be defined in terms of completion to an interval graph. One of the equivalent definitions of tree-depth is that it is the size of the largest clique in a completion to a trivially perfect graph. These graph classes form the following chain

$$\textbf{trivially perfect} \subset \textbf{interval} \subset \textbf{chordal},$$

corresponding to the parameters tree-depth, pathwidth, and treewidth.

However, while for the computation of treewidth and pathwidth there exist $\mathcal{O}^*(c^n)$, $c < 2$, time algorithms [7, 8, 11, 21], to the best of our knowledge no such algorithm for tree-depth has been known prior to our work. In this paper, we construct the first exact algorithm which for any input graph G computes its tree-depth in time $\mathcal{O}^*(c^n)$, $c < 2$. The running time of the algorithm is $\mathcal{O}^*(1.9602^n)$. The approach is based on the structural characteristics of the minimal forest that defines the tree-depth of the graph.

The rest of the paper is organized as follows. In Section 2 we give some basic definitions and preliminary combinatorial results on the minimal trees for tree-depth and in Section 3, based on the results from Section 2, we present the $\mathcal{O}(1.9602^n)$ time algorithm for tree-depth. Finally, in Section 4 we conclude with open problems. Due to space constraints the proofs of the Lemmata marked with $(\star)$ have been omitted.

2 Minimal Rooted Forests for Tree-Depth

2.1 Preliminaries

For a graph $G = (V, E)$, we use $V(G)$ to denote V and $E(G)$ to denote E. If $S \subseteq V(G)$ we denote by $G \setminus S$ the graph obtained from G after removing the vertices of S. In the case where $S = \{u\}$, we abuse notation and write $G \setminus u$ instead of $G \setminus \{u\}$. We denote by $G[S]$ the subgraph of G induced by the set S. For $S \subseteq V(G)$, the *open neighborhood* of S in G, $N_G(S)$, is the set $\{u \in G \setminus S \mid \exists v \in S : \{u, v\} \in E(G)\}$. Again, in the case where $S = \{v\}$ we abuse notation and write $N_G(v)$ instead of $N_G(\{v\})$. Given two vertices v and u we denote by $\mathbf{dist}_G(v, u)$ their distance in G. We use $\mathcal{C}(G)$ to denote the set of connected components of G.

---

[1] The $\mathcal{O}^*(\cdot)$ notation suppresses factors that are polynomial in the input size.

2.2 Tree-Depth

A *rooted forest* is a disjoint union of rooted trees. The *height* of a vertex x in a rooted forest F is the number of vertices of the path from the root (of the tree to which x belongs) to x and is denoted by $\mathbf{height}(x, F)$. The height of F is the maximum height of the vertices of F and is denoted by $\mathbf{height}(F)$. Let x, y be vertices of F. The vertex x is an *ancestor* of y if x belongs to the path linking y and the root of the tree to which y belongs. The *closure* $\mathbf{clos}(F)$ of a rooted forest F is the graph with vertex set $V(F)$ and edge set $\{\{x, y\} \mid x \text{ is an ancestor of } y \text{ in } F, x \neq y\}$. For every vertex y of F we denote by P_y the unique path linking y and the root of the tree to which y belongs, and denote by $p(y)$ the parent of y in F, i.e. the neighbor of y in P_y. Vertices whose parent is y are called the *children* of y. We call a vertex x of F a *branching point* if x is not a root of F and $\deg_F(x) > 2$ or if x is a root of F and $\deg_F(x) \geq 2$. For a vertex v of a rooted tree T, we denote by T_v the maximal subtree of T rooted in v. For example, if v is the root of T, then $T_v = T$.

Let G be a graph. The *tree-depth* of G, denoted $\mathbf{td}(G)$, is the least $k \in \mathbb{N}$ such that there exists a rooted forest F on the same vertex set as G such that $G \subseteq \mathbf{clos}(F)$ and $\mathbf{height}(F) = k$. Note that if G is connected then F must be a tree, and the tree-depth of a disconnected graph is the maximum of tree-depth among its connected components. Thus, when computing tree-depth we may focus on the case when G is connected and F is required to be a rooted tree.

With every rooted tree T of height h we associate a sequence $(t_1, t_2, t_3, \ldots)$, where $t_i = |\{v \mid \mathbf{height}(v, T) = i\}|$, $i \in \mathbb{N}$, that is, t_i is the number of vertices of the tree T of height i, $i \in \mathbb{N}$. Note that since T is finite, this sequence contains only finitely many non-zero values.

Let T_1 and T_2 be two rooted trees with heights h_1 and h_2, and corresponding sequences $(t_1^1, t_2^1, t_3^1, \ldots)$ and $(t_1^2, t_2^2, t_3^2, \ldots)$, respectively. We say that $T_1 \prec T_2$ if and only if there exists an $i \in \mathbb{N}$ such that $t_i^1 < t_i^2$ and $t_j^1 = t_j^2$, for every $j > i$. Note in particular that if $h_1 < h_2$, then taking $i = h_2$ in this definition proves that $T_1 \prec T_2$.

Definition 1. *Let G be a connected graph. A rooted tree T is* minimal *for G if*

1. *$V(T) = V(G)$ and $G \subseteq \mathbf{clos}(T)$, and*
2. *there is no tree T' such that $V(T') = V(G)$, $G \subseteq \mathbf{clos}(T')$, and $T' \prec T$.*

The next observation follows from the definitions of $\prec$ and of tree-depth.

Observation 1. *Let G be a connected graph and T be a rooted tree for G such that $V(T) = V(G)$, $G \subseteq \mathbf{clos}(T)$, and $\mathbf{height}(T) > \mathbf{td}(G)$. Then there exists a rooted tree T' such that $V(T') = V(G)$, $G \subseteq \mathbf{clos}(T')$, and $\mathbf{height}(T') < \mathbf{height}(T)$, and thus $T' \prec T$.*

The following combinatorial lemmata reveal the structures of minimal trees which will be handy in the algorithm.

Lemma 1 ($\star$). *Let T^1 be a rooted tree with root r, $v \in V(T^1)$, and T^* be a rooted tree with root r^* such that $T^* \prec T_v^1$. If T^2 is the rooted tree obtained from*

T^1 after considering the union of $T^1 \setminus V(T_v^1)$ with T^* and adding an edge between r^* and $p(v)$ (if $p(v)$ exists), then $T^2 \prec T^1$.

Lemma 2. *Let G be a connected graph. If T is a minimal tree for G with root r then for every $v \in V(T)$,*

1. *$G[V(T_v)]$ is connected,*
2. *T_v is a minimal tree for $G[V(T_v)]$, and*
3. *if $v' \in V(T_v)$ is a branching point with minimum $\mathbf{dist}_{T_v}(v, v')$ then $N_G(v) \cap V(T_u) \neq \emptyset$, for every child u of v'.*

Proof. We first prove (1). Assume in contrary that there exists a vertex $v \in V(T)$ such that the graph $G[V(T_v)]$ is not connected. Notice that we may choose v in such a way that $\mathbf{dist}_T(r, v)$ is maximum. We first exclude the case where $v = r$. Indeed, notice that if $v = r$, then $G[V(T_r)] = G$ is connected by the hypothesis. Thus, $v \neq r$. Notice also that if v is a leaf of T then T_v is the graph consisting of one vertex, so it is again connected. Therefore, v is not a leaf of T. Let $v_1, v_2, \ldots, v_p$ be the children of v. The choice of v (maximality of distance from r) implies that $G[V(T_{v_i})]$ is a connected component of $G[V(T_v)] \setminus v$, $i \in [p]$. Moreover, from the fact that $G[V(T_v)]$ is not connected, it follows that there exists at least one $i_0 \in [p]$ such that $N_G(v) \cap V(T_{v_{i_0}}) = \emptyset$. Let T' be the tree obtained from T by removing the edge $\{v, v_{i_0}\}$ and adding the edge $\{p(v), v_{i_0}\}$. Observe that $G \subseteq \mathbf{clos}(T')$. Moreover, notice that by construction of T', we may consider T' as the tree obtained from the union of $T \setminus V(T_{p(v)})$ with $T'_{p(v)}$ after adding the edge $\{p(v), p(p(v))\}$ (if $p(v) \neq r$). It is easy to see that $T'_{p(v)} \prec T_{p(v)}$. Therefore, from Lemma 1, we end up with a contradiction to the minimality of T.

To prove (2), we assume in contrary that there exists a vertex $v \in V(T)$ such that T_v is not a minimal tree for $G[V(T_v)]$. By the hypothesis that T is a minimal tree for G, it follows that $v \neq r$. As T_v is not a minimal tree for $G[V(T_v)]$, there exists a rooted tree T' with root r' such that $V(T') = V(T_v)$, $G[V(T_v)] \subseteq \mathbf{clos}(T')$, and $T' \prec T_v$. Let now T^* be the rooted tree obtained from the union of $T \setminus V(T_v)$ with T' after adding an edge between $p(v)$ and r'. Notice then that $G \subseteq \mathbf{clos}(T^*)$. Moreover, from Lemma 1, we get that $T^* \prec T$, a contradiction to the minimality of T.

We now prove (3). Let v be a vertex of T and v' be a branching point of T_v such that $\mathbf{dist}_{T_v}(v, v')$ is minimum, that is, v' is the highest branching point in T_v. Assume in contrary that there exists a child u of v' such that $N_G(v) \cap V(T_u) = \emptyset$. Let T' be the tree obtained from T by switching the position of the vertices v and v', where $T' = T$ if $v = v'$. Notice that $\mathbf{clos}(T) = \mathbf{clos}(T')$ and T and T' are isomorphic, hence T' is also a minimal tree for G. Moreover, children of v in T' are exactly children of v' in T. Observe also that if w is a child of v' in T, hence also a child of v in T', then $T_w = T'_w$ and $N_{G[V(T'_v)]}(V(T'_w)) \subseteq \{v\}$. As $N_G(v) \cap V(T_u) = \emptyset$, we obtain that $G[V(T'_v)]$ is not connected. However, T' is a minimal tree for G and therefore, from (1), $G[V(T'_v)]$ is connected, a contradiction. This completes the proof of the last claim and of the lemma. $\square$

3   Computing Tree-Depth

3.1   The Naive DP, and Pruning the Space of States

To construct our algorithm, we need an equivalent recursive definition of tree-depth.

Proposition 1 ([13]). *The* tree-depth *of a connected graph G is equal to*

$$\mathbf{td}(G) = \begin{cases} 1 & \text{if } |V(G)| = 1 \\ 1 + \min_{v \in V(G)} \max_{H \in \mathcal{C}(G \setminus v)} \mathbf{td}(H) & \text{otherwise} \end{cases} \qquad (1)$$

Proposition 1 already suggests a dynamic programming algorithm computing tree-depth of a given graph G in $\mathcal{O}^*(2^n)$ time. Assume without loss of generality that G is connected, as otherwise we may compute the tree-depth of each connected component of G separately. For every $X \subseteq V(G)$ such that $G[X]$ is connected, we compute $\mathbf{td}(G[X])$ using (1). Assuming that the tree-depth of all the connected graphs induced by smaller subsets of vertices has been already computed, computation of formula (1) takes polynomial time. Hence, if we employ dynamic programming starting with the smallest sets X, we can compute $\mathbf{td}(G)$ in $\mathcal{O}^*(2^n)$ time. Let us denote this algorithm by $\mathbb{A}_0$.

The reason why $\mathbb{A}_0$ runs in pessimistic $\mathcal{O}^*(2^n)$ time is that the number of subsets of $V(G)$ inducing connected subgraphs can be as large as $\mathcal{O}(2^n)$. Therefore, if we aim at reducing the time complexity, we need to prune the space of states significantly. Let us choose some ε, $0 < \varepsilon < \frac{1}{6}$, to be determined later, and let G be a connected graph on n vertices. We define the space of states $\mathcal{S}_\varepsilon$ as follows:

$$\mathcal{S}_\varepsilon = \{ S \subseteq V(G) \mid 1 \le |S| \le (\tfrac{1}{2} - \varepsilon) n \text{ and } G[S] \text{ is connected, or}$$
$$\exists X \subseteq V(G) : |X| \le (\tfrac{1}{2} - \varepsilon) n \text{ and } G[S] \in \mathcal{C}(G \setminus X) \}.$$

Observe that thus all the sets belonging to $\mathcal{S}_\varepsilon$ induce connected subgraphs of G. The subsets $S \in \mathcal{S}_\varepsilon$ considered in the first part of the definition will be called of the *first type*, and the ones considered in the second part will be called of the *second type*. Note that $V(G) \in \mathcal{S}_\varepsilon$ since it is a subset of second type for $X = \emptyset$.

Lemma 3 ($\star$). *If G is a graph on n vertices, then $|\mathcal{S}_\varepsilon| = \mathcal{O}^*\left(\binom{n}{(\frac{1}{2} - \varepsilon)n} \right)$. Moreover, $\mathcal{S}_\varepsilon$ may be enumerated in $\mathcal{O}^*\left(\binom{n}{(\frac{1}{2} - \varepsilon) n} \right)$ time.*

In our algorithms we store the family $\mathcal{S}_\varepsilon$ as a collection of binary vectors of length n in a prefix tree (a trie). Thus when constructing $\mathcal{S}_\varepsilon$ we can avoid enumerating duplicates, and then test belonging to $\mathcal{S}_\varepsilon$ in $O(n)$ time.

We now define the pruned dynamic programming algorithm $\mathbb{A}_\varepsilon$ that for every $X \in \mathcal{S}_\varepsilon$ computes value $\mathbf{td}_*(G[X])$ defined as follows:

$$\mathbf{td}_*(G[X]) = \begin{cases} 1 & \text{if } |X| = 1 \\ 1 + \min_{v \in X} \max_{H \in \mathcal{C}(G[X] \setminus v), \ V(H) \in \mathcal{S}_\varepsilon} \mathbf{td}_*(H) & \text{otherwise} \end{cases} \qquad (2)$$

We use the convention that $\mathbf{td}_*(G[X]) = +\infty$ if $X \notin \mathcal{S}_\varepsilon$. The algorithm $\mathbb{A}_\varepsilon$ can be implemented in a similar manner as $\mathbb{A}$ so that its running time is $\mathcal{O}^*(|\mathcal{S}_\varepsilon|)$. We consider sets from $\mathcal{S}_\varepsilon$ in increasing order of cardinalities (sorting $|\mathcal{S}_\varepsilon|$ with respect to cardinalities takes $\mathcal{O}^*(|\mathcal{S}_\varepsilon|)$ time) and simply apply formula (2). Note that computation of formula (2) takes polynomial time, since we need to consider at most n vertices v, and for every connected component $H \in \mathcal{C}(G \setminus v)$ we can test whether its vertex set belongs to $\mathcal{S}_\varepsilon$ in $\mathcal{O}(n)$ time.

For a set $S \in \mathcal{S}_\varepsilon$ and T being a minimal tree for $G[S]$, we say that T is *covered* by $\mathcal{S}_\varepsilon$ if $V(T_v) \in \mathcal{S}_\varepsilon$ for every $v \in S$. The following lemma expresses the crucial property of $\mathbf{td}_*$.

Lemma 4. *For any connected graph G and any subset $S \subseteq V(G)$, it holds that $\mathbf{td}_*(G[S]) \geq \mathbf{td}(G[S])$. Moreover, if $S \in \mathcal{S}_\varepsilon$ and there exists a minimal tree T for $G[S]$ that is covered by $\mathcal{S}_\varepsilon$, then $\mathbf{td}_*(G[S]) = \mathbf{td}(G[S])$.*

Proof. We first prove the first claim by induction with respect to the cardinality of S. If $\mathbf{td}_*(G[S]) = +\infty$ then the claim is trivial. Therefore, we assume that $S \in \mathcal{S}_\varepsilon$, there exists some $r \in S$ such that $\mathbf{td}_*(G[S]) = 1 + \max_{H \in \mathcal{C}(G[S] \setminus r)} \mathbf{td}_*(H)$, and $V(H) \in \mathcal{S}_\varepsilon$ for each $H \in \mathcal{C}(G[S] \setminus r)$. By the induction hypothesis, since $|V(H)| \leq |S|$ for each $H \in \mathcal{C}(G[S] \setminus r)$, we infer that $\mathbf{td}_*(H) \geq \mathbf{td}(H)$ for each $H \in \mathcal{C}(G[S] \setminus r)$. On the other hand, by (1) we have that $\mathbf{td}(G[S]) \leq 1 + \max_{H \in \mathcal{C}(G[S] \setminus r)} \mathbf{td}(H)$. Therefore,

$$\mathbf{td}(G[S]) \leq 1 + \max_{H \in \mathcal{C}(G[S] \setminus r)} \mathbf{td}(H)$$
$$\leq 1 + \max_{H \in \mathcal{C}(G[S] \setminus r)} \mathbf{td}_*(H) = \mathbf{td}_*(G[S]),$$

and the induction step follows.

We now prove the second claim, again by induction with respect to the cardinality of S. Let T be a minimal tree for $G[S]$ that is covered by $\mathcal{S}_\varepsilon$. Let r be the root of T and let $v_1, v_2, \ldots, v_p$ be the children of r in T. By (2) of Lemma 2, we have that T_{v_i} is a minimal tree for $G[V(T_{v_i})]$, for each $i \in [p]$. Moreover, since T was covered by $\mathcal{S}_\varepsilon$, then so does each T_{v_i}. By the induction hypothesis we infer that $\mathbf{td}_*(G[V(T_{v_i})]) = \mathbf{td}(G[V(T_{v_i})])$ for each $i \in [p]$, since $|V(T_{v_i})| < |S|$. Moreover, since T and each T_{v_i} are minimal, we have that

$$\mathbf{td}(G[S]) = \mathbf{height}(T) = 1 + \max_{i \in [p]} \mathbf{height}(T_{v_i}) = 1 + \max_{i \in [p]} \mathbf{td}(G[V(T_{v_i})])$$
$$= 1 + \max_{i \in [p]} \mathbf{td}_*(G[V(T_{v_i})]) \geq \mathbf{td}_*(G[S]).$$

The last inequality follows from the fact that, by (1) of Lemma 2, $G[V(T_{v_i})]$ are connected components of $G[S] \setminus r$ and moreover that their vertex sets belong to $\mathcal{S}_\varepsilon$. Hence, vertex r was considered in (2) when defining $\mathbf{td}_*(G[S])$. We infer that $\mathbf{td}(G[S]) \geq \mathbf{td}_*(G[S])$, and $\mathbf{td}(G[S]) \leq \mathbf{td}_*(G[S])$ by the first claim, so $\mathbf{td}_*(G[S]) = \mathbf{td}(G[S])$. $\qquad\square$

Lemma 4 implies that the tree-depth is already computed exactly for all connected subgraphs induced by significantly less than half of the vertices.

Corollary 1. *For any connected graph G on n vertices and any $S \in \mathcal{S}_\varepsilon$, if $|S| \leq (\frac{1}{2} - \varepsilon)n$, then $\mathbf{td}_*(G[S]) = \mathbf{td}(G[S])$.*

Proof. If T is a minimal tree for $G[S]$, then for every $v \in V(T)$, $G[V(T_v)]$ is connected by (1) of Lemma 2, and $|V(T_v)| \leq (\frac{1}{2} - \varepsilon)n$. Hence, for every $v \in V(T)$ it holds that $V(T_v) \in \mathcal{S}_\varepsilon$ and the corollary follows from Lemma 4. □

Finally, we observe that for any input graph G the algorithm $\mathbb{A}_\varepsilon$ already computes the tree-depth of G unless every minimal tree for G has a very special structure. Let T be a minimal tree for G. A vertex $v \in V(G)$ is called *problematic* if (i) $|V(T_v)| > (\frac{1}{2} - \varepsilon)n$, and (ii) $|V(P_{p(v)})| > (\frac{1}{2} - \varepsilon)n$. We say that a minimal tree T for G is *problematic* if it contains some problematic vertex.

Corollary 2. *For any connected graph G, if G admits a minimal tree that is not problematic, then $\mathbf{td}_*(G) = \mathbf{td}(G)$.*

Proof. We prove that any minimal tree T for G that is not problematic, is in fact covered by $\mathcal{S}_\varepsilon$. Then the corollary follows from Lemma 4.

Take any $v \in V(G)$; we need to prove that $V(T_v) \in \mathcal{S}_\varepsilon$. First note that $G[V(T_v)]$ is connected by (1) of Lemma 2. Hence if $|V(T_v)| \leq (\frac{1}{2} - \varepsilon) n$, then it trivially holds that $V(T_v) \in \mathcal{S}_\varepsilon$ by the definition of $\mathcal{S}_\varepsilon$. Otherwise we have that $|V(P_{p(v)})| \leq (\frac{1}{2} - \varepsilon) n$, since v is not problematic. Note then that $N_G(V(T_v)) \subseteq V(P_{p(v)})$ and so $G[V(T_v)]$ is a connected component of $V(G) \setminus V(P_{p(v)})$. Consequently, $V(T_v)$ is a subset of second type for $X = V(P_{p(v)})$. □

3.2 The Algorithm

Corollary 2 already restricts cases when the pruned dynamic program $\mathbb{A}_\varepsilon$ misses the minimal tree: this happens only when all the minimal trees for the input graph G are problematic. Therefore, it remains to investigate the structure of problematic minimal trees to find out, if some problematic minimal tree could have smaller height than the minimal tree computed by $\mathbb{A}_\varepsilon$.

Let G be the input graph on n vertices. Throughout this section we assume that G admits some problematic minimal tree T. Let v be a problematic vertex in T. Let moreover v' be the highest branching point in T_v (possibly $v' = v$ if v is already a branching point in T), or v' be the only leaf of T_v in case T_v does not contain any branching points. Let $Z = V(P_{v'})$; observe that since v is problematic, we have that $|Z| > (\frac{1}{2} - \varepsilon) n$. Let $Q_1, Q_2, \ldots, Q_a$ be all the subtrees of T rooted in $N_T(Z \setminus \{v'\})$, that is, in the children of vertices of $Z \setminus \{v'\}$ that do not belong to Z, and let $R_1, R_2, \ldots, R_b$ be the subtrees of T rooted in children of v'. Note that trees $Q_1, Q_2, \ldots, Q_a, R_1, R_2, \ldots, R_b$ are pairwise disjoint, and by the definition of a minimal tree we have that $N_G(V(Q_i)), N_G(V(R_j)) \subseteq Z$ for any $i \in [a]$, $j \in [b]$.

Let $Q = \bigcup_{i=1}^{a} V(Q_i)$ and $R = \bigcup_{j=1}^{b} V(R_j)$. For any problematic minimal tree T and a problematic vertex v in it, we say that v *defines* the sets $Z, Q_1, Q_2, \ldots, Q_a, R_1, R_2, \ldots, R_b, Q, R$ in T.

Observation 2. *If $b > 0$, then $Z = N_G(V(R_j) \cup Q)$ for any $j \in [b]$.*

Proof. As $N_G(V(Q_i)), N_G(V(R_j)) \subseteq Z$ for any $i \in [a]$, $j \in [b]$, we have that $Z \supseteq N_G(V(R_j) \cup \bigcup_{i=1}^a V(Q_i))$. We proceed to proving the reverse inclusion.

Take any $z \in Z$, and let z' be the highest branching point in T_z; note that z' is always defined since $b > 0$ and thus v' is a branching point. If $z' = v'$, then by (3) of Lemma 2 we infer that $z \in N_G(V(R_j))$, $j \in [b]$. Otherwise, we have that if $z' \in Z \setminus \{v'\}$. Since z' is a branching point, there exists some subtree Q_i rooted in a child of z'. We can again use (3) of Lemma 2 to infer that $z \in N_G(V(Q_i))$, so also $z \in N_G(Q)$.   □

Observe that if $b = 0$, then we trivially have that $Z = V(G) \setminus Q$.

Observation 3. $|Q| < 2\varepsilon n$.

Proof. Since v is problematic, we have that $|V(P_{p(v)})| > \left(\frac{1}{2} - \varepsilon\right) n$ and $|V(T_v)| > \left(\frac{1}{2} - \varepsilon\right) n$. Observe also that $V(P_{p(v)}) \cup V(T_v) = Z \cup R$ by the definition of Z. Since $V(P_{p(v)}) \cap V(T_v) = \emptyset$ we have that:

$$|Z \cup R| = |V(P_{p(v)}) \cup V(T_v)| > (1 - 2\varepsilon)n.$$

Since $Q = V(G) \setminus (Z \cup R)$, the claim follows.   □

Observation 4. $|R| < \left(\frac{1}{2} + \varepsilon\right) n - |Q|$.

Proof. Since $R = V(G) \setminus (Z \cup Q)$ and $|Z| > \left(\frac{1}{2} - \varepsilon\right) n$, we have that

$$|R| = |V(G) \setminus (Z \cup Q)| = n - |Z| - |Q| < \left(\frac{1}{2} + \varepsilon\right) n - |Q|.$$ □

Observation 5. *If $b > 0$, then $b \geq 2$ and* $\min_{j \in [b]} |V(R_j)| < \left(\frac{1}{4} + \frac{\varepsilon}{2}\right) n - \frac{|Q|}{2}$.

Proof. If $b > 0$ then v' is a branching point and it has at least two children. It follows that $b \geq 2$. For the second claim, observe that since $b \geq 2$ we have that $\min_{j \in [b]} |V(R_j)| \leq |R|/2$ and the claim follows from Observation 4.   □

We can proceed to the description of our main algorithm, denoted further $\mathbb{A}$. Similarly as before, without loss of generality let us assume that G is connected. First, the algorithm constructs the family $\mathcal{S}_\varepsilon$ using Lemma 3, and runs the algorithm $\mathbb{A}_\varepsilon$ on it. Note that these steps can be performed in time $\mathcal{O}^*\left(\binom{n}{(\frac{1}{2}-\varepsilon)n}\right)$. We can therefore assume that the value $\mathbf{td}_*(G[S])$ is computed for every $S \in \mathcal{S}_\varepsilon$, and in particular for $S = V(G)$.

Now the algorithm proceeds to checking whether a better problematic minimal tree T with problematic vertex v can be constructed. We adopt the notation introduced in the previous paragraphs for a problematic minimal tree T. We aim at identifying set Z and sets $V(Q_1), V(Q_2), \ldots, V(Q_a), V(R_1), V(R_2), \ldots, V(R_b)$. Without loss of generality assume that if $b > 0$, then $V(R_1)$ has the smallest cardinality among $V(R_1), V(R_2), \ldots, V(R_b)$, i.e., $|V(R_1)| \leq |V(R_2)|, \ldots, |V(R_b)|$. Let then $Y = Q \cup R_1$ if $b > 0$, and $Y = Q$ if $b = 0$.

The algorithm branches into at most $(n + 1)$ subbranches, in each fixing the expected cardinality y of Y. Note that by Observations 3 and 5 and the fact that $\varepsilon < \frac{1}{6}$ we may assume that $y < |Q| + \left(\frac{1}{4} + \frac{\varepsilon}{2}\right) n - \frac{|Q|}{2} = \frac{|Q|}{2} + \left(\frac{1}{4} + \frac{\varepsilon}{2}\right) n < \left(\frac{1}{4} + \frac{3\varepsilon}{2}\right) n$. Then the algorithm branches into $\binom{n}{y}$ subbranches, in each fixing a different subset of vertices of size smaller than y as the set Y. Note that sets $V(Q_1), V(Q_2), \ldots, V(Q_a), V(R_1)$ are then defined as vertex sets of connected components of $G[Y]$. The algorithm branches further into $(n + 1)$ cases. In one case the algorithm assumes that $b = 0$ and therefore concludes that $Q = Y$. In other cases the algorithm assumes that $b > 0$ and picks one of the components of $G[Y]$ assuming that its vertex set is $V(R_1)$, thus recognizing Q as $Y \setminus V(R_1)$, i.e., the union of vertex sets of remaining components of $G[Y]$.

In the case when $b = 0$ the algorithm concludes that $Z = V(G) \setminus Q$. In the cases when $b > 0$, the algorithm concludes that $Z = N_G(Y)$ using Observation 2. Having identified Z, the sets $V(R_1), V(R_2), \ldots, V(R_j)$ can be recognized as vertex sets of connected components of $V(G) \setminus (Z \cup Q)$. Observations 2, 3, and 5 ensure that for every problematic minimal tree T for G, there will be at least one subbranch where sets $Z, V(Q_1), V(Q_2), \ldots, V(Q_a), V(R_1), V(R_2), \ldots, V(R_b)$ are fixed correctly. Observe also that in each of at most $(n + 1)$ branches where y has been fixed, we produced at most $(n + 1) \cdot \binom{n}{y}$ subbranches. We perform also sanity checks: whenever any produced branch does not satisfy any of Observations 2, 3, 4 or 5, or the fact that $V(R_1)$ is a smallest set among $V(R_1), V(R_2), \ldots, V(R_j)$, we terminate the branch immediately.

The algorithm now computes $\mathbf{td}(G[V(Q_i)])$ and $\mathbf{td}(G[V(R_j)])$ for all $i \in [a]$, $j \in [b]$. Recall that by Corollary 1, for any set $X \subseteq V(G)$ such that $G[X]$ is connected and $|X| \leq \left(\frac{1}{2} - \varepsilon\right)n$, we have that $\mathbf{td}(G[X]) = \mathbf{td}_*(G[X])$, and hence the value $\mathbf{td}(G[X])$ has been already computed by algorithm $\mathbb{A}_\varepsilon$. Since $|Q| \leq 2\varepsilon n$ and $\varepsilon < \frac{1}{6}$, we infer that this is the case for every set $V(Q_i)$ for $i \in [a]$, and values $\mathbf{td}(G[V(Q_i)])$ are already computed. The same holds for every R_j assuming that $|V(R_j)| \leq \left(\frac{1}{2} - \varepsilon\right)n$.

Assume then that there exists some j_0 such that $|V(R_{j_0})| > \left(\frac{1}{2} - \varepsilon\right)n$, i.e., we have no guarantee that the algorithm $\mathbb{A}_\varepsilon$ computed $\mathbf{td}(G[V(R_{j_0})])$ correctly. Note that by Observation 4 and the fact that $\varepsilon < \frac{1}{6}$, there can be at most one such j_0. Furthermore, if this is the case, then by Observation 5 we have that $b \geq 2$ and $V(R_{j_0})$ cannot be the smallest among sets $V(R_1), V(R_2), \ldots, V(R_b)$; hence, $j_0 \neq 1$ and $V(R_{j_0}) \subseteq V(G) \setminus (Z \cup Y)$. Therefore, we must necessarily have that $y = |Y| \leq |V(G)| - |Z| - |V(R_{j_0})| < n - \left(\frac{1}{2} - \varepsilon\right)n - \left(\frac{1}{2} - \varepsilon\right)n = 2\varepsilon n$, and moreover $|V(R_{j_0})| \leq |V(G)| - |Z| - |Y| < n - \left(\frac{1}{2} - \varepsilon\right)n - y = \left(\frac{1}{2} + \varepsilon\right)n - y$.

Formally, if none of these assertions holds, the branch would be terminated by the sanity check. To compute $\mathbf{td}(G[V(R_{j_0})])$ we employ the naive dynamic programming routine on $G[V(R_{j_0})]$, i.e., algorithm $\mathbb{A}_0$. Observe, however, that in this application we do not need to recompute the values for subsets of size at most $\left(\frac{1}{2} - \varepsilon\right)n$, since the values for them were already computed by the algorithm $\mathbb{A}_\varepsilon$. Therefore, since $|R_{j_0}| \leq \left(\frac{1}{2} + \varepsilon\right)n - y$ and $\varepsilon < \frac{1}{6}$, the application of algorithm $\mathbb{A}$ takes at most $\mathcal{O}^*\left(\binom{\left(\frac{1}{2} + \varepsilon\right)n - y}{\left(\frac{1}{2} - \varepsilon\right)n}\right)$ time.

Summarizing, for every choice of y (recall that $y < \left(\frac{1}{4} + \frac{3\varepsilon}{2}\right)n$), the algorithm produced at most $(n+1) \cdot \binom{n}{y}$ branches, and in branches with $y < 2\varepsilon n$ it could have used extra $\mathcal{O}^*\left(\binom{(\frac{1}{2}+\varepsilon)n-y}{(\frac{1}{2}-\varepsilon)n}\right)$ time for computing values $\mathbf{td}(G[V(R_j)])$ whenever there was no guarantee that algorithm $\mathbb{A}_\varepsilon$ computed them correctly.

We arrive at the situation where in each branch the algorithm already identified set Z, sets $V(Q_1), V(Q_2), \ldots, V(Q_a), V(R_1), V(R_2), \ldots, V(R_b)$, and values $\mathbf{td}(G[V(Q_i)])$ and $\mathbf{td}(G[V(R_j)])$ for $i \in [a]$, $j \in [b]$. Note, however, that the algorithm does not have yet the full knowledge of the shape of tree T, because we have not yet determined in which order the vertices of Z appear on the path $P_{v'}$, and thus we do not know where the trees Q_i and R_j are attached to this path. Fortunately, it turns out that finding an optimum such ordering of vertices of Z is polynomial-time solvable.

For $i \in [a+b]$ let $M_i = Q_i$ if $i \le a$ and $M_i = R_{i-a}$ otherwise, and let $h_i = \mathbf{td}(G[V(M_i)])$. Note that since T is minimal, by (2) of Lemma 2 we have that $h_i = \mathbf{height}(M_i)$ for each $i \in [a+b]$. Let also $Z_i = N_G(V(M_i))$; note that since $G \subseteq \mathbf{clos}(T)$, we have that $Z_i \subseteq Z$. Let σ be an ordering of Z, i.e., σ is a bijective function from Z to $[|Z|]$. Finally, we define the *weight* of σ as follows:

$$\mu(\sigma) = \max\left(|Z|, \max_{i \in [a+b]} (\max(\sigma(Z_i)) + h_i)\right). \tag{3}$$

The following lemma is implied by the definitions of tree-depth and of measure μ, and is crucial for the final part of our algorithm.

Lemma 5 ($\star$). *Let G be the input graph, and let $Z, \{V(M_i)\}_{i \in [a+b]}$ be any partitioning of vertices of G such that $Z_i = N_G(V(M_i))$ is a subset of Z for any $i \in [a+b]$. Moreover, let $h_i = \mathbf{td}(G[V(M_i)])$ and for σ being an ordering of Z, let $\mu(\sigma)$ be defined by (3). Then $\mathbf{td}(G) \le \mu(\sigma)$ for any ordering σ of Z. However, if G admits a problematic minimal tree T and $Z, \{V(M_i)\}_{i \in [a+b]}$ are defined by any problematic vertex in this tree, then $\mathbf{td}(G) = \min_\sigma \mu(\sigma)$.*

We are left with the following scheduling problem. Given a set Z of size at most n, a family number of subsets $Z_i \subseteq Z$ for $i \in [a+b]$ and corresponding integers $h_i \le n$, we would like to compute the minimum possible $\mu(\sigma)$ among orderings σ of Z. Let this problem be called MINIMUM ORDERING WITH INDEPENDENT DELAYS (MOID, for short).

Lemma 6. MINIMUM ORDERING WITH INDEPENDENT DELAYS *is polynomial-time solvable.*

Proof. Observe that since $|Z| \le n$ and $h_i \le n$, for any ordering σ we have that $\mu(\sigma) \le 2n$. We therefore iterate through all possible values M from $|Z|$ to $2n$, and for each M we check whether there exists some σ with $\mu(\sigma) \le M$. The first M for which this test returns a positive outcome is equal to $\min_\sigma \mu(\sigma)$.

For a given M, construct an auxiliary bipartite graph H with Z on one side and $\{1, 2, \ldots, |Z|\}$ on the other side. We put an edge between an element z and an index j if and only if the following holds: for every Z_i to which z belongs, it holds that $j + h_i \leq M$. It is easy to verify that orderings σ of Z with $\mu(\sigma) \leq M$ correspond one-to-one to perfect matchings in H. Indeed, if we are given an ordering σ with $\mu(\sigma) \leq M$, then we have that for every $z \in Z$ and Z_i to which z belongs, it holds that $\sigma(z) + h_i \leq M$ by the definition of $\mu(\sigma)$. Hence, $\{z, \sigma(z)\}$ is an edge in H and $\{\{z, \sigma(z)\} \mid z \in Z\}$ is a perfect matching in H. On the other hand, if we are given a perfect matching $\{\{z, j_z\} \mid z \in Z\}$ in H, then we may define an ordering σ of Z by putting $\sigma(z) = j_z$. Then for every $z \in Z$ and Z_i to which z belongs, we have that $\{z, \sigma(z)\}$ is an edge in H and, consequently, $\sigma(z) + h_i \leq M$. As we chose z and Z_i arbitrarily, it follows that $\max_{i \in [a+b]} (\max(\sigma(Z_i)) + h_i) \leq M$ and so $\mu(\sigma) \leq M$.

Therefore, to solve the MOID problem it suffices to construct H in polynomial time and run any polynomial-time algorithm for finding a perfect matching in H.   □

We remark that MINIMUM ORDERING WITH INDEPENDENT DELAYS can be also solved in $\mathcal{O}(n + \sum_{i=1}^{a+b} |Z_i|)$ time using greedy arguments. Since we are not interested in optimizing polynomial factors, in the proof of Lemma 6 we used the more concise matching argument to keep the description simple. We leave finding a faster algorithm for MOID to the reader as an interesting exercise.

Concluding, in every subbranch algorithm $\mathbb{A}$ constructs an instance of MOID and solves it in polynomial time using the algorithm of Lemma 6. Lemma 5 ensures that none of the values found in subbranches will be larger than $\mathbf{td}(G)$, and that if G admits a problematic minimal tree T then $\mathbf{td}(G)$ will be found in at least one subbranch. Therefore, by Corollary 2 we can conclude the algorithm $\mathbb{A}$ by outputting the minimum of $\mathbf{td}_*(G)$, computed by $\mathbb{A}_\varepsilon$, and the values returned by subbranches.

Let us proceed with the analysis of the running time of algorithm $\mathbb{A}$. First, we have enumerated $\mathcal{S}_\varepsilon$ and run the algorithm $\mathbb{A}_\varepsilon$, which took

$$T_1(n) = \mathcal{O}^* \left(\binom{n}{(\frac{1}{2} - \varepsilon) n} \right)$$

time. Then we created a number of subbranches. For every subbranch with $y \geq 2\varepsilon n$ we have spent polynomial time, and the number of these subbranches is bounded by $(n+1)^2 \cdot \binom{n}{(\frac{1}{4} + \frac{3\varepsilon}{2})n}$ since $y < (\frac{1}{4} + \frac{3\varepsilon}{2}) n$ and $\varepsilon < \frac{1}{6}$. Hence, on these subbranches we spent

$$T_2(n) = \mathcal{O}^* \left(\binom{n}{(\frac{1}{4} + \frac{3\varepsilon}{2}) n} \right)$$

time in total. Finally, for every subbranch with $y < 2\varepsilon n$ we have spent at most $\mathcal{O}^*(\binom{(\frac{1}{2} + \varepsilon)n - y}{(\frac{1}{2} - \varepsilon)n})$ time. As the number of such branches is bounded by $(n+1) \cdot \binom{n}{y}$, the total time spent on these branches is

$$T_3(n) = \mathcal{O}^* \left(\max_{y < 2\varepsilon n} \left(\binom{n}{y} \cdot \binom{(\frac{1}{2} + \varepsilon)n - y}{(\frac{1}{2} - \varepsilon)n} \right) \right).$$

If we now let $\varepsilon = \frac{1}{10}$, then $T_1(n), T_2(n) = \mathcal{O}^*\left(\binom{n}{\frac{2}{5}n}\right) = \mathcal{O}^*(1.9602^n)$. It can be also easily shown that for any $y < \frac{1}{5}n$, it holds that $\binom{n}{y} \cdot \binom{\frac{3}{5}n-y}{\frac{2}{5}n} = \mathcal{O}^*(1.9602^n)$. To prove this, we can use the following simple combinatorial bound: $\binom{n_1}{k_1} \cdot \binom{n_2}{k_2} \le \binom{n_1+n_2}{k_1+k_2}$. This inequality can be proved by combinatorial interpretation as follows: every choice of k_1 elements from a set of size n_1 and of k_2 elements from a set of size n_2, defines uniquely a choice of $k_1 + k_2$ elements from the union of these sets, which is of size $n_1 + n_2$. Therefore, we obtain: $\binom{n}{y} \cdot \binom{\frac{3}{5}n-y}{\frac{2}{5}n} = \binom{n}{y} \cdot \binom{\frac{3}{5}n-y}{\frac{1}{5}n-y} \le \binom{\frac{8}{5}n-y}{\frac{1}{5}n} \le \binom{\frac{8}{5}n}{\frac{1}{5}n} = \mathcal{O}^*(1.828^n)$. Consequently, $T_1(n), T_2(n), T_3(n) = \mathcal{O}^*(1.9602^n)$, and the whole algorithm runs in $\mathcal{O}^*(1.9602^n)$ time.

4   Conclusion

In this work we gave the first exact algorithm computing the tree-depth of a graph faster than $\mathcal{O}^*(2^n)$. As Bodlaender et al. [3] observe, both pathwidth and treewidth can be reformulated as vertex ordering problems and thus computed by a simple dynamic programming algorithm similar to the classical Held-Karp algorithm in time $\mathcal{O}^*(2^n)$ [9]. For example, computing the optimum value of treewidth is equivalent to finding an elimination ordering which minimizes the sizes of cliques created during the elimination process. As far as tree-depth is concerned, Nešetřil and Ossona de Mendez [17] give an alternative definition of tree-depth in terms of weak-colorings, which in turn are defined also via vertex orderings; however, it is unclear whether this definition can be used for an algorithm working in $\mathcal{O}^*(2^n)$ time. Interestingly enough, for many of vertex ordering problems, like Hamiltonicity, treewidth, or pathwidth, an explicit algorithm working in time $\mathcal{O}^*(c^n)$ for some $c < 2$ can be designed, see [1, 7, 21]. On the other hand, for several other vertex ordering problems no such algorithms are known. The two natural problems to attack are (i) the computation of cutwidth, and (ii) the Minimum Feedback Arc Set in Digraph problem; see [3, 5] for definitions and details. It is known that the cutwidth of a graph can be computed in time $\mathcal{O}^*(2^t)$, where t is the size of a vertex cover in the graph [5]; thus the problem is solvable in time $\mathcal{O}^*(2^{n/2})$ on bipartite graphs. We leave existence of faster exponential algorithms for these problems as an open question.

References

1. Björklund, A.: Determinant sums for undirected hamiltonicity. In: Proceedings of the 51st Annual IEEE Symposium on Foundations of Computer Science (FOCS 2010), pp. 173–182. IEEE (2010)
2. Bodlaender, H.L., Deogun, J.S., Jansen, K., Kloks, T., Kratsch, D., Müller, H., Tuza, Z.: Rankings of graphs. SIAM J. Discrete Math. 11(1), 168–181 (1998)
3. Bodlaender, H.L., Fomin, F.V., Koster, A.M.C.A., Kratsch, D., Thilikos, D.M.: A note on exact algorithms for vertex ordering problems on graphs. Theory Comput. Syst. 50(3), 420–432 (2012)

 4. Bodlaender, H.L., Gilbert, J.R., Hafsteinsson, H., Kloks, T.: Approximating treewidth, pathwidth, frontsize, and shortest elimination tree. J. Algorithms 18(2), 238–255 (1995)
 5. Cygan, M., Lokshtanov, D., Pilipczuk, M., Pilipczuk, M., Saurabh, S.: On cutwidth parameterized by vertex cover. Algorithmica, 1–14 (2012)
 6. Deogun, J.S., Kloks, T., Kratsch, D., Müller, H.: On the vertex ranking problem for trapezoid, circular-arc and other graphs. Discrete Applied Mathematics 98(1-2), 39–63 (1999)
 7. Fomin, F.V., Kratsch, D., Todinca, I., Villanger, Y.: Exact algorithms for treewidth and minimum fill-in. SIAM J. Comput. 38(3), 1058–1079 (2008)
 8. Fomin, F.V., Villanger, Y.: Treewidth computation and extremal combinatorics. Combinatorica 32(3), 289–308 (2012)
 9. Held, M., Karp, R.M.: A dynamic programming approach to sequencing problems. Journal of SIAM 10, 196–210 (1962)
10. Katchalski, M., McCuaig, W., Seager, S.M.: Ordered colourings. Discrete Mathematics 142(1-3), 141–154 (1995)
11. Kitsunai, K., Kobayashi, Y., Komuro, K., Tamaki, H., Tano, T.: Computing directed pathwidth in $O(1.89^n)$ time. In: Thilikos, D.M., Woeginger, G.J. (eds.) IPEC 2012. LNCS, vol. 7535, pp. 182–193. Springer, Heidelberg (2012)
12. Kloks, T., Müller, H., Wong, C.K.: Vertex ranking of asteroidal triple-free graphs. Inf. Process. Lett. 68(4), 201–206 (1998)
13. Nešetřil, J., de Mendez, P.O.: Tree-depth, subgraph coloring and homomorphism bounds. Eur. J. Comb. 27(6), 1022–1041 (2006)
14. Nešetřil, J., de Mendez, P.O.: Grad and classes with bounded expansion I. Decompositions. Eur. J. Comb. 29(3), 760–776 (2008)
15. Nešetřil, J., de Mendez, P.O.: Grad and classes with bounded expansion II. Algorithmic aspects. Eur. J. Comb. 29(3), 777–791 (2008)
16. Nešetřil, J., de Mendez, P.O.: Grad and classes with bounded expansion III. Restricted graph homomorphism dualities. Eur. J. Comb. 29(4), 1012–1024 (2008)
17. Nešetřil, J., de Mendez, P.O.: Sparsity - Graphs, Structures, and Algorithms. Algorithms and combinatorics, vol. 28. Springer (2012)
18. Robertson, N., Seymour, P.D.: Graph Minors. XIII. The Disjoint Paths Problem. J. Comb. Theory, Ser. B 63(1), 65–110 (1995)
19. Robertson, N., Seymour, P.D.: Graph Minors. XX. Wagner's conjecture. J. Comb. Theory, Ser. B 92(2), 325–357 (2004)
20. Schäffer, A.A.: Optimal Node Ranking of Trees in Linear Time. Inf. Process. Lett. 33(2), 91–96 (1989)
21. Suchan, K., Villanger, Y.: Computing pathwidth faster than 2^n. In: Chen, J., Fomin, F.V. (eds.) IWPEC 2009. LNCS, vol. 5917, pp. 324–335. Springer, Heidelberg (2009)

Faster Exact Algorithms
for Some Terminal Set Problems

Rajesh Chitnis[1,*], Fedor V. Fomin[2], Daniel Lokshtanov[2], Pranabendu Misra[3],
M.S. Ramanujan[3], and Saket Saurabh[2,3]

[1] University of Maryland, USA
`rchitnis@cs.umd.edu`
[2] University of Bergen, Norway
`{fedor.fomin,daniello}@ii.uib.no`
[3] Institute of Mathematical Sciences, India
`{pranabendu,msramanujan,saket}@imsc.res.in`

Abstract. Many problems on graphs can be expressed in the following language: given a graph $G = (V, E)$ and a terminal set $T \subseteq V$, find a minimum size set $S \subseteq V$ which intersects all "structures" (such as cycles or paths) passing through the vertices in T. We call this class of problems as *terminal set problems*. In this paper we introduce a general method to obtain faster exact exponential time algorithms for many terminal set problems. More precisely, we show that

- NODE MULTIWAY CUT can be solved in time $O(1.4766^n)$.
- DIRECTED UNRESTRICTED NODE MULTIWAY CUT can be solved in time $O(1.6181^n)$.
- There exists a deterministic algorithm for SUBSET FEEDBACK VERTEX SET running in time $O(1.8980^n)$ and a randomized algorithm with expected running time $O(1.8826^n)$. Furthermore, SUBSET FEEDBACK VERTEX SET on chordal graphs can be solved in time $O(1.6181^n)$.
- DIRECTED SUBSET FEEDBACK VERTEX SET can be solved in time $O(1.9993^n)$.

A key feature of our method is that, it uses the existing best polynomial time, fixed parameter tractable and exact exponential time algorithms for the non-terminal version of the same problem (i.e. when $T = V$), as subroutines. Therefore faster algorithms for these special cases will imply further improvements in the running times of our algorithms. Our algorithms for NODE MULTIWAY CUT, and SUBSET FEEDBACK VERTEX SET on chordal graphs improve the current best algorithms for these problems and answers an open question posed in [15]. Furthermore, our algorithms for DIRECTED UNRESTRICTED NODE MULTIWAY CUT and DIRECTED SUBSET FEEDBACK VERTEX SET are the first exact algorithms improving upon the brute force $O^*(2^n)$-algorithms.

1 Introduction

The goal of the design of moderately exponential time algorithms for NP-complete problems is to establish algorithms for which the worst-case running time is provably

---

* Supported in part by NSF CAREER award 1053605, ONR YIP award N000141110662, DARPA/AFRL award FA8650-11-1-7162 and a Simons Award for Graduate Students in Theoretical Computer Science.

G. Gutin and S. Szeider (Eds.): IPEC 2013, LNCS 8246, pp. 150–162, 2013.

faster than the one of enumerating all prospective solutions, or loosely speaking, algorithms better than brute-force enumeration. For example, for NP-complete problems on graphs on n vertices and m edges whose solutions are either subsets of vertices or edges, the brute-force or trivial algorithms basically enumerate all subsets of vertices or edges. This mostly leads to algorithms of time complexity $O^*(2^n)$ or $O^*(2^m)$, based on whether we are enumerating vertices or edges[1]. Thus the goal of exact algorithms for graph problems is to improve upon the algorithms running in time $O^*(2^n)$ or $O^*(2^m)$. See the book [11] for an introduction to exact exponential algorithms.

One of the most well studied directions in exact algorithms is to delete vertices of the input graph such that the resulting graph satisfies some interesting properties. More precisely, a natural optimization problem associated with a graph class $\mathcal{G}$ is the following: given a graph G, what is the minimum number of vertices to be deleted from G to obtain a graph in $\mathcal{G}$? For example, when $\mathcal{G}$ is the class of empty graphs, forests or bipartite graphs, the corresponding problems are VERTEX COVER (VC), FEEDBACK VERTEX SET (FVS) and ODD CYCLE TRANSVERSAL (OCT), respectively. The best known algorithms for VC, FVS and OCT run in time $O^*(1.2108^n)$ [24], $O^*(1.7347^n)$ [12] and $O^*(1.4661^n)$ [20,24] respectively. The other problems in this class for which non-trivial exact algorithms are known include finding an induced r-regular subgraph [16], induced subgraph of bounded degeneracy [21] and induced subgraph of bounded treewidth [12].

In this paper we study another class of graph problems which we call as *terminal set problems*. In these problems, the input is a graph $G = (V, E)$ and a terminal set $T \subseteq V$, and the goal is to find a minimum size set $S \subseteq V$ that intersects certain structures such as cycles or paths passing through the vertices in T. In this paper we introduce a general method to obtain faster exact exponential time algorithms for many terminal set problems. The general algorithmic technique is the following. Let the size of the terminal set T be k. We first observe that the size of the optimum solution is at most k (or a function of k). Let S be an optimum solution to the problem and let $X = S \cap (V \setminus T)$. We guess X and delete it from G. Since $S \setminus X \subseteq T$, we create an auxiliary graph on T and determine the rest of the solution using either a known polynomial time algorithm, or a fixed parameter tractable algorithm, or a non-trivial exact algorithm for the non-terminal version (when $T = V$) of the same problem. We now provide a list of problems for which we give improved or new algorithms using our method together with a short overview of previous work on each application.

NODE MULTIWAY CUT *and* DIRECTED UNRESTRICTED NODE MULTIWAY CUT: A fundamental min-max theorem about connectivity in graphs is Menger's Theorem, which states that the maximum number of vertex disjoint paths between two vertices s and t, is equal to the minimum number of vertices whose removal separates these two vertices. Indeed, a maximum set of vertex disjoint paths between s, t and a minimum size set of vertices separating these two vertices can be computed in polynomial time. A known generalization of this theorem, commonly known as Mader's T-path Theorem [18] states that, given a graph G and a subset T of vertices, there are either k vertex disjoint paths with only the end points in T (such paths are called T-*paths*), or there is a vertex set of size at most $2k$ which intersects every T-path. Although computing a maximum set of vertex disjoint T-paths can be done in polynomial time by using

---

[1] Throughout this paper we use the O^* notation which suppresses polynomial factors.

matching techniques, the decision version of the dual problem of finding a minimum set of vertices that intersects every T-path is NP-complete for $|T| > 2$. Formally, this problem is the following classical NODE MULTIWAY CUT problem.

NODE MULTIWAY CUT (NMC)

Input: An undirected graph $G = (V, E)$ and a set of terminals $T = \{t_1, t_2, \ldots, t_k\}$.
Question: Find a set $S \subseteq V(G) \setminus T$ of minimum size such that $G \setminus S$ has no path between a t_i, t_j pair for any $i \neq j$.

This is a very well studied problem in terms of approximation, as well as parameterized algorithms [5,8,14,19]. A variant of this problem where S is allowed to intersect the set T, is known as UNRESTRICTED NODE MULTIWAY CUT (UNMC). The best known parameterized algorithm for NODE MULTIWAY CUT decides in time $O^*(2^\ell)$ whether there is a solution of size at most ℓ or not. Fomin et al. [10] designed an exact algorithm for UNMC running in time $O^*(1.8638^n)$. In this paper we design an algorithm with running time $O(1.4766^n)$ for both NMC and UNMC.

Next we consider the directed variant of NODE MULTIWAY CUT, namely DIRECTED NODE MULTIWAY CUT (DNMC) where the input is a directed graph and a set $T = \{t_1, \ldots, t_k\}$ of terminals and the objective is to find a set of minimum size which intersects every $t_i \to t_j$ path for every $t_i, t_j \in T$ with $i \neq j$. For the *unrestricted* version of this problem, namely DIRECTED UNRESTRICTED NODE MULTIWAY CUT (DUNMC), we design an exact algorithm with running time $O(1.6181^n)$.

SUBSET FEEDBACK VERTEX SET *and* DIRECTED SUBSET FEEDBACK VERTEX SET: An exact algorithm for FEEDBACK VERTEX SET (FVS) – finding a minimum sized vertex subset such that its removal results in an acyclic graph – remained elusive for several years. In a breakthrough paper Razgon [22] designed an exact algorithm for this problem running in time $O^*(1.8899^n)$. Later, Fomin et al. [9] building upon the work in [22] designed an exact algorithm for FVS running in time $O^*(1.7548^n)$. The current best known algorithm for this problem uses potential maximal clique machinery and runs in time $O^*(1.7347^n)$ [13]. Razgon studied the directed version of FVS and obtained an exact algorithm running in time $O^*(1.9977^n)$ [23]. This is the only known non-trivial exact algorithm for DIRECTED FEEDBACK VERTEX SET (DFVS). A natural generalization of the FEEDBACK VERTEX SET problem is when we only want to hit all the cycles passing through a specified set of terminals. This leads to the following problem.

SUBSET FEEDBACK VERTEX SET (SFVS)

Input: An undirected graph $G = (V, E)$, a set of terminals $T \subseteq V$ of size k
Question: Find a minimum set of vertices which hits every cycle passing through T

Fomin et al. [10] designed an algorithm for SFVS on general graphs which runs in time $O^*(1.8638^n)$. It is important to note that their algorithm not only finds a minimum sized solution, but also enumerates all minimal solutions in the same time. Using our methodology we design an algorithm for SFVS which runs in time $O^*(1.8980^n)$. However, if we are allow randomization then we can design an algorithm with an expected running time of $O^*(1.8826^n)$. Golovach et al. [15] initiated the study of exact algorithms for SFVS on special graph classes by giving an enumeration algorithm for SFVS on chordal graphs which runs in time $O^*(1.6708^n)$. They left it as an open question whether there exists algorithms for SFVS on chordal graphs (even on split graphs) which are faster than $O^*(1.6708^n)$. Though our algorithm using the described methodology for SFVS in general does not improve on the best known algorithm, it answers this question in the affirmative for SFVS on chordal graphs and we obtain an algorithm with running time $O(1.6181^n)$. More generally, our algorithm for SFVS runs in $O(1.6181^n)$ on any graph class $\mathcal{G}$ which is closed under vertex deletions and edge contractions, and where the weighted FVS problem can be solved in polynomial time. Finally, we also consider the directed variant of the SFVS problem, namely DIRECTED SUBSET FEEDBACK VERTEX SET (DSFVS), and obtain an algorithm with running time $O(1.9993^n)$.

2   Preliminaries

Let C be a cycle in a graph G. A *chord* of C is an edge $e \notin C$ which connects two vertices of C. A graph G is called a *chordal graph* if every cycle on four or more vertices has a *chord*.

Now we define the *contraction* of an edge or a subgraph in G. Let G be an undirected graph and let (u, v) be an edge in G. Let G' be the graph obtained from G in the following manner. We add a new vertex w. For every edge (u, z) where $z \neq v$ we add an edge (w, z), and for every edge (y, v) where $y \neq u$ we add an edge (y, w). Finally we delete the vertices u and v, and any parallel edges from the graph. We say that G' is obtained from G by *contracting* the edge (u, v). Let H be a subgraph of G. Consider the graph G' obtained from G by contracting every edge of H in an arbitrary order. We say that G' is obtained from G by contracting the subgraph H.

Now we define the *torso graph* of a subset of vertices in G. Let $G = (V, E)$ be an undirected graph and T and V' be subsets of V. We denote by $\texttt{torso}(T, V')$ the graph defined in the following manner. The vertex set of this graph is T and the edge set comprises of all pairs (t_i, t_j) such that there is a $t_i - t_j$ path in G whose internal vertices lie in $V' \setminus T$ or there is an edge $(t_i, t_j) \in E$.

3   NODE MULTIWAY CUT

In this section we design an exact algorithm for the NODE MULTIWAY CUT problem. We begin by giving an algorithm for unrestricted version of this problem.

3.1   UNRESTRICTED NODE MULTIWAY CUT

The following observation follows from the fact that the set of terminals in an instance of UNRESTRICTED NODE MULTIWAY CUT itself is a potential solution.

Observation 1. *Let* (G, T) *be an instance of* UNRESTRICTED NODE MULTIWAY CUT *and S be an optimum solution to this instance . Then $|S| \leq |T|$.*

Now we design an algorithm for UNRESTRICTED NODE MULTIWAY CUT using the FPT algorithm for NODE MULTIWAY CUT and Observation 1. This algorithm uses the FPT algorithm for multiway cut of Cygan et al.[8]. We will use this algorithm for the instances where k is"small".

Lemma 1. [$\star$][2] *Let (G, T) be an instance of* UNRESTRICTED NODE MULTIWAY CUT *where $|T| = k$. Then we can find an optimum solution to this instance in time $O^*(2^k)$.*

Next, we design another algorithm for UNRESTRICTED NODE MULTIWAY CUT which will be used for the instances where k is "large".

Lemma 2. *Let (G, T) be an instance of* UNRESTRICTED NODE MULTIWAY CUT *where $G = (V, E)$ and let S be an optimum solution to this instance. Let $X = S \cap (V \setminus T)$ and $Y = S \setminus X$. Then Y is a vertex cover of the graph $\texttt{torso}(T, V \setminus X)$. Conversely, if Y' is any vertex cover for the graph $\texttt{torso}(T, V \setminus X)$, then the set $X \cup Y'$ is a solution to this instance.*

Proof. We first show that Y is indeed a vertex cover of $G' = \texttt{torso}(T, V \setminus X)$. Let E' be the edge set of G'. Suppose that Y is not a vertex cover of G' and there is an edge $(t_i, t_j) \in E'$ which is not covered by Y. Observe that $(t_i, t_j) \notin E$, since this would contradict the assumption of S being a solution. Therefore, it must be the case that there is a path P between t_i and t_j in $G[V \setminus X]$ whose internal vertices are disjoint from T. Since this path is disjoint from both X and Y, it is also present in the graph $G \setminus S$, a contradiction. Hence, we conclude that Y is indeed a vertex cover of $\texttt{torso}(T, V \setminus X)$.

We now show that for any vertex cover Y' of G', the set $X \cup Y'$ is a solution to the instance (G, T) of UNRESTRICTED NODE MULTIWAY CUT. Suppose to the contrary that there is a vertex cover Y' of G' such that the set $S' = X \cup Y'$ is not a solution to the instance (G, T). That is, there is a t_i-t_j path in $G \setminus S'$ for some $t_i, t_j \in T$. Observe that this implies the existence of a $t_{i'}$-$t_{j'}$ path P for some $t_{i'}, t_{j'} \in T$ such that the internal vertices of P are disjoint from $T \cup S'$. Therefore the edge $(t_{i'}, t_{j'})$ is not covered by Y' in G', a contradiction. east one of t_i or t_j must be in the set Y'. $\square$

Using the above lemma and the FPT algorithm for Vertex Cover of Chen et al.[4], we are able to show the following lemma.

Lemma 3. *There is an algorithm that, given an instance $(G = (V, E), T)$ of* UNRESTRICTED NODE MULTIWAY CUT, *runs in time $O\left(1.7850^n \left(\frac{1.2738}{1.7850}\right)^k\right)$ and returns an optimum solution where $k = |T|$ and $n = |V|$.*

Proof. The description of the algorithm is as follows. For every $X \subseteq (V \setminus T)$ such that $|X| \leq k$, we construct the graph $G_X = \texttt{torso}(T, V \setminus X)$ and compute a minimum vertex cover Y_X for G_X. We compute the minimum vertex cover by using the FPT algorithm of [4], which runs in time $O^*(1.2738^\ell)$ where ℓ is the size of an optimum vertex cover. Finally, we return the set $X \cup Y_X$ which is a smallest solution over all choices of X. The correctness of this algorithm follows from Lemma 2.

---

[2] The proofs of results labeled with $\star$ are deferred to the full version of the paper due to space constraints.

In order to bound the running time of this algorithm, first observe that for each X, the set Y_X has size at most $k - |X|$. Therefore, the FPT algorithm we use to compute a minimum vertex cover of $\texttt{torso}(T, V \setminus X)$ runs in time $O^*(1.2738^{k-|X|})$. Summing over all choices of X, the time taken by our algorithm is upper bounded by

$$\sum_{x=0}^{k} \binom{n-k}{x} O^*(1.2738^{k-x})$$

$$= O^*(1.2738^k) \sum_{x=0}^{k} \binom{n-k}{x} \left(\frac{1}{1.2738}\right)^x$$

$$= O^*(1.2738^k) \left(1 + \frac{1}{1.2738}\right)^{n-k}$$

$$= O\left(1.7850^n \left(\frac{1.2738}{1.7850}\right)^k\right) \qquad \square$$

Now we are ready to prove the main theorem of this section.

Theorem 2. *There is an algorithm that, given an instance $(G = (V, E), T)$ of UNRESTRICTED NODE MULTIWAY CUT, runs in time $O(1.4766^n)$ and returns an optimum solution where $n = |V|$.*

Proof. Let (G, T) be the given instance of UNRESTRICTED NODE MULTIWAY CUT and $|T| = k$. Recall that we have described two different algorithms for UNRESTRICTED NODE MULTIWAY CUT. We now choose either of these algorithms based on the values of k and n. If $k \leq 0.5622n$, then we use the algorithm described in Lemma 1. In this case, the running time is upper bounded by $O^*(2^k) \leq O^*(2^{0.5622n}) \leq O(1.4766^n)$. If $k > 0.5622n$, then we use the algorithm described in Lemma 3. This algorithm runs in time $O\left(1.7850^n \left(\frac{1.2738}{1.7850}\right)^k\right)$ which is a decreasing function of k. Substituting $k = 0.5622n$, we get an upper bound on the running time as $O(1.4766^n)$. This completes the proof of the theorem. $\qquad \square$

3.2   NODE MULTIWAY CUT

In this subsection, we give an exact algorithm for the NODE MULTIWAY CUT problem. We start with the following observation which follows from the fact that any solution to an instance of NODE MULTIWAY CUT is disjoint from the set of terminals in the instance.

Observation 3. *Let (G, T) be an instance of NODE MULTIWAY CUT. If T is not an independent set in G, then there is no solution to the instance (G, T). Furthermore, if two terminals t_1 and t_2 have a common neighbor v, then v must be in every solution for the given instance.*

Due to Observation 3, we may assume that the terminal set is independent and the neighborhoods of the terminals in G are pairwise disjoint. This reduces the restricted NODE MULTIWAY CUT to the following generalization of the UNRESTRICTED NODE MULTIWAY CUT, also known as the GROUP MULTIWAY CUT problem.

GROUP MULTIWAY CUT

Input: An undirected graph $G = (V, E)$ and pairwise disjoint sets of terminals $\{T_1, T_2, \ldots, T_\ell\}$.

Question: Find a set $S \subseteq V(G)$ of minimum size such that $G \setminus S$ has no $u - v$ path for any $u \in T_i, v \in T_j$ and $i \neq j$.

We have the following theorem, whose proof is along the same lines as the proof of Theorem 2.

Theorem 4. [$\star$] *There is an algorithm that, given an instance $(G = (V, E), T_1, \ldots, T_\ell)$ of* GROUP MULTIWAY CUT, *runs in time $O(1.4766^n)$ and returns an optimum solution, where $n = |V|$.*

The following theorem follows from Theorem 4 and Observation 3.

Theorem 5. *There is an algorithm that, given an instance $(G = (V, E), T)$ of* NODE MULTIWAY CUT, *runs in time $O(1.4766^n)$ and returns an optimum solution, where $n = |V|$.*

4 DIRECTED UNRESTRICTED NODE MULTIWAY CUT

In this section, we consider the DIRECTED UNRESTRICTED NODE MULTIWAY CUT problem.

DIRECTED UNRESTRICTED NODE MULTIWAY CUT

Input: A directed graph $D = (V, A)$ and a set of terminals $T = \{t_1, t_2, \ldots, t_k\}$.

Question: Find a set $S \subseteq V$ of minimum size such that $G \setminus S$ has no $t_i \rightarrow t_j$ path for any $i \neq j$.

Since we consider the version where the terminals can be deleted, we have the following observation.

Observation 6. *Let (G, T) be an instance of* DIRECTED UNRESTRICTED NODE MULTIWAY CUT *and S be an optimum solution to this instance. Then, $|S| \leq |T|$.*

The proof of the next lemma is identical to the proof of Lemma 2 and therefore, we do not repeat it.

Lemma 4. *Let (D, T) be an instance of* DIRECTED UNRESTRICTED NODE MULTIWAY CUT *where $D = (V, A)$ and let S be an optimum solution to this instance. Let $X = S \cap (V \setminus T)$ and $Y = S \cap X$. Then Y is a vertex cover of the graph $\texttt{torso}(T, V \setminus X)$. Conversely if Y' is any vertex cover of the graph $\texttt{torso}(T, V \setminus X)$, then the set $X \cup Y'$ is a solution to this instance.*

Now we describe our algorithm for DIRECTED UNRESTRICTED NODE MULTIWAY CUT.

Theorem 7. DIRECTED UNRESTRICTED NODE MULTIWAY CUT *can be solved in time $O^*(1.6181^n)$.*

Proof. The description of the algorithm is as follows. For every $X \subseteq (V \setminus T)$ such that $|X| \le k$, we construct the graph $D_X = \texttt{torso}(T, V \setminus X)$ and compute a minimum vertex cover Y_X for D_X. We compute the minimum vertex cover by using the FPT algorithm of Chen et al. [4], which runs in time $O^*(1.2738^\ell)$ where ℓ is the size of an optimum vertex cover. Finally, we return the set $X \cup Y_X$ which is a smallest solution over all choices of X. The correctness of this algorithm follows from Lemma 4.

Let $\mathcal{T}$ be the running time of our algorithm. We have the following claim.

Claim. $\mathcal{T} = O(1.6181^n)$.

For every choice of X we run the FPT algorithm for vertex cover, which takes time $O^*(1.2738^{k-|X|})$. Therefore we have,

Proof

$$
\begin{aligned}
\mathcal{T} &= \sum_{x=0}^{k} \binom{n-k}{x} O^*(1.2738^{k-x}) \\
&= \sum_{x=0}^{k} \binom{n-k}{x} O^*(1.618^{k-x}) \\
&= O^*(1.618^k) \sum_{x=0}^{k} \binom{n-k}{x} \left(\frac{1}{1.618}\right)^x \\
&= O^*(1.618^k) \left(\frac{1}{1.618} + 1\right)^{n-k} \\
&= O^*(1.618^k) \times (1.6181)^{n-k} \\
&= O(1.6181^n) \qquad\qquad\qquad\qquad \square
\end{aligned}
$$

This completes the proof of the theorem. $\qquad\qquad\qquad\qquad\qquad\qquad\qquad\qquad \square$

5   SUBSET FEEDBACK VERTEX SET

In this section we design an exact algorithm for SUBSET FEEDBACK VERTEX SET. We actually design two different algorithms for the problem, and then use these two algorithms to construct our final exact algorithm.

Let (G, T) be the given instance of SUBSET FEEDBACK VERTEX SET. Recall that we are allowed to pick terminal vertices into a solution. The following observation follows from the fact that the set of terminals itself is a solution.

Observation 8. *Let (G, T) be an instance of* SUBSET FEEDBACK VERTEX SET, *and let S be an optimum solution to this instance. Then $|S| \le |T|$.*

Lemma 5.   *1. There is an algorithm that, given an instance $(G = (V, E), T)$ of* SUB-SET FEEDBACK VERTEX SET, *runs in time $O\left(1.2^n \left(\frac{5}{12}\right)^k\right)$ and returns an optimum solution where $k = |T|$ and $n = |V|$.*

2. *There is an algorithm that, given an instance $(G = (V, E), T)$ of* SUBSET FEED-
   BACK VERTEX SET, *runs in time* $O\left(2^n \left(\frac{1.7548}{2}\right)^k\right)$ *and returns an optimum solu-
   tion where* $k = |T|$ *and* $n = |V|$.

Proof. For every $X \subseteq (V \setminus T)$ such that $|X| \leq k$, let T_X be the set of terminals t such that, $G \setminus X$ contains a cycle passing through t which contains no other terminal vertex. Let G_X be the graph obtained from $G \setminus (X \cup T_X)$ by contracting every connected component of $G \setminus (T \cup X)^3$. Let Y_X be a minimum feedback vertex set for G_X containing only vertices of T.

For the first algorithm, we compute Y_X in the following manner. We assign a weight of $k+1$ to the vertices not in T and 1 to the vertices in T. We then use an FPT algorithm to compute a minimum feedback vertex set of G_X of weight at most k. We use the FPT algorithm of Chen et. al. [3] which runs in time $O^*(5^p)$, where p is the minimum weight of an feedback vertex set.

For the second algorithm we compute Y_X by computing a maximum induced forest of G_X which contains all the non-terminal vertices and taking its complement. Let F be the set of all non-terminal vertices in G_X and let $q = n - |F|$ be the number of terminal vertices in G_X. We use the algorithm of Fomin et al. [9](Section 3) on (G_X, F) and compute a maximum induced forest of G_X containing F. The arguments of Fomin et al. imply that the algorithm runs in time $O^*(1.7548^q)$.

Let $S_X = X \cup T_X \cup Y_X$. We compute S_X for every X and output the one with the smallest number of vertices as our solution.

Correctness. The correctness of both the algorithm follow from the following claims.

Claim. [$\star$] Let S be an optimum solution to the given instance of SUBSET FEEDBACK VERTEX SET and let $X = S \cap (V \setminus T)$. Let T_X be the set of terminals t such that, $G \setminus X$ contains a cycle passing through t which contains no other terminal vertices. Then $T_X \subseteq S$.

The above claim shows the correctness of adding T_X to the solution. The following lemma shows that once T_X is added to the solution, it suffices to compute a minimum feedback vertex set for the graph G_X.

Claim. [$\star$] Let S be an optimum solution to the given instance of SUBSET FEEDBACK VERTEX SET and let $X = S \setminus T$ and $Y = S \setminus X$. Furthermore, suppose that there are no cycles in G containing a unique vertex of T. Let G_X be obtained from $G \setminus X$ by contracting every connected component of $G \setminus (T \cup X)$. Then Y is a minimum feedback vertex set of G_X. Conversely if Y' is any feedback vertex set of G_X, then the set $X \cup Y'$ is a solution for the given instance of SUBSET FEEDBACK VERTEX SET.

Running Time. Let $\mathcal{T}_1$ be the running time of the first algorithm and, $\mathcal{T}_2$ be the running time of the second algorithms. The following two claims establish the running times of both the algorithms.

Claim. [$\star$] $\mathcal{T}_1 = O\left(1.2^n \times \left(\frac{5}{1.2}\right)^k\right)$.

---

[3] We can compute both T_X and G_X in polynomial time.

Claim. [⋆] $T_2 = O\left(2^n \times \left(\frac{1.7548}{2}\right)^k\right)$.

This completes the proof of the lemma.   □

Theorem 9. *There is an algorithm that, given an instance $(G = (V, E), T)$ of* SUBSET FEEDBACK VERTEX SET, *runs in time $O(1.9161^n)$ and returns an optimum solution where $n = |V|$.*

Proof. Let (G, T) be the given instance of SUBSET FEEDBACK VERTEX SET, where G is a graph on n vertices and $|T| = k$. Based on the values of n and k we run one of the two algorithms described above.

If $k \leq 0.32789n$, then we run the first algorithm described in Lemma 5. The running time is upper bounded by $O\left(1.2^n \times \left(\frac{5}{1.2}\right)^{0.32789n}\right) = O(1.9161^n)$. Otherwise if $k > 0.32789n$, then we run the second algorithm described in Lemma 5. This algorithm runs in time $O\left(2^n \times \left(\frac{1.7548}{2}\right)^k\right)$ which is a decreasing function of k. Substituting $k = 0.32789n$, we get an upper bound of $O(1.9161^n)$ on the running time in this case as well.   □

We remark that, there are faster FPT [2,7] and Exact algorithms [12] known for FVS on undirected graphs (we can modify these algorithms in order to handle input instances with undeletable vertices). If we use the fastest known deterministic and randomized algorithms, then we obtain the following theorem.

Theorem 10. SUBSET FEEDBACK VERTEX SET *can be solved in $O^*(1.8980^n)$ deterministic time, or in $O^*(1.8826^n)$ randomized time.*

5.1 SUBSET FEEDBACK VERTEX SET on Chordal Graphs

In this section we give an algorithm for SUBSET FEEDBACK VERTEX SET on chordal graphs which improves upon the previous best algorithm of Golovach et al. [15], and is much simpler. The main difference between this algorithm and the algorithm for SUBSET FEEDBACK VERTEX SET described earlier is that we use a polynomial time algorithm to solve weighted FEEDBACK VERTEX SET on chordal graphs ([6,25]), instead of an FPT or an exact algorithm. It is well known (see also [1]) that chordal graphs are closed under vertex deletions and edge contractions.

We are now ready to prove the main theorem of this section:

Theorem 11. *There is an algorithm that, given an instance $(G = (V, E), T)$ of* SUBSET FEEDBACK VERTEX SET *on Chordal Graphs, returns an optimum solution in $O(1.6181^n)$ time, where $n = |V|$.*

Proof. The algorithm is the same as the two algorithms described in Lemma 5 except that we use the polynomial time algorithm for FEEDBACK VERTEX SET on chordal graphs instead of the FPT or the exact exponential algorithm. For every choice of X, we compute T_X and G_X in polynomial time. Observe that the graph G_X is obtained from G by vertex deletions and edge contractions, implying that G_X is also a chordal graph. Assign weight 1 to each terminal vertex and weight $k + 1$ to each non-terminal vertex, and compute in polynomial time a minimum weight feedback vertex set Y_X of G_X using the result of Corneil and Fonlupt [6]. We now analyze the running time of our algorithm.

Let S be any optimum solution and let $X = S \setminus T$. Observe that $|X| \leq |S| \leq |T|$. The next lemma 6 shows that the number of choices of X is at most $O(1.6181^n)$.

Lemma 6. [$\star$] *Let V be a set of n elements and T is a subset of V. Then the number of distinct sets $X \subseteq V$ such that $S \cap T = \emptyset$ and $|X| \leq |T|$ is $O(1.6181^n)$.*

Since after choosing X we do only a polynomial time computation, the running time of our algorithm is $O(1.6181^n)$.   □

We remark that we can use the above method to obtain faster exact algorithm for SUBSET FEEDBACK VERTEX SET on other graph classes, such as AT-free graphs [17], which are closed under vertex deletions and edge contractions, and FEEDBACK VERTEX SET is solvable in polynomial time on them.

6   DIRECTED SUBSET FEEDBACK VERTEX SET

In this section we give an exact algorithm for the DIRECTED SUBSET FEEDBACK VERTEX SET problem running in time $O^*(1.9993^n)$. The problem is defined as follows.

DIRECTED SUBSET FEEDBACK VERTEX SET
Input: A directed graph $D = (V, A)$ and a set of terminal vertices T of size k.
Question: Find a minimum set of vertices in D which intersects every cycle in D which contains at least one vertex of T.

Next we observe the following property of directed graphs.

Observation 12. *Let $D = (V, A)$ be a directed graph. For any vertex $v \in V$, the following holds: v belongs to a closed walk in D if and only if v belongs to a cycle in D.*

Lemma 7. *Let $(D = (V, A), T)$ be an instance of DIRECTED SUBSET FEEDBACK VERTEX SET. Let S be an optimum solution to this instance and $X = S \setminus T, Y = S \setminus X$. Furthermore, suppose that every cycle in $D \setminus X$ that intersects T, contains at least two vertices of T. Then Y is a feedback vertex set in the graph $\texttt{torso}(T, V \setminus X)$ if and only if $X \cup Y$ is a subset feedback vertex set for the instance (D, T).*

Proof. Suppose $X \cup Y$ is a solution in D where $Y \subseteq T$. If Y is not a feedback vertex set in $D_X = \texttt{torso}(T, V \setminus X)$, then there is a cycle C_X in $D_X \setminus Y$. From C_X in D_X we can obtain a closed walk C' in D in the following manner. We replace every edge (t_i, t_j) of C_X which is not present in A, with a path P_{ij} from t_i to t_j in $D \setminus X$ whose internal vertices lie in $V \setminus (T \cup X)$. Therefore we get a closed walk C' in $D \setminus (X \cup Y)$ which contains a terminal. By Observation 12, there is a cycle in D which passes through a terminal in D, which is not covered by $X \cup Y$. This is a contradiction.

Conversely, let Y be a feedback vertex set in D_X, but $X \cup Y$ is not a solution in D. Then there is a cycle C in $D \setminus (X \cup Y)$ and note that this cycle contains at least two vertices of T. Further assume that C is the shortest such cycle. Observe that every minimal subpath P_{ij} of C from terminals t_i to t_j whose internal vertices lie in $V \setminus T$, implies an edge (t_i, t_j) in D_X. Therefore we can obtain a cycle C' in D_X from C by replacing the subpath P_{ij} with the edge (t_i, t_j), for every pair of terminals t_i, t_j in C. Observe that this cycle is not covered by Y. This is a contradiction.

This completes the proof of the lemma.   □

The following observation is immediate since the set T forms a potential solution.

Observation 13. *Let (D, T) be an instance of* DIRECTED SUBSET FEEDBACK VER- TEX SET *and let S be an optimum solution for this instance. Then, $|S| \leq |T|$.*

We are now ready to prove the main theorem of this section.

Theorem 14. [⋆] *There is an algorithm that, given an instance $(D = (V, A), T)$ of* DIRECTED SUBSET FEEDBACK VERTEX SET, *runs in time $O(1.9993^n)$ and returns an optimum solution where $n = |V|$.*

7 Conclusion

We introduced a methodology of obtaining non-trivial exact exponential algorithms for several *terminal set problems*. We conclude with open problems which seems to be evasive to our approach. Designing an algorithm faster than $O^*(2^n)$ for DIRECTED NODE MULTIWAY CUT remains an interesting question. Another interesting problem is SUBSET ODD CYCLE TRANSVERSAL, where the task is to find a vertex subset of minimum size hitting all cycles of odd length containing at least one terminal. Again, the problem is trivially solvable in $O^*(2^n)$ but no faster algorithm for this problem is known. We conclude by remarking that an approach based on our methodology might result in such an algorithm since ODD CYCLE TRANSVERSAL is solvable in time $O^*(1.4661^n)$ [20,24]. Finally designing an algorithm for MULTICUT on both undi- rected and directed graphs, faster than the trivial $O^*(2^n)$ algorithm, remains an inter- esting open problem.

References

1. Belmonte, R., Golovach, P.A., Heggernes, P., van 't Hof, P., Kamiński, M., Paulusma, D.: Finding contractions and induced minors in chordal graphs via disjoint paths. In: Asano, T., Nakano, S.-i., Okamoto, Y., Watanabe, O. (eds.) ISAAC 2011. LNCS, vol. 7074, pp. 110– 119. Springer, Heidelberg (2011)
2. Cao, Y., Chen, J., Liu, Y.: On feedback vertex set new measure and new structures. In: Ka- plan, H. (ed.) SWAT 2010. LNCS, vol. 6139, pp. 93–104. Springer, Heidelberg (2010)
3. Chen, J., Fomin, F.V., Liu, Y., Lu, S., Villanger, Y.: Improved algorithms for feedback vertex set problems. J. Comput. Syst. Sci. 74(7), 1188–1198 (2008)
4. Chen, J., Kanj, I.A., Xia, G.: Improved upper bounds for vertex cover. Theoretical Computer Science 411(40), 3736–3756 (2010)
5. Chen, J., Liu, Y., Lu, S.: An Improved Parameterized Algorithm for the Minimum Node Multiway Cut Problem. Algorithmica 55(1), 1–13 (2009)

6. Corneil, D., Fonlupt, J.: The complexity of generalized clique covering. Discrete Applied Mathematics 22(2), 109–118 (1988–1989)
7. Cygan, M., Nederlof, J., Pilipczuk, M., Pilipczuk, M., van Rooij, J., Wojtaszczyk, J.: Solving connectivity problems parameterized by treewidth in single exponential time. In: 2011 IEEE 52nd Annual Symposium on Foundations of Computer Science (FOCS), pp. 150–159. IEEE (2011)
8. Cygan, M., Pilipczuk, M., Pilipczuk, M., Wojtaszczyk, J.O.: On multiway cut parameterized above lower bounds. In: Marx, D., Rossmanith, P. (eds.) IPEC 2011. LNCS, vol. 7112, pp. 1–12. Springer, Heidelberg (2012)
9. Fomin, F.V., Gaspers, S., Pyatkin, A.V., Razgon, I.: On the minimum feedback vertex set problem: Exact and enumeration algorithms. Algorithmica 52(2), 293–307 (2008)
10. Fomin, F.V., Heggernes, P., Kratsch, D., Papadopoulos, C., Villanger, Y.: Enumerating minimal subset feedback vertex sets. In: Dehne, F., Iacono, J., Sack, J.-R. (eds.) WADS 2011. LNCS, vol. 6844, pp. 399–410. Springer, Heidelberg (2011)
11. Fomin, F.V., Kratsch, D.: Exact Exponential Algorithms, 1st edn. Springer-Verlag New York, Inc., New York (2010)
12. Fomin, F.V., Villanger, Y.: Finding induced subgraphs via minimal triangulations. In: 27th International Symposium on Theoretical Aspects of Computer Science (STACS), vol. 5, pp. 383–394. Schloss Dagstuhl–Leibniz-Zentrum fuer Informatik (2010)
13. Fomin, F.V., Villanger, Y.: Finding induced subgraphs via minimal triangulations. In: STACS, vol. 5, pp. 383–394 (2010)
14. Garg, N., Vazirani, V., Yannakakis, M.: Multiway Cuts in Directed and Node Weighted Graphs. In: Shamir, E., Abiteboul, S. (eds.) ICALP 1994. LNCS, vol. 820, pp. 487–498. Springer, Heidelberg (1994)
15. Golovach, P.A., Heggernes, P., Kratsch, D., Saei, R.: An exact algorithm for subset feedback vertex set on chordal graphs. In: Thilikos, D.M., Woeginger, G.J. (eds.) IPEC 2012. LNCS, vol. 7535, pp. 85–96. Springer, Heidelberg (2012)
16. Gupta, S., Raman, V., Saurabh, S.: Maximum r-regular induced subgraph problem: Fast exponential algorithms and combinatorial bounds. SIAM J. Discrete Math. 26(4), 1758–1780 (2012)
17. Kratsch, D., Müller, H., Todinca, I.: Feedback vertex set on at-free graphs. Discrete Applied Mathematics 156(10), 1936–1947 (2008)
18. Mader, W.: Über die Maximalzahl kreuzungsfreier H-Wege. Arch. Math (Basel) 31(4), 387–402 (1978/1979), http://dx.doi.org/10.1007/BF01226465
19. Marx, D.: Parameterized Graph Separation Problems. Theor. Comput. Sci. 351(3), 394–406 (2006)
20. Mishra, S., Raman, V., Saurabh, S., Sikdar, S.: König deletion sets and vertex covers above the matching size. In: Hong, S.-H., Nagamochi, H., Fukunaga, T. (eds.) ISAAC 2008. LNCS, vol. 5369, pp. 836–847. Springer, Heidelberg (2008)
21. Pilipczuk, M., Pilipczuk, M.: Finding a maximum induced degenerate subgraph faster than 2 n. In: Thilikos, D.M., Woeginger, G.J. (eds.) IPEC 2012. LNCS, vol. 7535, pp. 3–12. Springer, Heidelberg (2012)
22. Razgon, I.: Exact computation of maximum induced forest. In: Arge, L., Freivalds, R. (eds.) SWAT 2006. LNCS, vol. 4059, pp. 160–171. Springer, Heidelberg (2006)
23. Razgon, I.: Computing Minimum Directed Feedback Vertex Set in $O^*(1.9977^n)$. In: ICTCS, pp. 70–81 (2007)
24. Robson, J.M.: Algorithms for maximum independent sets. J. Algorithms 7(3), 425–440 (1986)
25. Yannakakis, M., Gavril, F.: The maximum k-colorable subgraph problem for chordal graphs. Information Processing Letters 24(2), 133–137 (1987)

Parameterized Algorithms for Modular-Width

Jakub Gajarský[1,*], Michael Lampis[2], and Sebastian Ordyniak[1,**]

[1] Faculty of Informatics, Masaryk University, Brno, Czech Republic
{gajarsky,ordyniak}@fi.muni.cz
[2] KTH Royal Institute of Technology, Stockholm, Sweden
mlampis@kth.se

Abstract. It is known that a number of natural graph problems which are FPT parameterized by treewidth become W-hard when parameterized by clique-width. It is therefore desirable to find a different structural graph parameter which is as general as possible, covers dense graphs but does not incur such a heavy algorithmic penalty.

The main contribution of this paper is to consider a parameter called modular-width, defined using the well-known notion of modular decompositions. Using a combination of ILP and dynamic programming we manage to design FPT algorithms for Coloring and Partitioning into paths (and hence Hamiltonian path and Hamiltonian cycle), which are W-hard for both clique-width and its recently introduced restriction, shrub-depth. We thus argue that modular-width occupies a sweet spot as a graph parameter, generalizing several simpler notions on dense graphs but still evading the "price of generality" paid by clique-width.

1   Introduction

The topic of this paper is the exploration of the algorithmic properties of some structural graph parameters. This area is typically dominated by an effort to achieve two competing goals: generality and algorithmic tractability. A good example of this tension is the contrast between treewidth and clique-width.

A large wealth of problems are known to be FPT when parameterized by treewidth [6,5,4]. One drawback of treewidth, however, is that this parameterization excludes a large number of interesting instances, since, in particular, graphs of small treewidth are necessarily sparse. The notion of clique-width (and its cousins rank-width [22] and boolean-width [7]) tries to ameliorate this problem by covering a significantly larger family of graphs, including many dense graphs. As it turns out though, the price one has to pay for this added generality is significant. Several natural problems which are known to be fixed-parameter tractable for treewidth become W-hard when parameterized by these measures [18,17,16].

---

* Research funded by the Czech Science Foundation under grant P202/11/0196
** Research funded by Employment of Newly Graduated Doctors of Science for Scientific Excellence (CZ.1.07/2.3.00/30.0009).

G. Gutin and S. Szeider (Eds.): IPEC 2013, LNCS 8246, pp. 163–176, 2013.

It thus becomes an interesting problem to explore the trade-offs offered by these and other graph parameters. More specifically, one may ask: is there a natural graph parameter which covers dense graphs but still allows FPT algorithms for the problems lost to clique-width? This is the main question motivating this paper. We first attempt to use the recently introduced notion of shrub-depth for this role [20]. Shrub-depth is a restriction of clique-width which shows some hope, since it has been used to obtain improved algorithmic meta-theorems. Unfortunately, as we will establish, the hardness constructions for COLORING and HAMILTONIAN PATH used in [18] go through with small modifications for this restricted parameter as well.

The main contribution of this paper is then the consideration of a parameter called modular-width which, we argue, nicely fills this niche. One way to define modular-width is by using the standard concept of modular decompositions (see e.g. [24]), as the maximum degree of the optimal modular decomposition tree. As a consequence, a graph's modular-width can be computed in polynomial time. Note that the concept of modular-width was already briefly considered in [8], but was then abandoned in that paper in favor of the more general clique-width. To the best of our knowledge, modular-width has not been considered as a parameter again, even though modular decompositions have found a large number of algorithmic applications, including in parameterized complexity (see [21] for a general survey and [26,10,1] for example applications in parameterized complexity).

We give here the first evidence indicating that modular-width is a structural parameter worthy of further study. On the algorithmic side, modular-width offers a significant advantage compared to clique-width, a fact we demonstrate by giving FPT algorithms for several variants of HAMILTONICITY and CHROMATIC NUMBER, all problems W-hard for clique-width. At the same time, we show that modular-width significantly generalizes several simpler parameters, such as neighborhood diversity [23] and twin-cover [19], which also allow FPT algorithms for these problems.

Our main algorithmic tool is a form of dynamic programming on the modular decomposition of the input graph. Unlike dynamic programming on the more standard tree decompositions, the main obstacle here is in combining the DP tables of the children of a node to compute the table for the node itself. This is in general a hard problem, but we show that it can sometimes be made tractable if every node of the decomposition has small degree, hence the parameterization by modular-width.

Even if the modular decomposition has small degree, combining the DP tables is still not necessarily a trivial problem. A second idea we rely on (in the case of HAMILTONICITY) is to use an Integer Linear Program, whose number of variables is bounded by the number of modules we are trying to combine. It is our hope that this technique, which seems quite general, will be applicable to other problems as well.

Full Version Proofs of statements marked with $(\star)$ are shortened or omitted due to space restrictions. Detailed proofs can be found in the full version, available at arxiv.org/abs/1308.2858.

2 Preliminaries

We use standard notation from graph theory as can be found in, e.g., [9]. Let G be a graph. We denote the vertex set of G by $V(G)$ and the edge set of G by $E(G)$. Let $X \subseteq V(G)$ be a set of vertices of G. The *subgraph of G induced by X*, denoted $G[X]$, is the graph with vertex set X and edges $E(G) \cap [X]^2$. By $G \setminus X$ we denote the subgraph of G induced by $V(G) \setminus X$. Similarly, for $Y \subseteq E(G)$ we define $G \setminus Y$ to be the subgraph of G obtained by deleting all edges in Y from G. For a graph G and a vertex $v \in V(G)$, we denote by $N_G(v)$ and $N_G[v]$ the open and closed neighborhood of v in G, respectively.

2.1 Considered Problems

We consider the following problems on graphs. Let G be a graph. A *coloring* of G is a function $\lambda : V(G) \to \mathbb{N}$ such that for every edge $\{u, v\} \in E(G)$ it holds that $\lambda(u) \neq \lambda(v)$. We denote by $\lambda(G)$ the set of colors used by the coloring λ, i.e., $\lambda(G) = \{\, \lambda(v) : v \in V(G) \,\}$, and by $\Lambda(G)$ the set of all colorings of G that use at most $|V(G)|$ colors. The *chromatic number* of G, denoted by $\chi(G)$, is the smallest number c such that G has a coloring λ with $|\lambda(G)| \leq c$.

GRAPH COLORING
Input: A graph G.
Question: Compute $\chi(G)$.

Let G be a graph. A partition of G into paths is a set of (vertex-)disjoint paths of G whose union contains every vertex of G. We denote by $\mathrm{ham}(G)$ the least integer p such that G has a partition into p paths.

PARTITIONING INTO PATHS
Input: A graph G.
Question: Compute $\mathrm{ham}(G)$.

HAMILTONIAN PATH
Input: A graph G.
Question: Does G have a Hamiltonian Path?

HAMILTONIAN CYCLE
Input: A graph G.
Question: Does G have a Hamiltonian Cycle?

2.2 Parameterized Complexity

Here we introduce the relevant concepts of parameterized complexity theory. For more details, we refer to text books on the topic [12,15,25]. An instance of a parameterized problem is a pair (I, k) where I is the main part of the instance, and k is the parameter. A parameterized problem is *fixed-parameter tractable* if instances (I, k) can be solved in time $f(k)|I|^c$, where f is a computable function of k, and c is a constant. FPT denotes the class of all fixed-parameter tractable problems. Hardness for parameterized complexity classes is based on *fpt-reductions*. A parameterized problem L is fpt-reducible to another parameterized problem L' if there is a mapping R from instances of L to instances of L' such that (i) $(I, k) \in L$ if and only if $(I', k') = R(I, k) \in L'$, (ii) $k' \le g(k)$ for a computable function g, and (iii) R can be computed in time $O(f(k)|I|^c)$ for a computable function f and a constant c. Central to the completeness theory of parameterized complexity is the hierarchy FPT $\subseteq$ W[1] $\subseteq$ W[2] $\subseteq \ldots$. Each intractability class W[t] contains all parameterized problems that can be reduced to a certain parameterized satisfiability problem under fpt-reductions.

2.3 Treewidth

The treewidth of a graph is defined using the following notion of a tree decomposition (see, e.g., [3]). A *tree decomposition* of an (undirected) graph $G = (V, E)$ is a pair (T, χ) where T is a tree and χ is a labeling function that assigns each tree node t a set $\chi(t)$ of vertices of the graph G such that the following conditions hold: (1) Every vertex of G occurs in $\chi(t)$ for some tree node t, (2) For every edge $\{u, v\}$ of G there is a tree node t such that $u, v \in \chi(t)$, and (3) For every vertex v of G, the tree nodes t with $v \in \chi(t)$ form a connected subtree of T. The *width* of a tree decomposition (T, χ) is the size of a largest bag $\chi(t)$ minus 1 among all nodes t of T. A tree decomposition of smallest width is *optimal*. The *treewidth* of a graph G is the width of an optimal tree decomposition of G.

2.4 Shrub-depth

The recently introduced notion of *shrub-depth* [20] is the "low-depth" variant of clique-width, similar to the role that tree-depth plays with respect to treewidth.

Definition 1. *We say that a graph G has a* tree-model *of m colors and depth $d \ge 1$ if there exists a rooted tree T (of height d) such that*

1. *the set of leaves of T is exactly $V(G)$,*
2. *the length of each root-to-leaf path in T is exactly d,*
3. *each leaf of T is assigned one of m colors (this is not a graph coloring, though),*
4. *and the existence of a G-edge between $u, v \in V(G)$ depends solely on the colors of u, v and the distance between u, v in T.*

The class of all graphs having a tree-model of m colors and depth d is denoted by $\mathcal{TM}_m(d)$.

Definition 2. *A class of graphs $\mathcal{G}$ has* shrub-depth *d if there exists m such that $\mathcal{G} \subseteq \mathcal{TM}_m(d)$, while for all natural m it is $\mathcal{G} \not\subseteq \mathcal{TM}_m(d-1)$.*

Note that Definition 2 is asymptotic as it makes sense only for infinite graph classes. Particularly, classes of shrub-depth 1 are known as the graphs of bounded *neighborhood diversity* in [23], i.e., those graph classes on which the twin relation on pairs of vertices (for a pair to share the same set of neighbors besides this pair) has a finite index.

2.5 Modular-Width

For our algorithms we consider graphs that can be obtained from an algebraic expression that uses the following operations:

(O1) create an isolated vertex;

(O2) the disjoint union of 2 graphs denoted by $G_1 \oplus G_2$, i.e., $G_1 \oplus G_2$ is the graph with vertex set $V(G_1) \cup V(G_2)$ and edge set $E(G_1) \cup E(G_2)$;

(O3) the complete join of 2 graphs denoted by $G_1 \otimes G_2$, i.e., $G_1 \otimes G_2$ is the graph with vertex set $V(G_1) \cup V(G_2)$ and edge set $E(G_1) \cup E(G_2) \cup \{\{v, w\} : v \in V(G_1) \text{ and } w \in V(G_2)\}$;

(O4) the substitution operation with respect to some graph G with vertices $v_1, \ldots, v_n$, i.e., for graphs $G_1, \ldots, G_n$ the *substitution* of the vertices of G by the graphs $G_1, \ldots, G_n$, denoted by $G(G_1, \ldots, G_n)$, is the graph with vertex set $\bigcup_{1 \leq i \leq n} V(G_i)$ and edge set $\bigcup_{1 \leq i \leq n} E(G_i) \cup \{\{u, v\} : u \in V(G_i) \text{ and } v \in V(G_j) \text{ and } \{v_i, v_j\} \in E(G)\}$. Hence, $G(G_1, \ldots, G_n)$ is obtained from G by substituting every vertex $v_i \in V(G)$ with the graph G_i and adding all edges between the vertices of a graph G_i and the vertices of a graph G_j whenever $\{v_i, v_j\} \in E(G)$.

Let A be an algebraic expression that uses only the operations (O1)–(O4). We define the *width* of A as the maximum number of operands used by any occurrence of the operation (O4) in A. It is well-known that the *modular-width* of a graph G, denoted mw(G), is the least integer m such that G can be obtained from such an algebraic expression of width at most m. Furthermore, an algebraic expression of width mw(G) can be constructed in linear time [27].

2.6 Integer Linear Programming

For our algorithms, we use the well-known result that INTEGER LINEAR PROGRAMMING is fixed-parameter tractable parameterized by the number of variables.

INTEGER LINEAR PROGRAMMING FEASIBILITY	**Parameter:** p
Input: A matrix $A \in \mathcal{Z}^{m \times p}$ and a vector $b \in \mathcal{Z}^m$.	
Question: Is there a vector $x \in \mathcal{Z}^p$ such that $Ax \leq b$?	

Proposition 1 ([14]). INTEGER LINEAR PROGRAMMING FEASIBILITY *is fixed-parameter tractable and can be solved in time* $O(p^{2.5p+o(p)} \cdot L)$ *where L is the number of bits in the input.*

3   Hardness for Problems on Shrub-depth

In this section we give evidence that the recently introduced parameter shrub-depth is not restrictive enough to obtain fixed-parameter algorithms for problems that are W[1]-hard on graphs of bounded cliquewidth. In particular, we show that GRAPH COLORING and HAMILTONIAN PATH are W[1]-hard parameterized by the number of colors (used in a tree-model of the input graph) on classes of graphs of shrub-depth 5. Note that restricting the shrub-depth means restricting the height of the tree-model that can be employed and for every restriction on the height of the tree-model the number of colors needed to model the graph gives a different parameter. In particular, if we restrict the shrub-depth to 1 the number of colors of the tree-model corresponds to the neighborhood diversity of a graph. This implies that GRAPH COLORING and HAMILTONIAN PATH become fixed-parameter tractable when parameterized by the number of colors (used in a tree-model of the input graph) on classes of graphs of shrub-depth 1 [23]. It is an interesting open question what is the least possible shrub-depth that allows for fixed-parameter algorithms for the problems GRAPH COLORING and HAMILTONIAN PATH.

Theorem 1 (⋆). GRAPH COLORING *parameterized by the number of colors (used in a tree-model of the input graph) is* W[1]-*hard on classes of graphs of shrub-depth* 5.

Theorem 2 (⋆). HAMILTONIAN PATH *parameterized by the number of colors (used in a tree-model of the input graph) is* W[1]-*hard on class of graphs of shrub-depth* 5.

4   Modular-Width and Other Parameters

In this section we study the relationships of modular-width, shrub-depth and other important width parameters. Of particular importance is the observation that modular-width generalizes the recently introduced parameters neighborhood diversity [23] and twin-cover [19]. Both of these parameters have been introduced to obtain FPT algorithms on dense graphs for problems that are hard for clique-width. Figure 1 summarizes these relationships. Most of these relationships are well-known or have recently been shown in [23,19,20,8]. Consequently, we only show the relationships whose proofs cannot been found anywhere else.

Theorem 3. *Let* G *be a graph. Then* $\mathrm{mw}(G) \leq \mathrm{nd}(G)$ *and* $\mathrm{mw}(G) \leq 2^{\mathrm{tc}(G)} + \mathrm{tc}(G)$. *Furthermore, both inequalities are strict, i.e., there are graphs with bounded modular-width and unbounded neighborhood diversity (or unbounded twin-cover number).*

Proof. Let G be a graph. Using the definition of neighborhood diversity from [23] it follows that G has a partition $\{V_1, \ldots, V_{\mathrm{nd}(G)}\}$ of its vertex set such that for every $1 \leq i \leq \mathrm{nd}(G)$ it holds that the graph $G[V_i]$ is either a clique or an independent set and for every $1 \leq i < j \leq \mathrm{nd}(G)$, either all vertices in

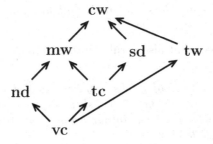

Fig. 1. Relationship between vertex cover (vc), neighborhood diversity (nd), twin-cover (tw), modular-width (mw), shrub-depth (sd), treewidth (tw), and clique-width (cw). Arrows indicate generalizations, e.g., modular-width generalizes both neighborhood diversity and twin-cover.

V_i are adjacent to all vertices of V_j or G contains no edges between vertices in V_i and vertices in V_j. Let G' be the graph with vertex set $v_1, \ldots, v_{\mathrm{nd}(G)}$ and an edge between v_i and v_j if and only if the graph G contains all edges between vertices in V_i and vertices in V_j. Then $G = G'(G[V_1], \ldots, G[V_{\mathrm{nd}(G)}])$. Furthermore, because for every $1 \leq i \leq \mathrm{nd}(G)$, V_i is either a clique or an independent set, we can obtain the graph $G[V_i]$ from an algebraic expression A_i that uses only the operations (O1)–(O3). Substituting the algebraic expressions A_i for every $G[V_i]$ into $G'(G[V_1], \ldots, G[V_{\mathrm{nd}(G)}])$ gives us the desired algebraic expression for G of width $\mathrm{nd}(G)$.

Let G be a graph. Using the definition of twin-cover from [19] it follows that there is a set C of at most $\mathrm{tc}(G)$ vertices of G such that every component C' of $G \setminus C$ is a clique and every vertex in C' is connected to the same vertices in C. Let $C_1, \ldots, C_l$ be sets of components of $G \setminus C$ such that 2 components of $G \setminus C$ are contained in the same set C_i if and only if their vertices have the same neighbors in C. Because there are at most $2^{|C|}$ possible such neighborhoods, we obtain that $l \leq 2^{|C|}$. Let G' be the graph with vertices $C \cup \{c_1, \ldots, c_l\}$ and edges $E(G[C]) \cup \{\, \{c_i, v_i\} : v_i \in N_G(C_i)\,\}$. Then $G = G'(v_1, \ldots, v_{|C|}, G[C_1], \ldots, G[C_l])$. Furthermore, because the graphs $G[C_i]$ are disjoint unions of cliques, we can obtain each of these graphs from an algebraic expression A_i that uses only the operations (O1)–(O3). Substituting the algebraic expressions A_i for every $G[C_i]$ into $G'(v_1, \ldots, v_{|C|}, G[C_1], \ldots, G[C_l])$ gives us the desired algebraic expression for G of width $2^{|C|} + |C| \leq 2^{\mathrm{tc}(G)} + \mathrm{tc}(G)$.

To see that both inequalities are strict we refer the reader to [20, Example 5.4 a)]. The example exhibits a family of co-graphs, i.e., graphs of modular-width 0, with unbounded neighborhood diversity and unbounded twin-cover number. □

The following theorem shows that modular-width and shrub-depth are orthogonal to each other.

Theorem 4 (⋆). *There are classes of graphs with unbounded modular-width and bounded shrub-depth and vice versa.*

The next theorem shows that also shrub-depth generalizes neighborhood diversity and twin-cover.

Theorem 5. *Let $\mathcal{G}$ be a class of graphs. If $\mathcal{G}$ has bounded neighborhood diversity or bounded twin-cover number then $\mathcal{G}$ has shrub-depth at most 2. Furthermore, there are classes of graphs that have unbounded neighborhood diversity and twin-cover number but shrub-depth 2.*

Proof. Suppose that $\mathcal{G}$ has bounded neighborhood diversity, i.e., there is some natural number c such that $\mathrm{nd}(G) \leq c$ for every $G \in \mathcal{G}$. We show that every graph $G \in \mathcal{G}$ has a tree-model of height at most 1 that uses at most $\mathrm{nd}(G) \leq c$ colors. Using the definition of neighborhood diversity from [23] it follows that G has a partition $\{V_1, \ldots, V_{\mathrm{nd}(G)}\}$ of its vertex set such that for every $1 \leq i \leq \mathrm{nd}(G)$ it holds that the graph $G[V_i]$ is either a clique or an independent set and for every $1 \leq i < j \leq \mathrm{nd}(G)$, either all vertices in V_i are adjacent to all vertices of V_j or G contains no edges between vertices in V_i and vertices in V_j. Then the tree-model for G consists of a root r and 1 leaf for every vertex v in G with color i if $v \in V_i$.

Now, suppose that $\mathcal{G}$ has bounded twin-cover, i.e., there is some natural number c such that $\mathrm{tc}(G) \leq c$ for every $G \in \mathcal{G}$. We show that every graph $G \in \mathcal{G}$ has a tree-model of height at most 2 that uses at most $2^{\mathrm{tc}(G)} + \mathrm{tc}(G) \leq 2^c + c$ colors. Using the definition of twin-cover from [19] it follows that there is a set $W = \{w_1, \ldots, w_{\mathrm{tc}(G)}\}$ of $\mathrm{tc}(G)$ vertices of G such that every component C' of $G \setminus W$ is a clique and every vertex in C' is connected to the same vertices in W. Let $C_1^1, \ldots, C_{p_1}^1, \ldots, C_1^l, \ldots, C_{p_l}^l$ be all the components of $G \setminus W$ such that 2 components $C_{i_1}^{j_1}$ and $C_{i_2}^{j_2}$ have the same neighborhood in W if and only if $j_1 = j_2$. Because there are at most $2^{|W|}$ possible such neighborhoods, we obtain that $l \leq 2^{|W|}$. We construct a tree-model of G as follows. We start with the root node r, which has 1 child, say C_i^j, for every component of $G \setminus W$, and 1 child (which is also a leaf of the tree-model), say w_i, with color $l + i$ for every $1 \leq i \leq |W|$. Finally, every node C_i^j has $|V(C_i^j)|$ children (which are also leaves of the tree-model) with color j. This finishes the construction of the tree-model for G.

To see that there are classes of graphs that have unbounded neighborhood diversity and twin-cover number but shrub-depth 2, consider the class $\mathcal{S}$ from the proof of Theorem 4. As shown in Theorem 4 $\mathcal{S}$ has unbounded modular-width but shrub-depth 2. It now follows from Theorem 3 that $\mathcal{S}$ has also unbounded neighborhood diversity and unbounded twin-cover number. □

5   Algorithms on Modular-Width

In this section we show that PARTITIONING INTO PATHS, HAMILTONIAN PATH, HAMILTONIAN CYCLE, and COLORING are fixed-parameter tractable parameterized by the modular-width of the input graph. Our algorithms use a bottom-up dynamic programming approach along the parse-tree of an algebraic expression as defined in Section 2.5. That is for every node of such a parse-tree we compute a solution (or a record representing a solution) given solutions (or records) for the children of the node in the parse-tree. The running time of our algorithms is then the number of nodes in the parse-tree times the maximum time spend at any node of the parse-tree. Because the number of nodes of a parse-tree is linear in the number of vertices of the created graph, it suffices to bound the maximum time spend at any node of the parse-tree. Furthermore, because the operations (O1)–(O3) can be replaced by one operation of type (O4) that uses at most 2 operands, we only need to bound the time spend to compute a record for the graph obtained by operation (O4). To avoid cumbersome run-time bounds we use the notation O^+ to suppress poly-logarithmic factors, i.e., we write $O^+(f)$ when we have $O(f \log^d f)$ for some constant d.

5.1   Coloring

This section is devoted to a proof of the following theorem. Recall the definition of GRAPH COLORING and related notions from Section 2.1.

Theorem 6. GRAPH COLORING *parameterized by the modular-width of the input graph is fixed-parameter tractable.*

As outlined above we only need to bound the time spend to compute a record for a node of type (O4) of the parse-tree. In the case of GRAPH COLORING a record is simply the chromatic number of the graph. Hence, we will have shown the theorem after showing the following lemma.

Lemma 1. *Let G be a graph with vertices $v_1, \ldots, v_n$, $G_1, \ldots, G_n$ be graphs, and $H := G(G_1, \ldots, G_n)$. Then $\chi(H)$ can be computed from $\chi(G_1), \ldots, \chi(G_n)$ in time $O^+(2^n n^2 \max_{1 \le i \le n} \chi(G_i))$.*

We will prove the lemma by reducing the coloring problem to the following problem.

MAX WEIGHTED PARTITION

Input: An n-element set N and functions $f_1, \ldots, f_k$ from the subsets of N to integers from the range $[-M, M]$.

Question: A k-partition $(S_1, \ldots, S_k)$ of N that maximizes $f_1(S_1) + \cdots + f_k(S_k)$.

Proposition 2. *([2, Theorem 4.])* MAX WEIGHTED PARTITION *can be solved in time $O^+(2^n k^2 M)$.*

To simplify the reduction to MAX WEIGHTED PARTITION we need the following Proposition and Lemma.

Proposition 3 ([23]). *Let G be a graph with vertices $v_1, \ldots, v_n$ and $s_1, \ldots, s_n$ be natural numbers. Then $\chi(G(K_{s_1}, \ldots, K_{s_n})) = \min_{\lambda \in \Lambda(G)}(\sum_{c \in \lambda(G)} \max\{ s_i : v_i \in \lambda^{-1}(c) \})$.*

Lemma 2 (⋆). *Let G be a graph with vertices $v_1, \ldots, v_n$, $G_1, \ldots, G_n$ be graphs, $H_K := G(K_{\chi(G_1)}, \ldots, K_{\chi(G_n)})$, and $H := G(G_1, \ldots, G_n)$. Then $\chi(H_K) = \chi(H)$.*

We can now proceed with a proof of Lemma 1.

Proof (of Lemma 1).
We reduce the coloring problem to the MAX WEIGHTED PARTITION problem as follows: We set $N := V(G)$ and $f_1(S) = \cdots = f_k(S) = -\max\{\chi(G_i) : v_i \in S\}$ for every subset S of N. It follows from Proposition 3 and Lemma 2 that the maximum weight of a partition of this instance corresponds to the chromatic number $\chi(H)$. Hence, the lemma follows from Proposition 2. □

5.2   Partitioning into Paths

This section is devoted to a proof of the following theorem. Recall the definition of PARTITIONING INTO PATHS and related notions from Section 2.1.

Theorem 7. PARTITIONING INTO PATHS *(and hence also* HAMILTONIAN PATH *and* HAMILTONIAN CYCLE*) parameterized by the modular width of the input graph is fixed-parameter tractable.*

As outlined above we only need to bound the time spend to compute a record for a node of type (O4) of the parse-tree. In the case of PARTITIONING INTO PATH a record of a graph G is the pair $(\mathrm{ham}(G), |V(G)|)$. Hence, we will have shown the theorem after showing the following lemma. From now on we will assume that G is a graph with vertices $v_1, \ldots, v_n$, $G_1, \ldots, G_n$ are graphs, $H = G(G_1, \ldots, G_n)$, and $m = |E(G)|$.

Lemma 3. *Given the graph G and the pairs*
$(\mathrm{ham}(G_1), |V(G_1)|), \ldots, (\mathrm{ham}(G_n), |V(G_n)|)$, then the pair $(\mathrm{ham}(H), |V(H)|)$ can be computed in time $O^+\left(\mathrm{ham}(H)\left((m+n)2^n + n\left(2(m+n)\right)^{5(m+n)+o(m+n)}\right)\right)$.

The remainder of this section is devoted to a proof of this lemma.

For a graph G and a positive integer i we define the graph $G \oplus i$ as the graph with vertex set $V(G) \cup \{1, \ldots, i\}$ and edge set $E(G) \cup \{\{v, j\} : v \in V(G)$ and $1 \le j \le i\}$, i.e., the graph $G \oplus i$ is obtained from G by adding i vertices and connect them to every vertex in G.

Proposition 4 (⋆). *Let G be a graph and*
$h(G) = \min\{i : G \oplus i$ has a Hamiltonian cycle$\}$. Then $\mathrm{ham}(G) = h(G)$.

A slightly less general version of the following lemma has already been proven in [23].

Lemma 4. *Let* HAMILTONIAN CYCLE *be the ILP with variables*
$\{e_{ij}, e_{ji} : \{v_i, v_j\} \in E(G)\}$ and constraints:

For every $1 \leq i \leq n$:

(1) $\sum_{j \in \{ l \,:\, v_l \in N_G(v_i) \}} e_{ij} = \sum_{j \in \{ l \,:\, v_l \in N_G(v_i) \}} e_{ji}$ *("incoming = outgoing")*

(2) $\sum_{j \in \{ l \,:\, v_l \in N_G(v_i) \}} e_{ij} \leq |V(G_i)|$ *(at most $|V(G_i)|$)*

(3) $\mathrm{ham}(G_i) \leq \sum_{j \in \{ l \,:\, v_l \in N_G(v_i) \}} e_{ij}$ *(at least $\mathrm{ham}(G_i)$)*

For every partition of $V(G)$ into vertex sets A and B:

(4) $\sum_{1 \leq i < j \leq n \,:\, \{v_i, v_j\} \in E(G) \wedge |e \cap A| = 1} e_{ij} + e_{ji} \geq 1$ *("connectivity")*

For every variable e_{ij}:

(5) $e_{ij} \geq 0$.

Then H has a Hamiltonian cycle if and only if the ILP HAMILTONIAN CYCLE is feasible. Furthermore, the size of the ILP is at most $O^+(m2^n)$ and it has $2m$ variables.

Proof. The size bound on the ILP HAMILTONIAN CYCLE is obvious. Suppose that H has a Hamiltonian cycle C. W.l.o.g. we can assume that C is directed. For every $\{v_i, v_j\} \in E(G)$ we set e_{ij} to be the number of arcs (x, y) in C such that $x \in V(G_i)$ and $y \in V(G_j)$ and similarly we set e_{ji} to be the number of arcs (x, y) in C such that $x \in V(G_j)$ and $y \in V(G_i)$. Then, because C is a Hamiltonian cycle of H, this assignment of e_{ij} and e_{ji} satisfies the constrains (1)–(5), as required.

For the reverse direction, suppose that the ILP HAMILTONIAN CYCLE is feasible and let β be an assignment of the variables e_{ij} and e_{ji} witnessing this. Let G' be the directed multigraph obtained from G by replacing every edge $\{v_i, v_j\}$ with $\beta(e_{ij})$ parallel arcs from v_i to v_j and $\beta(e_{ji})$ parallel arcs from v_j to v_i. Because of the constrains (1), (4), and (5), it follows that G' contains a directed eularian tour T, i.e., a closed directed walk that visits all the arcs of G' exactly once. Clearly, when fixing any vertex of G', the tour T defines an ordering of the arcs of G'. Let π be any such ordering of the arcs of G'. For every $1 \leq i \leq n$, let $\mathcal{P}_i = (P_1^i, \ldots, P_{p_i}^i)$ be a partition of G_i into p_i disjoint paths, where $p_i = \sum_{j \in \{ l \,:\, v_l \in N_G(v_i) \}} e_{ij}$. Because of the constrains (2) and (3) we know that such a partition exists for every $1 \leq i \leq n$. For every arc $a = (v_i, v_j)$ in T where a is the l-th arc leaving v_i in T and a is the l'-th arc entering v_j in T (according to the ordering π), we denote by $e(a)$ the edge of H from the second endpoint of P_l^i to the first endpoint of $P_{l'}^j$. Then the edges in $\{ e(a) : a \in T \}$ together with the edges of all the path $P_1^1, \ldots, P_{p_n}^n$ form a Hamiltonian cycle in H, as required.   □

Lemma 5. *Given the graph G and the pairs $(\mathrm{ham}(G_1), |V(G_1)|), \ldots,$ $(\mathrm{ham}(G_n), |V(G_n)|)$ it can be decided whether the graph H has a Hamiltonian cycle in time $O^+(m2^n + (2m)^{5m + o(m)}n)$.*

Proof. To decide whether the graph H has a Hamiltonian cycle we construct and solve the ILP HAMILTONIAN CYCLE from Lemma 4. The running time of this algorithm is then the time it takes to construct the ILP, i.e., $O^+(m2^n)$, plus the time needed to solve the ILP, i.e., $O^+((2m)^{5m + o(m)} \log(m2^n)) \in O^+((2m)^{5m + o(m)}n)$ using Proposition 1. This concludes the proof of the Lemma.   □

We are now ready to show Lemma 3.

Proof (Proof of Lemma 3). Clearly, $|V(H)| = \sum_{1 \leq i \leq n} |V(G_i)|$ so it remains to show how to compute ham(H). Because of Proposition 4 ham(H) is equal to the minimum positive integer $1 \leq l \leq |V(H)|$ such that the graph $H \oplus l$ has a Hamiltonian cycle. For every $1 \leq l \leq |V(H)|$ the graph $H \oplus l$ is equal to the graph $G'(G_1, \ldots, G_n, I_l)$ where G' is the graph obtained from G by adding one vertex v_{n+1} and making it adjacent to all vertices of G, and the graph I_l is the independent set on l vertices. Because ham$(I_l) = |V(I_l)| = l$ we can use Lemma 5 to decide whether the graph $H \oplus l$ has an Hamiltonian cycle in time $O(m'2^{n'} + (2m')^{5m'+o(m')}n')$ where $n' = n+1$ and $m' = m+n$. This concludes the proof of the lemma. □

6 Conclusion

We examined some of the algorithmic properties of modular-width, a natural structural parameter. Our results indicate that this is a notion which may be worthy of further study independently of its more famous cousin, clique-width, since it its decreased generality does offer some algorithmic pay-off.

As a direction for further research, it would be interesting to see if more problems which are hard for clique-width (or even for treewidth) become tractable for modular-width. Two prime suspects in this category are EDGE DOMINATING SET and PARTITION INTO TRIANGLES.

Beyond that, it would be interesting to investigate if the techniques of this paper can be further generalized, perhaps eventually leading to meta-theorem-like results. In particular, our ILP-based solution for HAMILTONICITY may be applicable (with some modifications) to other problems. One may ask: what properties must a problem possess for us to be able to give a straightforward DP algorithm that uses ILP to combine the tables?

The main property that a problem should satisfy for these ideas to apply is that the sets of partial solutions arising in the dynamic programming formulation should be *convex*. Convexity is important here, since we would like to be able to express the information contained in the DP tables using linear constraints, in order to use an ILP. Convexity was easy to establish in the case of PARTITION INTO PATHS and similar problems, since the set of feasible partial solutions is the set of integers k such that there exists a partition of a subgraph into k paths. If one knows the minimum feasible k, all larger integers are also feasible and this is trivially a convex set. The question then becomes, are there any other natural problems where convexity can be established (perhaps non-trivially) and used in this way?

References

1. Bessy, S., Paul, C., Perez, A.: Polynomial kernels for 3-leaf power graph modification problems. Discrete Applied Mathematics 158(16), 1732–1744 (2010)
2. Björklund, A., Husfeldt, T., Koivisto, M.: Set partitioning via inclusion-exclusion. SIAM J. Comput. 39(2), 546–563 (2009)

3. Bodlaender, H.L.: A tourist guide through treewidth. Acta Cybernetica 11(1–2), 1–22 (1993)
4. Bodlaender, H.L.: The algorithmic theory of treewidth. Electronic Notes in Discrete Mathematics 5, 27–30 (2000)
5. Bodlaender, H.L.: Treewidth: Characterizations, applications, and computations. In: Fomin, F.V. (ed.) WG 2006. LNCS, vol. 4271, pp. 1–14. Springer, Heidelberg (2006)
6. Bodlaender, H.L., Koster, A.M.C.A.: Combinatorial optimization on graphs of bounded treewidth. Comput. J. 51(3), 255–269 (2008)
7. Bui-Xuan, B.-M., Telle, J.A., Vatshelle, M.: Boolean-width of graphs. Theor. Comput. Sci. 412(39), 5187–5204 (2011)
8. Courcelle, B., Makowsky, J.A., Rotics, U.: Linear time solvable optimization problems on graphs of bounded clique-width. Theory of Computer Systems 33(2), 125–150 (2000)
9. Diestel, R.: Graph Theory, 2nd edn. Graduate Texts in Mathematics, vol. 173. Springer, New York (2000)
10. Dom, M., Guo, J., Hüffner, F., Niedermeier, R., Truß, A.: Fixed-parameter tractability results for feedback set problems in tournaments. In: Calamoneri, T., Finocchi, I., Italiano, G.F. (eds.) CIAC 2006. LNCS, vol. 3998, pp. 320–331. Springer, Heidelberg (2006)
11. Dom, M., Lokshtanov, D., Saurabh, S., Villanger, Y.: Capacitated domination and covering: A parameterized perspective. In: Grohe, M., Niedermeier, R. (eds.) IWPEC 2008. LNCS, vol. 5018, pp. 78–90. Springer, Heidelberg (2008)
12. Downey, R.G., Fellows, M.R.: Parameterized Complexity. Monographs in Computer Science. Springer, New York (1999)
13. Fellows, M.R., Fomin, F.V., Lokshtanov, D., Rosamond, F.A., Saurabh, S., Szeider, S., Thomassen, C.: On the complexity of some colorful problems parameterized by treewidth. Inf. Comput. 209(2), 143–153 (2011)
14. Fellows, M.R., Lokshtanov, D., Misra, N., Rosamond, F.A., Saurabh, S.: Graph layout problems parameterized by vertex cover. In: Hong, S.-H., Nagamochi, H., Fukunaga, T. (eds.) ISAAC 2008. LNCS, vol. 5369, pp. 294–305. Springer, Heidelberg (2008)
15. Flum, J., Grohe, M.: Parameterized Complexity Theory. Texts in Theoretical Computer Science. An EATCS Series, vol. XIV. Springer, Berlin (2006)
16. Fomin, F.V., Golovach, P.A., Lokshtanov, D., Saurabh, S.: Clique-width: on the price of generality. In: Mathieu, C. (ed.) Proceedings of the Twentieth Annual ACM-SIAM Symposium on Discrete Algorithms, SODA 2009, New York, NY, USA, January 4-6, pp. 825–834. SIAM (2009)
17. Fomin, F.V., Golovach, P.A., Lokshtanov, D., Saurabh, S.: Algorithmic lower bounds for problems parameterized with clique-width. In: Charikar, M. (ed.) Proceedings of the Twenty-First Annual ACM-SIAM Symposium on Discrete Algorithms, SODA 2010, Austin, Texas, USA, January 17-19, pp. 493–502. SIAM (2010)
18. Fomin, F.V., Golovach, P.A., Lokshtanov, D., Saurabh, S.: Intractability of clique-width parameterizations. SIAM J. Comput. 39(5), 1941–1956 (2010)
19. Ganian, R.: Twin-cover: Beyond vertex cover in parameterized algorithmics. In: Marx, D., Rossmanith, P. (eds.) IPEC 2011. LNCS, vol. 7112, pp. 259–271. Springer, Heidelberg (2012)
20. Ganian, R., Hliněný, P., Nešetřil, J., Obdržálek, J., Ossona de Mendez, P., Ramadurai, R.: When trees grow low: Shrubs and fast mso1. In: Rovan, B., Sassone, V., Widmayer, P. (eds.) MFCS 2012. LNCS, vol. 7464, pp. 419–430. Springer, Heidelberg (2012)

21. Habib, M., Paul, C.: A survey of the algorithmic aspects of modular decomposition. Computer Science Review 4(1), 41–59 (2010)
22. Oum, S.: Rank-width and vertex-minors. J. Comb. Theory, Ser. B 95(1), 79–100 (2005)
23. Lampis, M.: Algorithmic meta-theorems for restrictions of treewidth. Algorithmica 64(1), 19–37 (2012)
24. McConnell, R.M., Spinrad, J.: Modular decomposition and transitive orientation. Discrete Mathematics 201(1–3), 189–241 (1999)
25. Niedermeieri, R.: Invitation to Fixed-Parameter Algorithms. Oxford Lecture Series in Mathematics and its Applications. Oxford University Press, Oxford (2006)
26. Protti, F., da Silva, M.D., Szwarcfiter, J.L.: Applying modular decomposition to parameterized cluster editing problems. Theory of Computing Systems 44(1), 91–104 (2009)
27. Tedder, M., Corneil, D.G., Habib, M., Paul, C.: Simpler linear-time modular decomposition via recursive factorizing permutations. In: Aceto, L., Damgård, I., Goldberg, L.A., Halldórsson, M.M., Ingólfsdóttir, A., Walukiewicz, I. (eds.) ICALP 2008, Part I. LNCS, vol. 5125, pp. 634–645. Springer, Heidelberg (2008)

A Faster FPT Algorithm
for Bipartite Contraction[*]

Sylvain Guillemot and Dániel Marx

Institute for Computer Science and Control, Hungarian Academy of Sciences
(MTA SZTAKI), Budapest, Hungary

Abstract. The BIPARTITE CONTRACTION problem is to decide, given a
graph G and a parameter k, whether we can can obtain a bipartite graph
from G by at most k edge contractions. The fixed-parameter tractability
of the problem was shown by Heggernes et al. [13], with an algorithm
whose running time has double-exponential dependence on k. We present
a new randomized FPT algorithm for the problem, which is both conceptually simpler and achieves an improved $2^{O(k^2)}nm$ running time, i.e.,
avoiding the double-exponential dependence on k. The algorithm can be
derandomized using standard techniques.

1 Introduction

A graph modification problem aims at transforming an input graph into a graph
satisfying a certain property, by at most k operations. These problems are typically studied from the viewpoint of fixed-parameter tractability, where the goal
is to obtain an algorithm with running time $f(k)n^c$ (or *FPT algorithm*). Here,
$f(k)$ is a computable function depending only on the parameter k, which confines
the combinatorial explosion that is seemingly inevitable for an NP-hard problem. The most intensively studied graph modification problems involve vertex-
or edge-deletions as their base operation; fixed-parameter tractability has been
established for the problems of transforming a graph into a forest [11,8,3], a
bipartite graph [26,11,18,15,24], a chordal graph [20], a planar graph [22], or
an interval graph [27,2]. Results have also been obtained for problems involving
directed graphs [5] or group-labeled graphs [10,7].

Recently, there has been an interest in graph modification problems involving
edge contractions. These problems fall in the following general framework. Given
a graph property Π, the problem Π-CONTRACTION is to decide, for a graph G
and a parameter k, whether we can obtain a graph in Π, by starting from
G and performing at most k edge contractions. For each graph property Π
admitting a polynomial recognition algorithm, it is then natural to ask whether
Π-CONTRACTION admits an FPT algorithm. Such algorithms have been given

---

[*] Research supported by the European Research Council (ERC) grant
"PARAMTIGHT: Parameterized complexity and the search for tight complexity
results," reference 280152 and OTKA grant NK105645.

G. Gutin and S. Szeider (Eds.): IPEC 2013, LNCS 8246, pp. 177–188, 2013.

when Π is the class of paths, the class of trees [12], the class of planar graphs [9], or the class of bipartite graphs [13].

For the case of bipartite graphs, the problem is called BIPARTITE CONTRACTION, and Heggernes et al. [13] obtained an FPT algorithm with a running time double-exponential in k. The algorithm combines several tools from parameterized algorithmics, such as the irrelevant vertex technique, important separators, treewidth reduction, and Courcelle's theorem. In this note, we present a new FPT algorithm for the problem, which is both conceptually simpler and faster. Similar to the compression routine for ODD CYCLE TRANSVERSAL in [26], we reduce BIPARTITE CONTRACTION to several instances of an auxiliary cut problem. Our main effort is spent on obtaining an FPT algorithm for this cut problem. This is achieved by using the notion of important separators from [19], together with the randomized coloring technique introduced by Alon et al. [1]. We obtain the following result:

Theorem 1. BIPARTITE CONTRACTION *has a randomized* FPT *algorithm with running time* $2^{O(k^2)}nm$ *and a deterministic algorithm with running time* $2^{O(k^2)}n^{O(1)}$.

This paper is organized as follows. We first introduce the relevant notation and definitions in Section 2. We explain in Section 3 how BIPARTITE CONTRACTION can be reduced to several instances of a suitable cut problem called RANK-CUT. In Section 4, we define a constrained version of the RANK-CUT problem and show that it is polynomial-time solvable. In Section 5, we present a randomized reduction of RANK-CUT to its constrained version. Finally, in Section 6 we derandomize this reduction and we complete the proof of Theorem 1. Section 7 concludes the paper.

2   Preliminaries

Given a graph G, we let $V(G)$ and $E(G)$ denote its vertex set and edge set, respectively. We let $n = |V(G)|$ and $m = |E(G)|$. Given $X \subseteq V(G)$, we denote by $G[X]$ the subgraph of G induced by X, and we denote by $G \setminus X$ the subgraph of G induced by $V(G) \setminus X$. Given a set $F \subseteq E(G)$ of edges, we denote by $V(F)$ the endpoints of the edges in F, and we say that F *spans* the vertices in $V(F)$. Given $F \subseteq E(G)$, we denote by $G[F]$ the graph with vertex set $V(F)$ and edge set F; we denote by $G \setminus F$ the graph with vertex set $V(G)$ and edge set $E(G) \setminus F$. For an edge e, we denote by G/e the graph obtained by contracting edge e, that is, by removing the endpoints u and v of e and introducing a new vertex that is adjacent to every vertex that is adjacent to at least one of u or v. Given $F \subseteq E(G)$, we denote by G/F the graph obtained from G by contracting all the edges of F; it is easy to observe that the graph G/F does not depend on the order in which we perform the contractions.

Fix two disjoint subsets of vertices X, Y of a graph G. An (X, Y)-*walk* in G is a sequence $W = v_0 v_1 \ldots v_\ell$ of vertices such that $v_0 \in X, v_\ell \in Y$, and $v_i v_{i+1} \in E(G)$ for $0 \le i < \ell$; the *length* of W is ℓ, and we call W an (X, Y)-*path* in G if the vertices v_i are pairwise distinct. We will simply use the term "path" when the sets X, Y are irrelevant. An (X, Y)-*cut* in G is defined as a set $C \subseteq E(G)$ such that $G \setminus C$ has no (X, Y)-path; an (X, Y)-*separator* in G is defined as a set $S \subseteq V(G) \setminus (X \cup Y)$ such that $G \setminus S$ has no (X, Y)-path. Note that the (X, Y)-separator is by definition disjoint from X and Y. An (X, Y)-cut (resp. (X, Y)-separator) C is *inclusion-wise minimal* if no proper subset of C is an (X, Y)-cut (resp. (X, Y)-separator).

A *bipartite modulator* of G is a set $F \subseteq E(G)$ such that $G \setminus F$ is bipartite. The *rank* of a graph G is the number of edges of a spanning forest of G. The rank of a set $F \subseteq E(G)$ of edges, denoted by $r(F)$ is the rank of $G[F]$. As observed in [21], we can alternatively define BIPARTITE CONTRACTION as the following problem: given a graph G and an integer k, find a bipartite modulator F of G such that $r(F) \leq k$. We reproduce the proof here for completeness.

Lemma 2. *The following statements are equivalent:*

(i) *There exists a set $F \subseteq E(G)$ such that $|F| \leq k$ and G/F is bipartite;*
(ii) *there exists a set $F \subseteq E(G)$ such that $r(F) \leq k$ and $G \setminus F$ is bipartite.*

Proof. $(i) \Rightarrow (ii)$: Let F' denote the edges of G having both endpoints in a same connected component of $G[F]$. Observe that $r(F') \leq k$, as F contains a spanning forest of $G[F']$. We claim that $G \setminus F'$ is bipartite. Observe that the vertex set of each connected component of $G[F]$ is an independent set in $G \setminus F'$. Therefore, a proper 2-coloring of G/F can be turned into a proper 2-coloring of $G \setminus F'$ if we color every vertex in a connected component K of $G[F]$ by the color of the single vertex corresponding to K in G/F.

$(ii) \Rightarrow (i)$: Let us fix a proper 2-coloring of $G \setminus F$. We can assume that F is a minimal set of edges such that $G \setminus F$ is bipartite. Therefore, in each connected component of $G[F]$, every vertex has the same color in the 2-coloring of $G \setminus F$. Hence, contracting each connected component of F to a single vertex gives a bipartite graph. This graph can be obtained by contracting the edges of a spanning forest of F, which has $r(F) \leq k$ edges. $\qquad \square$

3   Reduction to a Cut Problem

We first define a *compression* version of the problem, named BIPARTITE CONTRACTION COMPRESSION: given a graph G, an integer k, and a set $X \subseteq V(G)$ with $|X| \leq 2k$ such that $G \setminus X$ is bipartite, is there a bipartite modulator F of G with $r(F) \leq k$? The following lemma establishes how a *compression routine* for the problem entails the fixed-parameter tractability of BIPARTITE CONTRACTION.

Lemma 3. *Suppose that* BIPARTITE CONTRACTION COMPRESSION *is solvable in time $T(k, n, m)$. Then* BIPARTITE CONTRACTION *is solvable in time $O(9^k knm + T(k, n, m))$.*

Proof. An instance $I = (G, k)$ of BIPARTITE COMPRESSION is solved the following way. First, we run the algorithm of Reed et al. [26] to look for a set $X \subseteq V(G)$ of size $\leq 2k$ such that $G \setminus X$ is bipartite; the running time of the algorithm is

$O(3^{2k}knm)$ (see also [18])[1]. If there is no such set, then we answer "no". Otherwise, we run the algorithm for BIPARTITE CONTRACTION COMPRESSION on (G, k, X). This takes time $O(9^k knm + T(k, n, m))$ as claimed. The correctness of this algorithm follows by observing that, if F is a solution for instance I of BIPARTITE CONTRACTION, then $X = V(F)$ is a set of size at most $2k$ such that $G \setminus X$ is bipartite.   □

In the rest of this section, we concentrate on the BIPARTITE CONTRACTION COMPRESSION problem. We solve the problem similarly to the compression routine for ODD CYCLE TRANSVERSAL [26]. First, we adapt the construction of [26] to the case of edge sets.

Suppose that we are given a graph G in which a bipartite modulator has to be found, along with a set $X \subseteq V(G)$ such that $G \setminus X$ is bipartite. We construct a graph G' as follows. Let S_1, S_2 be a bipartition of $G \setminus X$, and let $<$ be an arbitrary total ordering of $V(G)$. We let $V(G') = (V(G) \setminus X) \cup X'$ with $X' = \{x_1, x_2 : x \in X\}$, and $E(G') = E(G \setminus X) \cup \{uv_{3-i} : uv \in E, u \in S_i, v \in X\} \cup \{u_1 v_2 : uv \in E, u, v \in X, u < v\}$. Observe that G' is bipartite, with bipartition $S_1' = S_1 \cup \{x_1 : x \in X\}$ and $S_2' = S_2 \cup \{x_2 : x \in X\}$. Furthermore, if we identify x_1 with x_2 for every $x \in X$ in G', then we get the graph G; in particular, G and G' have the same number of edges.

Define a bijection $\Phi : E(G) \to E(G')$ which preserves each edge of $E(G \setminus X)$, maps each edge $uv \in E(G)$ ($u \in S_i, v \in X$) to the edge uv_{3-i}, and maps each edge $uv \in E(G)$ ($u, v \in X, u < v$) to the edge $u_1 v_2$. We say that a partition of X' into two sets X_A', X_B' is *valid* if for each $x \in X$, exactly one of $\{x_1, x_2\}$ is in X_A'. The following lemma is similar to Lemma 1 of [26].

Lemma 4. *For every $F \subseteq E(G)$, the following statements are equivalent:*

(i) $G \setminus F$ is bipartite,
(ii) there is a valid partition X_A', X_B' of X' such that $\Phi(F)$ is an (X_A', X_B')-cut in G'.

Proof. $(i) \Rightarrow (ii)$. Suppose that $G \setminus F$ is bipartite with bipartition V_1, V_2. Define the partition X_A', X_B' of X' such that: for $u \in X$, $u_1 \in X_A'$ iff $u \in V_1$. Observe that X_A', X_B' is a valid partition of X'. We claim that $C = \Phi(F)$ is an (X_A', X_B')-cut in G'. Towards a contradiction, assume that $G' \setminus C$ contains an X_A', X_B'-path P'. Suppose that the endpoints of P' are $u_i \in X_A', v_j \in X_B'$; then $u \in V_i, v \in V_{3-j}$. The path P' corresponds to a u, v-path P in $G \setminus F$, of the same length. If $j = i$, then u_i, v_j both belong to S_i', and we obtain that P is a path of even length between V_i and V_{3-i}, contradiction. If $j = 3 - i$, then $u_i \in S_i', v_j \in S_{3-i}'$, and we obtain that P is a path of odd length between V_i and V_i, contradiction. We conclude that C is an (X_A', X_B')-cut in G', as claimed.

---

[1] Very recently, two linear-time algorithms for ODD CYCLE TRANSVERSAL were announced [25,14]. However, using these linear-time algorithms here would not improve the overall asymptotic running time of our algorithm, as the dominating term comes from the RANK-CUT algorithm of Theorem 12.

$(ii) \Rightarrow (i)$. Suppose that $C \subseteq E(G')$ is an (X'_A, X'_B)-cut in G', for some valid partition X'_A, X'_B of X'. We claim that $F = \Phi^{-1}(C)$ is such that $G \setminus F$ is bipartite. We define a 2-coloring χ of $G \setminus F$ as follows: (1) If $u \in X$, then $\chi(u) = 1$ iff $u_1 \in X'_A$; (2) If $u \in V \setminus X$ is reachable from X'_A in $G' \setminus C$, then $\chi(u) = 1$ iff $u \in S_1$; (3) If $u \in V \setminus X$ is not reachable from X'_A in $G' \setminus C$, then $\chi(u) = 1$ iff $u \in S_2$. We verify that χ is a proper 2-coloring of $G \setminus F$. Consider an edge $uv \in E(G) \setminus F$, there are three cases. If $u, v \notin X$, then $u \in S_i, v \in S_{3-i}$; as u, v are either both in case (2) or both in case (3), it follows that $\chi(u) \neq \chi(v)$. If $u \in S_i, v \in X$, then uv_{3-i} is an edge of $G' \setminus C$; if $v_{3-i} \in X'_A$ then $\chi(v) = 3 - i$ by (1), and $\chi(u) = i$ by (2); if $v_{3-i} \in X'_B$ then $\chi(v) = i$ by (1), and $\chi(u) = 3 - i$ by (3). If $u, v \in X$ with $u < v$, then $u_1 v_2$ is an edge of $G' \setminus C$, and thus we have u_1, v_2 both in X'_A or both in X'_B, which implies that $\chi(u) \neq \chi(v)$. We conclude that $G \setminus F$ is bipartite, as claimed.   □

Lemma 4 turns the problem of finding a bipartite modulator into several instances of a cut problem (one for each valid partition). The same way as Lemma 2 shows the equivalence of BIPARTITE CONTRACTION with the problem of finding a bipartite modulator with a rank constraint, we show that Lemma 4 allows us to solve BIPARTITE CONTRACTION COMPRESSION by solving a cut problem with a rank constraint. However, there is a technical detail related to the fact that two vertices $x_1, x_2 \in X'$ correspond to each vertex $x \in X$ in the construction of G'; we need the following definition to deal with this issue. Let $M, F \subseteq E(G)$ be two subsets of edges. We define the M-rank $r_M(F)$ of F as the rank of the graph $G[F \cup M]/M$. Our auxiliary problem is defined as follows.

RANK-CUT
   Input: A graph G, an integer k, two sets $X, Y \subseteq V(G)$, and a set $M \subseteq E(G)$ with $|M| \leq 2k$
Question: Is there an (X, Y)-cut C in G such that $r_M(C) \leq k$?

The following simple observation will be useful later:

Lemma 5. *If $|M| \leq 2k$ and $r_M(C) \leq k$, then $C \cup M$ spans at most $6k$ vertices.*

Proof. Each contraction can decrease rank by at most one, hence the rank of $G[C \cup M]$ is at most $3k$. As $G[C \cup M]$ has no isolated vertices by definition, it follows that $G[C \cup M]$ has at most $6k$ vertices.   □

We now describe how an FPT algorithm for RANK-CUT yields an FPT algorithm for BIPARTITE CONTRACTION COMPRESSION; Sections 4–6 show the fixed-parameter tractability of RANK-CUT itself.

Lemma 6. *Suppose that RANK-CUT is solvable in time $T(k, n, m)$. Then BI-PARTITE CONTRACTION COMPRESSION is solvable in $O(4^k(T(k, n, m) + n + m))$ time.*

Proof. Consider an instance $I = (G, k, X)$ of BIPARTITE CONTRACTION COMPRESSION. From G and X, we construct the graph G' as described above Lemma 4. We let H be obtained from G' by adding the edge $x_1 x_2$ for every $x \in X$; let $M \subseteq E(H)$ be the set of these new edges.

We solve BIPARTITE CONTRACTION COMPRESSION by the following algorithm. For each valid partition X'_A, X'_B of X', we run the algorithm for RANK-CUT on the instance $I' = (H, k, X'_A, X'_B, M)$. We answer "yes" if and only if one of the instances I' was a yes-instance of RANK-CUT. Note that, as $|X| \leq 2k$ by assumption, we have $|M| \leq 2k$ and thus each instance I' is a valid instance of RANK-CUT. As there are $2^{|X|} \leq 4^k$ valid partitions of X', the claimed running time follows. We show that it correctly solves BIPARTITE CONTRACTION COMPRESSION.

Suppose that I admits a solution F with $r(F) \leq k$, then $G \setminus F$ is bipartite. Thus, by Lemma 4 there exists a valid partition X'_A, X'_B of X' such that $\Phi(F)$ is an (X'_A, X'_B)-cut in G'. Hence, $C = \Phi(F) \cup M$ is an (X'_A, X'_B)-cut in H, and since $H[C]/M$ is isomorphic to $G[F]$, we have $r_M(C) = r(F) \leq k$. It follows that C is a solution for RANK-CUT on the instance $I' = (H, X'_A, X'_B, M, k)$. Conversely, suppose that C is a solution of RANK-CUT on the instance $I' = (H, X'_A, X'_B, M, k)$, for some valid partition X'_A, X'_B of X'. Observe that $M \subseteq C$ (as each edge of M is between X'_A and X'_B), and that $C \setminus M$ is an (X'_A, X'_B)-cut in $H \setminus M = G'$. Thus, if we define $F = \Phi^{-1}(C \setminus M)$, we obtain that $G \setminus F$ is bipartite by Lemma 4. Observe that contracting the edges of M in $H[C]$ gives a graph isomorphic to $G[F]$. Therefore, $r(F) = r_M(C) \leq k$, and we conclude that F is a solution for the instance I. $\square$

4   Solving a Constrained Version of Rank-Cut

In this section, we introduce a constrained variant of RANK-CUT, and show its polynomial-time solvability. We give a randomized reduction of RANK-CUT to this variant in the next section. In the constrained problem, the cut has to be constructed as the union of disjoint components prescribed in the input:

CONSTRAINED RANK-CUT
   Input: A graph G, an integer k, two subsets $X, Y \subseteq V(G)$, a set $M \subseteq E(G)$, and a partition $P = (V_1, \ldots, V_\ell)$ of $V(G)$ such that

   (i) $G[V_i]$ is connected for every $1 \leq i \leq \ell$, and
   (ii) there is no edge of M between V_i and V_j for any $i \neq j$.

   Question: Is there a set $Z \subseteq \{1, \ldots, \ell\}$ such that $C_Z = \cup_{i \in Z} E(G[V_i])$ is an (X, Y)-cut in G with $r_M(C_Z) \leq k$?

Note that a V_i can consist of a single vertex, in which case $E(G[V_i]) = \emptyset$ and it does not matter if i is in Z or not. We show that the constrained problem can be reformulated as a weighted separator problem.

Lemma 7. CONSTRAINED RANK-CUT *can be solved in* $O(k(n + m))$ *time.*

Proof. Let $I = (G, k, X, Y, M, P)$ be an instance of CONSTRAINED RANK-CUT with $P = (V_1, \ldots, V_\ell)$. Starting with G, we build a weighted graph G' as follows:

- we remove the edges of $\cup_{i=1}^{\ell} E(G[V_i])$;
- we give an infinite weight to the vertices of $V(G)$;
- for each $1 \leq i \leq \ell$, we add a vertex v_i of weight $r_M(E(G[V_i]))$, and we make v_i adjacent to the vertices of V_i.

We answer "yes" if and only if G' has an (X, Y)-separator of weight at most k. We claim that this algorithm takes $O(k(n + m))$ time. First, observe that G' has at most $n + m$ edges: for each $i \in \{1, \ldots, \ell\}$, we replace the edges of $E(G[V_i])$ by a number of edges equal to $|V_i| \leq |E(G[V_i])| + 1$. As we are trying to find an (X, Y)-separator of weight at most k in G', we can accomplish this by performing at most k rounds of the Ford-Fulkerson max-flow min-cut algorithm, giving the running time $O(k(n + m))$.

Given a set $Z \subseteq \{1, \ldots, \ell\}$, let us define edge set $C_Z = \cup_{i \in Z} E(G[V_i])$ and vertex set $S_Z = \{v_i : i \in Z\}$. The following claim establishes the correctness of the algorithm.

> *Claim.* For any $Z \subseteq \{1, \ldots, \ell\}$,
> (i) $r_M(C_Z)$ equals the weight of S_Z;
> (ii) C_Z is an (X, Y)-cut in G iff S_Z is an (X, Y)-separator in G'.

To prove (i), note first that the vertex set of each connected component of $G[C_Z]$ is some V_i. Furthermore, as the two endpoints of each edge in M is in the same V_i, it is also true in the graph $G[C_Z \cup M]$ that the vertex set of each connected component is some V_i. Thus, each connected component of $G[C_Z \cup M]/M$ is obtained from a set V_i by identifying some vertices. We obtain that $r_M(C_Z) = \sum_{i \in Z} r_M(E(G[V_i]))$ equals the weight of S_Z.

To prove (ii), suppose that C_Z is an (X, Y)-cut in G; we need to show that S_Z is an (X, Y)-separator in G'. By way of contradiction, assume that G' contains an (X, Y)-path W avoiding S_Z. For each segment of W of the form $x v_i y$ with $x, y \in V_i, i \notin Z$, we replace it by an x, y-path in $G[V_i]$ (recall that the neighbors of v_i are in V_i). We obtain an (X, Y)-walk in G avoiding C_Z, a contradiction.

Conversely, suppose that S_Z is an (X, Y)-separator in G', and let us show that C_Z is an (X, Y)-cut in G. By way of contradiction, assume that G contains an (X, Y)-path W avoiding C_Z. Then W can be partitioned as $W_1 W_2 \ldots W_r$, where each W_j is a maximal subpath of W included in a set V_i (possibly, W_j contains a single vertex). Each W_j that has at least two vertices is an x, y-path included in a component V_i with $i \notin Z$; we replace it by a path of the form $x v_i y$, to obtain an (X, Y)-walk in G' avoiding S_Z, a contradiction. □

5   Reduction to the Constrained Version

In this section, we describe a randomized reduction mapping an instance $I = (G, k, X, Y, M)$ of RANK-CUT to an instance $I' = (G, k, X, Y, M, P)$ of CONSTRAINED RANK-CUT.

The first step of the reduction identifies a set of *relevant edges* $E_{\mathrm{rel}} \subseteq E(G)$ that spans a graph of bounded degree. It relies on the notion of important separators introduced in [19], which we recall now. Fix two disjoint sets $X, Y \subseteq V(G)$, and let S be an (X, Y)-separator in G. We denote by $\mathrm{Reach}_G(X, S)$ the set of vertices of G reachable from X in $G \backslash S$; note that $\mathrm{Reach}_G(X, S)$ is disjoint from Y. We say that S is an *important* (X, Y)-separator if (i) S is an inclusion-wise minimal (X, Y)-separator, (ii) there is no (X, Y)-separator S' with $|S'| \leq |S|$ and $\mathrm{Reach}_G(X, S) \subset \mathrm{Reach}_G(X, S')$. We have the following result:

Lemma 8 ([19,4,6]). *Let k be a nonnegative integer. There are at most 4^k important (X, Y)-separators of size $\leq k$, and they can be enumerated in time $O(4^k k(n + m))$.*

We now describe the construction of the set E_{rel}. Starting with G, we construct a graph G' by subdividing each edge e with a vertex z_e. Given two subsets $X, Y \subseteq V(G)$, we denote by $C_k(X, Y)$ the union of the important (X, Y)-separators of size at most k in the extended graph G'. As there are are at most 4^k such separators by Lemma 8, we have $|C_k(X, Y)| \leq k \cdot 4^k$. Given a vertex $u \in V(G)$, we denote by $E(u)$ the set of edges of G incident to u. We define the set $E_{\mathrm{rel}} \subseteq E(G)$ as follows: (i) for each $u \in V(G)$, let $E_{\mathrm{rel}}(u) = \{e \in E(u) : z_e \in C_{6k}(X, \{u\}) \cup C_{6k}(Y, \{u\})\}$; (ii) E_{rel} consists of those edges $uv \in E(G)$ such that $uv \in E_{\mathrm{rel}}(u) \cap E_{\mathrm{rel}}(v)$. By Lemma 8, E_{rel} can be constructed in time $O(4^{6k} k \cdot n(n + m))$, as we need to enumerate important separators for n vertices. Furthermore, the graph $G[E_{\mathrm{rel}}]$ has maximum degree $d = 12k \cdot 4^{6k}$, as each set $E_{\mathrm{rel}}(u)$ comes from the union of two sets $C_{6k}(X, \{u\})$ and $C_{6k}(Y, \{u\})$, each of which has size at most $6k \cdot 4^{6k}$. The interest of the set E_{rel} is that it contains any minimal solution for I.

Lemma 9. *Any minimal solution C of a* RANK-CUT *instance I is included in E_{rel}.*

Proof. We show that for every $e = uv \in C$, we have $e \in E_{\mathrm{rel}}(v)$; this will imply that $e \in E_{\mathrm{rel}}(u)$ by symmetry, and thus $e \in E_{\mathrm{rel}}$. As C is a minimal (X, Y)-cut, if we define U to be the set of vertices reachable from X in $G \setminus C$, then $X \subseteq U \subseteq V(G) \setminus Y$ holds and C is the set of edges with exactly one endpoint in U. Let C_X denote the endpoints of C inside U, and let C_Y denote the endpoints of C inside $V(G) \setminus U$. We suppose that $v \in C_Y$, as the case $v \in C_X$ is similar. Let us define the vertex set S of G' as $S = (C_Y \setminus v) \cup \{z_e : e \in C \cap E(v)\}$. We make the following observations:

- S is an (X, v)-separator in G', as each (X, v)-path in G either goes through $C_Y \setminus v$, or goes through an edge of C incident to v (note also that S is disjoint from $X \cup \{v\}$).
- $u \in \mathrm{Reach}_{G'}(X, S)$: as C is a minimal (X, Y)-cut, there has to be an (X, u)-path in G disjoint from C, that is, fully contained in U, which means that the corresponding path in G' avoids S.
- $|S| \leq 6k$. By Lemma 5, there are at most $6k$ vertices in C. Every vertex of C can appear in C_Y or can be adjacent to v, but not both. Therefore, each vertex spanned by C contributes at most one to S and $|S| \leq 6k$ follows.

By the definition of important separators, there exists an important (X, v)-separator S' in G' such that $\mathrm{Reach}_{G'}(X, S) \subseteq \mathrm{Reach}_{G'}(X, S')$ and $|S'| \leq |S|$. As z_e is adjacent to u and v, as $u \in \mathrm{Reach}_{G'}(X, S')$ and as S' is an (X, v)-separator in G', it follows that $z_e \in S'$. Now, $S' \subseteq C_{6k}(X, \{v\})$ implies that $z_e \in C_{6k}(X, \{v\})$, and we conclude that $e \in E_{\mathrm{rel}}(v)$.   □

We construct an instance I' of CONSTRAINED RANK-CUT from the instance I of RANK-CUT, by the following random process. Let $p = \frac{1}{6kd} = 2^{-O(k)}$. We color edges of $E_{\mathrm{rel}} \setminus M$ with color black with probability p, and with color red otherwise. Let E_b denote the set containing the edges of E_{rel} colored black, as well as the edges of M. Consider the subgraph G_b of G containing only the edges in E_b and let partition $P = (V_1, \ldots, V_{\ell'})$ represent the way the connected components of this subgraph partition $V(G)$ (note that P can have classes that contain only a single vertex). By definition, $G[V_i]$ is connected for every i and the two endpoints of each edge in M is in the same V_i. Therefore, the CONSTRAINED RANK-CUT instance $I' = (G, k, X, Y, M, P)$ is a valid instance, as it satisfies both (i) and (ii).

Lemma 10. *The following two statements hold:*

1. *If I is a no-instance of* RANK-CUT, *then I' is a no-instance of* CONSTRAINED RANK-CUT.
2. *If I is a yes-instance of* RANK-CUT, *then I' is a yes-instance of* CONSTRAINED RANK-CUT *with probability* $\geq 2^{-O(k^2)}$.

Proof. Clearly, if I' has a solution Z, then C_Z is a solution for instance I of RANK-CUT. Conversely, suppose that I has a minimal solution $C \subseteq E(G)$ with $r_M(C) \leq k$. Let $U_1, \ldots, U_{\ell'}$ denote the vertex sets of the connected components of $G[C \cup M]$ (note that this is not necessarily a partition of $V(G)$, as it is possible to have vertices that are not incident to any edge of $C \cup M$). Let F be a spanning forest of $G[C \cup M]$ containing as many edges of M as possible. Let $B = F \setminus M$; as all these edges are in C, we have $B \subseteq E_{\mathrm{rel}}$ by Lemma 9, and since we have $r_M(C) \leq k$ it follows that $|B| \leq k$. Let $R = \cup_{i=1}^{\ell'} E_{\mathrm{rel}}(U_i)$, where $E_{\mathrm{rel}}(U_i)$ denotes the set of edges in E_{rel} with exactly one endpoint in U_i. By Lemma 5, $C \cup M$ spans at most $6k$ vertices, thus $\sum_{i=1}^{\ell'} |U_i| \leq 6k$. As each vertex of $V(G)$ has at most d incident edges in E_{rel}, we have $|R| \leq d \sum_{i=1}^{\ell'} |U_i| \leq 6kd = 2^{O(k)}$. Now, (i) with probability at least $p^k = 2^{-O(k^2)}$, every edge of B is colored black, (ii) with probability at least $(1 - \frac{1}{6kd})^{6kd} \geq \frac{1}{4}$, every edge of R is colored red (indeed, the function $x \mapsto (1 - \frac{1}{x})^x$ is increasing on $[1, +\infty[$ and is thus $\geq \frac{1}{4}$ for $x \geq 2$). These two events are independent, as they involve disjoint sets of edges. Suppose that both events happen. Then, E_b contains all edges of F, but no edge of R. Consider the subgraph G_b of G containing only the edges in E_b and let partition $P = (V_1, \ldots, V_\ell)$ represent the way the connected components of this subgraph partition $V(G)$. Then every U_i is one class of this partition. Thus $C' = \cup_{i=1}^{\ell'} E(G[U_i])$ is a solution for instance I' (as $C' \supseteq C \cup M$ and $r_M(C') = r_M(C) \leq k$). We conclude that I' is a yes-instance with probability $\geq 2^{-O(k^2)}$.   □

From Lemmas 7 and 10, we obtain:

Theorem 11. RANK-CUT *has a randomized* FPT *algorithm with running time* $2^{O(k^2)}nm$.

Proof. Let $I = (G, k, X, Y, M)$ be an instance of RANK-CUT. We first remove all isolated vertices of G in $O(n + m)$ time, obtaining a graph G for which each connected component has at least two vertices, ensuring that $n + m = O(m)$. We then compute the set E_{rel} in time $O(4^{6k}knm)$, and we construct the instance I' of CONSTRAINED RANK-CUT by random selection as described above. This instance I' can be solved in time $O(k(n + m))$ by Lemma 7. By Lemma 10, the probability of a false "no" answer is at least $p_e = 2^{-O(k^2)}$. Thus repeating this process $\frac{1}{p_e} = 2^{O(k^2)}$ times yields a randomized FPT algorithm for RANK-CUT running in time $2^{O(k^2)}nm$ and having success probability $(1 - p_e)^{\frac{1}{p_e}} \geq \frac{1}{4}$.   □

6   Derandomization

We now derandomize the proofs of Lemma 10 and Theorem 11 using the standard technique of splitters. Given integers n, s, t, an (n, s, t)-*splitter* is a family $\mathcal{F}$ of functions $f : [n] \to [t]$ such that for every $S \subseteq [n]$ with $|S| = s$, there is a function of $\mathcal{F}$ that is injective of S. Naor et al. [23] give a deterministic construction of an (n, s, s^2)-splitter of size $O(s^6 \log s \log n)$. We can use this splitter construction to build a family of colorings of E_{rel} to replace the randomized selection of colors in Lemma 10. By setting the parameters appropriately, we can ensure that at least one coloring in the family has the property that every edge of B is colored black and every edge of R is colored red. The (n, s, s^2)-splitter of Naor et al. [23] can be constructed in polynomial time, but unfortunately the exact running time is not stated. Therefore, in the following theorem, we do not specify the polynomial factors of the running time.

Theorem 12. RANK-CUT *has a deterministic* FPT *algorithm with running time* $2^{O(k^2)}n^{O(1)}$.

Proof. Consider an instance $I = (G, k, X, Y, M)$ of RANK-CUT. We first construct the set E_{rel} as in Section 5, and we identify $E_{rel} \setminus M$ with the set $[m']$ where $m' = |E_{rel} \setminus M|$. Let $s = k + 4kd = 2^{O(k)}$. Using the result of [23], we construct an (m', s, s^2)-splitter $\mathcal{F}$ of size $O(s^6 \log s \log m)$. Instead of randomly coloring the elements of $E_{rel} \setminus M$, we go through the following deterministic family of colorings: for every $f \in \mathcal{F}$ and every subset $U \subseteq [s^2]$ of size at most k, we color $e \in E_{rel} \setminus M$ black if and only if $f(e) \in U$. For each such coloring, we perform the reduction to CONSTRAINED RANK-CUT as in Lemma 10 and then solve the instance using the algorithm of Lemma 7. We return "yes" if and only if at least one of the resulting CONSTRAINED RANK-CUT instances is a yes-instance.

It is clear that if one of the CONSTRAINED RANK-CUT instances is a yes-instance, then I is a yes-instance of RANK-CUT. Conversely, suppose that I is a yes-instance and let B and R be the set of edges defined in the proof of Lemma 10. As $|B| + |R| \leq s$, there is a function $f \in \mathcal{F}$ that is injective on $B \cup R$

and there is a set $U \subseteq [s^2]$ of size at most k such that $b \in B \cup R$ satisfies $b \in B$ if and only $f(b) \in U$. For this choice of f and U, the algorithm considers a coloring that colors B black and R red. Therefore, the reduction creates a yes-instance of CONSTRAINED RANK-CUT. $\square$

Theorems 11 and 12 respectively give randomized and deterministic FPT algorithms for RANK-CUT. Combining them with Lemmas 3 and 6, we obtain (i) a $2^{O(k^2)}nm$ randomized algorithm for BIPARTITE CONTRACTION, (ii) a $2^{O(k^2)}n^{O(1)}$ deterministic algorithm BIPARTITE CONTRACTION. This establishes Theorem 1 stated in the introduction.

7   Concluding Remarks

We have obtained a randomized $2^{O(k^2)}nm$ algorithm for BIPARTITE CONTRACTION. Can the dependence on k be improved? It seems plausible that the problem admits a $2^{O(k)}n^{O(1)}$ FPT algorithm, as such algorithms are known for EDGE BIPARTIZATION [11] as well as for other edge contraction problems [12]. We note that important separators are a common feature of [13] and of our algorithm, so they could be the key to further improvements.

Regarding kernelization, Heggernes et al. [13] asked whether BIPARTITE CONTRACTION has a polynomial kernel. While this question is still open, it is now known that ODD CYCLE TRANSVERSAL (and thus EDGE BIPARTIZATION) have randomized polynomial kernels [16]. As EDGE BIPARTIZATION reduces to BIPARTITE CONTRACTION, this raises the question whether the matroid-based techniques of [16,17] can be applied to the more general BIPARTITE CONTRACTION as well. The notion of rank in the RANK-CUT problem is the same as the notion of rank in graphic matroids, hence it is possible that the rank constraint can be incorporated into the arguments of [16,17] based on linear representation of matroids.

References

1. Alon, N., Yuster, R., Zwick, U.: Color-coding. J. ACM 42(4), 844–856 (1995)
2. Cao, Y., Marx, D.: Interval deletion is fixed-parameter tractable. CoRR, abs/1211.5933 (2012), Accepted to SODA 2014
3. Chen, J., Fomin, F.V., Liu, Y., Lu, S., Villanger, Y.: Improved algorithms for feedback vertex set problems. J. Comput. Syst. Sci. 74(7), 1188–1198 (2008)
4. Chen, J., Liu, Y., Lu, S.: An improved parameterized algorithm for the minimum node multiway cut problem. Algorithmica 55(1), 1–13 (2009)
5. Chen, J., Liu, Y., Lu, S., O'Sullivan, B., Razgon, I.: A fixed-parameter algorithm for the directed feedback vertex set problem. J. ACM 55(5) (2008)
6. Chitnis, R.H., Hajiaghayi, M., Marx, D.: Fixed-parameter tractability of directed multiway cut parameterized by the size of the cutset. SIAM Journal of Computing 42(4), 1674–1696 (2013), http://arxiv.org/abs/1110.0259

7. Cygan, M., Pilipczuk, M., Pilipczuk, M.: On Group Feedback Vertex Set Parameterized by the Size of the Cutset. In: Golumbic, M.C., Stern, M., Levy, A., Morgenstern, G. (eds.) WG 2012. LNCS, vol. 7551, pp. 194–205. Springer, Heidelberg (2012)

8. Dehne, F.K., Fellows, M.R., Langston, M.A., Rosamond, F.A., Stevens, K.: An $O(2^{O(k)}n^3)$ FPT algorithm for the undirected feedback vertex set problem. Theor. Comput. Syst. 41(3), 479–492 (2007)

9. Golovach, P.A., van 't Hof, P., Paulusma, D.: Obtaining Planarity by Contracting Few Edges. In: Rovan, B., Sassone, V., Widmayer, P. (eds.) MFCS 2012. LNCS, vol. 7464, pp. 455–466. Springer, Heidelberg (2012)

10. Guillemot, S.: FPT algorithms for path-transversals and cycle-transversals problems. Discrete Optimization 8(1), 61–71 (2011)

11. Guo, J., Gramm, J., Hüffner, F., Niedermeier, R., Wernicke, S.: Compression-based fixed-parameter algorithms for feedback vertex set and edge bipartization. J. Comput. Syst. Sci. 72(8), 1386–1396 (2006)

12. Heggernes, P., van 't Hof, P., Lévêque, B., Lokshtanov, D., Paul, C.: Contracting Graphs to Paths and Trees. In: Marx, D., Rossmanith, P. (eds.) IPEC 2011. LNCS, vol. 7112, pp. 55–66. Springer, Heidelberg (2012)

13. Heggernes, P., van't Hof, P., Lokshtanov, D., Paul, C.: Obtaining a Bipartite Graph by Contracting Few Edges. In: FSTTCS 2011, pp. 217–228 (2011)

14. Iwata, Y., Oka, K., Yoshida, Y.: Linear-time FPT algorithms via network flow. CoRR, abs/1307.4927 (2013), Accepted to SODA 2014

15. Kawarabayashi, K., Reed, B.A.: An (almost) Linear Time Algorithm for Odd Cycle Transversal. In: SODA 2010, pp. 365–378 (2010)

16. Kratsch, S., Wahlström, M.: Compression via Matroids: a Randomized Polynomial Kernel for Odd Cycle Transversal. In: SODA 2012, pp. 94–103 (2012)

17. Kratsch, S., Wahlström, M.: Representative sets and irrelevant vertices: New tools for kernelization. In: FOCS 2012, pp. 450–459 (2012)

18. Lokshtanov, D., Saurabh, S., Sikdar, S.: Simpler Parameterized Algorithm for OCT. In: Fiala, J., Kratochvíl, J., Miller, M. (eds.) IWOCA 2009. LNCS, vol. 5874, pp. 380–384. Springer, Heidelberg (2009)

19. Marx, D.: Parameterized graph separation problems. Theoretical Computer Science 351(3), 394–406 (2006)

20. Marx, D.: Chordal deletion is fixed-parameter tractable. Algorithmica 57(4), 747–768 (2010)

21. Marx, D., O'Sullivan, B., Razgon, I.: Finding small separators in linear time via treewidth reduction. ACM Transactions on Algorithms 9(4) (2013)

22. Marx, D., Schlotter, I.: Obtaining a planar graph by vertex deletion. Algorithmica 62(3–4), 807–822 (2012)

23. Naor, M., Schulman, L.J., Srinivasan, A.: Splitters and near-optimal derandomization. In: FOCS 1995, pp. 182–191 (1995)

24. Narayanaswamy, N., Raman, V., Ramanujan, M., Saurabh, S.: LP can be a cure for Parameterized Problems. In: STACS 2012, pp. 338–349 (2012)

25. Ramanujan, M.S., Saurabh, S.: Linear time parameterized algorithms via skew-symmetric multicuts. CoRR, abs/1304.7505 (2013), Accepted to SODA 2014

26. Reed, B.A., Smith, K., Vetta, A.: Finding odd cycle transversals. Oper. Res. Lett. 32(4), 299–301 (2004)

27. Villanger, Y., Heggernes, P., Paul, C., Telle, J.A.: Interval completion is fixed parameter tractable. SIAM J. Comput. 38(5), 2007–2020 (2009)

On the Ordered List Subgraph Embedding Problems

Olawale Hassan[1], Iyad Kanj[1], Daniel Lokshtanov[2], and Ljubomir Perković[1]

[1] School of Computing, DePaul University, Chicago, USA
oahassan@gmail.com, {ikanj,lperkovic}@cs.depaul.edu
[2] Department of Informatics, University of Bergen, Bergen, Norway
daniello@ii.uib.no

Abstract. In the parameterized ORDERED LIST SUBGRAPH EMBEDDING problem (p-OLSE) we are given two graphs G and H, each with a linear order defined on its vertices, a function L that associates with every vertex in G a list of vertices in H, and a parameter k. The question is to decide if we can embed (one-to-one) a subgraph S of G of k vertices into H such that: (1) every vertex of S is mapped to a vertex from its associated list, (2) the linear orders inherited by S and its image under the embedding are respected, and (3) if there is an edge between two vertices in S then there is an edge between their images. If we require the subgraph S to be embedded as an induced subgraph, we obtain the ORDERED LIST INDUCED SUBGRAPH EMBEDDING problem (p-OLISE). The p-OLSE and p-OLISE problems model various problems in Bioinformatics related to structural comparison/alignment of proteins.

We investigate the complexity of p-OLSE and p-OLISE with respect to the following structural parameters: the *width* Δ_L of the function L (size of the largest list), and the maximum degree Δ_H of H and Δ_G of G. We provide tight characterizations of the classical and parameterized complexity, and approximability of the problems with respect to the structural parameters under consideration.

1   Introduction

1.1   Problem Definition and Motivation

We consider the following problem that we refer to as the parameterized ORDERED LIST SUBGRAPH EMBEDDING problem, shortly (p-OLSE):

Given: Two graphs G and H with linear orders $\prec_G$ and $\prec_H$ defined on the vertices of G and H; a function $L : V(G) \longrightarrow 2^{V(H)}$; and $k \in \mathbb{N}$
Parameter: k
Question: Is there a subgraph S of G of k vertices and an injective map $\varphi : V(S) \longrightarrow V(H)$ such that: (1) $\varphi(u) \in L(u)$ for every $u \in S$; (2) for every $u, u' \in S$, if $u \prec_G u'$ then $\varphi(u) \prec_H \varphi(u')$; and (3) for every $u, u' \in S$, if $uu' \in E(G)$ then $\varphi(u)\varphi(u') \in E(H)$

G. Gutin and S. Szeider (Eds.): IPEC 2013, LNCS 8246, pp. 189–201, 2013.

The parameterized ORDERED LIST INDUCED SUBGRAPH EMBEDDING (p-OLISE) problem, in which we require the subgraph S to be embedded as an induced subgraph, is defined the same way as p-OLSE except that condition (3) is replaced with: for every $u, u' \in S$, $uu' \in E(G)$ if and only if $\varphi(u)\varphi(u') \in E(H)$. The optimization versions of p-OLSE and p-OLISE, denoted opt-OLSE and opt-OLISE, respectively, ask for a subgraph S of G with maximum number of vertices such that there exists a valid list embedding φ that embeds S into H.

The p-OLSE and p-OLISE problems have applications in the area of Bioinformatics because they provide a graph-theoretical model for numerous protein and DNA structural comparison problems (see [3,5,6,15]).

In this paper we investigate the complexity of p-OLSE and p-OLISE with respect to the following structural parameters: the *width* Δ_L of the function L (*i.e.*, the size of the largest list $|L(u)|$, for $u \in G$) and the maximum degree Δ_H of H and Δ_G of G. Restrictions on the structural parameters Δ_H, Δ_G and Δ_L are very natural in Bioinformatics. The parameters Δ_H and Δ_G model the maximum number of hydrophobic bonds that an amino acid in each protein can have; on the other hand, Δ_L is usually a parameter set by the Bioinformatics practitioners when computing the top few alignments of two proteins [15].

1.2 Previous Related Results

Goldman et al. [8] studied protein comparison problems using the notion of *contact maps*, which are undirected graphs whose vertices are linearly ordered. Goldman et al. [8] formulated the protein comparison problem as a CONTACT MAP OVERLAP problem, in which we are given two contact maps and we need to identify a subset of vertices S in the first contact map, a subset of vertices S' in the second with $|S| = |S'|$, and an order-preserving bijection $f : S \longrightarrow S'$, such that the number of edges in S that correspond to edges in S' is maximized. In [8], the authors proved that the CONTACT MAP OVERLAP problem is MAXSNP-complete, even when both contact maps have maximum degree one. The main difference between the CONTACT MAP OVERLAP problem and the opt-OLSE and opt-OLISE problems under consideration is that in opt-OLSE and opt-OLISE the function is restricted to mapping a vertex to one in its list, and the goal is to maximize the number of vertices not the number of edges that can be embedded.

The p-OLISE problem generalizes the LONGEST ARC-PRESERVING COMMON SUBSEQUENCE (LAPCS) problem, which is a well-studied problem (see [1,5,6,9,10,12]). In LAPCS, we are given two sequences S_1 and S_2 over a fixed alphabet, where each sequence has arcs/edges between its characters, and the problem is to compute a longest common subsequence of S_1 and S_2 that respects the arcs. The p-OLISE problem generalizes LAPCS since no restriction is placed on the size of the alphabet, and a vertex can be mapped to any vertex from its list. Consequently, the "positive" results obtained in this paper about p-OLISE and opt-OLISE apply directly to their corresponding versions of LAPCS; on the other hand, we are able to borrow the $W[1]$-hardness result from [6] to conclude the W-hardness results in Proposition 3 and Proposition 4. The LAPCS problem was introduced by [5,6] where it was shown to be

$W[1]$-complete (parameterized by the length of common subsequence sought) in the case when the arcs are *crossing*. Several works studied the complexity and approximation of LAPCS with respect to various restrictions on the types of the arcs (*e.g., nested, crossing*, etc.) [1,5,6,9,10,12]. The work in [1,9] considered the problem in the case of nested arcs parameterized by the total number of characters that need to be deleted from S_1 and S_2 to obtain the arc-preserving common subsequence. They showed that the problem is FPT with respect to this parameterization, and they also showed it to be FPT when parameterized by the length of the common subsequence in the case when the alphabet consists of four characters.

A slight variation of p-OLSE was considered in [3,15], where the linear order imposed on G and H was replaced with a partial order (directed acyclic graphs); the problem was referred to as the GRAPH EMBEDDING problem in [3] and as the GENERALIZED SUBGRAPH ISOMORPHISM problem in [15]. The aforementioned problems were mainly studied in [3,15] assuming no bound on Δ_H and Δ_G (*i.e.,* unbounded) and, not surprisingly, only hardness results were derived. In [15], a parameterized algorithm with respect to the treewidth of G and the map width Δ_L combined was given. Most of the hardness results in [3,15] were obtained by a direct reduction from the INDEPENDENT SET or CLIQUE problems. For example, it was shown in [3] that the problem of embedding the whole graph G into H is $\mathcal{NP}$-hard, but is in $\mathcal{P}$ if $\Delta_L = 2$. It was also shown that the problem of embedding a subgraph of G of order k into H is $W[1]$-complete even when $\Delta_L = 1$, and cannot be approximated to a ratio $n^{\frac{1}{2}-\varepsilon}$ unless $\mathcal{P} = \mathcal{NP}$; we borrow these two hardness results as they also work for p-OLSE and p-OLISE.

Finally, one can draw some similarities between p-OLISE and the celebrated SUBGRAPH ISOMORPHISM and GRAPH EMBEDDING problems. The main differences between p-OLISE and the aforementioned problems are: (1) in p-OLSE we have linear orders on G and H that need to be respected by the map sought, (2) we ask for an embedding of a subgraph of G rather than the whole graph G, and (3) each vertex must be mapped to a vertex from its list. In particular, requirement (1) above precludes the application of well-known (logic) meta-theorems (see [7]) to the restrictions of p-OLISE that are under consideration in this paper.

1.3   Our Results and Techniques

We draw a complete complexity landscape of p-OLSE and p-OLISE with respect to the computational frameworks of classical complexity, parameterized complexity, and approximation, in terms of the structural parameters Δ_H, Δ_G and Δ_L. Table 1 outlines the obtained results about p-OLSE and p-OLISE and their optimization versions. Note that even though our hardness results are for specific values of the parameters Δ_H, Δ_G, and Δ_L, these results certainly hold true for restrictions of the problems to instances in which the corresponding parameters are upper bounded by (or equal to — by adding dummy vertices) any constants larger than these specific values. Observe also that the results we obtain *completely and tightly* characterize the complexity (with respect to all frameworks under consideration) of the problems with respect to Δ_H, Δ_G and Δ_L (unbounded vs. bounded, and when applicable, for different specific values).

Section 2 presents various complexity and approximation results. The NP-hardness results are obtained by a reduction from the k-MULTI-COLORED INDE-PENDENT SET problem, and the $W[1]$-hardness results are obtained by tweaking the $W[1]$-hardness results given in the literature [3,6], or by simple known reductions from the INDEPENDENT SET problem. Section 3 presents FPT algorithms for various restrictions of p-OLSE and p-OLISE. The FPT results in Theorem 2 for p-OLSE, when $\Delta_L = O(1)$, $\Delta_G = O(1)$, and $\Delta_H = \infty$, are derived using the random separation method. This method is applied after transforming the problem — via reduction operations — to the INDEPENDENT SET problem on a graph composed of (1) a permutation graph and (2) a set of additional edges between the permutation graph vertices such that the number of additional edges incident to any vertex is at most a constant; Lemma 4 then shows that the INDEPENDENT SET problem on such graphs is FPT. On the other hand, the FPT results in Proposition 6, when $\Delta_H = 0$, $\Delta_G = O(1)$ (resp. $\Delta_G = 0$ and $\Delta_H = O(1)$ for p-OLISE by symmetry) and $\Delta_L = \infty$, are also derived using the random separation method, but the argument is simpler.

To cope with the W-hardness of p-OLSE in certain cases, we consider a different parameterization of the problem, namely the parameterization by the vertex cover number of G, and denote the associated problem by p-VC-OLSE. This parameterization is not interesting for p-OLISE since we proved that p-OLISE is NP-complete in the case when $\Delta_G = 0$, $\Delta_H = 1$ and $\Delta_L = 1$, and hence the problem is *para-NP-hard* with respect to this parameterization. Proposition 7 shows that p-VC-OLSE is $W[1]$-complete in the case when $\Delta_H = 1$, $\Delta_G = 1$ and Δ_L is unbounded (note that if either $\Delta_H = 0$ or $\Delta_G = 0$ then p-OLSE is FPT when $\Delta_L = \infty$). So we restrict our attention to the case when $\Delta_L = O(1)$, and show in this case that the problem is FPT even when both Δ_H and Δ_G are unbounded; the method relies on a bounded search tree approach, combined with the dynamic programming algorithm described in Proposition 1.

1.4   Background and Terminologies

We assume knowledge of the basic background about approximation algorithms.

Graphs. For a graph H we denote by $V(H)$ and $E(H)$ the set of vertices and edges of H, respectively. The *order* of a graph H is $|V(H)|$. For a set of vertices $S \subseteq V(H)$, we denote by $H[S]$ the subgraph of H induced by the vertices in S. For a subset of edges $E' \subseteq E(H)$, we denote by $H - E'$ the graph $(V(H), E(H) \setminus E')$. For a vertex $v \in H$, $N(v)$ denotes the set of neighbors of v in H. The *degree* of a vertex v in H, denoted $deg_H(v)$, is $|N(v)|$. A vertex v is *isolated* in H if $deg_H(v) = 0$. The *degree* of H, denoted $\Delta(H)$, is $\Delta(H) = \max\{deg_H(v) : v \in H\}$. An *Independent Set* of a graph H is a set of vertices I such that no two vertices in I are adjacent. A *vertex cover* of H is a set of vertices such that each edge in H is incident to at least one vertex in this set; we denote by $\tau(H)$ the cardinality of a minimum vertex cover of H. Let L and L' be two parallel lines in the plane.

Table 1. Classical, approximation, and parameterized complexity maps of p-OLSE and p-OLISE with respect to Δ_H, Δ_G and Δ_L. The inapproximability results are under the assumption that $\mathcal{P} \neq \mathcal{NP}$. The symbol ∞ stands for unbounded degree, and the results with the $O(1)$ upper bound on the degree hold true for *any* fixed degree.

p-OLSE	Δ_H	Δ_G	Δ_L	**Complexity**
Classical	∞	0	∞	$\mathcal{P}$
	0	1	1	$\mathcal{NP}$-complete
Approximation	∞	$O(1)$	∞	$\mathcal{APX}$-complete
	0	∞	1	not approximable to $n^{\frac{1}{2}-\varepsilon}$ (Thm. 0.4, [3])
Parameterized	∞	$O(1)$	$O(1)$	FPT
	1	1	∞	$W[1]$-complete
	0	∞	1	$W[1]$-complete (Thm. 0.3, [3])
	0	$O(1)$	∞	FPT
	∞	0	∞	FPT (even in $\mathcal{P}$)

p-OLISE	Δ_H	Δ_G	Δ_L	**Complexity**
Classical	0	0	∞	$\mathcal{P}$
	0	1	1	$\mathcal{NP}$-complete
	1	0	1	$\mathcal{NP}$-complete
Approximation	$O(1)$	$O(1)$	∞	$\mathcal{APX}$-complete
	0	∞	1	not approximable to $n^{\frac{1}{2}-\varepsilon}$ (Thm. 0.4, [3])
	∞	0	1	not approximable to $n^{\frac{1}{2}-\varepsilon}$
Parameterized	∞	0	1	$W[1]$-complete
	0	∞	1	$W[1]$-complete
	0	$O(1)$	∞	FPT
	$O(1)$	0	∞	FPT
	1	1	1	$W[1]$-complete

A *permutation graph* P is the intersection graph of a set of line segments D such that one endpoint of each of those segments lies on L and the other endpoint lies on L'.

Parameterized Complexity. A *parameterized problem* is a set of instances of the form (x, k), where $x \in \Sigma^*$ for a finite alphabet set Σ, and k is a non-negative integer called the *parameter*. A parameterized problem Q is *fixed parameter tractable* (FPT), if there exists an algorithm that on input (x, k) decides if (x, k) is a yes-instance of Q in time $f(k)n^{O(1)}$, where f is a computable function independent of $n = |x|$; we will denote by *fpt-time* a running time of the form $f(k)n^{O(1)}$. A hierarchy of fixed-parameter intractability, *the W-hierarchy* $\bigcup_{t \geq 0} W[t]$, was introduced based on the notion of *fpt-reduction*, in which the 0-th level $W[0]$ is the class FPT. It is commonly believed that $W[1] \neq$ FPT. The asymptotic notation $O^*()$ suppresses a polynomial factor in the input length.

We will denote an instance of p-OLSE or p-OLISE by the tuple $(G, H, \prec_G, \prec_H, L, k)$. We shall call an injective map φ satisfying conditions (1)-(3) in the definition of p-OLSE and p-OLISE (given in Section 1) for some subgraph S of G, a *valid list embedding*, or simply a *valid embedding*. Constraint (3) will be referred to as the *embedding constraint* (note that constraint (3) is different in the two problems). We define the *width* of L, denoted Δ_L, as $\max\{|L(v)| \mid v \in G\}$. It is often more convenient to view/represent the map L as a set of edges joining every vertex $u \in G$ to the vertices of H that are in $L(u)$.

2   Complexity Results

Consider the restrictions of the opt-OLSE and opt-OLISE problems to instances in which $\Delta_G = \Delta_H = 0$ (for p-OLSE we can even assume that $\Delta_H = \infty$ as the edges in H do not play any role when $\Delta_G = 0$). This version of the problem can be easily shown to be solvable in polynomial time by dynamic programming:

Proposition 1. *The opt-OLSE and opt-OLISE problems (and hence p-OLSE and p-OLISE) restricted to instances in which $\Delta_G = \Delta_H = 0$ are solvable in $O(|V(G)| \cdot |V(H)|)$ time (and hence are in $\mathcal{P}$).*

Proposition 1 will be useful for Proposition 2 and Theorem 3.

If $\Delta_G > 0$, the p-OLSE and p-OLISE problems become $\mathcal{NP}$-complete, even in the simplest case when $\Delta_G = 1$, $\Delta_H = 0$ and $\Delta_L = 1$. For p-OLISE, the same proof by symmetry shows the NP-completeness of the problem when $\Delta_H = 1$, $\Delta_G = 0$ and $\Delta_L = 1$ (this version of p-OLSE is in $\mathcal{P}$).

Theorem 1. *The p-OLSE and p-OLISE problems restricted to instances in which $\Delta_G = 1$, $\Delta_H = 0$ and $\Delta_L = 1$ are $\mathcal{NP}$-complete.*

Proposition 2. *The opt-OLSE problem restricted to instances in which $\Delta_G = O(1)$ has an approximation algorithm of ratio $(\Delta_G + 1)$, and the opt-OLISE problem restricted to instances in which $\Delta_G = O(1)$ and $\Delta_H = O(1)$ has an approximation algorithm of ratio $(\Delta_H + 1) \cdot (\Delta_G + 1)$.*

Proof. Let $(G, H, \prec_G, \prec_H, L)$ be an instance of opt-OLSE, and consider the following algorithm. Apply the dynamic programming algorithm in Proposition 1 to $(G, H, \prec_G, \prec_H, L)$ after removing the edges of G and the edges of H, and let S and φ be the subgraph and map obtained, respectively. Apply the following trivial approximation algorithm to compute an independent set I of S: pick a vertex v in S, include v in I, remove v and $N(v)$ from S, and repeat until S is empty. Return the subgraph $G[I] = I$, and the restriction of φ to I, φ_I. Clearly, we have $|I| \geq |V(S)|/(\Delta_G + 1)$.

Since φ is a valid list embedding of S with respect to G and H with their edges removed, and since I is an independent set of $G[S]$, it is clear that φ_I is a valid list embedding of $G[I]$ into H. Therefore, the algorithm is an approximation algorithm. Now let S_{opt} be an optimal solution of the instance. Clearly, we have $|V(S_{opt})| \leq |V(S)|$. Therefore, $|V(S_{opt})|/|I| \leq |V(S)|/|I| \leq (\Delta_G + 1)$.

For opt-OLISE, we follow the same steps to obtain I, and then we apply a similar approximation algorithm to H, to retain in I a subset whose image under φ_I is an independent set of H. At least $|I|/(\Delta_H + 1)$ vertices are retained.   □

The inapproximability results outlined in Table 1 follow by simple reductions from the MAXIMUM INDEPENDENT SET problem. The APX-hardness for the restrictions of opt-OLSE and opt-OLISE considered in Proposition 2 follow by a reduction from MAXIMUM INDEPENDENT SET on bounded-degree graphs. On the other hand, the inapproximability results for opt-OLSE and opt-OLISE when $\Delta_G = \infty$, $\Delta_L = 1$ and $\Delta_H = 0$ (and the symmetric case for opt-OLISE when

$\Delta_G = 0$, $\Delta_L = 1$ and $\Delta_H = \infty$) follow by a reduction from the (general) MAXIMUM INDEPENDENT SET that was given in [3] for a variation of p-OLSE.

Evans [6] proved that the LONGEST ARC-PRESERVING COMMON SUBSE-QUENCE (LAPCS) is $W[1]$-complete. LAPCS is a special case of p-OLISE, and it turns out that the reduction in [6] results in an instance that can be modeled by an instance of either p-OLSE or p-OLISE in which $\Delta_H = 1$, and $\Delta_G = 1$:

Proposition 3. *The p-OLSE and p-OLISE problems restricted to instances in which $\Delta_H = 1$ and $\Delta_G = 1$ are $W[1]$-complete.*

The reduction in [6] can also be tweaked to show that p-OLISE, restricted to instances in which $\Delta_H = O(1)$, $\Delta_G = O(1)$, and $\Delta_L = O(1)$ is $W[1]$-complete. To see this, observe first that the result would follow if we proved the $W[1]$-hardness of p-OLISE restricted to instances in which $\Delta_H = O(1)$, $\Delta_G = O(1)$, and the number of vertices in G that have the same vertex $v \in H$ in their list is also $O(1)$, since that would correspond to $\Delta_L = O(1)$ if we switched G and H (by symmetry). Using color-coding, the reduction in [6] can be tweaked to a Turing fpt-reduction in which $\Delta_G = 1$, $\Delta_H = 1$, and every vertex $v \in H$ appears in the list of exactly one vertex in G:

Proposition 4. *The p-OLISE problem restricted to instances in which $\Delta_H = 1$, $\Delta_G = 1$, and $\Delta_L = 1$ is $W[1]$-complete.*

The result below follows from a reduction given in [3] for a variant of p-OLSE:

Proposition 5. *([3], Theorem 0.3) The p-OLSE problem restricted to instances in which $\Delta_G = \infty$, $\Delta_H = 0$ and $\Delta_L = 1$ is $W[1]$-complete, and the p-OLISE problem restricted to instances in which $\Delta_G = \infty$ (resp. $\Delta_H = \infty$ by symmetry), $\Delta_H = 0$ (resp. $\Delta_G = 0$) and $\Delta_L = 1$ is $W[1]$-complete.*

3   FPT Results

We start by discussing the results for p-OLSE when both Δ_G and Δ_L are $O(1)$ (Δ_H may be unbounded). Let $(G, H, \prec_G, \prec_H, L, k)$ be an instance of p-OLSE in which both Δ_G and Δ_L are upper bounded by a fixed constant. Consider the graph $\mathcal{G}$ whose vertex-set is $V(G) \cup V(H)$ and whose edge-set is $E(G) \cup E(H) \cup E_L$, where $E_L = \{uv \mid u \in G, v \in H, v \in L(u)\}$; that is, $\mathcal{G}$ is the union of G and H plus the edges that represent the mapping L. We perform the following *splitting* operation on the vertices of $\mathcal{G}$ (see Figure 1 for illustration):

Definition 1. *Let u be a vertex in $\mathcal{G}$ and assume that $u \in G$ (the operation is similar when $u \in H$). Suppose that the vertices of G are ordered as $\langle u_1, \ldots, u_n \rangle$ with respect to $\prec_G$, and suppose that $u = u_i$, for some $i \in \{1, \ldots, n\}$. Let $e_1 = uv_1, \ldots, e_r = uv_r$ be the edges incident to u in E_L, and assume that $v_1 \prec_H v_2 \prec_H \ldots \prec_H v_r$. By splitting vertex u we mean: (1) replacing u in $\mathcal{G}$ with vertices $u_i^1, \ldots, u_i^r$ such that the resulting ordering of the vertices in G with respect to $\prec_G$ is $\langle u_1, \ldots, u_{i-1}, u_i^1, \ldots, u_i^r, u_{i+1}, \ldots, u_n \rangle$; (2) removing all the edges $e_1, \ldots, e_r$ from $\mathcal{G}$ and replacing them with the edges $u_i^1 v_r, u_i^2 v_{r-1}, \ldots, u_i^r v_1$; and (3) replacing every edge uu_j in G with the edges $u_i^s u_j$, for $s = 1, \ldots, r$.*

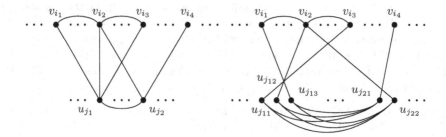

Fig. 1. Illustration of the splitting operation when applied to vertices u_{j_1} and u_{j_2}

Let $\mathcal{G}_{split}$ be the graph resulting from $\mathcal{G}$ by splitting every vertex in G and every vertex in H (in an arbitrary order), where G_{split} is the graph resulting from splitting the vertices in G and H_{split} that resulting from splitting the vertices of H. Let E_{split} be the set of edges having one endpoint in G_{split} and the other in H_{split}, $L_{split} : G_{split} \longrightarrow 2^{V(H_{split})}$ defined by $L_{split}(u) = \{v \mid uv \in E_{split}\}$ for $u \in V(G_{split})$, and let $\prec_{G_{split}}$ and $\prec_{H_{split}}$ be the orders on G_{split} and H_{split}, respectively, resulting from $\prec_G, \prec_H$ after the splitting operation. The following lemma can be easily verified:

Lemma 1. *The graph $\mathcal{G}_{split}$ satisfies the properties: (i) for every $u \in V(G_{split})$ we have $deg_{G_{split}}(u) \leq \Delta_L \cdot \Delta_G$;[1] (ii) in the graph $(V(G_{split}) \cup V(H_{split}), E_{split})$ every vertex has degree exactly 1 (in particular $|L_{split}(u)| = 1$ for every $u \in V(G_{split})$), and (iii) the instance $(G, H, \prec_G, \prec_H, L, k)$ is a yes-instance of p-OLSE if and only if $(G_{split}, H_{split}, \prec_{G_{split}}, \prec_{H_{split}}, L_{split}, k)$ is.*

Next, we perform the following operation, denoted **Simplify**, to $\mathcal{G}_{split}$. Since every vertex in $(V(G_{split}) \cup V(H_{split}), E_{split})$ has degree 1 by part (ii) of Lemma 1, if two vertices $u, u' \in G_{split}$ are such that either (1) $uu' \notin E(G_{split})$ but $vv' \in E(H_{split})$ or (2) both $uu' \in E(G_{split})$ and $vv' \in E(H_{split})$, where $\{v\} = L_{split}(u)$ and $\{v'\} = L_{split}(u')$, then we can remove edge vv' from $E(H_{split})$ in case (1) and we can remove both edges uu' and vv' in case (2) without affecting any embedding constraint. Without loss of generality, we will still denote by $(G_{split}, H_{split}, \prec_{G_{split}}, \prec_{H_{split}}, L_{split}, k)$ the resulting instances after the removal of the edges satisfying cases (1) and (2) above. Note that $E(H_{split}) = \emptyset$ at this point, and hence if $uu' \in E(G_{split})$ then no valid list embedding can be defined on a subset that includes both u and u'. Note also that $\mathcal{G}_{split} - E(G_{split})$ is a permutation graph P in which the vertices of G_{split} can be arranged on one line according to the order induced by $\prec_{G_{split}}$, and the vertices of H_{split} can be arranged on a parallel line according to the order induced by $\prec_{H_{split}}$. The vertex-set of P corresponds to the edges in E_{split}, and two vertices in P are adjacent if and only if their two corresponding edges cross. Note that two vertices in P correspond to two edges of the form $e = uv$ and $e' = u'v'$, where $u, u' \in G_{split}$ and $v, v' \in H_{split}$. Let $\mathcal{I}$ be the graph whose vertex-set is $V(P)$ and whose edge set

---

[1] Note that the degree of a vertex in H_{split} may be unbounded.

is $E(P) \cup E_c$, where $E_c = \{ee' \mid e, e' \in V(P), e = uv, e' = u'v', uu' \in E(G_{split})\}$ is the set of *conflict edges*; that is, $\mathcal{I}$ consists of the permutation graph P plus the set of conflict edges E_c, where each edge in E_c joins two vertices in P whose corresponding endpoints in G_{split} cannot both be part of a valid solution.

Lemma 2. *For every vertex $e \in \mathcal{I}$, the number of conflict edges incident to e in $\mathcal{I}$, denoted $deg_c(e)$, is at most $\Delta_L \cdot \Delta_G$.*

Lemma 3. *The instance $(G_{split}, H_{split}, \prec_{G_{split}}, \prec_{H_{split}}, L_{split}, k)$, and hence $(G, H, \prec_G, \prec_H, L, k)$, before* **Simplify** *is applied is a yes-instance of p-OLSE if and only if $\mathcal{I}$ has an independent set of size k.*

Proof. A size-k independent set I in $\mathcal{I}$ corresponds to a set of k edges $u_{i_1}v_{j_1}, \ldots, u_{i_k}v_{j_k}$ in G_{split} such that $u_{i_1} \prec_{G_{split}} \cdots \prec_{G_{split}} u_{i_k}$, $v_{j_1} \prec_{H_{split}}$ $\cdots \prec_{H_{split}} v_{j_k}$, and $S = G_{split}[\{u_{i_1}, \ldots, u_{i_k}\}]$ is a subgraph in G_{split} whose vertices form an independent set. Clearly, the embedding $\varphi(u_{i_s}) = \{v_{j_s}\}$, $s = 1, \ldots, k$, is a valid embedding that embeds S into H_{split} because it respects both $\prec_{G_{split}}, \prec_{H_{split}}$, and because it respects the embedding constraints. To see why the latter statement is true, note that, for any two vertices u_{i_s} and u_{i_r} $(r \neq s)$ in S, either there was no edge between u_{i_s} and u_{i_r} before the application of the operation **Simplify**, or there was an edge and got removed by **Simplify**, and in this case there must be an edge between v_{j_s} and v_{j_r} in H_{split}; in either case, φ respects the embedding constraints.

Conversely, let φ be a valid embedding that embeds a subgraph S of order k where $V(S) = \{u_{i_1}, \ldots, u_{i_k}\}$, and $\varphi(u_{i_s}) = v_{j_s}$, for $s = 1, \ldots, k$. We claim that the set $I = \{e_1 = u_{i_1}v_{j_1}, \ldots, e_k = u_{i_k}v_{j_k}\}$ is an independent set in $\mathcal{I}$. Since φ is a valid list embedding, no edge in P exists between any two vertices in I. Let $e_r = u_{i_r}v_{j_r}$ and $e_s = u_{i_s}v_{j_s}$ be two vertices in I, where $r \neq s$. If there is no edge between u_{i_r} and u_{i_s} in $E(G_{split})$, then no edge exists between e_s and e_r in $\mathcal{I}$. On the other hand, if there is an edge between u_{i_r} and u_{i_s} in $E(G_{split})$, then because φ is a valid embedding, there must be an edge between v_{i_r} and v_{i_s}. After applying **Simplify**, the edge between u_{i_r} and u_{i_s} will be removed, and hence no edge exists between e_s and e_r in $\mathcal{I}$. It follows that I is a size-k independent set in $\mathcal{I}$. □

Lemma 4. *Let $\mathcal{C}$ be a hereditary class of graphs on which the* INDEPENDENT SET *problem is solvable in polynomial time, and let $\Delta \geq 0$ be a fixed integer constant. Let $\mathcal{C}' = \{\mathcal{I} = (V(P), E(P) \cup E_c) \mid P \in \mathcal{C}, E_c \subseteq V(P) \times V(P)\}$, where at most Δ edges in E_c are incident to any vertex in $\mathcal{I}$. Assuming that a graph in $\mathcal{C}'$ is given as $(V(P), E(P) \cup E_c)$ (i.e., E_c is given), the* INDEPENDENT SET *problem is* FPT *on graphs in the class $\mathcal{C}'$.*

Proof. Let $(\mathcal{I} = (V(P), E(P) \cup E_c), k)$ be an instance of INDEPENDENT SET, where $\mathcal{I} \in \mathcal{C}'$. We use the random separation method introduced by Cai et al. [4]; this method can be de-randomized in fpt-time using the notion of universal sets and perfect hash functions [2,13,14].

We apply the random separation method to the subgraph $(V(P), E_C)$, and color the vertices in $V(P)$ with two colors, "green" and "red", randomly and independently. If I is an independent set of size k in $\mathcal{I}$, since there are at most Δ edges of E_c that are incident to any vertex in $\mathcal{I}$, the probability that all vertices in I are colored green and all their neighbors along the edges in E_c are colored red, is at least $2^{-k+\Delta k} = 2^{-(\Delta+1)k}$. Using universal sets and perfect hash functions, by trying FPT-many 2-colorings, if a size-k independent set exists, then there is a 2-coloring among the ones we try that will result in the independent set vertices being colored green, and all their neighbors along edges in E_c being colored red. Therefore, it suffices to determine, given a 2-colored graph $\mathcal{I}$, whether there is an independent set of size k consisting of green vertices whose neighbors along the edges in E_c are red vertices. We explain how to do so next.

Suppose that the vertices in $\mathcal{I}$ are colored green or red, and $\mathcal{I}_g$ be the subgraph of $\mathcal{I}$ induced by the green vertices, and $\mathcal{I}_r$ that induced by the red vertices. Notice that if there is an independent set I consisting of k green vertices whose neighbors along the edges in E_c are red, then for each vertex u in I, u is an isolated vertex in the graph $(V(\mathcal{I}_g), E_c)$. Moreover, since I is an independent set, then no edge in $E(P)$ exists between any two vertices in I. Therefore, if we form the subgraph $G_0 = (V_0, E_0)$, where V_0 is the set of vertices in $\mathcal{I}_g$ that are isolated with respect to the set of edges E_c, and E_0 is the set of edges in $E(P)$ whose both endpoints are in V_0, then I is an independent set in G_0. On the other hand, any independent set of G_0 is also an independent set of $\mathcal{I}$. Since G_0 is a subgraph of $P \in \mathcal{C}$ and $\mathcal{C}$ is hereditary, it follows that $G_0 \in \mathcal{C}$ and we can compute a maximum independent set I_{max} in G_0 in polynomial time. If $|I_{max}| \geq k$, then we accept the instance; otherwise, we try the next 2-coloring. If no 2-coloring results in an independent set of size at least k, we reject.   □

Theorem 2. *The p-OLSE problem restricted to instances in which $\Delta_G = O(1)$ and $\Delta_L = O(1)$ is FPT.*

Proof. Let $(G, H, \prec_G, \prec_H, L, k)$ be an instance of p-OLSE in which both Δ_G and Δ_L are upper bounded by a fixed constant. We form the graph $\mathcal{G}$ and perform the splitting operation described in Definition 1 to obtain the instance $(G_{split}, H_{split}, \prec_{G_{split}}, \prec_{H_{split}}, L_{split}, k)$. By Lemma 1, $(G, H, \prec_G, \prec_H, L, k)$ is a yes-instance of p-OLSE if and only if $(G_{split}, H_{split}, \prec_{G_{split}}, \prec_{H_{split}}, L_{split}, k)$ is. We apply the operation **Simplify** to the instance and construct the graph $\mathcal{I} = (V(P), E(P) \cup E_c)$ as described above, where P is a permutation graph. Note that the set of edges E_c is known. By Lemma 3, $(G, H, \prec_G, \prec_H, L, k)$ is a yes-instance of p-OLSE if and only if $\mathcal{I}$ has an independent set of size k. Since the INDEPENDENT SET problem is solvable in polynomial time on permutation graphs ([11]), the class of permutation graph is hereditary, and every vertex in $\mathcal{I}$ has at most $\Delta_L \cdot \Delta_G$ edges in E_c incident to it by Lemma 2, it follows from Lemma 4 that we can decide if $\mathcal{I}$ has an independent set of size k in fpt-time.   □

Unfortunately, the above result does not hold true for p-OLISE, even when $\Delta_G = O(1)$, $\Delta_H = O(1)$, and $\Delta_L = O(1)$.

From Proposition 3, we know that in the case when Δ_L is unbounded, and both $\Delta_G = 1$ and $\Delta_H = 1$, p-OLSE is W[1]-hard. The following proposition says that the condition $\Delta_H = 1$ is essential for this W-hardness result:

Proposition 6. *The p-OLSE and p-OLISE problems restricted to instances in which $\Delta_H = 0$, $\Delta_G = O(1)$ (resp. $\Delta_G = 0$ and $\Delta_H = O(1)$ for p-OLISE by symmetry) and $\Delta_L = \infty$ are FPT.*

Proof. We prove the result for p-OLSE. The proof is exactly the same for p-OLISE. The proof uses the random separation method, but is simpler than the proof of Lemma 4. Let $(G, H, \prec_G, \prec_H, L, k)$ be an instance of p-OLSE. Observe that if S is the solution that we are looking for then $G[S]$ must be an independent set since $\Delta_H = 0$. Use the random separation method to color G with green or red. Since $\Delta_G = O(1)$, if a solution S exists, then in fpt-time (deterministic) we can find a 2-coloring in which all vertices in S are green and their neighbors in G are red. So we can work under this assumption. Let G_g be the subgraph of G induced by the green vertices, and G_r that induced by the red vertices. Observe that any green vertex in G_g that is not isolated in G_g can be discarded by our assumption (since all neighbors of a vertex in S must be in G_r). Therefore, we can assume that G_g is an independent set. We can now compute a maximum cardinality subgraph of G_g that can be (validly) embedded into H using the dynamic programming algorithm in Proposition 1; if the subgraph has order at least k we accept; otherwise, we try another 2-coloring of G. If no 2-coloring of G results in a solution of size at least k, we reject. $\square$

4   Parameterization by the Vertex Cover Number

We study the parameterized complexity of p-OLSE parameterized by the size of a vertex cover ν in the graph G; we denote the corresponding problem with p-VC-OLSE. The reduction in [6], which can be used to prove the $W[1]$-hardness of p-OLSE when restricted to instances in which $\Delta_H \leq 1$, $\Delta_G \leq 1$ and Δ_L is unbounded, results in an instance in which the number of vertices in G, and hence $\tau(G)$, is upper bounded by a function of the parameter. Therefore:

Proposition 7. *p-VC-OLSE restricted to instances in which $\Delta_H = 1$ and $\Delta_G = 1$ is $W[1]$-complete.*

Therefore, we can focus our attention on studying the complexity of p-VC-OLSE restricted to instances in which $\Delta_L \in O(1)$.

Theorem 3. *p-VC-OLSE restricted to instances in which $\Delta_L = O(1)$ is FPT.*

Proof. Let $(G, H, \prec_G, \prec_H, L, k, \nu)$ be an instance of p-VC-OLSE, where k is the desired solution size and ν is the size of a vertex cover in G. In fpt-time (in ν) we can compute a vertex cover C of G of size ν (if no such vertex cover exists we reject). Let $I = V(G) \setminus C$, and note that I is an independent set of G. Suppose that the solution we are seeking (if it exists) is S, and the valid mapping of S is φ. Let $S_C = S \cap C$, and let φ_C be the restriction of φ to S_C. We enumerate each subset of C as S_C, enumerate each possible mapping from S_C to $L(S_C)$ as φ_C, and check the validity of φ_C. This enumeration takes $O^*((2\Delta_L)^\nu)$ time, which is fpt-time in ν. Therefore, we will assume that the solution intersects C at a known subset S_C, and that the restriction of φ to S_C is a valid map φ_C, and reject the instance if this assumption is proved to be wrong.

We remove all vertices in $C \setminus S_C$ from G and update L accordingly; without loss of generality, we will still use G to refer to the resulting graph whose vertex-set at this point is $I \cup S_C$. Let $u_{i_1}, \ldots, u_{i_r}$ be the vertices in S_C, where $u_{i_1} \prec_G \ldots \prec_G u_{i_r}$. Since φ_C is valid, we have $\varphi(u_{i_1}) \prec_H \ldots \prec_H \varphi(u_{i_r})$. We perform the following operation. For each vertex u in I and each vertex $v \in L(u)$, if setting $\varphi(u) = v$ violates the embedding constraint in the sense that either (1) there is a vertex $u_{i_j} \in S_C$ such that $uu_{i_j} \in E(G)$ but $v\varphi_C(u_{i_j}) \notin E(H)$ or (2) there is a vertex $u_{i_j} \in S_C$ such that $u \prec u_{i_j}$ (resp. $u_{i_j} \prec u$) but $\varphi_C(u_{i_j}) \prec v$ (resp. $v \prec \varphi_C(u_{i_j})$), then remove v from $L(u)$. Afterwards, partition the vertices in G into at most $r + 1$ intervals, $I_0, \ldots, I_r$, where I_0 consists of the vertices preceding u_{i_1} (with respect to $\preceq_G$), I_r consists of those vertices following u_{i_r}, and I_j consists of those vertices that fall strictly between vertices $u_{i_{j-1}}$ and u_{i_j}, for $j = 1, \ldots, r$. Similarly, partition H into $r + 1$ intervals, $I'_0, \ldots, I'_r$, where I'_0 consists of the vertices preceding $\varphi_C(u_{i_1})$ (with respect to $\prec_H$), I'_r those vertices following $\varphi_C(u_{i_r})$, and I'_j those vertices that fall strictly between vertices $\varphi_C(u_{i_{j-1}})$ and $\varphi_C(u_{i_j})$, for $j = 1, \ldots, r$. Clearly, any valid mapping φ that respects $\prec_G$ and $\prec_H$ must map vertices in the solution that belong to I_j to vertices in I'_j, for $j = 0, \ldots, r$, in a way that respects the restrictions of $\prec_G$ and $\prec_H$ on I_j and I'_j, respectively. On the other hand, since after the above operation every vertex u in I can be validly mapped to any vertex $v \in L(u)$, any injective mapping φ_j that maps a subset of vertices in I_j to a subset in I'_j in a way that respects the restrictions of $\prec_G$ and $\prec_H$ to I_j and I'_j, respectively, can be extended to a valid embedding whose restriction to S_C is φ_C. Therefore, our problem reduces to determining whether there exist injective maps φ_j, $j = 0, \ldots, r$, mapping vertices in I_j to vertices in I'_j, such that the total number of vertices mapped in the I_j's is $k - r$. Consider the subgraphs $G_j = G[I_j]$ and $H_j = H[I'_j]$, for $j = 0, \ldots, r$. Since G_j is an independent set, the presence of edges in H_j does not affect the existence of a valid list mapping from vertices in G_j to H_j, and hence those edges can be removed. Therefore, we can solve the opt-OLSE problem using Proposition 1 on the two graphs G_j and H_j to compute a maximum cardinality subset of vertices S_j in G_j that can be validly embedded into H_j via an embedding φ_j. If the union of the S_j's with S_C has cardinality at least k, we accept. If after all enumerations of S_C and φ_C we do not accept, we reject the instance. This algorithm runs in time $O^*((2\Delta_L)^\nu)$. $\qquad\square$

References

1. Alber, J., Gramm, J., Guo, J., Niedermeier, R.: Computing the similarity of two sequences with nested arc annotations. Theoretical Computer Science 312(2-3), 337–358 (2004)
2. Alon, N., Yuster, R., Zwick, U.: Color-coding. Journal of the ACM 42, 844–856 (1995)
3. Ashby, C., Huang, X., Johnson, D., Kanj, I., Walker, K., Xia, G.: New algorithm for protein structure comparison and classification. BMC Genomics 14(S2:S1) (2012)

4. Cai, L., Chan, S.M., Chan, S.O.: Random separation: A new method for solving fixed-cardinality optimization problems. In: Bodlaender, H.L., Langston, M.A. (eds.) IWPEC 2006. LNCS, vol. 4169, pp. 239–250. Springer, Heidelberg (2006)
5. Evans, P.: Algorithms and complexity for annotated sequence analysis. Technical report, Ph.D. thesis, University of Victoria (1999)
6. Evans, P.A.: Finding common subsequences with arcs and pseudoknots. In: Crochemore, M., Paterson, M. (eds.) CPM 1999. LNCS, vol. 1645, pp. 270–280. Springer, Heidelberg (1999)
7. Flum, J., Grohe, M.: Parameterized Complexity Theory. Springer, Berlin (2010)
8. Goldman, D., Istrail, S., Papadimitriou, C.H.: Algorithmic aspects of protein structure similarity. In: FOCS, pp. 512–522 (1999)
9. Gramm, J., Guo, J., Niedermeier, R.: On exact and approximation algorithms for distinguishing substring selection. In: Lingas, A., Nilsson, B.J. (eds.) FCT 2003. LNCS, vol. 2751, pp. 195–209. Springer, Heidelberg (2003)
10. Jiang, T., Lin, G., Ma, B., Zhang, K.: The longest common subsequence problem for arc-annotated sequences. J. Discrete Algorithms 2(2), 257–270 (2004)
11. Kim, H.: Finding a maximum independent set in a permutation graph. Information Processing Letters 36(1), 19–23 (1990)
12. Lin, G.-H., Chen, Z.-Z., Jiang, T., Wen, J.: The longest common subsequence problem for sequences with nested arc annotations. Journal of Computer and System Sciences 65(3), 465–480 (2002)
13. Naor, M., Schulman, L.J., Srinivasan, A.: Splitters and near-optimal derandomization. In: FOCS, pp. 182–191 (1995)
14. Schmidt, J., Siegel, A.: The spatial complexity of oblivious k-probe hash functions. SIAM Journal on Computing 19(5), 775–786 (1990)
15. Song, Y., Liu, C., Huang, X., Malmberg, R., Xu, Y., Cai, L.: Efficient parameterized algorithms for biopolymer structuresequence alignment. IEEE/ACM Trans. Comput. Biology Bioinform. 3(4), 423–432 (2006)

A Completeness Theory
for Polynomial (Turing) Kernelization[*]

Danny Hermelin[1], Stefan Kratsch[2,**], Karolina Sołtys[3],
Magnus Wahlström[4,***], and Xi Wu[5]

[1] Ben Gurion University of the Negev, Be'er Sheva, Israel
hermelin@bgu.ac.il
[2] Technical University Berlin, Germany
stefan.kratsch@tu-berlin.de
[3] Max Planck Institute for Informatics, Saarbrücken, Germany
ksoltys@mpi-inf.mpg.de
[4] Royal Holloway, University of London, England
Magnus.Wahlstrom@rhul.ac.uk
[5] University of Wisconsin Madison, Madison, USA
xiwu@cs.wisc.edu

Abstract. The framework of Bodlaender et al. (ICALP 2008, JCSS 2009) and Fortnow and Santhanam (STOC 2008, JCSS 2011) allows us to exclude the existence of polynomial kernels for a range of problems under reasonable complexity-theoretical assumptions. However, some issues are not addressed by this framework, including the existence of Turing kernels such as the "kernelization" of LEAF OUT BRANCHING(k) into a disjunction over n instances each of size poly(k). Observing that Turing kernels are preserved by polynomial parametric transformations (PPTs), we define two *kernelization hardness* hierarchies by the PPT-closure of problems that seem fundamentally unlikely to admit efficient Turing kernelizations. This gives rise to the MK- and WK-hierarchies which are akin to the M- and W-hierarchies of ordinary parameterized complexity. We find that several previously considered problems are complete for the fundamental hardness class WK[1], including MIN ONES d-SAT(k), BINARY NDTM HALTING(k), CONNECTED VERTEX COVER(k), and CLIQUE parameterized by $k \log n$. We conjecture that no WK[1]-hard problem admits a polynomial Turing kernel. Our hierarchy subsumes an earlier hierarchy of Harnik and Naor (FOCS 2006, SICOMP 2010) that, from a parameterized perspective, is restricted to classical problems parameterized by witness size. Our results provide the first natural complete problems for, e.g., their class VC_1; this had been left open.

1 Introduction

Kernelization, or data reduction, is a central concept in parameterized complexity, and has important applications outside this field as well. Roughly speaking, a kernelization

---

[*] Due to space restrictions many proofs are deferred to the full version [20].

[**] Main work done while supported by the Netherlands Organization for Scientific Research (NWO), project "KERNELS: Combinatorial Analysis of Data Reduction".

[***] Main work done while the author was at the Max Planck Institute for Informatics.

G. Gutin and S. Szeider (Eds.): IPEC 2013, LNCS 8246, pp. 202–215, 2013.
© Springer International Publishing Switzerland 2013

algorithm reduces an instance of a given parameterized problem to an equivalent instance of size $f(k)$, where k is the parameter of the input instance. Appropriately, the function $f()$ is referred to as the *size* of the kernel. A kernel with a good size guarantee is very useful – whether one wants to solve a problem exactly, or apply heuristics, or compute an approximation, it never hurts to first apply the kernelization procedure.[1] It can also be seen more directly as instance compression, e.g., for storing a problem instance for the future use; see Harnik and Naor [18]. The common milestone for an efficient kernelization is a *polynomial kernel*, i.e., a kernel with a polynomial size guarantee. Several significant kernelization results can be found in the literature, sometimes using non-trivial mathematical tools; see, e.g., the $2k$-vertex kernel for VERTEX COVER [28], the $\mathcal{O}(k^2)$ kernel for FEEDBACK VERTEX SET [30], and the recent randomized polynomial kernel for ODD CYCLE TRANSVERSAL [26].

Fairly recently, work by Bodlaender et al. [4] together with a result of Fortnow and Santhanam [17] provided the first technique to rule out the existence of any polynomial kernel for certain problems, assuming that NP $\not\subseteq$ coNP/poly (and PH does not collapse [31]). A series of further papers have applied this framework to concrete problems and developed it further, e.g., [13,12,5,11,21,22,14].

However, there are relaxed notions of efficient kernelization which are not ruled out by any existing work, but which would still be useful in practice and interesting from a theoretical point of view. Almost immediately after the appearance of the above lower bounds framework, the question was raised whether there were notions of "cheating" kernels. For example for the problem k-PATH which could circumvent the above lower bounds by producing Turing kernels instead of standard many-one kernelizations [3]. Not long afterwards, the first example of such a cheating kernel appeared: Binkele-Raible et al. [2] showed that the k-LEAF OUT-BRANCHING problem (given a directed graph G and an integer k, does G contain a directed tree with at least k leaves?) does not admit a polynomial kernel unless NP $\not\subseteq$ coNP/poly, but does admit one (with $\mathcal{O}(k^3)$ vertices) if the root of the tree is fixed, implying a Turing kernel in the form of a disjunction over n instances, each of size polynomial in k. There are also simpler problems sharing the same behavior; for example, the problem of CLIQUE parameterized by maximum degree is trivially compatible with the lower-bound frameworks, implying that it has no polynomial many-one kernel unless NP $\not\subseteq$ coNP/poly, but admits a very simple Turing kernel into n instances of k vertices (by taking the neighborhood of each vertex). We will call such a disjunctive Turing kernel an *OR-kernel*. We observe that many of the positive aspects of standard (many-one) polynomial kernels are preserved by OR-kernels, or even generally Turing kernels; in particular the algorithmic consequences (e.g., a Turing kernel with polynomial individual instance sizes for a problem in NP implies an algorithm with a running time of $2^{k^{\mathcal{O}(1)}} n^{\mathcal{O}(1)}$, same as for a polynomial many-one kernel).

The question of the extent to which such Turing kernels exist is theoretically very interesting and one of the most important problems in the field. Some restricted forms of Turing kernels, e.g., polynomial AND-kernels, can already be excluded by the existing framework, as they are special cases of polynomial kernels which may use co-nondeterministic polynomial time (cf. [12,22,26]). However, for OR-kernels or Turing

---

[1] This is assuming that the kernel preserves solution values, which most do.

kernels the current framework does not apply (as also witnessed by the k-LEAF OUT-BRANCHING and CLIQUE by max degree problems). It is also unclear if the framework could be adapted to deal with them. Instead, we take an approach common in complexity theory, namely that of defining an appropriate notion of hardness, and studying problems that are complete under this notion. We start from a set of problems for which we conjecture that none of them has a polynomial Turing kernel, and show that they are equivalent under PPT-reductions (which preserve existence of polynomial Turing kernels). The result is a robust class of hardness of Turing kernelization, dubbed WK[1], whose complete problems include central problems from different areas of theoretical computer science. While we have no concrete evidence that our conjecture holds, we feel that the abundance of WK[1]-complete problems, where Turing-kernelization is found for none of them, might suggest its validity.

WK[1]-hard problems. A cornerstone problem of WK[1] is the k-step halting problem for non-deterministic Turing machines parameterized by $k \log n$. To see why this is a powerful problem, and why an efficient Turing kernel would seem unlikely, consider the k-clique problem. Given a graph G with n vertices and an integer k, it is easy to construct a Turing machine which checks in $\text{poly}(k)$ non-deterministic steps whether G contains a k-clique (by using a number of states polynomial in n). On the other hand, an OR-kernel for the problem (or more generally, a polynomial Turing kernel) would require reducing k-clique to a polynomial number of questions of size polynomial in $k \log n$ (e.g., $\text{poly}(n)$ induced subgraphs of G, each of size $\text{poly}(k \log n)$, which are guaranteed to cover any k-clique of G). In fact, the above-mentioned halting problem captures not only clique, but all problems where a witness of t bits can be verified in $\text{poly}(t, \log n)$ time (e.g., SUBGRAPH ISOMORPHISM).

Other WK[1]-complete problems include MIN ONES d-SAT, the problem of finding a satisfying assignment with at most k true variables for a d-CNF formula, parameterized by the solution size k; HITTING SET parameterized by the number of sets or hyperedges m; and CONNECTED VERTEX COVER parameterized by the solution size k. Of these, we in particular want to single out MIN ONES d-SAT, which captures all minimization problems for which the consistency of a solution can be locally verified (by looking at combinations of d values at a time). For example, this includes the $\mathcal{H}$-FREE EDGE MODIFICATION(k) problems, where $\mathcal{H}$ is a finite, fixed set of forbidden induced subgraphs, and the goal is to remove or add k edges in the input graph in order to obtain a graph with no induced subgraph in $\mathcal{H}$ [8].

Extending the hardness class WK[1], we also define a hierarchy of hardness classes WK[t] and MK[t] for $t \geq 1$, mirroring the W- and M-hierarchies of traditional parameterized complexity; see [16]. We note that there are also strong similarities to the work of Harnik and Naor [18], in particular to the VC-hierarchy (which is defined around the notion of *witness length* for problems in NP). However, the notion of a parameter seems more general and robust than witness length; consider for example the volume of work in FPT on structural parameters such as treewidth. We also feel that the connections to the traditional FPT hardness classes (see Section 3) flesh out and put into context Harnik and Naor's work, and the link to the Turing kernel question adds interest to the separation question. Still, the main focus of our work is the hardness class WK[1].

We hope that our conjecture, that WK[1] does not have polynomial Turing kernels, will inspire other researchers to revisit the kernelization properties of problems which have been shown not to admit standard polynomial kernels unless PH collapses, but for which hardness for the above-mentioned class is less obvious. In particular, we leave open the WK[1]-hardness of k-PATH, the problem for which the existence of Turing kernels was originally asked in [3].

2 Preliminaries

We begin our discussion by formally defining some of the main concepts used in this paper, and by introducing some terminology and notation that will be used throughout. All problem definitions are deferred to the full version of this paper [20]. We use $[n]$ to denote the set of integers $\{1, \ldots, n\}$.

Definition 1 (Kernelization). *A* kernelization algorithm, *or, in short, a* kernel *for a parameterized problem $L \subseteq \Sigma^* \times \mathbb{N}$ is a polynomial-time algorithm that on a given input $(x, k) \in \Sigma^* \times \mathbb{N}$ outputs a pair $(x', k') \in \Sigma^* \times \mathbb{N}$ such that $(x, k) \in L \Leftrightarrow (x', k') \in L$, and $|x'| + k' \leq f(k)$ for some function f. The function f above is referred to as the* size *of the kernel.*

In other words, a kernel is a polynomial-time reduction from a problem to itself that compresses the problem instance to a size depending only on the parameter. If the size of a kernel for L is polynomial, we say that L has a *polynomial kernel*. In the interest of robustness and ease of presentation, we relax the notion of kernelization to allow the output to be an instance of a different problem. This has been referred to as a generalized kernelization [4] or bikernelization [1]. The class of all parameterized problems with polynomial kernels in this relaxed sense is denoted by PK.

Definition 2 (Turing Kernelization). *A* Turing kernelization *for a parameterized problem $L \subseteq \Sigma^* \times \mathbb{N}$ is a polynomial-time algorithm with oracle access to a parameterized problem L' that can decide whether an input (x, k) is in L using queries of size bounded by $f(k)$, for some computable function f. The function f is referred to as the* size *of the kernel.*

If the size is polynomial, we say that L has a *polynomial Turing kernel*. The class of all parameterized problems with polynomial Turing kernels is denoted by Turing-PK.

Definition 3 (Polynomial Parametric Transformations [7]). *Let L_1 and L_2 be two parameterized problems. We write $L_1 \leq_{ppt} L_2$ if there exists a polynomial time computable function $\Psi : \{0, 1\}^* \times \mathbb{N} \to \{0, 1\}^* \times \mathbb{N}$ and a constant $c \in \mathbb{N}$, such that for all $(x, k) \in \Sigma^* \times \mathbb{N}$, if $(x', k') = \Psi(x, k)$ then $(x, k) \in L_1 \iff (x', k') \in L_2$, and $k' \leq ck^c$. The function Ψ is called a* polynomial parameter transformation *(PPT for short). If $L_1 \leq_{ppt} L_2$ and $L_2 \leq_{ppt} L_1$ we write $L_1 \equiv_{ppt} L_2$.*

Proposition 1. *Let L_1, L_2, and L_3 be three parameterized problems.*

- *If $L_1 \leq_{ppt} L_2$ and $L_2 \leq_{ppt} L_3$ then $L_1 \leq_{ppt} L_3$.*
- *If $L_1 \leq_{ppt} L_2$ and $L_2 \in$ PK (resp. Turing-PK) then $L_1 \in$ PK (resp. Turing-PK).*

We denote parameterizations in parentheses after the problem name, for example, CLIQUE$(k \log n)$. In this example, k is the solution size, and n the size of the input. (Recall that CLIQUE(k) is one of the fundamental hard problems for parameterized complexity, and unlikely to admit a kernel of any size [16]; however, under a parameter $p = k \log n$ it has a trivial kernel of size 2^p.)

Note that, if a problem Q is solvable in $2^{k^{\mathcal{O}(1)}} n^{\mathcal{O}(1)}$ time, then Q$(k) \equiv_{ppt}$ Q$(k \log n)$.

3   The WK- and MK-Hierarchies

In this section we introduce our hierarchies of inefficient kernelizability, the MK- and WK-hierarchies. Relations to the so-called *VC-hierarchy* of Harnik and Naor [18] are discussed in Section 3.1. To begin with, for $t \geq 0$ and $d \geq 1$, we inductively define the following classes $\Gamma_{t,d}$ and $\Delta_{t,d}$ of formulas following [16]:

$$\Gamma_{0,d} := \{\lambda_1 \wedge \cdots \wedge \lambda_c : c \in [d] \text{ and } \lambda_1, \ldots, \lambda_c \text{ are literals}\},$$
$$\Delta_{0,d} := \{\lambda_1 \vee \cdots \vee \lambda_c : c \in [d] \text{ and } \lambda_1, \ldots, \lambda_c \text{ are literals}\},$$
$$\Gamma_{t+1,d} := \{\textstyle\bigwedge_{i \in I} \delta_i : I \text{ is a finite non-empty set and } \delta_i \in \Delta_{t,d} \text{ for all } i \in I\},$$
$$\Delta_{t+1,d} := \{\textstyle\bigvee_{i \in I} \gamma_i : I \text{ is a finite non-empty set and } \gamma_i \in \Gamma_{t,d} \text{ for all } i \in I\}.$$

Thus, $\Gamma_{1,3}$ is the set of all 3-CNF formulas, and $\Gamma_{2,1}$ is the set of all CNF formulas. Given a class Φ of propositional formulas, we let $\Phi^+, \Phi^- \subseteq \Phi$ denote the restrictions of Φ to formulas containing only positive and negative literals, respectively. For any given Φ, we define two parameterized problems:

- Φ-WSAT$(k \log n)$ is the problem of determining whether a formula $\phi \in \Phi$ with n variables has a satisfying assignment of Hamming weight exactly k, parameterized by $k \log n$.
- Φ-SAT(n) is the problem of determining whether a formula $\phi \in \Phi$ with n variables is satisfiable, parameterized by n.

In particular, we will be interested in $\Gamma_{t,d}$-WSAT$(k \log n)$ and $\Gamma_{t,d}$-SAT(n).

We now reach our class definitions. For a parameterized problem $L \subseteq \Sigma^* \times \mathbb{N}$, we let $[L]_{\leq_{ppt}}$ denote the closure of L under polynomial parametric transformations. That is, $[L]_{\leq_{ppt}} := \{L' \subseteq \Sigma^* \times \mathbb{N} : L' \leq_{ppt} L\}$.

Definition 4. *Let $t \geq 1$ be an integer. The classes* WK$[t]$ *and* MK$[t]$ *are defined by*

- WK$[t] := \bigcup_{d \in \mathbb{N}} [\Gamma_{t,d}\text{-WSAT}(k \log n)]_{\leq_{ppt}}$.
- MK$[t] := \bigcup_{d \in \mathbb{N}} [\Gamma_{t,d}\text{-SAT}(n)]_{\leq_{ppt}}$.

The naming of the classes in our hierarchies comes from the close relationship of the MK- and WK-hierarchies to the M- and W-hierarchies of traditional parameterized complexity [16]. Roughly speaking, WK$[t]$ and MK$[t]$ are reparameterizations by a factor of $\log n$ (or log of the instance size) of the traditional parameterized complexity classes W$[t]$ and M$[t]$ (although W$[t]$ and M$[t]$ are closed under FPT reductions, which may use superpolynomial time in k).

There are also close connections to the so-called *subexponential time* S-hierarchy (see [16, Chapter 16]); specifically, S$[t]$ and MK$[t]$ are defined from the same starting problems, using closures under different types of reduction; see also Cygan et al. [10]. We further note that [10] asked as an open problem, a reduction which in our terms would go from an MK[2]-complete problem to one in WK[1], and our work suggests the difficulty of producing one (see Theorem 2).

We show the following complete problems for our hierarchy.

Theorem 1 ($\bigstar^2$). *Let $t \geq 1$. The following hold.*

- $\Gamma_{1,2}^{-}$-WSAT($k \log n$) *is* WK[1]-*complete.*
- $\Gamma_{t,1}^{-}$-WSAT($k \log n$) *is* WK[t]-*complete for odd $t > 1$.*
- $\Gamma_{t,1}^{+}$-WSAT($k \log n$) *is* WK[t]-*complete for even $t > 1$.*
- $\Gamma_{1,d}$-SAT(n) *is* MK[1]-*complete for every $d \geq 3$.*
- $\Gamma_{t,1}$-SAT(n) *is* MK[t]-*complete for $t \geq 2$.*

Theorem 1 above shows that the traditional problems used for showing completeness in the W- and M-hierarchies have reparameterized counterparts which are complete for our hierarchy. The theorem is proven using a set of PPTs from the specific class-defining problems to the corresponding target problem in the theorem. Our main contribution is a PPT for the first item, for which previous proofs used FPT-time reductions. The remaining items are either easy or well-known.

We now proceed to show the class containments in our hierarchy. The main containments are as follows.

Theorem 2 ($\bigstar$). MK[1] $\subseteq$ WK[1] $\subseteq$ MK[2] $\subseteq$ WK[2] $\subseteq$ MK[3] $\subseteq \cdots \subseteq$ EXPT.

We also study a few further particular classes. First, let PK$_{\text{NP}}$ denote the class of parameterized problems with polynomial kernels whose output problem lies in NP. We have the following relationship.

Lemma 1 ($\bigstar$). MK[1] $=$ PK$_{\text{NP}}$.

Proof (sketch). If a problem has a polynomial kernel within NP, then we may first kernelize, then reduce to 3-SAT by the NP-completeness of the latter. Conversely, any problem in MK[1] has a PPT to d-SAT(n) for some constant d, which (as the latter has bounded size) forms a kernelization within NP. $\qquad\square$

Next, for a problem L, we define a parameterized problem OR(L)(ℓ) where the input is a set of instances $x_1, \ldots, x_t$ of L, each of length at most ℓ, and the task is to decide whether $x_i \in L$ for at least one instance x_i.

Lemma 2 ($\bigstar$). *Let L be an* NP-*complete language. Then* MK[1] $\subseteq$ [OR(L)(ℓ)]$_{\leq_{ppt}}$ $\subseteq$ WK[1], *where the first inclusion is strict unless* NP $\subseteq$ coNP/poly, *and the second is strict unless every problem in* WK[1] *has a polynomial OR-kernel.*

Proof (sketch). The first containment is trivial; the second can be given via reduction to the BINARY NDTM HALTING problem (see Section 4). The first consequence follows from Fortnow and Santhanam [17]; the second follows since OR(L) has an OR-kernel, and OR-kernels are preserved by PPTs. $\qquad\square$

---

[2] The proofs for lemmas and theorems denoted by a star are deferred to the full version [20].

For AND(L), and problems with Turing kernels more generally, no similar containment is known. However, our hierarchy can still be useful for AND-compositional problems, in showing them to be *hard* for some level.

3.1 Comparison with the VC-Hierarchy

We now discuss the relations between the VC-hierarchy of Harnik and Naor [18] and the MK- and WK-hierarchies defined in this paper. Let us review some definitions. An NP-*language* L is for the purposes of this section defined by a pair (R_L, k), where $R_L(\cdot, \cdot)$ is a polynomial-time computable relation and $k(x) = |x|^{\mathcal{O}(1)}$ a polynomial-time computable function, and $x \in L$ for an instance x if and only if there is some string y with $|y| \leq k(x)$, such that $R_L(x, y)$ holds. The string y is called the *witness* for x. This naturally defines a parameterization of L, with parameter $k(x)$; the resulting problem is FPT (with running time $O^*(2^k)$). We refer to this parameterized problem as the *direct parameterization* of (R_L, k). Harnik and Naor consider the feasibility of a (possibly probabilistic) *compression* of an instance x of L into length poly($k(x)$), in a sense essentially equivalent to our (relaxed) notion of kernelization; see [18] for details.

Before we give the technical results, let us raise two points. First, Harnik and Naor deal solely with problems where the parameter is the length $k(x)$ of a witness. Although this always defines a valid parameter (as we have seen), it is not clear that every reasonable parameterization of an NP-problem can be interpreted as a witness in this sense. Second, although Harnik and Naor are rather lax with the choice of witness (frequently letting it go undefined), we want to stress that the choice of witness can have a big impact on the kernelization complexity of a problem. Consider as an example the case of HITTING SET (treated later in this paper). Let n denote the number of vertices of an instance, m the number of edges, and k the upper bound on solution size; note that the total coding size is $\mathcal{O}(mn)$. We will find that the problem is WK[1]-complete when parameterized by m, MK[2]-complete under the parameter n, and WK[2]-complete under the parameter $k \log n$. The two latter both represent plausible choices of witnesses; a witness of length m is less obvious (or natural), but there is a simple witness of length $m \log k \leq m \log m$ obtained by describing a partition of the edges into k sets. One may also consider structural parameters such as treewidth (e.g., of the bipartite vertex/edge incidence graph), for which a corresponding short witness seems highly unlikely (but which can be shown to be MK[2]-hard). That said, it is not hard to see that every problem in the WK- and MK-hierarchies has a corresponding witness, by the PPT-reduction that proves its class membership.

For reasons of space, we give only a brief summary of the results; see the full version [20] for more details.

Theorem 3 (★). *Let L be an NP-language defined by (R_L, k), and let $\mathcal{Q}$ be its direct parameterization. The following hold.*

1. *L is contained in (hard for, complete for) $\mathcal{VC}_{or}$ if and only if $\mathcal{Q}$ is contained in (hard for, complete for) the PPT-closure of OR(3-SAT).*
2. *L is contained in $\mathcal{VC}_1$ ($\mathcal{VC}_1$-hard, $\mathcal{VC}_1$-complete) if and only if $\mathcal{Q}$ is contained in WK[1] (WK[1]-hard, WK[1]-complete).*

3. *L is contained in $\mathcal{VC}_t$ ($\mathcal{VC}_t$-hard, $\mathcal{VC}_t$-complete) for $t > 1$ if and only if $\mathcal{Q}$ is contained in* $\mathrm{MK}[t]$ *($\mathrm{MK}[t]$-hard, $\mathrm{MK}[t]$-complete).*

Thus, we answer a question left open in [18], of finding a natural problem complete for $\mathcal{VC}_1$ (as well as some minor questions about specific problem placements).

4   Complete Problems for WK[1]

In this section we show that several natural problems are complete for our fundamental hardness class WK[1] and thus exemplify its robustness. Our starting point will be CLIQUE($k \log n$) which is clearly equivalent to $\Gamma_{1,2}^-$-WSAT($k \log n$); the latter is WK[1]-complete by Theorem 1.

Theorem 4. CLIQUE($k \log n$) *is complete for* WK[1].

4.1   Basic Problems

This section establishes the following theorem that covers some basic problems which will be convenient for showing WK[1]-hardness and completeness for other problems. Standard many-one polynomial kernels for these problems were excluded in previous work [4,24,25,13].

Theorem 5. *The following problems are all complete for* WK[1]:

- BINARY NDTM HALTING(k) *and* NDTM HALTING($k \log n$).
- MIN ONES d-SAT(k) *for* $d \geq 3$, *with at most k true variables.*
- HITTING SET(m) *and* EXACT HITTING SET(m), *with m sets.*
- SET COVER(n) *and* EXACT SET COVER(n), *with n elements.*

The following colorful variants are helpful for our reductions.

Lemma 3 ([15,13]). *The following equivalences hold.*

- MULTICOLORED CLIQUE($k \log n$) $\equiv_{ppt}$ CLIQUE($k \log n$).
- MULTICOLORED HITTING SET(m) $\equiv_{ppt}$ HITTING SET(m).

We now proceed with the reductions. For many problems, we will find it convenient to show hardness by reduction from EXACT HITTING SET(m) or HITTING SET(m); hence we begin by showing the completeness of these problems. We give a chain of reductions from CLIQUE($k \log n$), via EXACT HITTING SET(m) and NDTM HALTING($k \log n$), and back to CLIQUE($k \log n$), closing the cycle. After this we will treat the HITTING SET(m) problem. Note that NDTM HALTING($k \log n$) and BINARY NDTM HALTING(k) (the problem restricted to machines with a binary tape alphabet) are easily PPT-equivalent.

Lemma 4 ($\bigstar$). MULTICOLORED CLIQUE($k \log n$) $\leq_{ppt}$ EXACT HITTING SET(m).

Proof (sketch). Let the input be a graph $G = (V, E)$ with coloring function $c : V \to [k]$. Assume $V = [n]$, and let $b_\ell(v)$ be the ℓ:th bit in the binary expansion of v. We create a set family $\mathcal{F} \subseteq 2^E$ with sets $F_{i,j} = \{uv \in E : c(u) = i, c(v) = j\}$ for $1 \le i < j \le k$, and $F_{i,j,j',\ell} = \{uv \in E : c(u) = i, c(v) = j, b_\ell(u) = 1\} \cup \{uv \in E : c(u) = i, c(v) = j', b_\ell(u) = 0\}$ for all color pairs (i, j) and (i, j') (e.g., for $i, j, j' \in [k]$ with $i \ne j, j'$ and $j < j'$), $1 \le \ell \le \lceil \log n \rceil$. By the first set family, exactly one edge per color class is selected in a solution; the second family ensures that the selections are consistent (i.e., for each color class i, all incident edges are incident on the same vertex u). □

Lemma 5 (★). EXACT HITTING SET$(m) \le_{ppt}$ NDTM HALTING$(k \log n)$.

Proof (sketch). Let $\mathcal{F} = \{F_1, \ldots, F_m\}$ be a set family on universe $U = [n]$; we will create a NDTM with tape alphabet $[n]$ which verifies whether $\mathcal{F}$ has an exact hitting set. We may assume $n \le 2^m$. In the first phase, for each $i \in [m]$ we put the ID of a member u_i of F_i in cell i; in subsequent phases, we verify that no set F_i is hit twice (e.g., if $u_j \ne u_i$, then $u_j \notin F_i$). This is easily done in poly(m) steps, by a machine with poly$(n + m)$ states; thus $k \log n = m^{O(1)}$ and we are done. □

Lemma 6 (★). BINARY NDTM HALTING$(k) \le_{ppt}$ MIN ONES 3-SAT(k).

Proof (sketch). Let M be a Turing machine with ℓ transitions in the state diagram. We will create a 3-CNF formula ϕ that has a satisfying assignment of Hamming weight at most k', $k' = k^{O(1)}$, if and only if M accepts within k steps. We use variables $M_{e,t}, e \in [\ell], t \in [k]$, designating that M uses transition e of the state diagram as the t:th execution step, along with auxiliary variables tracing the machine state, head position, and tape contents. By using the log-cost selection formulas of [25], we can ensure that exactly one variable $M_{e,t}$ is true for each t, at the cost of an extra solution weight of $k \log \ell$. Given this, the consistency of an assignment can be enforced using only local (3-ary) conditions (details omitted). We also require that the final state is accepting. With some care, one can ensure that every satisfying assignment has the same weight $k' = \text{poly}(k, \log \ell)$. Since the problem is FPT, we may assume $\log \ell \le k$, thus we have polynomial reduction time and parameter growth, i.e., a PPT. □

The following lemma is a direct consequence of Theorem 1 in Section 3.

Lemma 7 (★). MIN ONES d-SAT$(k) \le_{ppt}$ CLIQUE$(k \log n)$ *for every fixed d.*

The remaining problems in Theorem 5 for which we need to show completeness are HITTING SET(m), SET COVER(n), and EXACT SET COVER(n). Since it is well known that HITTING SET$(m) \equiv_{ppt}$ SET COVER(n) and EXACT HITTING SET$(m) \equiv_{ppt}$ EXACT SET COVER(n), we finish the proof of Theorem 5 by proving WK[1]-completeness for HITTING SET(m).

Lemma 8 (★). HITTING SET(m) *is* WK[1]*-complete.*

4.2 Further Problems

We briefly list further problems which we have shown to be WK[1]-complete or WK[1]-hard. LOCAL CIRCUIT SAT$(k \log n)$ was defined by Harnik and Naor, for defining their VC-hierarchy [18]. The remaining proofs are adaptations of lower bounds proofs by Bodlaender *et al.* [7] and Dom *et al.* [13].

Theorem 6 (★). *The following problems are* WK[1]*-complete.*

- LOCAL CIRCUIT SAT($k \log n$).
- MULTICOLORED PATH(k) *and* DIRECTED MULTICOLORED PATH(k).
- MULTICOLORED CYCLE(k) *and* DIRECTED MULTICOLORED CYCLE(k).
- CONNECTED VERTEX COVER(k).
- CAPACITATED VERTEX COVER(k).
- STEINER TREE($k + t$), *for solution size k and t terminals.*
- SMALL SUBSET SUM(k) *(see [13] for parameter definition).*
- UNIQUE COVERAGE(k), *where k is the number of items to be covered.*

Theorem 7 (★). *The following problems are* WK[1]*-hard.*

- DISJOINT PATHS(k) *and* DISJOINT CYCLES(k).

5   Problems in Higher Levels

In this section we investigate the second level of the MK- and WK-hierarchies, and present some complete and hard problems for these classes.

MK[2]-complete problems. According to Theorem 1, MK[2] is the PPT-closure of the classical CNF satisfiability problem where the parameter is taken to be the number of variables in the input formula. The PPT-equivalence of this problem to HITTING SET(n) and SET COVER(m) is well known.

Theorem 8. HITTING SET(n) *and* SET COVER(m) *are complete for* MK[2].

Heggernes et al. [19] consider RESTRICTED PERFECT DELETION($|X|$) and RE-STRICTED WEAKLY CHORDAL DELETION($|X|$), where the input is a graph G, a set X of vertices of G such that $G - X$ is perfect (resp. weakly chordal), and an integer k, and the task is to select at most k vertices $S \subseteq X$ such that $G - S$ is perfect (resp. weakly chordal). We get the following corollary from Theorem 8 and PPTs given in [19].

Corollary 1. RESTRICTED PERFECT DELETION($|X|$) *and* RESTRICTED WEAKLY CHORDAL DELETION($|X|$) *are hard for* MK[2].

WK[2]-complete problems. Due to space limitations we only state the following completeness and hardness results for WK[2] and defer the proofs to the full version.

Theorem 9 (★). *The following problems are complete for* WK[2]:

- HITTING SET($k \log n$) *and* SET COVER($k \log m$).
- DOMINATING SET($k \log n$) *and* INDEPENDENT DOMINATING SET($k \log n$).
- STEINER TREE($k \log n$)

From Theorem 9, we immediately get the following corollary via PPTs by Loksh-tanov [27] and Heggernes *et al.* [19].

Corollary 2. *The following problems are all hard for* WK[2]:

- WHEEL-FREE DELETION($k \log n$).
- PERFECT DELETION($k \log n$).
- WEAKLY CHORDAL DELETION($k \log n$).

For the first four problems in Theorem 9 the results follow easily (see [20]), so let us focus on the more interesting case of STEINER TREE($k \log n$). While WK[2]-hardness for this problem follows immediately from *e.g.* the PPT from HITTING SET($k \log n$) given in [13], showing membership in WK[2] is more challenging. To facilitate this and other non-trivial membership proofs, we consider the issue of a machine characterization of WK[2], similarly to the WK[1]-complete NDTM HALTING($k \log n$) problem. The natural candidate would be MULTI-TAPE NDTM HALTING($k \log n$), as this same problem with parameter k is W[2]-complete [9]. However, while the problem with parameter $k \log n$ is easily shown to be WK[2]-hard, we were so far unable to show WK[2]-membership. On the other hand, the following extension of a single-tape non-deterministic Turing machine leads to a WK[2]-complete problem, which we name NDTM HALTING WITH FLAGS.

Definition 5. *A (single-tape, non-deterministic) Turing machine with flags is a standard (single-tape, non-deterministic) Turing machine which in addition to its working tape has access to a set F of flags. Each state transition of the Turing machine has the ability to read and/or write a subset of the flags. A transition that reads a set $S \subseteq F$ of flags is only applicable if all flags in S are set. A transition that writes a set $S \subseteq F$ of flags causes every flag in S to be set. In the initial state, all flags are unset. Note that there is no operation to reset a flag.*

Theorem 10. NDTM HALTING WITH FLAGS($k \log n$) *is* WK[2]*-complete.*

Proof. Showing WK[2]-hardness is easy by reduction from HITTING SET($k \log n$). In fact, the hitting set instance can be coded directly into the flags, without any motion of the tape head – simply construct a machine that non-deterministically makes k non-writing transitions, each corresponding to including a vertex in the hitting set, followed by one verification step. The machine has m flags, one for every set in the instance, and a step corresponding to selecting a vertex v activates all flags corresponding to sets containing v. Finally, the step to the accepting state may only be taken if all flags are set. By assuming $\log m \leq k \log n$ (or else solving the instance exactly) we get a PPT.

Showing membership in WK[2] can be done by translation to $\Gamma_{2,1}$-WSAT($k \log n$). The transition is similar to that in Lemma 6. The only complication is to enforce consistency of transitions which read and write sets of flags, but this is easily handled. Let $M_{e,t}$ signify that step number t of the machine follows edge e of the state diagram (as in Lemma 6). If transition e has a flag f as a precondition, then we simply add a clause

$$(\neg M_{e,t} \vee M_{e_{i_1},1} \vee \ldots \vee M_{e_{i_m},1} \vee \ldots \vee M_{e_{i_1},t-1} \vee \ldots \vee M_{e_{i_m},t-1}),$$

where $e_{i_1}, \ldots, e_{i_m}$ is an enumeration of all transitions in the state diagram which set flag f. The rest of the reduction proceeds without difficulty. □

Lemma 9 (★). STEINER TREE($k \log n$)$\leq_{ppt}$NDTM HALTING WITH FLAGS($k \log n$).

6   Discussion

We have defined a hierarchy of PPT-closed classes, akin to the M- and W-hierarchy of parameterized intractability, in order to build up a completeness theory for polynomial (Turing) kernelization. The fundamental hardness class is called WK[1] and we conjecture that no WK[1]-hard problem admits a polynomial Turing kernelization. At present, the state of the art in lower bounds for kernelization does not seem to provide a way to connect this conjecture to standard complexity assumptions. However, there is collective evidence by a wealth of natural problems that are complete for WK[1] and for which polynomial Turing kernels seem unlikely. (Recall that admittance of Turing kernels is preserved by PPTs and hence a single polynomial Turing kernel would transfer to all WK[1] problems.) Of course, our examples provide only a partial image of the WK[1] landscape. For example, the various kernelizability dichotomies that have been shown for CSP problems [25,23] can be shown to imply dichotomies between problems with polynomial kernels and WK[1]-complete problems (and in some cases the third class of W[1]-hard problems). We take this as further evidence of the naturalness of the class.

On the more structural side, we have discussed the relation to the earlier VC-hierarchy of Harnik and Naor [18] which, from our perspective, is restricted to NP-problems parameterized by witness size. Under this interpretation their hierarchy folds into ours, with the levels of their hierarchy mapping to a subset of the levels of our hierarchy.

Many questions remain. One is the WK[1]-hardness of PATH(k) and CYCLE(k); for these problems, we have only lower bound proofs in the framework of Bodlaender *et al.* [4], leaving the question of Turing kernels open. There are also several problems, including the work on structural graph parameters by Bodlaender, Jansen, and Kratsch, *e.g.* [6], which we have not investigated. It is also unknown whether MULTI-TAPE NDTM HALTING($k \log n$) is in WK[2]. Furthermore, it would be interesting to know some natural parameterized problems which are WK[2]-complete under a standard parameter (e.g., k rather than $k \log n$).

Still, the main open problem is to provide (classical or parameterized) complexity theoretical implications of polynomial Turing kernelizations for WK[1]. More modest variants of this goal include excluding only OR-kernels, and/or considering the general problem of Turing machine acceptance parameterized by witness length.

References

1. Alon, N., Gutin, G., Kim, E.J., Szeider, S., Yeo, A.: Solving max-r-sat above a tight lower bound. Algorithmica 61(3), 638–655 (2011)
2. Binkele-Raible, D., Fernau, H., Fomin, F.V., Lokshtanov, D., Saurabh, S., Villanger, Y.: Kernel(s) for problems with no kernel: On out-trees with many leaves. ACM Transactions on Algorithms 8(4), 38 (2012)
3. Bodlaender, H.L., Demaine, E.D., Fellows, M.R., Guo, J., Hermelin, D., Lokshtanov, D., Müller, M., Raman, V., van Rooij, J., Rosamond, F.A.: Open problems in parameterized and exact computation – IWPEC 2008. Technical Report UU-CS-2008-017, Dept. of Informatics and Computing Sciences, Utrecht University (2008)
4. Bodlaender, H.L., Downey, R.G., Fellows, M.R., Hermelin, D.: On problems without polynomial kernels. Journal of Computer and System Sciences 75(8), 423–434 (2009)

5. Bodlaender, H.L., Jansen, B.M.P., Kratsch, S.: Cross-composition: A new technique for kernelization lower bounds. In: Proc. of the 28th international Symposium on Theoretical Aspects of Computer Science (STACS), pp. 165–176 (2011)
6. Bodlaender, H.L., Jansen, B.M.P., Kratsch, S.: Kernel bounds for path and cycle problems. In: Marx, D., Rossmanith, P. (eds.) IPEC 2011. LNCS, vol. 7112, pp. 145–158. Springer, Heidelberg (2012)
7. Bodlaender, H.L., Thomassé, S., Yeo, A.: Kernel bounds for disjoint cycles and disjoint paths. Theor. Comput. Sci. 412(35), 4570–4578 (2011)
8. Cai, L.: Fixed-parameter tractability of graph modification problems for hereditary properties. Information Processing Letters 58(4), 171–176 (1996)
9. Cesati, M.: The Turing way to parameterized complexity. J. Comput. Syst. Sci. 67(4), 654–685 (2003)
10. Cygan, M., Dell, H., Lokshtanov, D., Marx, D., Nederlof, J., Okamoto, Y., Paturi, R., Saurabh, S., Wahlström, M.: On problems as hard as CNF-SAT. In: IEEE Conference on Computational Complexity, pp. 74–84 (2012)
11. Dell, H., Marx, D.: Kernelization of packing problems. In: Rabani [29], pp. 68–81
12. Dell, H., van Melkebeek, D.: Satisfiability allows no nontrivial sparsification unless the polynomial-time hierarchy collapses. In: Proc. of the 42th Annual ACM Symposium on Theory of Computing (STOC), pp. 251–260 (2010)
13. Dom, M., Lokshtanov, D., Saurabh, S.: Incompressibility through colors and IDs. In: Albers, S., Marchetti-Spaccamela, A., Matias, Y., Nikoletseas, S., Thomas, W. (eds.) ICALP 2009, Part I. LNCS, vol. 5555, pp. 378–389. Springer, Heidelberg (2009)
14. Drucker, A.: New limits to classical and quantum instance compression. In: FOCS, pp. 609–618. IEEE Computer Society (2012)
15. Fellows, M.R., Hermelin, D., Rosamond, F.A., Vialette, S.: On the parameterized complexity of multiple-interval graph problems. Theoretical Computer Science 410(1), 53–61 (2009)
16. Flum, J., Grohe, M.: Parameterized Complexity Theory. Springer-Verlag New York, Inc. (2006)
17. Fortnow, L., Santhanam, R.: Infeasibility of instance compression and succinct PCPs for NP. J. Comput. Syst. Sci. 77(1), 91–106 (2011)
18. Harnik, D., Naor, M.: On the compressibility of NP instances and cryptographic applications. SIAM Journal on Computing 39(5), 1667–1713 (2010)
19. Heggernes, P., van't Hof, P., Jansen, B.M.P., Kratsch, S., Villanger, Y.: Parameterized complexity of vertex deletion into perfect graph classes. In: Owe, O., Steffen, M., Telle, J.A. (eds.) FCT 2011. LNCS, vol. 6914, pp. 240–251. Springer, Heidelberg (2011)
20. Hermelin, D., Kratsch, S., Soltys, K., Wahlström, M., Wu, X.: Hierarchies of inefficient kernelizability. CoRR, abs/1110.0976 (2011)
21. Hermelin, D., Wu, X.: Weak compositions and their applications to polynomial lower bounds for kernelization. In: Rabani [29], pp. 104–113
22. Kratsch, S.: Co-nondeterminism in compositions: A kernelization lower bound for a Ramsey-type problem. In: Rabani [29], pp. 114–122
23. Kratsch, S., Marx, D., Wahlström, M.: Parameterized complexity and kernelizability of max ones and exact ones problems. In: Hliněný, P., Kučera, A. (eds.) MFCS 2010. LNCS, vol. 6281, pp. 489–500. Springer, Heidelberg (2010)
24. Kratsch, S., Wahlström, M.: Two edge modification problems without polynomial kernels. In: Chen, J., Fomin, F.V. (eds.) IWPEC 2009. LNCS, vol. 5917, pp. 264–275. Springer, Heidelberg (2009)
25. Kratsch, S., Wahlström, M.: Preprocessing of min ones problems: A dichotomy. In: Abramsky, S., Gavoille, C., Kirchner, C., Meyer auf der Heide, F., Spirakis, P.G. (eds.) ICALP 2010. LNCS, vol. 6198, pp. 653–665. Springer, Heidelberg (2010)

26. Kratsch, S., Wahlström, M.: Compression via matroids: a randomized polynomial kernel for odd cycle transversal. In: Rabani [29], pp. 94–103
27. Lokshtanov, D.: Wheel-free deletion is W[2]-hard. In: Grohe, M., Niedermeier, R. (eds.) IWPEC 2008. LNCS, vol. 5018, pp. 141–147. Springer, Heidelberg (2008)
28. Nemhauser, G.L., Trotter Jr., L.E.: Vertex packings: Structural properties and algorithms. Mathematical Programming 8(2), 232–248 (1975)
29. Rabani, Y. (ed.): Proceedings of the Twenty-Third Annual ACM-SIAM Symposium on Discrete Algorithms, SODA 2012. SIAM, Kyoto (2012)
30. Thomassé, S.: A $4k^2$ kernel for feedback vertex set. ACM Transactions on Algorithms 6(2) (2010)
31. Yap, C.-K.: Some consequences of non-uniform conditions on uniform classes. Theor. Comput. Sci. 26, 287–300 (1983)

On Sparsification for Computing Treewidth*

Bart M.P. Jansen

University of Bergen, Norway
bart.jansen@ii.uib.no

Abstract. We investigate whether an n-vertex instance (G, k) of TREE-WIDTH, asking whether the graph G has treewidth at most k, can efficiently be made sparse without changing its answer. By giving a special form of OR-cross-composition, we prove that this is unlikely: if there is an $\epsilon > 0$ and a polynomial-time algorithm that reduces n-vertex TREEWIDTH instances to equivalent instances, of an arbitrary problem, with $\mathcal{O}(n^{2-\epsilon})$ bits, then NP $\subseteq$ coNP/poly and the polynomial hierarchy collapses to its third level.

Our sparsification lower bound has implications for structural parameterizations of TREEWIDTH: parameterizations by measures that do not exceed the vertex count, cannot have kernels with $\mathcal{O}(k^{2-\epsilon})$ bits for any $\epsilon > 0$, unless NP $\subseteq$ coNP/poly. Motivated by the question of determining the optimal kernel size for TREEWIDTH parameterized by vertex cover, we improve the $\mathcal{O}(k^3)$-vertex kernel from Bodlaender et al. (STACS 2011) to a kernel with $\mathcal{O}(k^2)$ vertices. Our improved kernel is based on a novel form of *treewidth-invariant set*. We use the q-expansion lemma of Fomin et al. (STACS 2011) to find such sets efficiently in graphs whose vertex count is superquadratic in their vertex cover number.

1 Introduction

The task of preprocessing inputs to computational problems to make them less dense, called *sparsification*, has been studied intensively due to its theoretical and practical importance. Sparsification, and more generally, preprocessing, is a vital step in speeding up resource-demanding computations in practical settings. In the context of theoretical analysis, the *sparsification lemma* due to Impagliazzo et al. [21] has proven to be an important asset for studying subexponential-time algorithms. The work of Dell and van Melkebeek [15] on sparsification for SATISFIABILITY has led to important advances in the area of kernelization lower bounds. They proved that for all $\epsilon > 0$ and $q \geq 3$, assuming NP $\not\subseteq$ coNP/poly, there is no polynomial-time algorithm that maps an instance of q-CNF-SAT on n variables to an equivalent instance on $\mathcal{O}(n^{q-\epsilon})$ bits — not even if it is an instance of a *different* problem.

This paper deals with sparsification for the task of building minimum-width tree decompositions of graphs, or, in the setting of decision problems, of determining whether the treewidth of a graph G is bounded by a given integer k.

---

* This work was supported by ERC Starting Grant 306992 "Parameterized Approximation".

G. Gutin and S. Szeider (Eds.): IPEC 2013, LNCS 8246, pp. 216–229, 2013.
© Springer International Publishing Switzerland 2013

Preprocessing procedures for TREEWIDTH have been studied in applied [10,11,26] and theoretical settings [3,7]. A team including the current author obtained [7] a polynomial-time algorithm that takes an instance (G, k) of TREEWIDTH, and produces in polynomial time a graph G' such that $\mathrm{TW}(G) \leq k$ if and only if $\mathrm{TW}(G') \leq k$, with the guarantee that $|V(G')| \in \mathcal{O}(\mathrm{VC}^3)$ (VC denotes the size of a smallest vertex cover of the input graph). A similar algorithm was given that reduces the vertex count of G' to $\mathcal{O}(\mathrm{FVS}^4)$, where FVS is the size of a smallest feedback vertex set in G. Hence polynomial-time data reduction can compress TREEWIDTH instances to a number of vertices polynomial in their vertex cover (respectively feedback vertex) number. On the other hand, the natural parameterization of TREEWIDTH is trivially AND-compositional, and therefore does not admit a polynomial kernel unless NP $\subseteq$ coNP/poly [3,17]. These results give an indication of how far the vertex count of a TREEWIDTH instance can efficiently be reduced in terms of various measures of its complexity. However, they do not tell us anything about the question of *sparsification*: can we efficiently make a TREEWIDTH instance less dense, without changing its answer?

Our Results. Our first goal in this paper is to determine whether nontrivial sparsification is possible for TREEWIDTH instances. As a simple graph G on n vertices can be encoded in n^2 bits through its adjacency matrix, TREEWIDTH instances consisting of a graph G and integer k in the range $[1 \ldots n]$ can be encoded in $\mathcal{O}(n^2)$ bits. We prove that it is unlikely that this trivial sparsification scheme for TREEWIDTH can be improved significantly: if there is a polynomial-time algorithm that reduces TREEWIDTH instances on n vertices to equivalent instances of an arbitrary problem, with $\mathcal{O}(n^{2-\epsilon})$ bits, for some $\epsilon > 0$, then NP $\subseteq$ coNP/poly and the polynomial hierarchy collapses [27]. We prove this result by giving a particularly efficient form of OR-cross-composition [9]. We embed the OR of t n-vertex instances of an NP-complete graph problem into a TREEWIDTH instance with $\mathcal{O}(n\sqrt{t})$ vertices. The construction is a combination of three ingredients. We carefully inspect the properties of Arnborg et al.'s [1] NP-completeness proof for TREEWIDTH to obtain an NP-complete source problem called COBIPARTITE GRAPH ELIMINATION that is amenable to composition. Its instances have a restricted form that ensures that good solutions to the composed TREEWIDTH instance cannot be obtained by combining partial solutions to two different inputs. Then, like Dell and Marx [14], we use the layout of a $2 \times \sqrt{t}$ table to embed t instances into a graph on $\mathcal{O}(n^{\mathcal{O}(1)}\sqrt{t})$ vertices. For each way of choosing a cell in the top and bottom row, we embed one instance into the edge set induced by the vertices representing the two cells. Finally, we use ideas employed by Bodlaender et al. [8] in the superpolynomial lower bound for TREEWIDTH parameterized by the vertex-deletion distance to a clique: we compose the input instances of COBIPARTITE GRAPH ELIMINATION into a cobipartite graph to let the resulting TREEWIDTH instance express a logical OR, rather than an AND. Our proof combines these three ingredients with an intricate analysis of the behavior of elimination orders on the constructed instance. As the treewidth of the constructed cobipartite graph equals its pathwidth [24], the obtained sparsification lower bound for TREEWIDTH also applies to PATHWIDTH.

Our sparsification lower bound has immediate consequences for parameterizations of TREEWIDTH by graph parameters that do not exceed the vertex count, such as the vertex cover number or the feedback vertex number. Our result shows the impossibility of obtaining kernels of bitsize $\mathcal{O}(k^{2-\epsilon})$ for such parameterized problems, assuming NP $\not\subseteq$ coNP/poly. The kernel for TREEWIDTH parameterized by vertex cover (TREEWIDTH [VC]) obtained by Bodlaender et al. [6] contains $\mathcal{O}(\text{VC}^3)$ vertices, and therefore has bitsize $\Omega(\text{VC}^4)$. Motivated by the impossibility of obtaining kernels with $\mathcal{O}(\text{VC}^{2-\epsilon})$ bits, and with the aim of developing new reduction rules that are useful in practice, we further investigate kernelization for TREEWIDTH [VC]. We give an improved kernel based on *treewidth-invariant sets*: independent sets of vertices whose elimination from the graph has a predictable effect on its treewidth. While finding such sets seems to be hard in general, we show that the q-expansion lemma, previously employed by Thomassé [25] and Fomin et al. [19], can be used to find them when the graph is large with respect to its vertex cover number. The resulting kernel shrinks TREEWIDTH instances to $\mathcal{O}(\text{VC}^2)$ vertices, allowing them to be encoded in $\mathcal{O}(\text{VC}^3)$ bits. Thus we reduce the gap between the upper and lower bounds on kernel sizes for TREEWIDTH [VC]. Our new reduction rule for TREEWIDTH [VC] relates to the old rules like the crown-rule for k-VERTEX COVER relates to the high-degree Buss-rule [12]: by exploiting local optimality considerations, our reduction rule does not need to know the value of k.

Related Work. While there is an abundance of superpolynomial kernel lower bounds, few superlinear lower bounds are known for problems admitting polynomial kernels. There are results for hitting set problems [15], packing problems [14,20], and for domination problems on degenerate graphs [13].

2   Preliminaries

Parameterized Complexity and Kernels. A parameterized problem $\mathcal{Q}$ is a subset of $\Sigma^* \times \mathbb{N}$. The second component of a tuple $(x, k) \in \Sigma^* \times \mathbb{N}$ is called the *parameter* [16,18]. The set $\{1, 2, \ldots, n\}$ is abbreviated as $[n]$. For a finite set X and integer i we use $\binom{X}{i}$ to denote the collection of size-i subsets of X.

Definition 1 (Generalized kernelization). *Let* $\mathcal{Q}, \mathcal{Q}' \subseteq \Sigma^* \times \mathbb{N}$ *be parameterized problems and let* $h \colon \mathbb{N} \to \mathbb{N}$ *be a computable function. A* generalized kernelization *for* $\mathcal{Q}$ *into* $\mathcal{Q}'$ *of size* $h(k)$ *is an algorithm that, on input* $(x, k) \in \Sigma^* \times \mathbb{N}$, *takes time polynomial in* $|x| + k$ *and outputs an instance* (x', k') *such that:*

- *$|x'|$ and k' are bounded by $h(k)$.*
- *$(x', k') \in \mathcal{Q}'$ if and only if $(x, k) \in \mathcal{Q}$.*

The algorithm is a kernelization, *or in short a* kernel, *for* $\mathcal{Q}$ *if* $\mathcal{Q}' = \mathcal{Q}$. *It is a* polynomial (generalized) kernelization *if* $h(k)$ *is a polynomial.*

Cross-Composition. To prove our sparsification lower bound, we use a variant of cross-composition tailored towards lower bounds on the degree of the polynomial in a kernel size bound. The extension is discussed in the journal version [9] of the extended abstract on cross-composition [6].

Definition 2 (Polynomial equivalence relation). *An equivalence relation* $\mathcal{R}$ *on* Σ^* *is called a* polynomial equivalence relation *if the following conditions hold:*

1. *There is an algorithm that given two strings* $x, y \in \Sigma^*$ *decides whether* x *and* y *belong to the same equivalence class in time polynomial in* $|x| + |y|$.
2. *For any finite set* $S \subseteq \Sigma^*$ *the equivalence relation* $\mathcal{R}$ *partitions the elements of* S *into a number of classes that is polynomially bounded in the size of the largest element of* S.

Definition 3 (Cross-composition). *Let* $L \subseteq \Sigma^*$ *be a language, let* $\mathcal{R}$ *be a polynomial equivalence relation on* Σ^*, *let* $\mathcal{Q} \subseteq \Sigma^* \times \mathbb{N}$ *be a parameterized problem, and let* $f \colon \mathbb{N} \to \mathbb{N}$ *be a function. An* OR-cross-composition *of* L *into* $\mathcal{Q}$ *(with respect to* $\mathcal{R}$*) of cost* $f(t)$ *is an algorithm that, given* t *instances* $x_1, x_2, \ldots, x_t \in \Sigma^*$ *of* L *belonging to the same equivalence class of* $\mathcal{R}$, *takes time polynomial in* $\sum_{i=1}^{t} |x_i|$ *and outputs an instance* $(y, k) \in \Sigma^* \times \mathbb{N}$ *such that:*

- *The parameter* k *is bounded by* $\mathcal{O}(f(t) \cdot (\max_i |x_i|)^c)$, *where* c *is some constant independent of* t.
- $(y, k) \in \mathcal{Q}$ *if and only if there is an* $i \in [t]$ *such that* $x_i \in L$.

Theorem 1 ([9, Theorem 6]). *Let* $L \subseteq \Sigma^*$ *be a language, let* $\mathcal{Q} \subseteq \Sigma^* \times \mathbb{N}$ *be a parameterized problem, and let* d, ϵ *be positive reals. If* L *is NP-hard under Karp reductions, has an* OR-*cross-composition into* $\mathcal{Q}$ *with cost* $f(t) = t^{1/d + o(1)}$, *where* t *denotes the number of instances, and* $\mathcal{Q}$ *has a polynomial (generalized) kernelization with size bound* $\mathcal{O}(k^{d - \epsilon})$, *then* $NP \subseteq coNP/poly$.

Graphs. All graphs we consider are finite, simple, and undirected. An undirected graph G consists of a vertex set $V(G)$ and an edge set $E(G) \subseteq \binom{V(G)}{2}$. The open neighborhood of a vertex v in graph G is denoted $N_G(v)$, while its closed neighborhood is $N_G[v]$. The open neighborhood of a set $S \subseteq V(G)$ is $N_G(S) := \bigcup_{v \in S} N_G(v) \setminus S$, while the closed neighborhood is $N_G[S] := N_G(S) \cup S$. If $S \subseteq V(G)$ then $G[S]$ denotes the subgraph of G induced by S. We use $G - S$ to denote the graph $G[V(G) \setminus S]$. A graph is *cobipartite* if its edge-complement is bipartite. Equivalently, a graph G is cobipartite if its vertex set can be partitioned into two sets X and Y, such that both $G[X]$ and $G[Y]$ are cliques. A matching M in a graph G is a set of edges whose endpoints are all distinct. The endpoints of the edges in M are *saturated* by the matching. For disjoint subsets A and B of a graph G, we say that A has a perfect matching into B if there is a matching that saturates $A \cup B$ such that each edge in the matching has exactly one endpoint in each set. If $\{u, v\}$ is an edge in graph G, then *contracting* $\{u, v\}$ *into* u is the operation of adding edges between u and $N_G(v)$ while removing v. A graph H is a *minor* of a graph G, if H can be obtained from a subgraph of G by edge contractions.

Treewidth and Elimination Orders. While treewidth [2] is commonly defined in terms of tree decompositions, for our purposes it is more convenient to work with an alternative characterization in terms of *elimination orders*. *Eliminating* a vertex v in a graph G is the operation of removing v while completing its open neighborhood into a clique, i.e., adding all missing edges between neighbors of v. An elimination order of an n-vertex graph G is a permutation $\pi \colon V(G) \to [n]$ of its vertices. Given an elimination order π of G, we obtain a series of graphs by consecutively eliminating $\pi^{-1}(1), \ldots, \pi^{-1}(n)$ from G. The *cost* of eliminating a vertex v according to the order π, is the size of the *closed neighborhood* of v at the moment it is eliminated. The *cost of π on G*, denoted $c_G(\pi)$, is defined as the maximum cost over all vertices of G.

Theorem 2 ([2, Theorem 36]). *The treewidth of a graph G is exactly one less than the minimum cost of an elimination order for G.*

Lemma 1 ([4, Lemma 4], cf. [23, Lemma 6.13]). *Let G be a graph containing a clique $B \subseteq V(G)$, and let $A := V(G) \setminus B$. There is a minimum-cost elimination order π^* of G that eliminates all vertices of A before eliminating any vertex of B.*

Following the notation employed by Arnborg et al. [1] in their NP-completeness proof, we say that a *block* in a graph G is a maximal set of vertices with the same closed neighborhood. An elimination order π for G is *block-contiguous* if for each block $S \subseteq V(G)$, it eliminates the vertices of S contiguously. The following observation implies that every graph has a block-contiguous minimum-cost elimination order.

Observation 1. *Let G be a graph containing two adjacent vertices u, v such that $N_G[u] \subseteq N_G[v]$. Let π be an elimination order of G that eliminates v before u, and let the order π' be obtained by updating π such that it eliminates u just before v. Then the cost of π' is not higher than the cost of π.*

3   Sparsification Lower Bound for Treewidth

In this section we give the sparsification lower bound for TREEWIDTH. We phrase it in terms of a kernelization lower bound for the parameterization by the number of vertices, formally defined as follows.

> n-TREEWIDTH
> **Input:** An integer n, an n-vertex graph G, and an integer k.
> **Parameter:** The number of vertices n.
> **Question:** Is the treewidth of G at most k?

The remainder of this section is devoted to the proof of the following theorem.

Theorem 3. *If n-TREEWIDTH admits a (generalized) kernel of size $\mathcal{O}(n^{2-\epsilon})$, for some $\epsilon > 0$, then $NP \subseteq coNP/poly$.*

We prove the theorem by cross-composition. We therefore first define a suitable source problem for the composition in Section 3.1, give the construction of the composed instance in Section 3.2, analyze its properties in Section 3.3, and finally put it all together in Section 3.4. The proofs of statements marked with a star ($\bigstar$) are deferred to the full version [22] of this work due to space restrictions.

3.1 The Source Problem

The sparsification lower bound for TREEWIDTH will be established by cross-composing the following problem into it.

COBIPARTITE GRAPH ELIMINATION
Input: A cobipartite graph G with partite sets A and B, and a positive integer k, such that the following holds: $|A| = |B|$, $|A|$ is even, $k < \frac{|A|}{2}$, and A has a perfect matching into B.
Question: Is there an elimination order for G of cost at most $|A| + k$?

The NP-completeness proof extends the completeness proof for TREEWIDTH [1].

Lemma 2 (★). COBIPARTITE GRAPH ELIMINATION *is NP-complete.*

3.2 The Construction

We start by defining an appropriate polynomial equivalence relationship $\mathcal{R}$. Let all malformed instances be equivalent under $\mathcal{R}$, and let two valid instances of COBIPARTITE GRAPH ELIMINATION be equivalent if they agree on the sizes of the partite sets and on the value of k. This is easily verified to be a polynomial equivalence relation.

Now we define an algorithm that combines a sequence of equivalent inputs into a small output instance. As a constant-size NO-instance is a valid output when the input consists of solely malformed instances, in the remainder we assume that the inputs are well-formed. By duplicating some inputs, we may assume that the number of input instances t is a square, i.e., $t = r^2$ for some integer r. An input instance can therefore be indexed by two integers in the range $[r]$. Accordingly, let the input consist of instances $(G_{i,j}, A_{i,j}, B_{i,j}, k_{i,j})$ for $i, j \in [r]$, that are equivalent under $\mathcal{R}$. Thus the number of vertices is the same over all partite sets; let this be $n = |A_{i,j}| = |B_{i,j}|$ for all $i, j \in [r]$. Similarly, let k be the common target value for all inputs. For each partite set $A_{i,j}$ and $B_{i,j}$ in the input, label the vertices arbitrarily as $a_{i,j}^1, \ldots, a_{i,j}^n$ (respectively $b_{i,j}^1, \ldots, b_{i,j}^n$). We construct a cobipartite graph G' that expresses the OR of all the inputs, as follows.

1. For $i \in [r]$ make a vertex set A_i' containing n vertices $\hat{a}_i^1, \ldots, \hat{a}_i^n$.
2. For $i \in [r]$ make a vertex set B_i' containing n vertices $\hat{b}_i^1, \ldots, \hat{b}_i^n$.
3. Turn $\bigcup_{i \in [r]} A_i'$ into a clique. Turn $\bigcup_{i \in [r]} B_i'$ into a clique.
4. For each pair i, j with $i, j \in [r]$, we embed the adjacency of $G_{i,j}$ into G' as follows: for $p, q \in [n]$ make an edge $\{\hat{a}_i^p, \hat{b}_j^q\}$ if $\{a_{i,j}^p, b_{i,j}^q\} \in E(G_{i,j})$.

It is easy to see that at this point in the construction, graph G' is cobipartite. For any $i, j \in [r]$ the induced subgraph $G'[A_i' \cup B_j']$ is isomorphic to $G_{i,j}$ by mapping $\hat{a}_i^\ell$ to $a_{i,j}^\ell$ and $\hat{b}_j^\ell$ to $b_{i,j}^\ell$. As $G_{i,j}$ has a perfect matching between $A_{i,j}$ and $B_{i,j}$ by the definition of COBIPARTITE GRAPH ELIMINATION, this implies that G' has a perfect matching between A_i' and B_j' for all $i, j \in [r]$. These properties will be maintained during the remainder of the construction.

5. For each $i \in [r]$, add the following vertices to G':
   - n *checking vertices* $C_i' = \{c_i^1, \ldots, c_i^n\}$, all adjacent to B_i'.
   - n *dummy vertices* $D_i' = \{d_i^1, \ldots, d_i^n\}$, all adjacent to $\bigcup_{j \in [r]} A_j'$ and to C_i'.
   - $\frac{n}{2}$ *blanker vertices* $X_i' = \{x_i^1, \ldots, x_i^{n/2}\}$, all adjacent to A_i'.
6. Turn $\bigcup_{i \in [r]} A_i' \cup C_i'$ into a clique A'. Turn $\bigcup_{i \in [r]} B_i' \cup D_i' \cup X_i'$ into a clique B'.

The resulting graph G' is cobipartite with partite sets A' and B'. Define $k' := 3rn + \frac{n}{2} + k$. Observe that $|A'| = 2rn$ and that $|B'| = 2rn + \frac{rn}{2}$. Graph G' can easily be constructed in time polynomial in the total size of the input instances.

Intuition. Let us discuss the intuition behind the construction before proceeding to its formal analysis. To create a composition, we have to relate elimination orders in G' to those for input graphs $G_{i,j}$. All adjacency information of the input graphs $G_{i,j}$ is present in G'. As A' is a clique in G', by Lemma 1 there is a minimum-cost elimination order for G' that starts by eliminating all of B'. But when eliminating vertices of some B_{j*}' from G', they interact simultaneously with all sets A_i' ($i \in [r]$), so the cost of those eliminations is not directly related to the cost of elimination orders of a particular instance $G_{i*,j*}$. We therefore want to ensure that low-cost elimination orders for G' first "blank out" the adjacency of B' to all but one set A_{i*}', so that the cost of afterwards eliminating B_{j*}' tells us something about the cost of eliminating $G_{i*,j*}'$. To blank out the other adjacencies, we need earlier eliminations to make B' adjacent to all vertices of $\bigcup_{i \in [r] \setminus \{i*\}} A_i'$. These adjacencies will be created by eliminating the *blanker* vertices. For an index $i \in [r]$, vertices in X_i' are adjacent to A_i' and all of B'. Hence eliminating a vertex in X_i' indeed blanks out the adjacency of B' to A_i'. The weights of the various groups (simulated by duplicating vertices with identical closed neighborhoods) have been chosen such that low-cost elimination orders of G' starting with B', have to eliminate $r - 1$ blocks of blankers $X_{i_1}', \ldots, X_{i_{r-1}}'$ before eliminating any other vertex of B'. This creates the desired blanking-out effect. The checking vertices C_i' ($i \in [r]$) enforce that after eliminating $r-1$ blocks of blankers, an elimination order cannot benefit by mixing vertices from two or more sets $B_i', B_{i'}'$: each set B_i' from which a vertex is eliminated, introduces new adjacencies between B' and C_i'. Finally, the dummy vertices are used to ensure that after one set $B_i' \cup D_i'$ is completely eliminated, the cost of eliminating the remainder is small because $|B'|$ has decreased sufficiently.

3.3   Properties of the Constructed Instance

The following type of elimination orders of G' will be crucial in the proof.

Definition 4. *Let $i^*, j^* \in [r]$. An elimination order π' of G' is (i^*, j^*)-canonical if π' eliminates $V(G)$ in the following order:*

1. *first all blocks of blanker vertices X_i' for $i \in [r] \setminus \{i^*\}$, one block at a time,*
2. *then the vertices of B_{j*}', followed by dummies D_{j*}', followed by blankers X_{i*}',*
3. *alternatingly a block B_i' followed by the corresponding dummies D_i', until all remaining vertices of $\bigcup_{i \in [r]} B_i' \cup D_i'$ have been eliminated,*
4. *and finishes with the vertices $\bigcup_{i \in [r]} A_i' \cup C_i'$ in arbitrary order.*

Lemma 3 shows that the crucial part of a canonical elimination order is its behavior on B'_{j*}.

Lemma 3 (★). *Let π' be an (i^*, j^*)-canonical elimination order for G'.*

1. *No vertex that is eliminated before the first vertex of B'_{j*} costs more than $3rn$.*
2. *When a vertex of $D'_{j*} \cup X'_{i*}$ is eliminated, its cost does not exceed $3rn + \frac{n}{2}$.*
3. *No vertex that is eliminated after X'_{i*} costs more than $3rn$.*

The next lemma links this behavior to the cost of a related elimination order for G_{i^*, j^*}. Some terminology is needed. Consider an (i^*, j^*)-canonical elimination order π' for G', and an elimination order π for G_{i^*, j^*} that eliminates all vertices of B_{i^*, j^*} before any vertex of A_{i^*, j^*}. By numbering the vertices in B_{i^*, j^*} (a partite set of G_{i^*, j^*}) from 1 to n, we created a one-to-one correspondence between B_{i^*, j^*} and B'_{j*}, the first set of non-blanker vertices eliminated by π'. Hence we can compare the relative order in which vertices of B_{i^*, j^*} are eliminated in π and π'. If both π and π' eliminate the vertices of B_{i^*, j^*} in the same relative order, then we say that the elimination orders *agree on B_{i^*, j^*}*.

Lemma 4. *Let π' be an (i^*, j^*)-canonical elimination order of G'. Let π be an elimination order for G_{i^*, j^*} that eliminates all vertices of B_{i^*, j^*} before any vertex of A_{i^*, j^*}. If π' and π agree on B_{i^*, j^*}, then $c_{G'}(\pi') = 3rn + \frac{n}{2} - n + c_{G_{i^*, j^*}}(\pi)$.*

Proof. Consider the graph G'_B obtained from G' by performing the eliminations according to π' until we are about to eliminate the first vertex of B'_{j*}. By Definition 4 this means that all blocks of blankers X'_j for $j \neq j^*$ have been eliminated, and no other vertices. Using the construction of G' it is easy to verify that these eliminations have made all remaining vertices of B' adjacent to $\bigcup_{i \in [r] \setminus \{i^*\}} A'_i$, and that no new adjacencies have been introduced to $\bigcup_{i \in [r]} C'_i$ or to A'_{i*}. Graph $G'[A'_{i*} \cup B'_{j*}]$ was initially isomorphic to G_{i^*, j^*} by the obvious isomorphism based on the numbers assigned to the vertices. As no vertex adjacent to A'_{i*} has been eliminated yet, this also holds for $G'_B[A'_{i*} \cup B'_{j*}]$.

Consider what happens when eliminating the first vertex v' of B'_{j*} according to π'. Let $v \in B_{i^*, j^*}$ be the corresponding vertex in G_{i^*, j^*}. By the fact that the elimination orders agree, v is the first vertex of B_{i^*, j^*} to be eliminated under π.

The set $N_{G'_B}[v']$ contains C'_{j*}, $\bigcup_{j \neq j^*} B'_j \cup D'_j$, $\bigcup_{i \neq i^*} A'_i$, X'_{i*}, D'_{j*}, and the vertices of $G'[A'_{i*} \cup B'_{j*}]$ that correspond exactly to $N_{G_{i^*, j^*}}[v]$ by the isomorphism. So the cost of eliminating v' from G' exceeds the cost of eliminating v from G_{i^*, j^*} by exactly $|C'_{j*}| + |\bigcup_{j \neq j^*} B'_j \cup D'_j| + |\bigcup_{i \neq i^*} A'_i| + |X'_{i*}| + |D'_{j*}| = n + 2(r-1)n + (r-1)n + \frac{n}{2} + n = 3rn + \frac{n}{2} - n$. Now observe that by the isomorphism, eliminating v' from G' has exactly the same effect on the neighborhoods of B'_{j*} into A'_{i*}, as eliminating v from G_{i^*, j^*} has on the neighborhoods of B_{i^*, j^*} into A_{i^*, j^*}. Thus after one elimination, the remaining vertices of $A'_{i*} \cup B'_{j*}$ and $A_{i^*, j^*} \cup B_{i^*, j^*}$ induce subgraphs of G' and G_{i^*, j^*} that are isomorphic. Hence we may apply the same argument to the next vertex that is eliminated. Repeating this argument we establish that for each vertex in B'_{j*}, its elimination from G' costs exactly $3rn + \frac{n}{2} - n$ more than the corresponding elimination in G_{i^*, j^*}.

Now consider the cost of π on G_{i^*,j^*}: it is at least $n+1$, as the first vertex to be eliminated is adjacent to all of B_{i^*,j^*} (the graph is cobipartite) and to at least one vertex of A_{i^*,j^*} (since the COBIPARTITE GRAPH ELIMINATION instance G_{i^*,j^*} has a perfect matching between its two partite sets). After all vertices of B_{i^*,j^*} have been eliminated from G_{i^*,j^*}, the remaining vertices cost at most n; there are at most n vertices left in the graph at that point. Hence the cost of π on G_{i^*,j^*} is determined by the cost of eliminating B_{i^*,j^*}. For each vertex from that set that is eliminated, π' incurs a cost exactly $3rn + \frac{n}{2} - n$ higher. Hence $c_{G'}(\pi')$ is at least $(3rn + \frac{n}{2} - n) + (n+1) = 3rn + \frac{n}{2} + 1$. By Lemma 3 the cost that π' incurs before eliminating the first vertex of B'_{j^*} is at most $3rn$, the cost of eliminating $D'_{j^*} \cup X'_{i^*}$ is at most $3rn + \frac{n}{2}$, and the cost incurred after eliminating the last vertex of B'_{j^*} is at most $3rn$. Hence the cost of π' is determined by the cost of eliminating the vertices of B'_{j^*}. As this is exactly $3rn + \frac{n}{2} - n$ more than the cost of π on G_{i^*,j^*}, this proves the lemma. $\square$

The last technical step of the proof is to show that if G' has an elimination order of cost at most k', then it has such an order that is canonical.

Lemma 5 ($\bigstar$). *If G' has an elimination order of cost at most k', then there are indices $i^*, j^* \in [r]$ such that G' has an (i^*, j^*)-canonical elimination order of cost at most k'.*

3.4   Proof of Theorem 3

Having analyzed the relationship between elimination orders for G' and for the input graphs $G_{i,j}$ $(i,j \in [r])$, we can complete the proof. By combining the previous lemmata it is easy to show that G' acts as the logical OR of the inputs.

Lemma 6 ($\bigstar$). *G' has an elimination order of cost $\leq k' \Leftrightarrow$ there are $i,j \in [r]$ such that $G_{i,j}$ has an elimination order of cost $\leq n + k$.*

Lemma 7. *There is an OR-cross-composition of COBIPARTITE GRAPH ELIMINATION into n-TREEWIDTH of cost $\sqrt{t}$.*

Proof. In Section 3.2 we gave a polynomial-time algorithm that, given instances $(G_{i,j}, A_{i,j}, B_{i,j}, k_{i,j})$ of COBIPARTITE GRAPH ELIMINATION that are equivalent under $\mathcal{R}$ for $i,j \in [r]$, constructs a cobipartite graph G' with partite sets A' and B', and an integer k'. By Lemma 6 the resulting graph G' has an elimination order of cost k' if and only if there is a YES-instance among the inputs. By the correspondence between treewidth and bounded-cost elimination orders of Theorem 2, this shows that G' has treewidth at most $k'-1$ if and only if there is a YES-instance among the inputs. The polynomial equivalence relationship ensured that all partite sets of all inputs have the same number of vertices. For partite sets of size n, the constructed graph G' satisfies $|A'| = 2rn$ and $|B'| = \frac{5rn}{2}$. The number of vertices in G' is $n' = \frac{9rn}{2}$. Consider the n-TREEWIDTH instance $(G', n', k'-1)$. It expresses the logical OR of a series of $r^2 = t$ COBIPARTITE GRAPH ELIMINATION instances using a parameter value of $\frac{9n\sqrt{t}}{2} \in \mathcal{O}(n\sqrt{t})$. Hence the algorithm gives an OR-cross-composition of COBIPARTITE GRAPH ELIMINATION into n-TREEWIDTH of cost $\sqrt{t}$. $\square$

Theorem 3 follows from the combination of Lemma 7, Lemma 2, and Theorem 1. Since the pathwidth of a cobipartite graph equals its treewidth [24] and the graph formed by the cross-composition is cobipartite, the same construction gives an OR-cross-composition of bounded cost into n-PATHWIDTH.

Corollary 1. *If n-PATHWIDTH admits a (generalized) kernel of size $\mathcal{O}(n^{2-\epsilon})$, for some $\epsilon > 0$, then $NP \subseteq coNP/poly$.*

4   Quadratic-Vertex Kernel for Treewidth [VC]

In this section we present an improved kernel for TREEWIDTH [VC], which is formally defined as follows.

TREEWIDTH [VC]
Input: A graph G, a vertex cover $X \subseteq V(G)$, and an integer k.
Parameter: $|X|$.
Question: Is the treewidth of G at most k?

Our kernelization revolves around the following notion.

Definition 5. *Let G be a graph, let T be an independent set in G, and let $\hat{G}_T$ be the graph obtained from G by eliminating T; the order is irrelevant as T is independent. Then T is a* treewidth-invariant set *if for every $v \in T$, the graph $\hat{G}_T$ is a minor of $G - \{v\}$.*

Lemma 8. *If T is a treewidth-invariant set in G and $\Delta := \max_{v \in T} \deg_G(v)$, then $\mathrm{TW}(G) = \max(\Delta, \mathrm{TW}(\hat{G}_T))$.*

Proof. We prove that $\mathrm{TW}(G)$ is at least, and at most, the claimed amount.

($\geq$). As $\hat{G}_T$ is a minor of G, we have $\mathrm{TW}(G) \geq \mathrm{TW}(\hat{G}_T)$ (cf. [2]). If $\mathrm{TW}(\hat{G}_T) \geq \Delta$ then this implies the inequality. So assume that $\Delta > \mathrm{TW}(\hat{G}_T)$. Let $v \in T$ have degree Δ. By assumption, $\hat{G}_T$ is a minor of $G - \{v\}$. It contains all vertices of $N_G(v)$ since T is an independent set. As $N_G(v)$ is a clique in $\hat{G}_T$, there is a series of minor operations in $G - \{v\}$ that turns $N_G(v)$ into a clique. Performing these operations on G rather than $G - \{v\}$ results in a clique on vertex set $N_G[v]$ of size $\deg_G(v) + 1 = \Delta + 1$: the set $N_G(v)$ is turned into a clique, and v remains unchanged. Hence G has a clique with $\Delta + 1$ vertices as a minor, which is known to imply (cf. [2]) that its treewidth is at least Δ.

($\leq$). Consider an optimal elimination order $\hat{\pi}$ for $\hat{G}_T$, which costs $\mathrm{TW}(\hat{G}_T) + 1$ by Theorem 2. Form an elimination order π for G by first eliminating all vertices in T in arbitrary order, followed by the remaining vertices in the order dictated by $\hat{\pi}$. Consider what happens when eliminating the graph G in the order given by π. Each vertex $v \in T$ that is eliminated incurs cost $\deg_G(v) + 1 \leq \Delta + 1$: as T is an independent set, eliminations before v do not affect v's neighborhood. Once all vertices of T have been eliminated, the resulting graph is identical to $\hat{G}_T$, by definition. As π matches $\hat{\pi}$ on the vertices of $V(G) \backslash T$, and $\hat{\pi}$ has cost $\mathrm{TW}(\hat{G}_T) + 1$, the total cost of elimination order π on G is $\max(\Delta + 1, \mathrm{TW}(\hat{G}_T) + 1)$. By Theorem 2 this completes this direction of the proof.   □

Lemma 8 shows that when a treewidth-invariant set is eliminated from a graph, its treewidth changes in a controlled manner. To exploit this insight in a kernelization algorithm, we have to find treewidth-invariant sets in polynomial time. While it seems difficult to detect such sets in all circumstances, we show that the q-expansion lemma can be used to find a treewidth-invariant set when the size of the graph is large compared to its vertex cover number. The following auxiliary graph is needed for this procedure.

Definition 6. *Given a graph G with a vertex cover $X \subseteq V(G)$, we define the bipartite non-edge connection graph $H_{G,X}$. Its partite sets are $V(G) \setminus X$ and $\binom{X}{2} \setminus E(G)$, with an edge between a vertex $v \in V(G) \setminus X$ and a vertex $x_{\{p,q\}}$ representing $\{p,q\} \in \binom{X}{2} \setminus E(G)$ if $v \in N_G(p) \cap N_G(q)$.*

For disjoint vertex subsets S and T in a graph G, we say that S *is saturated by q-stars into T* if we can assign to every $v \in S$ a subset $f(v) \subseteq N_G(v) \cap T$ of size q, such that for any pair of distinct vertices $u, v \in S$ we have $f(u) \cap f(v) = \emptyset$. Observe that an empty set can trivially be saturated by q-stars.

Lemma 9 ($\star$). *Let (G, X, k) be an instance of* TREEWIDTH *[VC]. If $H_{G,X}$ contains a set $T \subseteq V(G) \setminus X$ such that $S := N_{H_{G,X}}(T)$ can be saturated by 2-stars into T, then T is a treewidth-invariant set.*

q-Expansion Lemma ([19, Lemma 12]). *Let q be a positive integer, and let m be the size of a maximum matching in a bipartite graph H with partite sets A and B. If $|B| > m \cdot q$ and there are no isolated vertices in B, then there exist nonempty vertex sets $S \subseteq A$ and $T \subseteq B$ such that S is saturated by q-stars into T and $S = N_H(T)$. Furthermore, S and T can be found in time polynomial in the size of H by a reduction to bipartite matching.*

Theorem 4. TREEWIDTH *[VC] has a kernel with $\mathcal{O}(|X|^2)$ vertices that can be encoded in $\mathcal{O}(|X|^3)$ bits.*

Proof. Given an instance (G, X, k) of TREEWIDTH [VC], the algorithm constructs the non-edge connection graph $H_{G,X}$ with partite sets $A = \binom{X}{2} \setminus E(G)$ and $B = V(G) \setminus X$. We attempt to find a treewidth-invariant set $T \subseteq B$. If B has an isolated vertex v, then by definition of $H_{G,X}$ the set $N_G(v)$ is a clique implying that $\{v\}$ is treewidth-invariant. If B has no isolated vertices, we apply the q-expansion lemma with $q := 2$ to attempt to find a set $S \subseteq A$ and $T \subseteq B$ such that S is saturated by 2-stars into T and $S = N_{H_{G,X}}(T)$. Hence such a set T is treewidth-invariant by Lemma 9. If we find a treewidth-invariant set T:

- If $\max_{v \in T} \deg_G(v) \geq k + 1$ then we output a constant-size NO-instance, as Lemma 8 then ensures that $\mathrm{TW}(G) \geq \deg_G(v) > k$.
- Otherwise we reduce to $(\hat{G}_T, X, k)$ and restart the algorithm.

Each iteration takes polynomial time. As the vertex count decreases in each iteration, there are at most n iterations until we fail to find a treewidth-invariant set. When that happens, we output the resulting instance. The q-expansion lemma ensures that at that point, $|B| \leq 2m$, where m is the size of a maximum matching in $H_{G,X}$. As m cannot exceed the size of the partite set A, which is bounded by $\binom{|X|}{2}$ as there cannot be more non-edges in a set of size $|X|$, we find that $|B| \leq 2\binom{|X|}{2}$ upon termination. As vertex set B of the graph $H_{G,X}$ directly corresponds to $V(G) \setminus X$, this implies that G has at most $|X| + 2\binom{|X|}{2}$ vertices after exhaustive reduction. Thus the instance that we output has $\mathcal{O}(|X|^2)$ vertices. We can encode it in $\mathcal{O}(|X|^3)$ bits: we store an adjacency matrix for $G[X]$, and for each of the $\mathcal{O}(|X|^2)$ vertices v in $V(G) \setminus X$ we store a vector of $|X|$ bits, indicating for each $x \in X$ whether v is adjacent to it. $\qquad\square$

5   Conclusion

In this paper we contributed to the knowledge of sparsification for TREEWIDTH by establishing lower and upper bounds. Our work raises a number of questions.

We showed that TREEWIDTH and PATHWIDTH instances on n vertices are unlikely to be compressible into $\mathcal{O}(n^{2-\epsilon})$ bits. Are there natural problems on general graphs that do allow (generalized) kernels of size $\mathcal{O}(n^{2-\epsilon})$? Many problems admit $\mathcal{O}(k)$-vertex kernels when restricted to *planar* graphs [5], which can be encoded in $\mathcal{O}(k)$ bits by employing succinct representations of planar graphs. Obtaining subquadratic-size compressions for NP-hard problems on classes of potentially *dense* graphs, such as unit-disk graphs, is an interesting challenge.

In Section 4 we gave a quadratic-vertex kernel for TREEWIDTH [VC]. While the algorithm is presented for the decision problem, it is easily adapted to the optimization setting (cf. [11]). The key insight for our reduction is the notion of treewidth-invariant sets, together with the use of the q-expansion lemma to find them when the complement of the vertex cover has superquadratic size. A challenge for future research is to identify treewidth-invariant sets that are not found by the q-expansion lemma; this might decrease the kernel size even further. As the sparsification lower bound proves that TREEWIDTH [VC] is unlikely to admit kernels of bitsize $\mathcal{O}(|X|^{2-\epsilon})$, while the current kernel can be encoded in $\mathcal{O}(|X|^3)$ bits, an obvious open problem is to close the gap between the upper and the lower bound. Does TREEWIDTH [VC] have a kernel with $\mathcal{O}(|X|)$ vertices? If not, then is there at least a kernel with $\mathcal{O}(|X|^2)$ rather than $\mathcal{O}(|X|^3)$ edges?

For PATHWIDTH [VC], a kernel with $\mathcal{O}(|X|^3)$ vertices is known [8]. Can this be improved to $\mathcal{O}(|X|^2)$ using an approach similar to the one used here? The obvious pathwidth-analogue of Lemma 8 fails, as removing a low-degree simplicial vertex may decrease the pathwidth of a graph. Finally, one may consider whether the ideas of the present paper can improve the kernel size for TREEWIDTH parameterized by a feedback vertex set [7].

References

1. Arnborg, S., Corneil, D.G., Proskurowski, A.: Complexity of finding embeddings in a k-tree. SIAM J. Algebra. Discr. 8, 277–284 (1987), doi:10.1137/0608024
2. Bodlaender, H.L.: A partial k-arboretum of graphs with bounded treewidth. Theor. Comput. Sci. 209(1-2), 1–45 (1998), doi:10.1016/S0304-3975(97)00228-4

3. Bodlaender, H.L., Downey, R.G., Fellows, M.R., Hermelin, D.: On problems without polynomial kernels. J. Comput. Syst. Sci. 75(8), 423–434 (2009), doi:10.1016/j.jcss.2009.04.001
4. Bodlaender, H.L., Fomin, F.V., Koster, A.M.C.A., Kratsch, D., Thilikos, D.M.: On exact algorithms for treewidth. In: Azar, Y., Erlebach, T. (eds.) ESA 2006. LNCS, vol. 4168, pp. 672–683. Springer, Heidelberg (2006)
5. Bodlaender, H.L., Fomin, F.V., Lokshtanov, D., Penninkx, E., Saurabh, S., Thilikos, D.M. (Meta) Kernelization. In: Proc. 50th FOCS, pp. 629–638 (2009), doi:10.1109/FOCS.2009.46
6. Bodlaender, H.L., Jansen, B.M.P., Kratsch, S.: Cross-composition: A new technique for kernelization lower bounds. In: Proc. 28th STACS, pp. 165–176 (2011), doi:10.4230/LIPIcs.STACS.2011.165
7. Bodlaender, H.L., Jansen, B.M.P., Kratsch, S.: Preprocessing for treewidth: A combinatorial analysis through kernelization. In: Aceto, L., Henzinger, M., Sgall, J. (eds.) ICALP 2011, Part I. LNCS, vol. 6755, pp. 437–448. Springer, Heidelberg (2011)
8. Bodlaender, H.L., Jansen, B.M.P., Kratsch, S.: Kernel bounds for structural parameterizations of pathwidth. In: Fomin, F.V., Kaski, P. (eds.) SWAT 2012. LNCS, vol. 7357, pp. 352–363. Springer, Heidelberg (2012)
9. Bodlaender, H.L., Jansen, B.M.P., Kratsch, S.: Kernelization lower bounds by cross-composition. CoRR, abs/1206.5941, arXiv:1206.5941 (2012)
10. Bodlaender, H.L., Koster, A.M.C.A.: Safe separators for treewidth. Discrete Math. 306(3), 337–350 (2006), doi:10.1016/j.disc.2005.12.017
11. Bodlaender, H.L., Koster, A.M.C.A., van den Eijkhof, F.: Preprocessing rules for triangulation of probabilistic networks. Comput. Intell. 21(3), 286–305 (2005), doi:10.1111/j.1467-8640.2005.00274.x
12. Buss, J.F., Goldsmith, J.: Nondeterminism within P. SIAM J. Comput. 22(3), 560–572 (1993), doi:10.1137/0222038
13. Cygan, M., Grandoni, F., Hermelin, D.: Tight kernel bounds for problems on graphs with small degeneracy. CoRR, abs/1305.4914, arXiv:1305.4914 (2013)
14. Dell, H., Marx, D.: Kernelization of packing problems. In: Proc. 23rd SODA, pp. 68–81 (2012)
15. Dell, H., van Melkebeek, D.: Satisfiability allows no nontrivial sparsification unless the polynomial-time hierarchy collapses. In: Proc. 42nd STOC, pp. 251–260 (2010), doi:10.1145/1806689.1806725
16. Downey, R., Fellows, M.R.: Parameterized Complexity. Monographs in Computer Science. Springer, New York (1999)
17. Drucker, A.: New limits to classical and quantum instance compression. In: Proc. 53rd FOCS, pp. 609–618 (2012), doi:10.1109/FOCS.2012.71
18. Flum, J., Grohe, M.: Parameterized Complexity Theory. Springer-Verlag New York, Inc. (2006)
19. Fomin, F.V., Lokshtanov, D., Misra, N., Philip, G., Saurabh, S.: Hitting forbidden minors: Approximation and kernelization. In: Proc. 28th STACS, pp. 189–200 (2011), doi:10.4230/LIPIcs.STACS.2011.189
20. Hermelin, D., Wu, X.: Weak compositions and their applications to polynomial lower bounds for kernelization. In: Proc. 23rd SODA, pp. 104–113 (2012)
21. Impagliazzo, R., Paturi, R., Zane, F.: Which problems have strongly exponential complexity? J. Comput. Syst. Sci. 63(4), 512–530 (2001), doi:10.1006/jcss.2001.1774
22. Jansen, B.M.P.: On sparsification for computing treewidth. CoRR, abs/1308.3665, arXiv:1308.3665 (2013)

23. Jansen, B.M.P.: The Power of Data Reduction: Kernels for Fundamental Graph Problems. PhD thesis, Utrecht University, The Netherlands (2013)
24. Möhring, R.H.: Triangulating graphs without asteroidal triples. Discrete Appl. Math. 64(3), 281–287 (1996), doi:10.1016/0166-218X(95)00095-9
25. Thomassé, S.: A $4k^2$ kernel for feedback vertex set. ACM Trans. Algorithms 6(2) (2010), doi:10.1145/1721837.1721848
26. van den Eijkhof, F., Bodlaender, H.L., Koster, A.M.C.A.: Safe reduction rules for weighted treewidth. Algorithmica 47(2), 139–158 (2007), doi:10.1007/s00453-006-1226-x
27. Yap, C.-K.: Some consequences of non-uniform conditions on uniform classes. Theor. Comput. Sci. 26, 287–300 (1983), doi:10.1016/0304-3975(83)90020-8

The Jump Number Problem:
Exact and Parameterized

Dieter Kratsch[1,*] and Stefan Kratsch[2,**]

[1] LITA, Université de Lorraine, Metz, France
dieter.kratsch@univ-lorraine.fr
[2] Technische Universität Berlin, Germany
stefan.kratsch@tu-berlin.de

Abstract. The JUMP NUMBER problem asks to find a linear extension of a given partially ordered set that minimizes the total number of jumps, i.e., the total number of consecutive pairs of elements that are incomparable originally. The problem is known to be NP-complete even on posets of height one and on interval orders. It has also been shown to be fixed-parameter tractable. Finally, the JUMP NUMBER problem can be solved in time $\mathcal{O}^*(2^n)$ by dynamic programming.

In this paper we present an exact algorithm to solve JUMP NUMBER in $\mathcal{O}(1.8638^n)$ time. We also show that the JUMP NUMBER problem on interval orders can be solved by an $\mathcal{O}(1.7593^n)$ time algorithm, and prove fixed-parameter tractability in terms of width w by an $\mathcal{O}^*(2^w)$ time algorithm. Furthermore, we give an almost-linear kernel for JUMP NUMBER on interval orders for parameterization by the number of jumps.

1 Introduction

A *partially ordered set* $P = (X, \prec_P)$ consists of a finite set X and an irreflexive, antisymmetric, and transitive binary relation $\prec_P$ defined on X. We call P a *poset* and $\prec_P$ its *partial order*; we often omit the subscript P. We denote the size of X by n. We say that u and v are comparable if $u \prec_P v$ or $v \prec_P u$, otherwise u and v are incomparable. The *comparability graph* $G(P)$ has vertex set X and two of its vertices are adjacent if and only if they are comparable in P. A set of pairwise comparable elements of P is called a *chain*, and a set of pairwise incomparable elements of P is called an *anti-chain*. The *height* of a poset P, denoted $h(P)$, is the maximum size of a chain in P minus one. The *width* of a poset P, denoted $w(P)$, is the maximum size of an anti-chain in P.

A total order L on X: $v_1 \prec_L v_2 \prec_L \ldots \prec_L v_n$ is a *linear extension of the poset* $P = (X, \prec_P)$ if for all $u, v \in X$: $u \prec_P v$ implies $u \prec_L v$. Let L: $v_1 \prec_L v_2 \prec_L \ldots \prec_L v_n$ be any linear extension of P. Then for all $i \in \{1, 2, \ldots, n-1\}$, we call (v_i, v_{i+1}) a *jump of* L if v_i and v_{i+1} are incomparable with respect to P, i.e., $v_i \not\prec_P v_{i+1}$. Otherwise when $v_i \prec_P v_{i+1}$, we call (v_i, v_{i+1}) a *bump of* L. The *number of jumps of a linear extension L of P*, denoted by $s(P, L)$, is the

---

* Supported by ANR Blanc AGAPE (ANR-09-BLAN-0159-03).
** Supported by the DFG Emmy Noether-program, project PREMOD, KR 4286/1.

G. Gutin and S. Szeider (Eds.): IPEC 2013, LNCS 8246, pp. 230–242, 2013.
© Springer International Publishing Switzerland 2013

total number of jumps of L, i.e., $s(P, L) = |\{i : (v_i, v_{i+1})$ is a jump$\}|$. Similarly $b(P, L)$ is defined as the total *number of bumps of L*. Clearly, for every linear extension L of P we have $s(P, L) + b(P, L) = n - 1$ since any pair (v_i, v_{i+1}) is either a jump or a bump of L. The *jump number of a poset P*, denoted $s(P)$, is the minimum number of jumps taken over all linear extensions of P. A linear extension L of a poset P is called *jump-optimal* if $s(P, L) = s(P)$. The *bump number of a poset P*, denoted $b(P)$, is the minimum number of bumps taken over all linear extensions of P.

The JUMP NUMBER *problem.* The JUMP NUMBER problem asks to compute for a poset $P = (X, \prec_P)$ the minimum number of jumps $s(P)$ in any linear extension L, or to find a jump-optimal linear extension L of P. The complexity of the JUMP NUMBER problem has been studied extensively in the eighties and nineties in poset theory and poset algorithmics [1–4]. Let us summarize the most important results. The problem was shown to be NP-hard by Pulleyblank [5]; see also [6]. Pulleyblank's proof also shows that the problem is NP-hard for posets of height one (see also [7]). Furthermore, JUMP NUMBER is known to be NP-hard on interval orders [8], but was long known to admit a 3/2-approximation algorithm [8–10]; this was recently improved by Krysztowiak [11] to factor 1.484. It is polynomial time solvable on various classes of orders among them N-free orders and 2-dimensional orders of height one [12, 13]. Finally let us mention that contrary to the JUMP NUMBER problem the BUMP NUMBER problem asking to minimize the total number of bumps over all linear extensions of a poset P is polynomial time solvable [14].

The JUMP NUMBER problem has already been studied from a parameterized point of view. First El-Zahar and Schmerl [15] gave a polynomial-time algorithm for fixed number of jumps (an XP algorithm). Then McCartin established an FPT algorithm that runs in time $\mathcal{O}(k^2 k! n)$ [16]. This is a very interesting result, however it does not imply an exact exponential algorithm faster than the trivial one based on listing all linear extensions of P and running in time $\mathcal{O}^*(n!)$.

Our work. We study the JUMP NUMBER problem on general posets and on interval orders. Let us mention that interval orders form a class of orders which has been studied extensively in poset theory and that there is a monograph on interval orders [17]. We present exact algorithms which improve significantly upon the standard $\mathcal{O}^*(2^n)$ time dynamic programming algorithm. Both exact algorithms are essentially refinements of this dynamic programming algorithm. The first one solves the JUMP NUMBER problem on general posets in $\mathcal{O}(1.8638^n)$ time. The second algorithm solves the JUMP NUMBER problem on interval orders in $\mathcal{O}(1.7593^n)$ time. As a byproduct, we obtain single-exponential FPT-algorithms for the JUMP NUMBER problem on interval orders when parameterized either by width or by the number of jumps. For the latter parameter we also establish an almost linear kernel, i.e., we can efficiently reduce to $\mathcal{O}(k)$ elements and bit-size $\mathcal{O}(k \log k)$.

Organization. Section 2 presents our exact algorithm for JUMP NUMBER on general posets. In Section 3 we give a faster exact algorithm for the restricted case of interval orders and a single-exponential algorithm for JUMP NUMBER on interval orders with respect to parameter width. In Section 4 we show the almost linear kernelization for JUMP NUMBER on interval orders. For completeness, Sections 5 and 6 recall the basic $\mathcal{O}^*(2^n)$ dynamic programming algorithm and some useful properties of binomial coefficients. Section 7 concludes the paper.

2   Computing the Jump Number in $\mathcal{O}(1.8638^n)$ Time

In this section we present an exact algorithm to compute the jump number of any given poset in time $\mathcal{O}(1.8638^n)$ based on refinements of a well-known dynamic programming approach.

Various problems on graphs ask to find a permutation of the elements of the ground-set minimizing or maximizing some function depending on the permutation. They can often be solved using dynamic programming over subsets, which typically yields an $\mathcal{O}^*(2^n)$ algorithm. This algorithm design technique goes back to the sixties and the Held-Karp algorithm for the Traveling Salesman Problem (see [18]); it can also be applied to problems on posets. Doing this, the JUMP NUMBER problem can be solved in time $\mathcal{O}^*(2^n)$ using dynamic programming over subsets improving upon trivial enumeration of all linear extensions in $\mathcal{O}^*(n!)$ time. For some of these permutation problems improvements over the $\Theta^*(2^n)$ worst case running time of the general approach have been achieved. The number of such problems is small, among them exact computation of treewidth and pathwidth [19, 20]. An outstanding result of this type is Björklund's $\mathcal{O}(1.657^n)$ time randomized algorithm for HAMILTONIAN CIRCUIT [21].

Recently, Cygan et al. [22] obtained such a result by showing that a precedence constraint scheduling problem called SCHED can be solved in time $\mathcal{O}^*((2-\epsilon)^n)$ answering a question of Woeginger [23]; in fact their ϵ is very small: $\epsilon \approx 10^{-15}$. Note that their problem can be seen as a poset problem, namely searching for a linear extension that minimizes a function. Thus, it comes as no surprise that the starting point for their algorithm, an observation about the relation of matchings in the comparability graph versus the number of initial sets, is also useful for the JUMP NUMBER problem; we explain this in the following paragraph.

Let us shortly recall the part of [22] important to us. The authors start with an $\mathcal{O}^*(2^n)$ dynamic programming algorithm for SCHED. Its running time is essentially given by its number of subproblems. Each useful subproblem is called an initial set, namely a set of the first elements of a linear extension already fixed during the dynamic programming. Clearly, the number of initial sets is at most 2^n. Achieving a better upper bound on the number of initial sets provides a refined running time bound for the dynamic programming algorithm. Since the algorithm constructs only linear extensions of P, and since $u \prec_P v$ implies that u occurs before v in any linear extension of P, there is no initial set containing v but not u. This is used to achieve an upper bound on the number of initial sets as follows. Let M be a maximum matching of the comparability graph $G(P)$ which can be computed in polynomial time [24]. Then for all $\{u, v\} \in M$ it holds that u and v are comparable in P, and thus if for example $u \prec_P v$ then no initial set contains v but not u. Consequently the number of initial sets is at most $2^{n-2|M|} \cdot 3^{|M|}$. Hence if M is large, i.e., $|M| \geq cn$ for large enough c, to be fixed later, then the running time will be $\mathcal{O}^*((2 - \epsilon)^n)$ for some $\epsilon > 0$.

This refined analysis via initial sets also applies to JUMP NUMBER (and to various other problems). In the case of a small maximum matching problem-specific ideas are required. For JUMP NUMBER it turns out that there is a single algorithm that handles both cases, i.e., small and large maximum matchings, and that explores the above number of subproblems when M is large; if M is small then more subproblems can be omitted, as we shall see.

Let us fix some poset $P = (X, \prec_P)$ and let M be a maximum matching of $G(P)$ with $|M| = cn$; note that $0 \leq c \leq 0.5$. Let W denote the elements of X that are not matched by M and let $U := X \setminus W = V(M)$. Clearly, W is an independent set in $G(P)$ and hence no two elements of W are comparable. Note that $|W| = n - 2cn$. Let $L = v_1 \prec_L v_2 \prec_L \ldots \prec_L v_n$ be a linear extension of P with minimum total number of jumps, and thus with maximum total number of bumps. Let $s(L, P) = s(P) = k$, thus $b(L, P) = n - k - 1$. The jumps of L give a chain partition of L into $C_1, C_2, \ldots C_{k+1}$ where for all i,

$$C_i = v_{t_i} \prec_P v_{t_i+1} \prec_P \ldots \prec_P v_{t_i+r_i}$$

is a chain of P, and (v_{t_i-1}, v_{t_i}) is a jump for all $i \in \{2, 3, \ldots, k+1\}$. Since W is an anti-chain of P every chain C_i contains at most one element of W.

Note that only non-trivial chains, i.e., those containing at least two elements, of the chain partition of L contain bumps. Thus to count the total number of bumps of L it suffices to consider only the non-trivial chains. Furthermore, every W-vertex in a non-trivial chain of the partition $C_1, C_2, \ldots, C_{k+1}$ is adjacent to a vertex of U in G. Thus these edges form a matching in G. Since M is a *maximum* matching this implies that at most $|M| \leq cn$ vertices of W belong to non-trivial chains. Hence it suffices to compute the maximum number of bumps on posets $P[W' \cup U]$ over all $W' \subseteq W$ satisfying $|W'| \leq |M| \leq cn$. Note that we can always add the ignored elements of $W \setminus W'$ while retaining a linear extension with at least the same number of bumps: For $w \in W \setminus W'$ consider adding it at the latest possible spot in a given linear extension, i.e., right before the first element x with $w \prec_P x$. This has two immediate consequences. First, all predecessors of w must also occur before x by transitivity, so there are no conflicts with this position of w in the linear extension. Second, we create at least the bump (w, x) and remove at most one bump before x; no decrease in total bumps. (If there is no such x then we insert at the end. Clearly, no conflicts are created and no bumps are removed.) Hence, the maximum number of bumps found for $P[W' \cup U]$ indeed transfers to P. From this the correct (minimum) jump number of P can be derived.

Note that, when computing the jump number of any $P[W' \cup U]$, we again benefit from the fact that the matching M limits the possible initial sets. Thus, the number of necessary subproblems can be bounded by the number of choices for $W' \in \binom{W}{\leq cn}$, i.e., the number of subsets of W of size up to cn, times $3^{|M|} = 3^{cn}$. Thus we get an upper bound of

$$\binom{n-2cn}{\le cn} \cdot 3^{cn}.$$

If $0.5 \ge c \ge 0.25$ then this can be bounded by 1.8613^n as follows.

$$\binom{n-2cn}{\le cn} \cdot 3^{cn} \le 2^{n-2cn} \cdot 2^{\log_2 3 \cdot cn} = 2^{n-2cn+\log_2 3 \cdot cn} = 2^{n \cdot (1-(2-\log_2 3)c)}$$

Now, we use $c \ge 0.25$ and $(2 - \log_2 3) > 0$.

$$\le 2^{n \cdot (1-(2-\log_2 3)0.25)} = \left(2^{0.5+0.25 \log_2 3}\right)^n < 1.8613^n$$

Otherwise, if $0 \le c \le 0.25$, then $0 \le c \le 0.5 \cdot (1-2c)$, i.e., $cn \le 0.5 \cdot (n-2cn)$, and hence Lemmas 3 and 4 (1) apply. Thus, to maximize over all values $0 \le c \le 0.25$ up to polynomial factor, we compute the maximum value of the function

$$f(c) = \frac{(1-2c)^{(1-2c)}}{c^c \cdot (1-3c)^{(1-3c)}} \cdot 3^c$$

The maximum is achieved for $c_0 \approx 0.2405117\ldots$ and implies a running time of $\mathcal{O}(1.8638^n)$. Together, both cases yield the following result.

Theorem 1. JUMP NUMBER *can be solved in time* $\mathcal{O}(1.8638^n)$.

3   Algorithms for Interval Orders

A partially ordered set $P = (X, \prec_P)$ is an *interval order* if there is a collection $\mathcal{I}$ of intervals of the real line and a bijection assigning to each element $v \in X$ an interval $\mathcal{I}(v)$ such that for all $x, y \in X$:

$$x \prec_P y \Leftrightarrow \mathcal{I}(x) < \mathcal{I}(y),$$

where $\mathcal{I}(x) < \mathcal{I}(y)$ means that interval $\mathcal{I}(x)$ appears to the left of interval $\mathcal{I}(y)$. Then $\mathcal{I}$ is called an interval model of P. We may assume that all endpoints of intervals in $\mathcal{I}$ are pairwise different and that $\{1, 2, \ldots, 2n\}$ is the set of all endpoints. Furthermore, for each interval i we denote its left endpoint by l_i and its right endpoint by r_i. With this notation we have $\mathcal{I}(x) < \mathcal{I}(y)$ if and only if $r_{\mathcal{I}(x)} < l_{\mathcal{I}(y)}$.

It is a well-known property of interval orders that they have at most n maximal anti-chains [7]. This can easily be seen as follows. The comparability graph $G(P)$ of an interval order P with interval model $\mathcal{I}$ is the complement of an interval graph with interval model $\mathcal{I}$. Consequently a maximal anti-chain of P is a maximal clique of the interval graph $\overline{G(P)}$, and an interval graph has at most n maximal cliques. Thus P has at most n maximal anti-chains.

Lemma 1. *An interval order P on n elements and of width $w(P)$ has at most $n \cdot 2^{w(P)}$ initial sets.*

Proof. Let I be an initial set of poset P. Then, as we have discussed, $x \in I$ and $y \prec x$ implies $y \in I$. Hence $x \in I$ implies that all predecessors of x also belong to I. Moreover, any initial set I of a poset P can be described as a union of predecessor sets: $I = I(W) = \mathrm{Pred}(W) = \cup_{x \in W}\{y : y \prec x\}$ where W is the set of all maximal elements in the poset induced by I. Hence W is an anti-chain of P, and thus there is a bijection between the initial sets of P and the anti-chains of P.

Let P be any interval order. Since an interval order has at most n maximal anti-chains [7], and each of them has size at most $w(P)$ and thus at most $2^{w(P)}$ subsets, it follows that P has at most $n \cdot 2^{w(P)}$ anti-chains and thus $n \cdot 2^{w(P)}$ initial sets.   □

Lemma 1 implies that JUMP NUMBER parameterized by width is fixed-parameter tractable on interval orders by a single-exponential algorithm: Indeed, it suffices to run the basic dynamic programming algorithm restricted to all sub-problems that correspond to one of the $\mathcal{O}(n \cdot 2^w)$ initial sets, giving time $\mathcal{O}^*(2^w)$. Furthermore, since $s(P) \geq w(P) - 1$ for every poset P, this immediately implies a single-exponential algorithm to solve JUMP NUMBER parameterized by number of jumps.

Theorem 2. JUMP NUMBER *on interval orders can be solved in time* $\mathcal{O}^*(2^w)$ *where w denotes the width of the given poset. It can be decided in time* $\mathcal{O}^*(2^k)$ *whether the jump number of an interval order is at most k.*

Now let us describe our exact algorithm. If $w(P) \leq cn$ for sufficiently small c then the exact exponential algorithm is using the $\mathcal{O}^*(2^w)$ algorithm which takes time $\mathcal{O}^*(2^{cn})$. This happens if $c \leq c_0$ for some $c_0 \geq 0.8$ that we will fix later.

Now let us consider the case $w(P) = cn$ for large c, i.e., $c \geq c_0$. Let W be a maximum anti-chain of P, which can be computed in polynomial time [25]. Then like in the exact exponential algorithm for general posets of the previous section, such large anti-chains do not contribute a lot to the number of bumps in an optimal linear extension. Let L be a linear extension with minimum total number of jumps, respectively maximum total number of bumps. Suppose L has k jumps then L generates a chain partition $(C_1, C_2, \ldots, C_{k+1})$. Since W is an anti-chain each chain C_i contains at most one element of W.

Each bump of L requires at least one incident element of $U := X \setminus W$ since W is an anti-chain. Thus, since $|U| = (1-c) \cdot n$ there are at most $(1-c) \cdot n$ bumps, and accordingly at most that many elements of W are involved in bumps. Therefore the algorithm tries all sets W' of at most $(1-c)n$ elements of W, and for each possible choice of W' and each subset $U' \subseteq U$ the dynamic programming algorithm computes the maximum number of bumps in any linear extension of $P[(U' \cup W')]$. The largest number b of bumps achieved over all choices of W' is the maximum number of bumps in a linear extension of P, and thus $s(P) = n - 1 - b$. The algorithm considers $\binom{cn}{\leq (1-c)n} \cdot 2^{(1-c)n}$ subproblems. Since $|W| = cn \geq c_0 n \geq 0.8n$ we have $(1-c)n \leq 0.5cn$ and Lemma 4 (1) applies. Thus, we get a bound of $\mathcal{O}^*(\binom{cn}{(1-c)n} \cdot 2^{(1-c)n})$ for the runtime in this case.

To balance the two running times $\mathcal{O}^*(2^{cn})$ and $\mathcal{O}^*(\binom{cn}{(1-c)n} \cdot 2^{(1-c)n})$ we use Lemma 3 and obtain $c_0 \approx 0.81493469\ldots$ and an overall running time of $\mathcal{O}(1.7593^n)$.

Theorem 3. *The* JUMP NUMBER *problem on interval orders is solvable in time* $\mathcal{O}(1.7593^n)$.

4   Linear Kernel for Jump Number on Interval Orders

In this section we present a kernelization for JUMP NUMBER on interval orders which reduces any given instance (P, k) to an equivalent instance with $\mathcal{O}(k)$ elements and bit-size $\mathcal{O}(k \log k)$. It is known that for any interval order $(X, \prec)$ the successor sets $\mathrm{Succ}(x) := \{y \in X \mid x \prec y\}$ are comparable with respect to set inclusion, i.e., for any two $x, y \in X$ we have $\mathrm{Succ}(x) \subseteq \mathrm{Succ}(y)$ or $\mathrm{Succ}(y) \subseteq \mathrm{Succ}(x)$. The same is true for the sets $\mathrm{Pred}(x) := \{y \mid y \prec x\}$ of predecessors. (In fact, either property is an alternative characterization of interval orders.) It is known that any linear order $\prec_L$ of X such that

$$x \prec_L y :\Leftarrow \begin{cases} \mathrm{Succ}(x) \supsetneq \mathrm{Succ}(y), \\ \mathrm{Succ}(x) = \mathrm{Succ}(y) \text{ and } \mathrm{Pred}(x) \supsetneq \mathrm{Pred}(y), \end{cases}$$

and breaking ties arbitrarily when $\mathrm{Succ}(x) = \mathrm{Succ}(y)$ and $\mathrm{Pred}(x) = \mathrm{Pred}(y)$ is a linear extension of the interval order $(X, \prec)$; cf. [26]; it is called *(decreasing)* Succ-*order*.

To obtain the kernelization we use results due to Felsner [26] who gave lower bounds on the jump number of an interval order $(X, \prec)$ in terms of parameters derived from the decreasing Succ-order. Given an interval order $P = (X, \prec)$ let $x_1, \ldots, x_n$ denote the obtained ordering. Since the sets $\mathrm{Succ}(x)$ are well-ordered we get that $i < j$ implies $\mathrm{Succ}(x_i) \supseteq \mathrm{Succ}(x_j)$. Felsner [26] describes three different cases for the behavior of any pair x_i, x_{i+1} in the ordering (note that necessarily $\mathrm{Succ}(x_i) \supseteq \mathrm{Succ}(x_{i+1})$):

1. $\mathrm{Succ}(x_i) = \mathrm{Succ}(x_{i+1})$: This is called an α-*jump at* x_i and α denotes the number of such jumps. Note that x_i and x_{i+1} must be incomparable; else one would appear in the successor set of the other causing disequality.
2. $\mathrm{Succ}(x_i) \supsetneq \mathrm{Succ}(x_{i+1})$ and $x_{i+1} \notin \mathrm{Succ}(x_i)$: This is called a β-*jump at* x_i and β also denotes the number of such jumps. Again the elements must be incomparable: We have $x_i \not\prec x_{i+1}$ since $x_{i+1} \notin \mathrm{Succ}(x_i)$, and $x_{i+1} \prec x_i$ would imply $x_i \in \mathrm{Succ}(x_{i+1}) \subsetneq \mathrm{Succ}(x_i) \not\ni x_i$.
3. If we have neither an α- nor a β-jump then it follows that $\mathrm{Succ}(x_i) \supsetneq \mathrm{Succ}(x_{i+1})$ and $x_{i+1} \in \mathrm{Succ}(x_i)$. Clearly, $x_i \prec x_{i+1}$.

Theorem 4 (Felsner [26]). *For any interval order* P, *the minimum number of jumps in any linear extension of* P *is at least*

$$\max\left\{\alpha, \alpha + \frac{\beta - \alpha}{3}\right\}.$$

It is an immediate corollary that **yes**-instances (P, k) for JUMP NUMBER on interval orders have $\alpha \leq k$ and $\beta \leq 3k$. Thus in order to obtain a kernelization it suffices to reduce the number of times that the third case happens. The following lemma gives a reduction rule for this. (The statement is equivalent to requiring three consecutive elements in Succ-order that have neither α- nor β-jumps.)

Lemma 2. *Let $P = (X, \prec)$ be an interval order and let $x_1, \ldots, x_n$ be an ordering of X according to decreasing Succ-order. Let $i \in \{1, \ldots, n-3\}$ such that*

1. $x_i \prec x_{i+1} \prec x_{i+2} \prec x_{i+3}$ *and*
2. $\mathrm{Succ}(x_i) \supsetneq \mathrm{Succ}(x_{i+1}) \supsetneq \mathrm{Succ}(x_{i+2}) \supsetneq \mathrm{Succ}(x_{i+3})$.

Then P has the same jump number as $P' := P - \{x_{i+2}\}$.

Proof. It is well-known that the jump number cannot increase by deleting elements: If we delete an element that has two incident bumps in a linear extension, then (by transitivity) its two neighbors still have an incident bump after deletion of the element; thus no jump is inserted. If there was at least one incident jump to begin with then this balances the possible jump between the former neighbors (after deletion of the element). Thus the jump number of P' is at most the jump number of P.

For the converse consider a linear extension L' of P'. We will show that x_{i+2} can be inserted into L' such that we get a linear extension L of P with at most the same number of jumps. Obviously we have to insert x_{i+2} somewhere between x_{i+1} and x_{i+3}. We have to check that this causes no violation of comparabilities between x_{i+2} and elements $x \in X \setminus \{x_i, \ldots, x_{i+3}\}$, and that the number of jumps does not increase.

1. Let $x \in \{x_1, \ldots, x_{i-1}\}$. Clearly, $\mathrm{Succ}(x) \supseteq \mathrm{Succ}(x_i)$. Since $x_i \prec x_{i+1}$ we get $x_{i+1} \in \mathrm{Succ}(x_i) \subseteq \mathrm{Succ}(x)$ and hence $x \prec x_{i+1}$. Thus there can be no violated comparabilities between x_{i+2} and elements $x \in \{x_1, \ldots, x_{i-1}\}$ when we insert x_{i+2} after x_{i+1}. Furthermore, such x are not relevant regarding preservation of the number of jumps: They appear before x_{i+1} in L' and we insert x after it.
2. Let $x \in \{x_{i+4}, \ldots, x_n\}$. Clearly, $\mathrm{Succ}(x_{i+3}) \supseteq \mathrm{Succ}(x)$. We have to consider the three possibilities for the $\prec$-relation between x and x_{i+2}:
   (a) If $x \prec x_{i+2}$ then $x_{i+2} \in \mathrm{Succ}(x) \subseteq \mathrm{Succ}(x_{i+3})$. However, this means that $x_{i+3} \prec x_{i+2}$ and contradicts antisymmetry of P.
   (b) If $x_{i+2} \prec x$ then we get $x_{i+1} \prec x_{i+2} \prec x$. Thus in L' the element x_{i+1} must precede x. Consequently, placing x_{i+2} after x_{i+1} and before any x with $x_{i+2} \prec x$ causes no comparability violation.
   (c) If neither $x \prec x_{i+2}$ nor $x_{i+2} \prec x$ then there is no comparability and hence also no possible violation.

We know now that inserting x_{i+2} after x_{i+1} and right before the first element x with $x_{i+2} \prec x$ does not cause any violation of comparabilities. Put differently, none of the elements succeeding x_{i+2} can appear before x_{i+1} since they are all successors of x_{i+1} too. We will now have to check that, additionally, this placement also does not increase the number of jumps. Clearly, additional jumps can only happen adjacent to x_{i+2}. Furthermore, $x_{i+2} \prec x$ is a bump. Thus, the only bad case for us would be if the old linear extension L' has a bump (x', x) and thus $x' \prec x$ but our insertion gives $x' \prec_L x_{i+2} \prec_L x$ with $x' \not\prec x_{i+2}$ and $x_{i+2} \prec x$ i.e., a jump and a bump in the new linear extension. Let us fix such choices of x and x' and discuss this case in detail.

We know from our previous discussion that $x, x' \notin \{x_1, \ldots, x_{i-1}\}$ since otherwise we would have $x \prec x_{i+1} \prec x_{i+2}$ (contradicting $x_{i+2} \prec x$) or $x' \prec x_{i+1} \prec x_{i+2}$ (contradicting $x' \not\prec x_{i+2}$). By the same argument we get $x, x' \notin \{x_i, x_{i+1}\}$ since either of that would precede x_{i+2} in P. Thus $x, x' \in \{x_{i+3}, \ldots, x_n\}$.

1. If $x' = x_{i+3}$ then $x_{i+2} \prec x_{i+3} = x'$; a contradiction to our choice of x as the first element in the linear extension that succeeds x_{i+2} in P: Since we assumed $x' \prec x$ it must appear before x in L' (and it also succeeds x_{i+2} in the present case); we would have chosen x' over x.

2. If $x' \in \{x_{i+4}, \ldots, x_n\}$ then $\mathrm{Succ}(x_{i+3}) \supseteq \mathrm{Succ}(x')$. Since $x' \prec x$ we have $x \in \mathrm{Succ}(x') \subseteq \mathrm{Succ}(x_{i+3})$. This however means that $x_{i+3} \prec x$ which contradicts our choice of x: It must appear before x in the linear extension and it does succeed x_{i+2} too.

Summarizing, we may conclude that no such choice of x and x' is possible. Thus, placing x_{i+2} after x_{i+1} and right before the first element that succeeds x_{i+2} in P causes no additional jumps and does not violate comparabilities. Hence, under the assumptions of the lemma the partial orders P and $P' := P - \{x_{i+2}\}$ have the same jump number. □

We are now able to complete the kernelization.

Theorem 5. JUMP NUMBER *on interval orders admits a kernelization with a linear number of elements, i.e., given an instance (P, k) where P is an interval order we get an equivalent instance with at most $9k + 3$ elements. This can be encoded in bit-size $\mathcal{O}(k \log k)$.*

Proof. Let (P, k) be an input for JUMP NUMBER where $P = (X, \prec)$ is an interval order. We use the implicit reduction rule of Lemma 2: Order the elements according to decreasing Succ-order: $x_1, \ldots, x_n$. For $i \in \{1, \ldots, n-3\}$, if there are no α- or β-jumps at x_i, x_{i+1}, and x_{i+2} then

1. $x_i \prec x_{i+1} \prec x_{i+2} \prec x_{i+3}$ and
2. $\mathrm{Succ}(x_i) \supsetneq \mathrm{Succ}(x_{i+1}) \supsetneq \mathrm{Succ}(x_{i+2}) \supsetneq \mathrm{Succ}(x_{i+3})$.

Thus, by Lemma 2 we may delete x_{i+2} from P without changing its jump number. Exhaustive application of this rule (possibly recomputing the Succ-order) can be performed in polynomial time.

Afterwards, at any three consecutive elements x_i, x_{i+1}, and x_{i+2} for $i \in \{1, \ldots, n-3\}$ in the sorted order there is at least one α- or β-jump. Thus, the total number of elements is at most $3(\alpha + \beta) + 3$, where α and β denote the number of α- and β-jumps, respectively. (Note that the $+3$ accounts for element number n and the possibility of having exactly every third element be an α- or β-jump.) By Theorem 4 due to Felsner [26] we have that the minimum number of jumps is at least

$$\max\left\{\alpha, \alpha + \frac{\beta - \alpha}{3}\right\}.$$

Thus, if this value exceeds k then we may safely return a **no**-instance (or simply answer **no**). Otherwise, we can bound $\alpha + \beta$, i.e., the total number of α- and β-jumps, as follows.

$$\alpha + \beta \leq 2\alpha + \beta = 3\alpha + \beta - \alpha = 3\left(\alpha + \frac{\beta - \alpha}{3}\right) \leq 3k$$

This is tight for $\alpha = 0$ and $\beta = 3k$. We get a bound of $3(\alpha + \beta) + 3 \leq 9k + 3$ elements in instances that are reduced as outlined above.

Finally, let us address the claimed bit size of $\mathcal{O}(k \log k)$. Recall the ordering by decreasing value of $\mathrm{Succ}(x)$, and let $x_1, x_2, \ldots, x_\ell$ be such an ordering for a reduced instance with $\ell \leq 9k + 3$. Clearly, $\mathrm{Succ}(x_1) \supseteq \mathrm{Succ}(x_2) \supseteq \ldots \supseteq \mathrm{Succ}(x_\ell)$. Note that for each element $x \in X$ there is a largest index $i(x) \in \{1, \ldots, \ell\}$ such that $x \in \mathrm{Succ}(x_{i(x)})$. Storing this index in $\mathcal{O}(\log k)$ bits for each of the ℓ elements lets us retrieve the sequence of successor sets and hence all comparabilities. This takes total size $\mathcal{O}(k \log k)$. This completes the proof. $\square$

5   The $\mathcal{O}^*(2^n)$ Dynamic Programming Algorithm

For the sake of completeness we present here the algorithm to compute the jump number of a given poset $P = (X, \prec_P)$ in time $\mathcal{O}^*(2^n)$ which is based on a standard dynamic programming over all subsets of X. Let us recall that the task is to compute the smallest possible number of jumps taken over all linear extension of P.

To describe the algorithm we use the following notation. The subposet of P induced by a subset $A \subseteq X$ is denoted by $P[A]$. The set of all maximal elements of a subposet $P[A]$ is denoted by $\mathrm{MAX}(A)$. We also use a binary-valued function $j(u, v)$ which is equal to 1 if u and v are not comparable and equal to 0 otherwise.

The algorithm computes the jump number of P by solving subproblems denoted by $s[A, v]$ for all $A \subseteq X$ and all $v \in \mathrm{MAX}(A)$. Thereby $s[A, v]$ is the smallest number of jumps in any linear extension of $P[A]$ having v as last element. The computation is based on the following recurrence:

$$s[A, v] = \begin{cases} \min_{u \in \mathrm{MAX}(A \setminus \{v\})} \left(s[A \setminus \{v\}, u] + j(u, v)\right) & \text{if } v \in \mathrm{MAX}(A), \\ \infty & \text{else.} \end{cases}$$

Initially, the algorithm sets $s[\{a\}, a] = 0$ for all elements a of P (in fact, we will see in a moment that this will only be needed for minimal elements). Then the above recurrence is applied to all subproblems $s[A, v]$ with $|A| \geq 2$ in increasing order of the size of A. Finally $s(P) = \min_{v \in X} s[X, v]$.

The number of subproblems to be solved is at most $n2^n$ and the running time is $\mathcal{O}(n^2 2^n)$. Without further efforts in the construction and analysis of the dynamic programming algorithm its running time is $\mathcal{O}^*(2^n)$ for every input.

Let us recall the observation that the dynamic programming can be restricted to initial sets. Consider the case of a set $A \subseteq X$ that is not initial and let $v \in X \setminus A$ such that $v \prec_P u$ for some $u \in A$. Now, when we take the recurrence for $A' = A \cup \{v\}$ we see that v is not maximal in A' since $v \prec_P u$; thus we get $s[A', v] = \infty$. By extension this means that the partial solution for A cannot contribute to the final outcome for $s[X, \cdot]$. Thus, we may safely skip all A that are not initial sets (often also called *down sets*).

As a technical detail, to benefit from the small number of initial sets the algorithm needs to enumerate all those sets with polynomial delay. (We would not want to check all subsets for whether they are down sets/initial sets.) This is straightforward and we omit a discussion at this point.

6   Binomial Coefficients

The following formula is crucial for the computation of the running times of our exact exponential algorithms. It is an immediate consequence of a related formula in [18, p.39].

Lemma 3. *Let α and β reals such that $1 \geq \alpha \geq \beta \geq 0$ and let n be a positive integer. Then the following holds*

$$\binom{\alpha n}{\beta n} \leq \left(\frac{\alpha^\alpha}{\beta^\beta (\alpha - \beta)^{\alpha - \beta}} \right)^n.$$

The following lemma is an immediate consequence of basic properties of binomial coefficients, i.e., the well-known monotonicity and the fact that the maximum of $\binom{s}{t}$ is approximately 2^s and is achieved for $t = \lfloor s/2 \rfloor$.

Lemma 4. *Let α and β reals such that $1 \geq \alpha \geq \beta \geq 0$ and let n be a positive integer.*

$$\binom{\alpha n}{\leq \beta n} = \sum_{i=0}^{\beta n} \binom{\alpha n}{i} \leq n \cdot \binom{\alpha n}{\beta n} \qquad \text{if } \beta < \alpha/2 \qquad (1)$$

$$\binom{\alpha n}{\leq \beta n} = \sum_{i=0}^{\beta n} \binom{\alpha n}{i} \leq 2^{\alpha n} \qquad \text{if } \beta \geq \alpha/2 \qquad (2)$$

7   Conclusion

We have studied exact algorithms for the JUMP NUMBER problem. For the general case on arbitrary posets we provide a $\mathcal{O}(1.8638^n)$ time algorithm improving over the $\mathcal{O}^*(2^n)$ dynamic programming over (all) subsets algorithm. For interval orders we improve the runtime to $\mathcal{O}(1.7593^n)$. Along the way we prove that

JUMP NUMBER is fixed-parameter tractable with runtime $\mathcal{O}^*(2^w)$ on interval orders when parameterized by the width w of the poset (the length of its longest anti-chain). Since $w - 1$ is a lower bound on the jump number, this immediately gives time $\mathcal{O}^*(2^k)$ parameterized by the number of jumps. This motivates two questions: (1) Does the fixed-parameter tractability by width extend also to general posets? The algorithm of Colbourn and Pulleyblank [27] runs in polynomial time for fixed width; in other words, JUMP NUMBER is in XP with respect to width. (2) Can JUMP NUMBER be solved in time $\mathcal{O}^*(c^k)$ in general?

Extending our results for JUMP NUMBER on interval orders, we provided a polynomial kernelization with respect to the target number k of jumps. Concretely, we reduce to $9k + 3$ elements and bit-size $\mathcal{O}(k \log k)$. Naturally, this brings up the open question regarding the possibility of a polynomial kernelization for JUMP NUMBER on general posets. We note that parameterization by width, unsurprisingly, does not admit a polynomial kernelization, similar to most other problems parameterized by "width-parameters" (e.g., treewidth, maximum degree); here is a sketch of a composition: Take t instances of JUMP NUMBER with width w each and asking for jump number at most k. Furthermore, using a result of Kratsch et al. [28], we can start from the improvement version where all these instances are **yes** for $k + 1$, without requiring NP-hardness of this variant. (Essentially, it suffices that we could solve JUMP NUMBER by queries to this improvement variant.) Now, make elements x and y comparable, i.e., $x \prec y$, if x is in an instance with smaller number than y. It is easy to see that linear extensions of the obtained order have to keep the elements of each instance consecutive: Indeed, every element must precede all elements from instances with larger numbers. It follows that jumps are only incurred among elements from the same instance and no jumps between them can be saved in comparison to any linear extension of only these elements. Thus, if all instances are **no** and hence need $k + 1$ jumps then the total is $t \cdot (k + 1)$. If at least one instance is **yes**, then we get a total at most $t \cdot (k + 1) - 1$. Thus asking for at most $t \cdot (k + 1) - 1$ jumps implements the OR of the instances. The width of the combined instance is w since anti-chains can only have elements from a single instance.

References

1. Sysło, M.M.: Minimizing the jump number for partially-ordered sets: a graph-theoretic approach, ii. Discrete Mathematics 63(2-3), 279–295 (1987)
2. Sysło, M.M.: An algorithm for solving the jump number problem. Discrete Mathematics 72(1-3), 337–346 (1988)
3. Müller, H.: Alternating cycle-free matchings. Order 7(1), 11–21 (1990)
4. Reuter, K.: The jump number and the lattice of maximal antichains. Discrete Mathematics 88(2-3), 289–307 (1991)
5. Pulleyblank, W.R.: On minimizing setups in precedence constrained scheduling (1981) (Unpublished manuscript)
6. Bouchitté, V., Habib, M.: NP-completeness properties about linear extensions. Order 4(2), 143–154 (1987)
7. Trotter, W.T.: Combinatorics and partially ordered sets: Dimension theory. John Hopkins University Press (2001)

8. Mitas, J.: Tackling the jump number of interval orders. Order 8(2), 115–132 (1991)
9. Felsner, S.: A 3/2-approximation algorithm for the jump number of interval orders. Order 6(4), 325–334 (1990)
10. Sysło, M.M.: The jump number problem on interval orders: A 3/2 approximation algorithm. Discrete Mathematics 144(1-3), 119–130 (1995)
11. Krysztowiak, P.: Improved approximation algorithm for the jump number of interval orders. Electronic Notes in Discrete Mathematics 40, 193–198 (2013)
12. Steiner, G., Stewart, L.K.: A linear time algorithm to find the jump number of 2-dimensional bipartite partial orders. Order 3(4), 359–367 (1987)
13. Sysło, M.M.: Minimizing the jump number for partially ordered sets: A graph-theoretic approach. Order 1(1), 7–19 (1984)
14. Habib, M., Möhring, R., Steiner, G.: Computing the bump number is easy. Order 5(2), 107–129 (1988)
15. El-Zahar, M.H., Schmerl, J.H.: On the size of jump-critical ordered sets. Order 1(1), 3–5 (1984)
16. McCartin, C.: An improved algorithm for the jump number problem. Inf. Process. Lett. 79(2), 87–92 (2001)
17. Fishburn, P.C.: Interval orders and interval graphs. Wiley (1985)
18. Fomin, F.V., Kratsch, D.: Exact Exponential Algorithms. Springer (2010)
19. Fomin, F.V., Villanger, Y.: Finding induced subgraphs via minimal triangulations. In: Marion, J.Y., Schwentick, T. (eds.) STACS. LIPIcs, vol. 5, pp. 383–394. Schloss Dagstuhl - Leibniz-Zentrum fuer Informatik (2010)
20. Suchan, K., Villanger, Y.: Computing pathwidth faster than 2^n. In: Chen, J., Fomin, F.V. (eds.) IWPEC 2009. LNCS, vol. 5917, pp. 324–335. Springer, Heidelberg (2009)
21. Björklund, A.: Determinant sums for undirected hamiltonicity. In: FOCS, pp. 173–182. IEEE Computer Society (2010)
22. Cygan, M., Pilipczuk, M., Pilipczuk, M., Wojtaszczyk, J.O.: Scheduling partially ordered jobs faster than 2^n. In: Demetrescu, C., Halldórsson, M.M. (eds.) ESA 2011. LNCS, vol. 6942, pp. 299–310. Springer, Heidelberg (2011)
23. Woeginger, G.J.: Exact Algorithms for NP-Hard Problems: A Survey. In: Jünger, M., Reinelt, G., Rinaldi, G. (eds.) Combinatorial Optimization (Edmonds Festschrift). LNCS, vol. 2570, pp. 185–207. Springer, Heidelberg (2003)
24. Micali, S., Vazirani, V.V.: An $O(\sqrt{|V|}|E|)$ algorithm for finding maximum matching in general graphs. In: FOCS, pp. 17–27. IEEE Computer Society (1980)
25. Spinrad, J.P.: Efficient graph representations. American Mathematical Society (2003)
26. Felsner, S.: Interval Orders: Combinatorial Structure and Algorithms. PhD thesis, Technical University Berlin (1992)
27. Colbourn, C.J., Pulleyblank, W.R.: Minimizing setups in ordered sets of fixed width. Order 1(3), 225–229 (1985)
28. Kratsch, S., Pilipczuk, M., Rai, A., Raman, V.: Kernel lower bounds using co-nondeterminism: Finding induced hereditary subgraphs. In: Fomin, F.V., Kaski, P. (eds.) SWAT 2012. LNCS, vol. 7357, pp. 364–375. Springer, Heidelberg (2012)

On the Hardness of Eliminating Small Induced Subgraphs by Contracting Edges

Daniel Lokshtanov[1], Neeldhara Misra[2], and Saket Saurabh[1,3]

[1] University of Bergen, Norway
daniello@ii.uib.no
[2] Indian Institute of Science, India
neeldhara@csa.iisc.ernet.in
[3] Institute of Mathematical Sciences, India
saket@imsc.res.in

Abstract. Graph modification problems such as vertex deletion, edge deletion or edge contractions are a fundamental class of optimization problems. Recently, the parameterized complexity of the CONTRACTIBILITY problem has been pursued for various specific classes of graphs. Usually, several graph modification questions of the deletion variety can be seen to be FPT if the graph class we want to delete into can be characterized by a finite number of forbidden subgraphs. For example, to check if there exists k vertices/edges whose removal makes the graph C_4-free, we could simply branch over all cycles of length four in the given graph, leading to a search tree with $O(4^k)$ leaves. Somewhat surprisingly, we show that the corresponding question in the context of contractibility is in fact W[2]-hard. An immediate consequence of our reductions is that it is W[2]-hard to determine if at most k edges can be contracted to modify the given graph into a chordal graph. More precisely, we obtain following results:

- C_ℓ-FREE CONTRACTION is W[2]-hard if $\ell \geq 4$ and FPT if $\ell \leq 3$.
- P_ℓ-FREE CONTRACTION is W[2]-hard if $\ell \geq 5$ and FPT if $\ell \leq 4$,
  where P_ℓ denotes a path on ℓ vertices.

We believe that this opens up an interesting line of work in understanding the complexity of contractibility from the perspective of the graph classes that we are modifying into.

1 Introduction

Graph modification problems constitute a broad and fundamental class of graph optimization problems. Typically, we are interested in knowing if a given input graph G is "close enough" to a graph H or a graph in a class of graphs $\mathcal{H}$. In the latter case, the goal is usually to see if G can be easily morphed into a graph with a certain property, and the class $\mathcal{H}$ is used to describe the said property [3]. Some of the most prevalent notions of closeness are defined in terms of vertex or edge deletion, or edge contraction. For example, when defined in terms of vertex deletion, one might ask if at most k vertices can be deleted to make the graph edgeless (here we are modifying into the class of empty graphs), and this is the classic VERTEX COVER problem.

G. Gutin and S. Szeider (Eds.): IPEC 2013, LNCS 8246, pp. 243–254, 2013.

In this work, we will restrict ourselves to the context of contractibility questions, and in particular, we would be contracting into graph classes that are described in terms of their induced forbidden subgraphs. In a $\mathcal{H}$-CONTRACTIBILITY problem, given a graph G and a positive integer k, the objective is to check if there exists a subset of at most k edges which, if contracted, lead to a graph in $\mathcal{H}$. Such questions are usually NP-complete on general graphs, and have recently received a lot of attention in the context of parameterized complexity. For example, it is known that the BIPARTITE CONTRACTION problem is FPT, and this is the contraction analog of EDGE BIPARTIZATION, which is the fundamental and well-studied question of whether k edges can be removed to make a given graph bipartite [9,6]. This result involved an interesting combination of techniques, including iterative compression, important separators, and irrelevant vertices. Also, the problems of determining if k edges can be contracted to obtain a tree, or a path, are known to be FPT using a non-trivial application of color coding [7]. The PLANAR CONTRACTION problem was also shown to be FPT recently [5], again using irrelevant vertex techniques combined with an application of Courcelle's theorem.

Questions of contractibility have been investigated quite extensively when the input graph is restricted to being chordal, usually yielding polynomial time algorithms (see, for instance, [8,2]). However, the natural question of CHORDAL CONTRACTION, while known to be NP-complete [1], remains un-investigated in the parameterized context. Before considering algorithms for CHORDAL CONTRACTION, we first explored the apparently easier question of contracting edges to obtain a C_4-free graph, that is, a graph with no induced cycles of length four. Notice that the vertex-deletion analog of this question is almost trivial from a parameterized point of view: we could simply branch over all cycles of length four in the given graph, leading to a search tree with $O(4^k)$ leaves. This is true of most problems which require us to "hit" a constant number of constant-sized forbidden subgraphs using a constrained budget. However, when we ask the same question in the context of contraction, the scenario is dramatically different: it is no longer true that a copy of a forbidden object can only be destroyed by edges that form the object — rather, edges contracted from "outside" the copy could also contribute towards its elimination. Therefore, the number of choices for branching is no longer obviously bounded. In fact, we find that the C_4-FREE CONTRACTION question turns out to be W[2]-hard, which we find rather surprising, considering the finite nature of the forbidden subgraph characterization of the graph class that we are interested in contracting to.

It turns out that our reduction also implies the hardness of CHORDAL CONTRACTION. On a closely related note, we show that the P_i-FREE CONTRACTION problem is also W[2]-hard. On the positive side, we show that it is FPT to determine if k edges can be contracted so that the resulting graph is a complete graph. In this case, the forbidden subgraph is just a single non-edge or an induced path on two edges. Further, we remark that it is easily checked that K_i-FREE CONTRACTION is FPT by the search tree technique. In this case, since the forbidden object, being a complete graph, cannot be "destroyed from outside", the branching is exhaustive.

The reason for describing the graph class $\mathcal{H}$ in terms of its forbidden subgraphs is to open up questions regarding a general characterization of the parameterized complexity of the problem in terms of the forbidden subgraphs, possibly analogous to the theorem of Asano and Hirata [1]. In this work, our goal is to motivate and initiate a study in this direction, by providing somewhat unexpected answers to a few specific cases.

Our Contributions. Let $\mathcal{H}$ be a graph class that has a forbidden induced subgraph characterization, and let $\mathcal{F}$ be the forbidden induced subgraphs for $\mathcal{H}$. Then, the $\mathcal{H}$ CONTRACTION question, or equivalently the $\mathcal{F}$-FREE CONTRACTION problem, is the following.

$\mathcal{F}$-FREE CONTRACTION   **Parameter:** k
Input: A graph $G = (V, E)$ and a positive integer k
Question: Is there a subset of at most k edges such that G/F has no induced copies of graphs $H \in \mathcal{F}$?

The C_ℓ-FREE CONTRACTION problem is known to be NP-complete. [1] for all fixed integer $\ell \geq 3$. We show, by a simple reduction from the HITTING SET problem, that the C_ℓ-FREE CONTRACTION problem is W[2]-hard for $\ell \geq 4$. Consequently, we establish that CHORDAL CONTRACTION is W[2]-hard. Further, we show that P_γ-FREE CONTRACTION is W[2]-hard for all $\gamma \geq 5$, while contracting to K_i-free graphs (for $i \geq 3$) and cliques turn out to be FPT.

The paper is organized as follows. After introducing some notation and preliminary notions in Section 2, we turn to the reductions. We first show that the C_4-FREE CONTRACTION problem is W[2]-hard, and subsequently describe a generalization. This is followed by the reduction for P_γ-FREE CONTRACTION. We conclude with the tractable cases and suggestions for future directions.

2 Preliminaries

In this section we state some basic definitions related to parameterized complexity and graph theory, and give an overview of the notation used in this paper. Our notation for graph theoretic notions is standard and follows Diestel [4]. We summarize some of the frequently used concepts here. For a finite set V, a pair $G = (V, E)$ such that $E \subseteq V^2$ is a graph on V. The elements of V are called *vertices*, while pairs of vertices (u, v) such that $(u, v) \in E$ are called *edges*. We also use $V(G)$ and $E(G)$ to denote the vertex set and the edge set of G, respectively. In the following, let $G = (V, E)$ and $G' = (V', E')$ be graphs, and $U \subseteq V$ some subset of vertices of G. Let G' be a subgraph of G. If E' contains all the edges $\{u, v\} \in E$ with $u, v \in V'$, then G' is an *induced subgraph* of G, *induced by* V', denoted by $G[V']$. For any $U \subseteq V$, $G \setminus U = G[V \setminus U]$. For $v \in V$, $N_G(v) = \{u \mid (u, v) \in E\}$.

The *contraction* of edge xy in G removes vertices x and y from G, and replaces them by a new vertex, which is made adjacent to precisely those vertices that were adjacent to at least one of the vertices x and y. A graph G is *contractible* to a graph H, or H-*contractible*, if H can be obtained from G by a sequence of edge contractions. Equivalently, G is H-contractible if there is a surjection $\varphi : V(G) \to V(H)$, with $W(h) = \{v \in V(G) \mid \varphi(v) = h\}$ for every $h \in V(H)$, that satisfies the following three conditions: **(1)** for every $h \in V(H)$, $W(h)$ is a connected set in G; **(2)** for every pair $h_i, h_j \in V(H)$, there is an edge in G between a vertex of $W(h_i)$ and a vertex of $W(h_j)$ if and only if $h_i h_j \in E(H)$; **(3)** $\mathcal{W} = \{W(h) \mid h \in V(H)\}$ is a partition of $V(G)$. We say that $\mathcal{W}$ is an H-*witness structure* of G, and the sets $W(h)$, for $h \in V(H)$, are called *witness sets* of $\mathcal{W}$. It is easy to see that if we contract every edge $uv \in E(G)$, such that u and v belong to the same witness set, then we obtain a graph isomorphic to H. Hence G is H-contractible if and only if it has an H-witness structure.

A path is a sequence of vertices $v_1, v_2, \ldots, v_r$ such that $(v_i, v_i + 1) \in E$ for all $1 \le i \le r - 1$. A cycle is a sequence of vertices $v_1, v_2, \ldots, v_r$ such that $(v_i, v_i + 1) \in E$ for all $1 \le i \le r - 1$, and $(v_r, v_1) \in E$. A graph is said to be *chordal*, or *triangulated* if it has no induced cycles of length four or more.

Parameterized Complexity. A parameterized problem is denoted by a pair $(Q, k) \subseteq \Sigma^* \times \mathbb{N}$. The first component Q is a classical language, and the number k is called the parameter. Such a problem is *fixed–parameter tractable* (FPT) if there exists an algorithm that decides it in time $O(f(k)n^{O(1)})$ on instances of size n. Next we define the notion of parameterized reduction.

Definition 1. *Let A, B be parameterized problems. We say that A is (uniformly many:1)* **fpt-reducible** *to B if there exist functions $f, g : \mathbb{N} \to \mathbb{N}$, a constant $\alpha \in \mathbb{N}$ and an algorithm Φ which transforms an instance (x, k) of A into an instance $(x', g(k))$ of B in time $f(k)|x|^\alpha$ so that $(x, k) \in A$ if and only if $(x', g(k)) \in B$.*

A parameterized problem is considered unlikely to be fixed-parameter tractable if it is $W[i]$-hard for some $i \ge 1$. To show that a problem is $W[2]$-hard, it is enough to give a parameterized reduction from a known $W[2]$-hard problem. Throughout this paper we follow this recipe to show a problem $W[2]$-hard.

3   Hardness of Contraction Problems

In this section we address the parameterized complexity of C_j-FREE CONTRACTION, CHORDAL CONTRACTION and P_j-FREE CONTRACTION. All the reductions are from the HITTING SET problem, and have a similar underlying flavor. We would begin by creating a separate induced instance of a forbidden object for every set in the universe. Then we will typically have edges corresponding to the elements in the universe, and the edges are placed to ensure that contracting them will "kill" exactly those forbidden objects that correspond to the sets that the element belongs to. Often, this is achieved with the following wireframe: we anchor all the edges corresponding to vertices of the universe to a common vertex, and let the forbidden object "dangle" from the same vertex. Now, to encode the instance, we add edges between the free end of the edges that correspond to the vertices of the universe and a suitably chosen vertex of the relevant forbidden objects. We would expect that this generic idea is realized in different ways depending on what the forbidden objects are. In the rest of this section, we will describe two instances of specific reductions in detail, formalizing the ideas described above.

3.1   Contracting to C_ℓ-free Graphs

Our first exploration is to do with the problem of contracting to graphs that contain no induced cycles of length ℓ. In the interest of exposition, we begin by explaining the reduction for the case of reducing to C_4-free graphs. Since it turns out that the reduced instance has no longer induced cycles, this reduction already implies the hardness of contracting k edges to obtain a chordal graph. We will subsequently describe an easy generalization of the construction.

C_4-FREE CONTRACTION   **Parameter:** k
Input: A graph $G = (V, E)$ and a positive integer k
Question: Is there a subset of at most k edges such that G/F has no induced cycles of length four?

We reduce from the HITTING SET problem. Let $(U, \mathcal{F})$ be an instance of HITTING SET, where $U = \{x_1, x_2, \ldots, x_n\}$ and $\mathcal{F} = \{S_1, S_2, \ldots, S_m\}$, where each $S_i \subseteq U$. We denote the reduced instance to be constructed by $G = (V, E)$. The vertex set consists of a special central vertex, denoted by g, one vertex for each element $x_i \in U$, denoted by u_i, and three vertices for every set S_i in the family $\mathcal{F}$, denoted by a_i, b_i, c_i. We now describe the edges. The central vertex is adjacent to every vertex other than $\{c_i \mid 1 \leq i \leq m\}$. We impose a clique on the vertices that correspond to elements of the universe. Next, we add the edges $(a_i c_i)$ and $(b_i c_i)$ for every $1 \leq i \leq m$. Finally, for every $x_i \in S_j$, we add the edge (u_i, c_j). This completes the construction. Formally, the instance is given as follows (also see Figure 1). $V := \{g\} \cup \{u_i \mid 1 \leq i \leq n\} \cup \left(\bigcup_{1 \leq i \leq m} \{a_i, b_i, c_i\} \right)$ and

$$E := \left(\bigcup_{1 \leq i \leq n, 1 \leq j \leq m} \{(g, u_i), (g, a_j), (g, b_j)\} \right) \cup \left(\bigcup_{1 \leq j \leq m} \{(c_j, a_j), (c_j, b_j)\} \right)$$
$$\cup \{(u_i, u_j) \mid 1 \leq i \neq j \leq n\} \cup \{(u_i, c_j) \mid 1 \leq i \leq n, 1 \leq j \leq m, \text{ and } x_i \in S_j\}$$

We begin by identifying the induced cycles of length four in the graph G. This will help us in showing the correctness of the reduction.

Proposition 1. *The only induced cycles of length four in the graph G are formed by the vertex sets given below:*

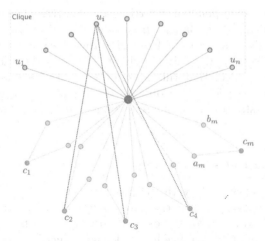

Fig. 1. The construction for reducing to C_4-free graphs. In this example, the adjacencies corresponding to the HITTING SET are illustrated for the element x_i, which is assumed to belong to the sets S_2, S_3 and S_4.

- $\{g, a_i, c_i, b_i\}$, for all $1 \leq i \leq m$,
- $\{u_i, g, a_j, c_j\}$, for all $x_i \in S_j$, and
- $\{u_i, g, b_j, c_j\}$, for all $x_i \in S_j$.

Proof. Clearly, for all $1 \leq i \leq m$, the vertices $\{g, a_i, b_i, c_i\}$ induce a four-cycle, and for all $x_i \in S_j$, the vertices $\{u_i, g, t, c_j\}$ (where t is either a_j or b_j) induce a four-cycle as well. Assume, for the sake of contradiction, that there exists an induced four-cycle other than the ones accounted for, with the vertex set $C := \{w, x, y, z\}$. Let T denote the vertex subset $\{g\} \cup \{u_i \mid 1 \leq i \leq n\}$. Note that $|C \cap T| \leq 2$, since $G[T]$ is a clique, and $G[C]$ is an induced cycle of length four. Notice that $G \setminus T$ is acyclic, so C intersects T in either one or two vertices.

First, consider the case when $|T \cap C| = 1$, and without loss of generality, let $T \cap C = \{w\}$. Suppose $w \neq g$. Then $w = u_i$ for some $1 \leq i \leq n$. Notice that u_i is adjacent to vertices $N_i := \{c_j \mid x_i \in S_j\}$. However, it is easily checked that no two vertices in N_i share a common neighbor in $G \setminus T$. Indeed, for $1 \leq p \neq q \leq m$, $N_{G \setminus T}(c_p) = \{b_p, a_p\}$ and $N_{G \setminus T}(c_q) = \{b_q, a_q\}$. Therefore, $N(x) \cap N(y) \cap G \setminus T = \emptyset$ for all $x, y \in N_i$, and w cannot be extended to an induced four-cycle from vertices in $G \setminus T$. On the other hand, let $w = g$. Then, let the neighbors of w in the four-cycle C are x and z. Clearly, $x := a_j$ or $x := b_j$, for some $1 \leq j \leq m$. Without loss of generality, let $x := a_j$. Now, $z \neq b_j$, since in this case, the unique cycle that w, x and z can be completed to is already accounted for. Thus, $z := v_\ell^{(a)}$ or $z := v_\ell^{(b)}$ for some $\ell \neq j$. Again, in this case, z and x share no common neighbors in $G \setminus T$, and we are done.

The second case is when $|T \cap C| = 2$. Again, without loss of generality, let $T \cap C = \{w, x\}$. First, consider the situation when $w \neq g$ and $x \neq g$. Let $w = u_p$ and $x = u_q$. For w and x to be part of an induced four-cycle, w and x need to have private neighbors in $G \setminus T$ that are adjacent. However, it is easy to verify that $N(u_p) \cup N(u_q)$ in $G \setminus T$ is an independent set. Therefore, there is no way of extending this choice of w and x to a four-cycle. Finally, suppose $w = g$, and let $x = u_p$. Every neighbor of u_p is c_i for some i and every neighbor of g lies in $\{a_j, b_j \mid 1 \le j \le m\}$. The only possibilities for forming induced four-cycles arise from choosing $c_i \in N(u_p)$ and either a_j or b_j with $j = i$. However, note that all of these cycles have been accounted for in the statement of the proposition. This completes the proof.   □

We now turn to the correctness of the reduction.

Lemma 1. *The graph G described as above is a* YES-*instance of* C_4-FREE CONTRACTION *if, and only if, $(U, \mathcal{F})$ is a* YES-*instance of* HITTING SET.

Proof. First, suppose $(U, \mathcal{F})$ is a YES-instance of HITTING SET, and let $S \subseteq U$ be a solution. Consider the edges corresponding to S in G, that is, let F be defined as $\{(g, u_i) \mid$ for all $u_i \in S\}$. We claim that G/F has no induced cycles of length four. Clearly, the proposed solution has the appropriate size, since we are picking one edge corresponding to every element of the hitting set, which is assumed to have size at most k. We now argue that the suggested set indeed forms a solution. First, notice that when the edge (g, u_i) is contracted, g becomes adjacent to every c_j for which $x_i \in S_j$. Since we are contracting vertices that form a hitting set, notice that for every $1 \le j \le m$, the edge (g, c_j) is present in G/F. By Proposition 1, the only induced four-cycles that need to be killed are as follows:

- $\{g, a_i, c_i, b_i\}$, for all $1 \le i \le m$,
- $\{u_i, g, a_j, c_j\}$, for all $x_i \in S_j$, and
- $\{u_i, g, b_j, c_j\}$, for all $x_i \in S_j$.

Notice that the edge (g, c_j) is a chord with respect to all these cycles, and this completes the argument in the forward direction.

In the reverse direction, suppose we have a subset of k edges, say F, such that G/F has no induced cycles of length four. We first argue that there exists a solution F that does not use any edge from the C_4 corresponding to the sets. Suppose F contains an edge e that is of the form (g, a_j) or (g, b_j). Clearly, contracting such an edge only affects the cycle $\{g, a_j, c_j, b_j\}$. Let x_i be any element of S_j. Consider the set $F^\star$ given by $F \setminus \{e\} \cup \{(g, u_i)\}$. It is easy to see that $F^\star$ is also a solution, since $G/F^\star$ has a chord in the cycle $\{g, a_j, c_j, b_j\}$. A similar argument shows that if F contains an edge of the form (a_j, c_j) or (b_j, c_j), then it can be replaced with an appropriately chosen edge of the form (g, u_i).

Finally, if F contains an edge e of the form (u_i, c_j), then notice that the only four-cycles of G that become triangulated in $G/\{e\}$ are: $\{g, a_j, c_j, b_j\}$, $\{u_i, g, a_j, c_j\}$, and $\{u_i, g, b_j, c_j\}$. All of these cycles also become triangulated when the edge (u_i, g) is contracted instead. Therefore, in this case also, we note that the set $F^\star$ given by $F \setminus \{e\} \cup \{(g, u_i)\}$ is also a solution.

Let $T^\star$ denote the set $\{u_1, \ldots, u_n\}$. By above arguments we have shown that there exists a solution F that is contained in the clique formed on $T^\star \cup \{g\}$. We are now ready to describe a hitting set S of size at most k. Let $\mathcal{W}$ be a G/F-witness structure of G and let $W(g)$ be the witness set that contains the global vertex g. Observe that since $G[W(g)]$ is connected we have that the $|W(g)| \leq k + 1$. We take S as $W(g) \setminus S$. Clearly, the size of S is at most k. It is also straightforward to see that S forms a hitting set. Indeed, consider any set $S_j \in \mathcal{F}$. Now consider the four-cycle given by $\{g, a_j, c_j, b_j\}$. Since it is triangulated, it must be the case that there is a $x_i \in S_i$ for which $u_i \in W(g)$, and hence $x_i \in S$. This concludes the reverse direction of the reduction.   □

From Lemma 1, and the hardness of the HITTING SET problem, we have the following:

Theorem 1. *The C_4-FREE CONTRACTION problem is* W[2]-*hard when parameterized by the size of the solution.*

Notice that in the analysis of Proposition 1, it is evident that the graph has no induced cycles of length five or more. Therefore, exactly the same arguments can be used to derive the fact that the problem of CHORDAL CONTRACTION, where we ask if k edges can be contracted to make the input graph chordal, is W[2]-hard when parameterized by k.

Corollary 1. *The CHORDAL CONTRACTION problem is* W[2]-*hard when parameterized by the size of the solution.*

Now we consider the C_ℓ-FREE CONTRACTION problem for $i \geq 5$. Notice that if we replace the cycles of length four with cycles of length ℓ in the reduction above, and make the vertices in u_i adjacent to the $\lfloor (\ell/2) \rfloor^{th}$ vertex in the cycle, then our claims follow by very similar arguments. We describe the construction and because of the similarity of the arguments defer the details of the correctness to the full version of this paper.

As before, let $(U, \mathcal{F})$ be an instance of HITTING SET, where $U = \{x_1, x_2, \ldots, x_n\}$ and $\mathcal{F} = \{S_1, S_2, \ldots, S_m\}$, where each $S_i \subseteq U$. We denote the reduced instance to be constructed by $G = (V, E)$. The vertex set consists of a special central vertex, denoted by g, one vertex for each element $x_i \in U$, denoted by u_i, and $(\ell - 1)$ vertices for every set S_i in the family $\mathcal{F}$, denoted by $a_i^1, a_i^2, \ldots, a_i^{\ell-1}$.

We now describe the edges. The central vertex is adjacent to the vertices u_i and a_j^1, $a_j^{\ell-1}$, for $1 \leq i \leq n$ and $1 \leq j \leq m$. We impose a clique on the vertices that correspond to elements of the universe. Next, we add the edges (g, a_i^1), $(g, a_i^{\ell-1})$ and (a_i^j, a_i^{j+1}) for every $1 \leq i \leq m$ and $1 \leq j \leq \ell - 2$. Finally, for every $x_i \in S_j$, we add the edge $(u_i, a_j^{\lfloor \ell/2 \rfloor})$. This completes the construction.

The proof of correctness is along the same lines as for the case of C_4-free contraction. In fact, for values of $\ell \geq 6$, there will be exactly m induced cycles of length ℓ in the graph G, as the cycles that use g, u_i and half of a cycle formed by a-vertices will not be of the requisite length, so the case analysis for the analog of Proposition 1 only simplifies. The detailed arguments are deferred to avoid repetition. This discussion brings us to the following theorem.

Theorem 2. *The C_ℓ-FREE CONTRACTION problem, for all fixed integer $\ell \geq 4$, is* W[2]-hard *when parameterized by the size of the solution.*

3.2   Contracting to P_γ-free Graphs

For the purposes of our discussion in this section, a path of length γ is a path on γ vertices and $(\gamma - 1)$ edges. For the problem of contracting to graphs that have no induced paths of length γ or longer (for $\gamma \geq 5$) we give describe two different reductions depending on the parity of γ. For the cases when $\gamma \leq 4$, in the next section, we describe approaches to FPT algorithms.

P_γ-FREE CONTRACTION   **Parameter:** k
Input: A graph $G = (V, E)$ and a positive integer k
Question: Is there a subset of at most k edges such that G/F has no induced paths of length γ?

The Case of Odd-Length Paths. We first describe the reduction for the case when γ is odd. Again, we reduce from HITTING SET. Let $(U, \mathcal{F})$ be an instance of HITTING SET, where $U = \{x_1, x_2, \ldots, x_n\}$ and $\mathcal{F} = \{S_1, S_2, \ldots, S_m\}$, where each $S_i \subseteq U$. We denote the reduced instance to be constructed by $G = (V, E)$. The vertex set consists of a special central vertex, denoted by g, one vertex for each element $x_i \in U$, denoted by u_i, and $(\gamma - 1)$ vertices for every set S_i in the family $\mathcal{F}$, denoted by $a_i^1, a_i^2, \ldots, a_i^{\lfloor (\gamma/2) \rfloor}, b_i^1, b_i^2, \ldots, b_i^{\lfloor (\gamma/2) \rfloor}$. For readability, we use ℓ to denote $\lfloor \gamma/2 \rfloor$. Also, let $T = \{u_1, \ldots, u_n\}$ denote the subset of vertices corresponding to the elements of the universe, and for every $1 \leq i \leq m$, denote the sets $\{a_i^1, a_i^2, \ldots, a_i^\ell\}$ $\{b_i^1, b_i^2, \ldots, b_i^\ell\}$ by A_i and B_i, respectively.

We now describe the edges. To begin with, we impose a clique on $T \cup \{g\}$. Next, add edges to ensure that the sets A_i and B_i induce paths of lengths ℓ, starting at a_i^1 and b_i^1, respectively. Further, we make the central vertex g adjacent to a_i^1 and b_i^1 for all $1 \leq i \leq m$. Notice that there is now an induced path of length γ starting at a_i^ℓ, going via g and ending at b_i^ℓ for all $1 \leq i \leq m$. To encode the hitting set structure, for every $x_i \in S_j$, make u_i adjacent to all vertices in $A_j \cup B_j$.

We now turn to the correctness of the reduction.

Lemma 2. *Let γ be a fixed odd integer ≥ 5. The graph G described as above is a* YES-*instance of P_γ-FREE CONTRACTION if, and only if, $(U, \mathcal{F})$ is a* YES-*instance of* HITTING SET.

Proof. First, suppose $(U, \mathcal{F})$ is a YES-instance of HITTING SET, and let $S \subseteq U$ be a solution. Consider the edges corresponding to S in G, that is, let F be defined as $\{(g, u_i) \mid$ for all $x_i \in S\}$. We claim that G/F has no induced paths of length γ. Clearly, the proposed solution has the appropriate size, since we are picking one edge corresponding to every element of the hitting set, which is assumed to have size at most k.

We now argue that the suggested set indeed forms a solution. For the sake of contradiction, let P be a path of length γ in G/F. First, note that g is a global vertex in G/F and therefore P does not contain g. Notice that $G \setminus (T \cup \{g\})$ is a disjoint union of paths of length ℓ, induced by the sets A_i, B_i, $1 \leq i \leq m$. This implies that P must contain at least one vertex from T, since $\ell < \gamma$. However, since T induces a clique, P can use at most two vertices from T. Finally, since $\gamma \geq 5$, we conclude that P must contain at least two vertices from one of the paths induced by A_i or B_i.

Let the path P be given by the sequence $p_1, p_2, \ldots, p_5$. Without loss of generality, let $p_1, p_2 \notin T \cup \{g\}$ (if either or both of them belong to $T \cup \{g\}$, then the last two vertices do not belong to $T \cup \{g\}$ and the path can be considered backwards). Note that both p_1 and p_2 belong to the same component of $G \setminus (T \cup \{g\})$, in other words, they belong to A_j or B_j for some $1 \leq j \leq m$. Let p_t be the nearest vertex along P such that $p_t \in T$. Now, the vertex p_t is evidently adjacent to both p_1 and p_2, creating a triangle in an induced path, which is a contradiction.

In the reverse direction, suppose we have a subset of k edges, say F, such that G/F has no induced paths of length γ. We now propose a hitting set S based on the edges in F. Let $\mathcal{W}$ be a G/F-witness structure of G and let $W(v)$ denote the witness set that contains the vertex v. We first consider the witness set of the global vertex g. For every $1 \leq i \leq n$ such that $W(g)$ contains the vertex u_i, include x_i in S. For every $1 \leq j \leq m$ such that $W(g)$ contains a vertex from $(A_j \cup B_j)$, choose an arbitrary element of the set S_j in S. Further, for every $v \notin T \cup \{g\}$, if $W(v)$ contains a vertex $u_i \in T$, include x_i in S.

We first reason that the size of the set thus described is at most k. Let λ be the number of vertices $v \notin T \cup \{g\}$ for which $W(v)$ included a vertex from T. Then, it is easy to see that:

$$\left(\sum_{v \in G \setminus (T \cup \{g\})} |W(v)| \right) + |W(g)| \leq k + 1 + \lambda.$$

Since we incorporate, from the witness sets $W(g)$ and $W(v)$, no elements corresponding to g or v (respectively), the number elements that feature in S is at most k.

We now argue that S is indeed a hitting set for $(U, \mathcal{F})$. In particular, we claim that if S_i is a set that is not hit by S, then $G[A_i \cup B_i \cup \{g\}]$ is an induced path of length γ in G/F, which would be the desired contradiction. Indeed, consider $G[A_i \cup B_i \cup \{g\}]$. For any vertex v in $(A_i \cup B_i)$, the edge (g, v) was not contracted (otherwise we would have included an element from S_i in S by construction). On the other hand, none of the vertices of T corresponding to elements contained in S_i were contracted to v, by the assumption that S does not hit S_i. All remaining vertices in $W(g)$ come from $A_j \cup B_j$ for $j \neq i$ and none of these vertices are adjacent to any of the vertices in $A_i \cup B_i$. Finally, we also know that if the witness sets $W(v)$ corresponding to $v \in A_i \cup B_i$ included vertices from T, then they must necessarily contain vertices corresponding to elements in S_i. But this would again contradict our assumption that S_i is not hit by S. The implication of this is that for all $v \in A_i \cup B_i$, the witness sets $W(v)$ do not contain any elements other than v. Therefore, the path $G[A_i \cup B_i \cup \{g\}]$ remains an induced path in G/F. This concludes the reverse direction of the reduction.   □

The Case of Even-Length Paths. We now describe the reduction for the case when γ is even. As in the case when γ was odd, we reduce from HITTING SET. Let $(U, \mathcal{F})$ be an instance of HITTING SET, where $U = \{x_1, x_2, \ldots, x_n\}$ and $\mathcal{F} = \{S_1, S_2, \ldots, S_m\}$, where each $S_i \subseteq U$. We denote the reduced instance to be constructed by $G = (V, E)$. The vertex set consists of a special central vertex, denoted by g, one vertex for each element $x_i \in U$, denoted by u_i, and $(\gamma - 3)$ vertices for every set S_i in the family $\mathcal{F}$, denoted by $a_i^1, a_i^2, \ldots, a_i^{\gamma-3}$. For readability, we use ℓ to denote $(\gamma - 3)$. Also, let $T = \{u_1, \ldots, u_n\}$ denote the subset of vertices corresponding to the elements of the universe, and for every $1 \le i \le m$, denote the sets $\{a_i^1, a_i^2, \ldots, a_i^\ell\}$ by A_i. Finally, introduce $2(k+1)$ additional vertices denoted by $\{g_1, \ldots, g_{k+1}, g_1', \ldots, g_{k+1}'\}$. This vertices in this set are sometimes referred to as *guard* vertices.

We now describe the edges. To begin with, we impose a clique on $T \cup \{g\}$. Next, add edges to ensure that the sets A_i induce paths of lengths ℓ, starting at a_i^1. Also add the edges (g_i, g_i') for all $1 \le i \le k+1$. Further, we make the central vertex g adjacent to a_i^1 for all $1 \le i \le m$ and g_i for all $1 \le i \le k+1$. Notice that there is now an induced path of length γ starting at a_i^ℓ, going via g and ending at g_j', for all $1 \le i \le m$ and all $1 \le j \le k+1$. To encode the hitting set structure, for every $x_i \in S_j$, make u_i adjacent to all vertices in A_j.

We now turn to the correctness of the reduction.

Lemma 3. [⋆] *Let γ be a fixed even integer ≥ 6. The graph G described as above is a YES-instance of P_γ-FREE CONTRACTION if, and only if, $(U, \mathcal{F})$ is a YES-instance of HITTING SET.*

Notice that the results above holds for $\gamma \ge 5$, and the cases when $\gamma \le 4$ are shown to be tractable in the next section. To conclude, from Lemmas 2 and 3, and the hardness of the HITTING SET problem, we have the following:

Theorem 3. *The P_γ-FREE CONTRACTION problem is W[2]-hard for all fixed integers $\gamma \ge 5$ when parameterized by the size of the solution.*

4   A Few Tractable Cases

In this section we give FPT algorithm for a few cases of $\mathcal{F}$-FREE CONTRACTION – namely K_ℓ-FREE CONTRACTION for every fixed integer $\ell \ge 3$, P_3-FREE CONTRACTION and P_4-FREE CONTRACTION. The last two problems can be shown to be FPT by arguments based on the MSO-expressibility of the problem and the fact that P_4- and P_3-free graphs have bounded rankwidth. In summary, we show the following.

Theorem 4. [⋆] *For every fixed integer $\ell \ge 3$, K_ℓ-FREE CONTRACTION is FPT. Also, the problems P_3-FREE CONTRACTION and P_4-FREE CONTRACTION are FPT.*

5 Future Directions

In this paper we initiated the study of $\mathcal{F}$-FREE CONTRACTION problem and answered questions when $\mathcal{F}$ consisted of a fixed cycle or a path of a particular length. An interesting, and potentially challenging, question would be to characterization the parameterized complexity of $\mathcal{F}$-FREE CONTRACTION in terms of properties of the forbidden subgraphs $\mathcal{F}$. On the other hand, it will also be interesting examine if there are subclasses of graphs on which the problems of C_j-FREE CONTRACTION (for $j \geq 4$) and P_j-FREE CONTRACTION (for $j \geq 5$) admit FPT algorithms while being NP-complete.

Acknowledgments. The authors would like to thank Chengwei Guo and Leizhen Cai for a careful reading of the paper and communicating a gap in the proof of Theorem 3, which has since been fixed.

References

1. Asano, T., Hirata, T.: Edge-contraction problems. J. Comput. Syst. Sci. 26(2), 197–208 (1983)
2. Belmonte, R., Heggernes, P., van 't Hof, P.: Edge contractions in subclasses of chordal graphs. Discrete Applied Mathematics 160(7-8), 999–1010 (2012)
3. Cai, L.: Fixed-parameter tractability of graph modification problems for hereditary properties. Inf. Process. Lett. 58(4), 171–176 (1996)
4. Diestel, R.: Graph Theory, 3rd edn. Springer, Heidelberg (2005)
5. Golovach, P.A., van 't Hof, P., Paulusma, D.: Obtaining planarity by contracting few edges. Theor. Comput. Sci. 476, 38–46 (2013)
6. Guillemot, S., Marx, D.: A faster fpt algorithm for bipartite contraction. CoRR, abs/1305.2743 (2013)
7. Heggernes, P., van 't Hof, P., Lévêque, B., Lokshtanov, D., Paul, C.: Contracting graphs to paths and trees. In: Parameterized and Exact Computation - 6th International Symposium, IPEC, pp. 55–66 (2011)
8. Heggernes, P., van 't Hof, P., Lévêque, B., Paul, C.: Contracting chordal graphs and bipartite graphs to paths and trees. Electronic Notes in Discrete Mathematics 37, 87–92 (2011)
9. Heggernes, P., van 't Hof, P., Lokshtanov, D., Paul, C.: Obtaining a bipartite graph by contracting few edges. In: Foundations of Software Technology and Theoretical Computer Science, FSTTCS, pp. 217–228 (2011)

Hardness of r-DOMINATING SET on Graphs of Diameter $(r + 1)$

Daniel Lokshtanov[1], Neeldhara Misra[2], Geevarghese Philip[3],
M.S. Ramanujan[4], and Saket Saurabh[1,4]

[1] University of Bergen, Norway
daniello.ii.uib.no
[2] Indian Institute of Science, India
neeldhara@csa.iisc.ernet.in
[3] Max-Planck-Institut für Informatik, Germany
gphilip@mpi-inf.mpg.de
[4] Institute of Mathematical Sciences, India
{msramanujan,saket}@imsc.res.in

Abstract. The DOMINATING SET problem has been extensively studied in the realm of parameterized complexity. It is one of the most common sources of reductions while proving the parameterized intractability of problems. In this paper, we look at DOMINATING SET and its generalization r-DOMINATING SET on graphs of bounded diameter in the realm of parameterized complexity. We show that DOMINATING SET remains W[2]-hard on graphs of diameter 2, while r-DOMINATING SET remains W[2]-hard on graphs of diameter $r+1$. The lower bound on the diameter in our intractability results is the best possible, as r-DOMINATING SET is clearly polynomial time solvable on graphs of diameter at most r.

1 Introduction

In the DOMINATING SET problem, we are given a graph G and a non-negative integer k, and the objective is to check if G contains a set of k vertices whose closed neighborhood contains all the vertices of G. In its generalization, r-DOMINATING SET, we are given a graph G and a non-negative integer k, and the question is whether G contains a set of k vertices such that every vertex of G is at distance at most r from one of these vertices. DOMINATING SET, together with its numerous variants, is one of the most classic and well-studied problems in algorithms and combinatorics [12].

A considerable part of the algorithmic study on this NP-complete problem has been focused on the design of parameterized algorithms. Formally, a *parameterization* of a problem is assigning an integer k to each input instance and a parameterized problem is *fixed-parameter tractable* (FPT) if there is an algorithm that solves the problem in time $f(k) \cdot |I|^{O(1)}$, where $|I|$ is the size of the input and f is an arbitrary computable function depending only on the parameter k. Just as NP-hardness is used as evidence that a problem probably is not polynomial time solvable, there exists a hierarchy of complexity classes

G. Gutin and S. Szeider (Eds.): IPEC 2013, LNCS 8246, pp. 255–267, 2013.
© Springer International Publishing Switzerland 2013

above FPT, and showing that a parameterized problem is hard for one of these classes is considered evidence that the problem is unlikely to be fixed-parameter tractable. The main classes in this hierarchy are:

$$FPT \subseteq W[1] \subseteq W[2] \subseteq \cdots \subseteq W[P] \subseteq XP$$

The principal analogue of the classical intractability class NP is W[1], which is a strong analogue, because a fundamental problem complete for W[1] is the k-STEP HALTING PROBLEM FOR NONDETERMINISTIC TURING MACHINES (with unlimited nondeterminism and alphabet size) — this completeness result provides an analogue of Cook's Theorem in classical complexity. In particular this means that an FPT algorithm for any W[1]-hard problem would yield a $O(f(k)n^c)$ time algorithm for k-STEP HALTING PROBLEM FOR NONDETERMINISTIC TURING MACHINES. A convenient source of W[1]-hardness reductions is provided by the result that CLIQUE is complete for W[1]. Other highlights of the theory include that DOMINATING SET, by contrast, is complete for W[2]. We refer to the following books for further details on parameterized complexity theory [8, 9, 13].

In general, DOMINATING SET and r-DOMINATING SET are W[2]-complete and therefore do not admit FPT algorithms unless an unexpected collapse occurs among certain parameterized complexity classes. However, there are interesting graph classes where FPT algorithms do exist for the DOMINATING SET problem. The project of widening the horizon where such algorithms exist spanned a multitude of ideas that made DOMINATING SET the testbed for some of the most cutting-edge techniques of parameterized algorithm design. For example, the initial study of parameterized subexponential algorithms for DOMINATING SET on planar graphs [2, 4, 10] resulted in the creation of bidimensionality theory, characterizing a broad range of graph problems that admit efficient approximate schemes or FPT algorithms on an equally broad range of graphs [5–7].

In this paper, we look at the effect of *diameter* on the parameterized complexity of DOMINATING SET and r-DOMINATING SET. In other words we study DOMINATING SET and r-DOMINATING SET on graphs of bounded diameter. We show that DOMINATING SET remains W[2]-complete on graphs of diameter 2, while r-DOMINATING SET remains W[2]-complete on graphs of diameter $r + 1$. The lower bound on the diameter in our intractability results is the best possible, as any graph with diameter at most r has an r-dominating of set of size exactly 1. The DOMINATING SET problem on split graphs was shown to be NP-complete in [1] and W[2]-hard in [14], while in [3], DOMINATING SET was shown to be NP-complete on graphs of diameter 2. In this paper, we demonstrate a reduction from the DOMINATING SET problem on split graphs to the DOMINATING SET problem on graphs of diameter 2, showing the W[2]-hardness of the problem on this graph class. Furthermore, this reduction will also demonstrate that CONNECTED DOMINATING SET is both NP-hard and W[2]-hard on graphs of diameter 2. We then extend these reductions in a non-trivial way to prove the classical as well as the parameterized intractability of generalizations of these problems. Our hardness reduction for r-DOMINATING SET on graphs of diameter $r + 1$ for $r \geq 2$ starts with a hypercube of diameter $r + 1$ and then embeds the input graph in this hypercube. The hard part of the reduction is the reverse direction where we need to argue that given a r-dominating set of the reduced graph we can obtain a dominating set of the input graph. We believe that our reduction strategy will be useful in other situations also.

2 Preliminaries

A parameterized problem is denoted by a pair $(Q, k) \subseteq \Sigma^* \times \mathbb{N}$. The first component Q is a classical language, and the number k is called the parameter. Such a problem is *fixed–parameter tractable* (FPT) if there exists an algorithm that decides it in time $O(f(k)n^{O(1)})$ on instances of size n. Next we define the notion of parameterized reduction.

Definition 1. *Let A, B be parameterized problems. We say that A is (uniformly many:1)* **fpt-reducible** *to B if there exist functions $f, g : \mathbb{N} \to \mathbb{N}$, a constant $\alpha \in \mathbb{N}$ and an algorithm Φ which transforms an instance (x, k) of A into an instance $(x', g(k))$ of B in time $f(k)|x|^\alpha$ so that $(x, k) \in A$ if and only if $(x', g(k)) \in B$.*

A parameterized problem is considered unlikely to be fixed-parameter tractable if it is W[i]-hard for some $i \geq 1$. To show that a problem is W[2]-hard, it is enough to give a parameterized reduction from a known W[2]-hard problem. Throughout this paper we follow this recipe to show a problem W[2]-hard. In fact, in this paper, all our reductions will run in polynomial time. Since this will be easy to see, we will not explicitly mention the time complexity of our reductions.

A *split graph* is a graph whose vertex set can be parititioned into two parts, one of which is a (maximal) clique and the other is an independent set. For any two vertices u and v, we let $d(u, v)$ denote the length of the shortest path between the vertices. Then the *diameter* of the graph $G = (V, E)$ is $max_{u,v \in V} d(u, v)$. In other words, the diameter of a graph is the length of the longest among all the shortest paths in the graph. For $S \subseteq V$, $G[S]$ denotes the graph *induced* by S in G. The vertex set of $G[S]$ is S, and the edge set is $\{(u, v) \mid u \in S, v \in S \text{ and } (u, v) \in E\}$. The r-*neighborhood* of a vertex v is the set of all vertices that are at distance at most r from v. The r-neighborhood of a vertex is denoted by $N^r(v)$, and the *closed* r-neighborhood of a vertex, given by $N^r(v) \cup \{v\}$, is denoted by $N^r[v]$. The r-neighborhood of a subset of vertices S is $\cup_{v \in S} N^r(v)$, and is denoted by $N^r(S)$. Likewise, the closed r-neighborhood of a subset of vertices S is $N^r(S) \cup S$, and is denoted by $N^r[S]$. We say that a vertex v is *global* to a set S of vertices if v is adjacent to every vertex in S. The *hamming distance* between two n-length strings is the number of positions at which the two string differ.

The DOMINATING SET and CONNECTED DOMINATING SET problems are defined as follows:

DOMINATING SET   **Parameter:** k
Input: A graph $G = (V, E)$, and an integer k.
Question: Does G have a subset S of at most k vertices such that for every $v \in V$, either $v \in S$, or there exists u such that $u \in S$ and $(u, v) \in E$?

CONNECTED DOMINATING SET   **Parameter:** k
Input: A graph $G = (V, E)$, and an integer k.
Question: Does G have a subset S of at most k vertices such that for every $v \in V$, either $v \in S$, or there exists u such that $u \in S$ and $(u, v) \in E$, and $G[S]$ is connected?

The DOMINATING SET and the CONNECTED DOMINATING SET problems are fundamental NP-complete [11], and W[2]-complete problems [8]. The r-DOMINATING SET and CONNECTED r-DOMINATING SET problems are defined below, and are also known to be NP-complete and W[2]-hard for every fixed constant r.

r-DOMINATING SET   **Parameter:** k
Input: A graph $G = (V, E)$, and an integer k.
Question: Does G have a subset S of at most k vertices such that for every $v \in V$, $v \in N^r[S]$?

CONNECTED r-DOMINATING SET   **Parameter:** k
Input: A graph $G = (V, E)$, and an integer k.
Question: Does G have a subset S of at most k vertices such that for every $v \in V$, $v \in N^r[S]$ and $G[S]$ is connected?

3   W-Hardness Of DOMINATING SET on Graphs of Diameter Two

In this section we show that DOMINATING SET remains W[2]-hard on split graphs of diameter 2.

Theorem 1. DOMINATING SET *is* $W[2]$-*hard on split graphs of diameter* 2.

Proof. We demonstrate this by a parameterized reduction from DOMINATING SET on connected split graphs. Let $G = (V, E)$ be a split graph, where $V = I \uplus C$ with $G[C]$ being a clique and $G[I]$, an independent set and let (G, k) be an instance of DOMINATING SET. We first make the following claim regarding dominating sets of G.

Claim. If (G, k) is a YES instance, then there exists a dominating set of size at most k that does not intersect I.

Proof. Since (G, k) is a YES instance of DOMINATING SET, G admits some subset S of size at most k that dominates all vertices in G. If $S \cap I = \emptyset$, then we are done. Suppose that this is not the case, and consider the set R obtained from S, by replacing every $v \in S \cap I$ with some $u \in N(v)$. Clearly, R is no larger than S and $R \cap I = \emptyset$. It is also easy to see that R is a dominating set of G:

- every vertex in the clique is dominated by R since $R \cap C \neq \emptyset$,
- any vertex $v \in I \setminus (S \cap I)$ is dominated by some vertex in $S \cap R$, since S was a dominating set, and vertices in $S \cap I$ cannot dominate the vertex under consideration,
- any vertex in $S \cap I$ is dominated by R, by construction.

This completes the proof of the claim.   □

We now proceed to the reduction. We will construct a split graph $H = (V_H = C' \uplus I', E_H)$. Recall that we desire H to be a graph of diameter 2. To this end, we obtain H from G by "replacing" the vertices of C with $\binom{|C|}{2}$ vertices, that is, H has one vertex for every pair of vertices in the clique partition of G. The adjacencies are as expected: a vertex corresponing to a pair of vertices is adjacent to the union of the neighborhoods of the original vertices. Finally, we induce a clique on the newly added vertices. Formally,

- $I' = I$,
- $C' = \{v[i,j] \mid i,j \in C, i \neq j\}$
- $(u, v[i,j]) \in E_H$ if, and only if, either $(u,i) \in E$ or $(u,j) \in E$,
- $(v[i,j], v[k,l]) \in E_H$ for all $(v[i,j], v[k,l]) \in \binom{C'}{2}$. Here, $\binom{C'}{2}$ is the family of two sized subsets of C'. This makes the set C' a clique, and hence H is indeed a split graph.

We now claim that (H, k) is a YES instance of DOMINATING SET if and only if $(G, 2k)$ is a YES instance. Since it is easily checked that H is a split graph and has diameter 2, the statement of the lemma will follow.

Indeed, let $S = \{u_1, u_2, \ldots, u_r\}$ be a dominating set of G of size at most $2k$. Notice that we can assume $S \cap I = \emptyset$ (see claim 3). Also, without loss of generality, we assume that r is even. Then, we claim that the set $R = \{v[u_1, u_2], v[u_3, u_4], \ldots v[u_{r-1}, u_r]\}$ is a dominating set of H, of size at most k. It is evident that all vertices in C' are dominated by R. Let $v \in I'$, and let $u_i \in S$ be such that $(u, v) \in E$ (notice that such a choice of u always exists, since S is – by assumption – a dominating set of G). But, since either $v[u_i, u_{i+1}]$ or $v[u_{i-1}, u_i]$ is contained in R, the vertex v is also dominated by R in the graph H.

On the other hand, let $R = \{v[u_1, u_2], v[u_3, u_4], \ldots v[u_{r-1}, u_r]\}$ be a dominating set of H of size at most k. Again, by claim 3 (which applies since we have that H is also a split graph), we may assume that $R \cap I' = \emptyset$. We claim that the set $S = \{u_1, u_2, \ldots, u_r\}$ is a dominating set of G of size at most $2k$. Clearly, $|S| \leq 2r \leq 2k$ and all vertices in C are dominated by S. Now, consider a vertex $v \in I$, and let $v[u_i, u_{i+1}] \in R$ be such that $(v[u_i, u_{i+1}], v) \in E_H$ (notice that such a vertex always exists, since R is – by assumption – a dominating set of H). Since both u_i and u_{i+1} are in S, v is also dominated by S. This completes the proof of the theorem.   □

4 W-Hardness Of r-DOMINATING SET on Graphs of Diameter $(r + 1)$

In this section, we describe a reduction from the DOMINATING SET problem on split graphs of diameter two to the r-DOMINATING SET problem on graphs of diameter $(r + 1)$ for $r \geq 2$.

The Construction. Let (G, k) be an instance of DOMINATING SET, where $G = (V, E)$ is a split graph of diameter two with $V = (I \uplus C)$. The independent set and clique of the split partition are given by I and C respectively.

We first describe an intermediate graph $G' = (V', E')$ that will serve as an wireframe for the construction. Let $\alpha := 4kr|I| + |C|$. The vertex set of G' comprises of words of length $(r + 1)$ over the alphabet $\{1, \ldots, \alpha\}$, and the edges are between vertices whose corresponding words differ in exactly one position.

Formally, we define $V' := \{1, \ldots, \alpha\}^{r+1}$. For every $u, v \in V'$, let $\delta(u, v)$ be the number of positions in which the strings u and v differ. In other words, $\delta(u, v)$ is the hamming distance between the strings u and v. We therefore have $E' = \{(u, v) \mid \delta(u, v) = 1\}$. This completes the description of G'.

We abuse language and speak of the hamming distance between two vertices to refer to the hamming distance between the strings corresponding to the vertices in question. Also, for a vertex v and $1 \leq i \leq r + 1$, we will use $v[i] \in [\alpha]$ to denote the value of the ith position in the string corresponding to v (sometimes also referred to as the ith coordinate).

It turns out that in G', the distance between a pair of vertices corresponds exactly to the hamming distances between them. We formalize this in the observation below, where we show that for any vertex v in V', the vertices of distance at most d from v in G' are precisely the vertices whose hamming distance from v is at most d.

Lemma 1. *For every vertex $v \in V'$, for every $d > 0$, the set*

$$N^d(v) = \{u \mid \delta(u, v) \leq d\}.$$

Proof. The proof is by induction on d. In the base case, for $d = 1$, the claim follows by the definition of adjacencies in G'. For the induction step, let $d > 0$, and assume that the claim holds for all $d^* < d$. Let $u_1, \ldots, u_t$ be the vertices at distance $(d - 1)$ from v. Consider $N^d(v) = \bigcup_{u \in N^{d-1}(v)} N[u]$. By the induction hypothesis, we have that $N^{d-1}(v) = \{u \mid \delta(u, v) \leq d - 1\}$. Therefore,

$$N^d(v) = \bigcup_{\{u \, : \, \delta(u,v) \leq d-1\}} \bigcup_{\{w \, : \, \delta(w,u) \leq 1\}} \{w\} = \{u \mid \delta(u, v) \leq d\}.$$

This completes the proof of the claim. $\square$

Notice that the distance between any pair of vertices in G' is at most $(r + 1)$. By Lemma 1, we also have that the distance between the vertices $(i, i, \ldots, i)$ and $(j, j, \ldots, j)$ is $r + 1$ for any $i, j \in [\alpha], i \neq j$. It follows that G' has diameter $r + 1$.

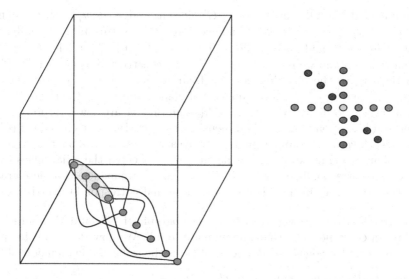

Fig. 1. (a) An illustration of the construction for $r = 2$ where the graph G is embedded along the diagonal of the bottom face of the cube. (b) An illustration of the adjacencies in G'. The red, blue and green vertices are the vertices adjacent to the yellow vertex in G'.

We are now ready to incorporate an encoding of G in the reduction. It is useful to think of V' as points inside an $(r+1)$-dimensional hypercube with sides of length α. We will focus on the plane obtained by setting all but first two coordinates to 1 and embed the graph G here in a way that does not decrease the diameter of the entire graph, and at the same time encodes a dominating set of G as an r-dominating set of the newly constructed graph and vice versa. We now formalize this intuition.

Recall that $(I \uplus C)$ is the split partition of the instance G. Let $p := |I|$ and $q := |C|$. Begin by labelling the vertices in I as $\{v_1, \ldots, v_p\}$ and those in C as $\{u_1, \ldots, u_q\}$. Let $\beta(i) = 4kr \cdot (i-1)$ and $\gamma = (4kr) \cdot p$. Furthermore, we use $\overline{1}_i$ to refer to the tuple $(1, \ldots, 1)$ of length i. We exclude the subscript when the length of the tuple is clear from the context. Before we go further, we collect the definitions of α, β and γ for easy reference.

- $\alpha := 4krp + q$, $\beta(i) := 4kr \cdot (i-1)$ and $\gamma := (4kr) \cdot p$.

Define the set $P_2 := \{(i, j, \overline{1}) \mid 1 \leq i, j \leq \alpha\}$ and let R denote the remaining vertices in V', that is, $R := V' \setminus P_2$. Let $D_2 \subset P_2$ denote the "diagonal" entries of P_2, that is, $D_2 = \{(i, i, \overline{1}) \mid 1 \leq i \leq \alpha\}$. We now establish the following correspondence between vertices of G and the vertices of D_2:

- For each vertex $v_\ell \in I$, the $4kr$ vertices $-$ $(\beta(\ell)+1, \beta(\ell)+1, \overline{1}), \ldots, (\beta(\ell+1), \beta(\ell+1), \overline{1})$ in G' all correspond to v_ℓ and we refer to this set as $\mathcal{I}_\ell$.
- For each vertex $u_i \in C$, the vertex $(\gamma+i, \gamma+i, \overline{1})$ corresponds to u_i and we refer to this vertex as $u_i^\star$.

We now add the following edges to G'. For each edge $(v_\ell, u_j) \in E$ such that $v_\ell \in I$ and $u_j \in C$, we make u_j^* adjacent to every vertex in $\mathcal{I}_\ell$. Finally, we consider the set P_2 and make a clique on the set $P_2 \setminus (\bigcup_{\ell=1}^{p} \mathcal{I}_\ell)$. This completes the construction and we refer to the graph thus constructed as $G'' = (V'', E'')$.

To tie back to the intuition described earlier, note that we considered the points of V' that lie on the two-dimensional plane obtained by the restriction of the last $(r-1)$ coordinates to $(1, 1, \ldots, 1)$ (recall that we are now interpreting the elements of V' as points in $(r+1)$-dimensional space). Here, we embedded $(4kr)$ copies of each vertex in I and a single copy of each vertex in C along the diagonal of this plane (see Figure 1). Following this, we replicated the adjacencies of G between the corresponding vertices in G' and finally, we made a complete graph on all the vertices in this plane except for those that correspond to vertices of I.

Diameter Bound. Notice that G' is a subgraph of G'', and therefore, the diameter of G'' is no more than the diameter of G'. We now show that in spite of the newly added edges, the diameter of G'' is the same as the diameter of G'.

Lemma 2. *The diameter of the graph G'' is $r + 1$.*

Proof. We show that the distance between the vertices $u = (\alpha, \alpha, \ldots, \alpha)$ and $v = (\alpha - 1, \alpha - 1, \ldots, \alpha - 1)$ in G'' is $r + 1$ which would imply the claim. Suppose, for the sake of contradiction, that that there is a path L of length at most r from u to v. Since such a path does not exist in G', this path must contain an edge from $E'' \setminus E'$. Since every edge in $E'' \setminus E'$ is contained in P_2, the path L has a non-trivial intersection with P_2. Since $u, v \notin P_2$, L begins and ends outside P_2. We let u' be the first vertex of P_2 on L and let v' be the last vertex of P_2 on L. Note that $u \neq u' \neq v' \neq v$.

Let L_u be the subpath of L from u to u', L_v be the subpath of L from v' to v. Clearly, L_u and L_v are also paths in G'. Note that the length of L_u is at least the length of a shortest path from u to u', and the length of L_v is at least the length of a shortest path from v' to v. Since L_u and L_v lie entirely outside P_2, the lengths of these shortest paths are the same in G'' and G'. This implies (using Lemma 1) that L_u and L_v both have length at least $(r-1)$, which implies that L has length at least $2(r-1) + 1$. Since $r \geq 2$ we have that $2r - 1$ can not be less than r. Thus we get our desired contradiction. $\qquad\square$

Correctness of the Reduction. We now turn to the correctness of the reduction. In the forward direction, consider a dominating set Z of size at most k for G. We have already seen that we may assume that $Z \subseteq C$, without loss of generality. Consider the set $\mathcal{C}_Z := \{u_j^* \mid u_j \in Z\}$. We claim that $\mathcal{C}_Z$ is an r-dominating set for G''.

Clearly, every vertex in P_2 is at a distance of at most 1 from $\mathcal{C}_Z$. Now, consider any vertex $v := (a_1, a_2, a_3, \ldots, a_{r+1}) \in R$. By Lemma 1, the vertex $(a_1, a_2, \overline{1}) \in P_2$ is at a distance of at most $(r-1)$ from v, and consequently at a distance of at most r from $\mathcal{C}_Z$. Hence, $\mathcal{C}_Z$ is indeed an r-dominating set for G''.

Conversely, consider a set Z of size at most k which is an r-dominating set for G''. In this direction, we will have to work our way from $Z \subset V''$ to a subset of P_2, and eventually to a subset of D_2 that will lead us to a correspondence between the vertices of Z and vertices of C in G. In the process, we will ensure that the vertices specified by the correspondence dominate I, using the fact that Z was a r-dominating set in G''.

The multiple copies of vertices in I will now be helpful in identifying vertices of Z that lie in P_2. To see this informally, fix $v_\ell \in I$, and consider $\mathcal{I}_\ell$. We will first show the presence of a "large" set such that any vertex outside P_2 can r-dominate a limited number of vertices in in this set. In fact, it will follow from the choice of α that even k vertices from outside P_2 cannot r-dominate this set. Therefore, for every $\ell \in [p]$, there must be a vertex from P_2 that belongs to Z to witness the r-domination of this large set. When these vertices correspond to $u_j^\star$ for some j, then the correspondence with a vertex in C is direct. In the other cases, it will turn out that the vertex in question dominates copies of at most two distinct vertices of I. In this situation, we will be able to identify an appropriate vertex from C to map to, using the fact that G has diameter two.

We now turn to a formal argument. To begin with, in the following observation we show that for any $v_\ell \in I$, there is an index j_ℓ in the range $[\beta(\ell) + 1, \beta(\ell + 1)]$ such that the dominating set does not contain any vertex of the form $(j_\ell, *, *, \ldots, *)$, or $(*, j_\ell, *, \ldots, *)$.

Observation 1. *For every $v_\ell \in I$, there is an index j_ℓ such that $\beta(\ell) + 1 \leq j_\ell \leq \beta(\ell + 1)$ and Z does not contain a vertex of the form $(j_\ell, \overline{x})$ or $(t, j_\ell, \overline{y})$ for any $\overline{x} \in [\alpha]^r$, $\overline{y} \in [\alpha]^{r-1}$ and $1 \leq t \leq \alpha$.*

Proof. Let $Z = \{z_1, z_2, \ldots, z_k\}$. Let $Z_{12} \subseteq [\alpha]$ be the set of all values in the first two coordinates of vertices in Z. Recall that for $v \in V''$, we let $v[i] \in [\alpha]$ denote the value of the ith co-ordinate of v. Then, we have:

$$Z_{12} := \{z[1] \mid z \in Z\} \cup \{z[2] \mid z \in Z\}.$$

Notice that $|Z_{12}| \leq 2k$ and the range of ℓ is at least $4kr$, and the observation follows by a simple application of the pigeon-hole principle.   □

Now consider the vertices in the dominating set that lie outside P_2, that is, $Z_R = Z \cap R$. Further, fix a vertex $v_\ell \in I$, and consider j_ℓ given by Observation 1 above. Consider the set of all vertices of G that are obtained by restricting the first two coordinates to (j_ℓ, j_ℓ). Formally, we let $\mathcal{T}_\ell = \bigcup_{\overline{a}} (j_\ell, j_\ell, \overline{a})$. Notice that no vertex in Z is contained in this set. To begin with, we will account for how many vertices of $\mathcal{T}_\ell$ can be r-dominated by a vertex in Z_R.

Lemma 3. *The r-neighborhood of any vertex in Z_R intersects $\mathcal{T}_\ell$ in at most $2\alpha^{r-2}$ vertices.*

Proof. Let $v \in Z_R$. We will prove the claim by identifying a suitably large set of vertices in $\mathcal{T}_\ell$ that are at distance $(r+1)$ from v. A natural candidate would be the vertices in $\mathcal{T}_\ell$ which are outside P_2 and at hamming distance $(r+1)$ from v. For technical reasons, we will consider this set but with the additional property that a *particular coordinate* is not equal to 1. Since v is not in P_2 there exists a coordinate $t \in \{3, \ldots, r+1\}$ such that $v[t] \neq 1$. Formally,

$$\mathcal{D}_\ell^v := \{u \mid u \in \mathcal{T}_\ell, u[t] \neq 1, \text{ and } u[i] \neq v[i] \text{ for all } 1 \leq i \leq r+1\}.$$

We first claim that no vertex from $\mathcal{D}_\ell^v$ lies in the r-neighborhood of v. Indeed, suppose not. Let $u \in \mathcal{D}_\ell^v$. For the sake of contradiction, consider any path L of length at most r from u to v. Note that such a path does not exist in G' (by Lemma 1 and the choice of u and v), this path must contain an edge from $E'' \setminus E'$. Since every edge in $E'' \setminus E'$ is contained in P_2, the path L has a non-trivial intersection with P_2. Since, by definition, $u, v \notin P_2$, L begins and ends outside P_2. We let u' be the first vertex of P_2 on L and let v' be the last vertex of P_2 on L. Note that $u \neq u' \neq v' \neq v$.

Let L_u be the subpath of L from u to u', L_v be the subpath of L from v' to v. Clearly, L_u and L_v are also paths in G'. Note that the length of L_u is at least the length of a shortest path from u to u', and the length of L_v is at least the length of a shortest path from v' to v. Since L_u and L_v lie entirely outside P_2, the lengths of these shortest paths are the same in G'' and G', and in particular, are equal to the hamming distances between the corresponding vertices. Let $p(u)$ and $p(v)$ denote, respectively, the set of positions where the last $(r-1)$ coordinates of u (respectively, v) differ from $\bar{1}_{r-1}$. Note that the tth position belongs to $p(u) \cap p(v)$. Also, every position that is not in $p(u)$ is in $p(v)$ – this is simply because u and v differ at every coordinate. Therefore, $|p(u)| + |p(v)| \geq (r-1) + 1 = r$.

Now, using Lemma 1, we have that the length of L_u is at least $|p(u)|$ and the length of L_v is at least p_v. Therefore, we have that L has length at least $r+1$ (since L uses at least one edge inside P_2), and this is the desired contradiction.

We have that among the vertices in $\mathcal{T}_\ell$ the vertices from $\mathcal{D}_\ell$ are not within the r-neighborhood of v. Note that $|\mathcal{T}_\ell| = \alpha^{(r-1)}$, and it is easy to see that $|\mathcal{D}_\ell^v| = (\alpha - 1)^{(r-1)} - (\alpha - 1)^{(r-2)}$. Thus, the intersection of the r-neighborhood of v with $\mathcal{T}_\ell$ is at most:

$$X := \alpha^{(r-1)} - [(\alpha - 1)^{(r-1)} - (\alpha - 1)^{(r-2)}]$$

Consider the term $\alpha^{(r-1)} - (\alpha - 1)^{(r-1)}$. Let $\lambda := (\alpha - 1)$.

$$\begin{aligned}
(\lambda + 1)^{(r-1)} - \lambda^{(r-1)} &= \left(\sum_{j=0}^{r-1} \binom{r-1}{j} \lambda^j \right) - \lambda^{(r-1)} \\
&= \sum_{j=0}^{r-2} \binom{r-1}{j} \lambda^j \\
&\leq \sum_{j=0}^{r-2} \binom{r-2}{j} \lambda^j = \lambda^{(r-2)} = (\alpha - 1)^{(r-2)}
\end{aligned}$$

Now we have:

$$X \le (\alpha - 1)^{(r-2)} + (\alpha - 1)^{(r-2)} \le 2(\alpha - 1)^{r-2} \le 2\alpha^{r-2},$$

which is the desired conclusion.   □

Next, we consider the vertices in the dominating set that are in P_2, but do not one-dominate the j_ℓ^{th} copy of v_j. In other words, we are concerned with vertices that are non-adjacent to $(j_\ell, j_\ell, \bar{1})$. Again, we will account for how much of $\mathcal{T}_\ell$ can be r-dominated by such vertices, and this observation will be analogous to the previous lemma.

Lemma 4. *Let $\mathcal{T}_\ell$ be defined as before. The r-neighborhood of any vertex in P_2 which is non-adjacent to $(j_\ell, j_\ell, \bar{1})$ intersects $\mathcal{T}_\ell$ in at most α^{r-2} vertices.*

Proof. Let $v \in P_2$ be a vertex that is not adjacent to $(j_\ell, j_\ell, \bar{1})$. Notice that by definition, $v[1] \ne j_\ell$ and $v[2] \ne j_\ell$. Consider $\mathcal{S}_\ell^v \subseteq \mathcal{T}_\ell$ defined as the set of vertices whose *hamming distance* from v is equal to $(r+1)$. We claim that for any vertex $u \in \mathcal{S}_\ell^v$, the distance between v and u in G'' is $(r+1)$. Indeed, consider any path from v to u. Since $v \in P_2$ and $u \notin P_2$, we let w be the last vertex on this path that belongs to P_2. If the distance from v to w is at least two, then we claim that the length of the path is at least $(r+1)$. This is because $w \in P_2$, implying that the hamming distance between w and v is at least $(r-1)$. (Recall that v and w have $\bar{1}_{r-1}$ on the last $r-1$ coordinates and the hamming distance between v and u is equal to $(r+1)$.) Since the subpath of L from w to u lies entirely outside P_2, the distance between w and u is equal to the hamming distance. Consequently, as long as the distance between v and w is at least two, we are done.

On the other hand, suppose the distance between v and w is one, that is, $w \in N(v) \cap P_2$. Since v is not adjacent to $(j_\ell, j_\ell, \bar{1})$, it follows that the hamming distance between w and u is in fact r, and therefore, the length of the path between w and u is r, for the same reasons as before. The only remaining case is when the path between v and u uses no edges in P_2, but in this case, the path is at least as long as the hamming distance between v and u, which is $(r+1)$ by choice of u. Therefore, we conclude that the length of the shortest path between v and any vertex in $\mathcal{S}_\ell^v$ is $r+1$. Since $|\mathcal{S}_\ell^v| = (\alpha - 1)^{(r-1)}$, the computation from the proof of Lemma 3 can be used to derive the desired conclusion.   □

Let Z_2 be the set of vertices of $Z \cap P_2$ which are non-adjacent to $(j, j, \bar{1})$. By Lemma 3 and Lemma 4, Z_R and Z_2 can together r-dominate at most $3k\alpha^{r-2}$ vertices. Since $|\mathcal{T}_\ell| = \alpha^{r-1} > 3k\alpha^{r-2}$, there is a vertex in $Z \cap P_2$ which is adjacent to $(j, j, \bar{1})$.

For every independent set vertex v_i, let j_i be the index with all the nice properties. For each i, let $(x_i, y_i, \bar{1})$ be a vertex in $P_2 \cap Z$ which is adjacent to $(j_i, j_i, \bar{1})$. Let $Y \subseteq (Z \cap P_2)$ be those vertices of Z in P_2 which are adjacent to $(j_i, j_i, \bar{1})$ for some i.

We now define a mapping $f : Y \to C$ as follows. Consider a vertex $(x_i, y_i, \bar{1}) \in Y$.

- If $x_i = y_i$, then, since vertices corresponding to the independent set vertices are independent in G'', the vertex $(x_i, x_i, \overline{1})$ corresponds to a vertex $v_c \in C$ and we set $f(x_i, y_i, \overline{1}) = v_c$.
- If $(x_i, x_i, \overline{1})$ and $(y_i, y_i, \overline{1})$ correspond to vertices v_a and v_b respectively where $v_a, v_b \in I$ and $v_c \in C$ is a vertex adjacent to both v_a and v_b in G (such a vertex always exists since G has diameter 2), then we set $f(x_i, y_i, \overline{1}) = v_c$.
- If $(x_i, x_i, \overline{1})$ corresponds to a vertex $v_a \in I$ and $(y_i, y_i, \overline{1})$ corresponds to a vertex $v_b \in C$, then we set $f(x_i, y_i, \overline{1}) = v_c$ where $v_c \in C$ is a vertex adjacent to v_a in G.

Lemma 5. *The set $f(Y)$ is a dominating set of size at most k for the graph G.*

Proof. Since $Y \subseteq Z$, Y has size at most k. Furthermore, the mapping f is clearly surjective, which implies that $|f(Y)| \leq k$. It remains to show that $f(Y)$ is a dominating set of G. Consider a vertex $v_i \in I$. We have already shown that there is a j_i and a vertex $u = (x_i, y_i, \overline{1}) \in Z$ such that u is adjacent to $(j_i, j_i, \overline{1})$. Furthermore, observe that the vertex $f(x_i, y_i, \overline{1})$ is by definition adjacent to v_i. Therefore $f(Y)$ dominates v_i and by the same argument, every vertex in I. Since $f(Y) \subseteq C$ and it is non-empty, the vertices in C are dominated as well. This completes the proof of the claim. $\square$

Thus we obtain the following theorems.

Theorem 2. *For all fixed $r \geq 1$, r-DOMINATING SET is $W[2]$-hard on graphs of diameter $(r+1)$.*

We note that, in all our reductions, without loss of generality, the r-dominating set in the reduced instances is connected. Hence, these reductions also prove $W[2]$-hardness of the *connected* variants of r-DOMINATING SET.

Theorem 3. *For all fixed $r \geq 1$, CONNECTED r-DOMINATING SET is $W[2]$-hard on graphs of diameter $(r+1)$.*

5   Conclusions

It is an interesting open problem to investigate if there are problems that are FPT on graphs of bounded diameter, even if they are W-hard on general graphs.

References

1. Bertossi, A.A.: Dominating sets for split and bipartite graphs. Information Processing Letters 19, 37–40 (1984)
2. Alber, J., Bodlaender, H.L., Fernau, H., Kloks, T., Niedermeier, R.: Fixed parameter algorithms for dominating set and related problems on planar graphs. Algorithmica 33, 461–493 (2002)

3. Ambalath, A.M., Balasundaram, R., Rao H., C., Koppula, V., Misra, N., Philip, G., Ramanujan, M.S.: On the kernelization complexity of colorful motifs. In: Raman, V., Saurabh, S. (eds.) IPEC 2010. LNCS, vol. 6478, pp. 14–25. Springer, Heidelberg (2010)

4. Demaine, E.D., Fomin, F.V., Hajiaghayi, M., Thilikos, D.M.: Fixed-parameter algorithms for (k, r)-center in planar graphs and map graphs. ACM Trans. Algorithms 1, 33–47 (2005)

5. Demaine, E.D., Fomin, F.V., Hajiaghayi, M., Thilikos, D.M.: Subexponential parameterized algorithms on bounded-genus graphs and H-minor-free graphs. J. ACM 52, 866–893 (2005)

6. Demaine, E.D., Hajiaghayi, M.: The bidimensionality theory and its algorithmic applications. The Computer Journal 51, 332–337 (2007)

7. Dorn, F., Fomin, F.V., Lokshtanov, D., Raman, V., Saurabh, S.: Beyond bidimensionality: Parameterized subexponential algorithms on directed graphs. In: Proceedings of the 27th International Symposium on Theoretical Aspects of Computer Science (STACS 2010). LIPIcs, vol. 5, pp. 251–262. Schloss Dagstuhl - Leibniz-Zentrum fuer Informatik (2010)

8. Downey, R.G., Fellows, M.R.: Parameterized Complexity. Springer (1998)

9. Flum, J., Grohe, M.: Parameterized Complexity Theory. Springer (2006)

10. Fomin, F.V., Thilikos, D.M.: Dominating sets in planar graphs: Branch-width and exponential speed-up. SIAM J. Comput. 36, 281–309 (2006)

11. Garey, M.R., Johnson, D.S.: Computers and Intractability: A Guide to the Theory of NP-Completeness. W. H. Freeman & Co., New York (1979)

12. Haynes, T.W., Hedetniemi, S.T., Slater, P.J.: Fundamentals of domination in graphs. Marcel Dekker Inc., New York (1998)

13. Niedermeier, R.: Invitation to fixed-parameter algorithms. Oxford Lecture Series in Mathematics and its Applications, vol. 31. Oxford University Press, Oxford (2006)

14. Raman, V., Saurabh, S.: Short cycles make W-hard problems hard: FPT algorithms for W-hard problems in graphs with no short cycles. Algorithmica 52, 203–225 (2008)

Amalgam Width of Matroids

Lukáš Mach[1,*] and Tomáš Toufar[2]

[1] DIMAP and Department of Computer Science,
University of Warwick, Coventry, United Kingdom
lukas.mach@gmail.com
[2] Computer Science Institute, Faculty of Mathematics and Physics,
Charles University, Prague, Czech Republic
tomas.toufar@gmail.com

Abstract. We introduce a new matroid width parameter based on the operation of *matroid amalgamation* called *amalgam-width*. The parameter is linearly related to branch-width on finitely representable matroids, while still allowing algorithmic applications on non-representable matroids (which is not possible for branch-width). In particular, any property expressible in the monadic second order logic can be decided in linear time for matroids with bounded amalgam-width. We also prove that the Tutte polynomial can be computed in polynomial time for matroids with bounded amalgam-width.

1 Introduction

It is well known that many NP-hard graph problems can be solved efficiently when restricted to trees or to graphs with bounded tree-width. Research of this phenomenon culminated in proving a celebrated theorem of Courcelle [2], which asserts that any graph property expressible in the monadic second order (MSO) logic can be decided in linear time for graphs of bounded tree-width. Such properties include, among many others, 3-colorability. There are several other width parameters for graphs with similar computational properties, e.g., boolean-width [1] and clique-width [5].

In this work, we study matroids, which are combinatorial structures generalizing the notions of graphs and linear independence. Although the tree-width for matroids can be defined [6], a more natural width parameter for matroids is branch-width. This is due to the fact that the branch-width of graphs can be introduced without referring to vertices, which are not explicitly available when working with (graphic) matroids. We postpone the formal definition of branch-width to Section 2 and just note that the branch-width of a matroid or a graph is linearly related to its tree-width.

---

* This author has received funding from the European Research Council under the European Union's Seventh Framework Programme (FP7/2007-2013)/ERC grant agreement no. 259385 and from the student project GAUK no. 592412 when being a PhD student at Computer Science Institute, Faculty of Mathematics and Physics, Charles University, Prague, Czech Republic.

G. Gutin and S. Szeider (Eds.): IPEC 2013, LNCS 8246, pp. 268–280, 2013.

It is natural to ask to what extent the above-mentioned algorithmic results for graphs have their counterparts for matroids. However, most width parameters (including branch-width) do not allow corresponding extension for general matroids without additional restrictions. Although computing decompositions of nearly optimal width is usually still possible (see [16, 17]), the picture becomes more complicated for deciding properties. Extensions to finitely representable matroids are feasible but significant obstacles emerge for non-representable matroids. This indicates a need for a width parameter reflecting the complex behavior of matroids that are not finitely representable.

Let us be more specific with the description of the state of the art for matroids. On the positive side, the analogue of Courcelle's theorem was proven by Hliněný [8] in the following form:

Theorem 1. *[8, Theorem 6.1] Let* **F** *be a finite field,* φ *be a fixed MSO formula and* $t \in \mathbf{N}$*. Then there is a fixed parameter algorithm deciding* φ *on* **F**-*represented matroids of branch-width bounded from above by* t*.*

However, as evidenced by several negative results, a full generalization of the above theorem to all matroids is not possible: Seymour [21] has shown that there is no sub-exponential algorithm testing whether a matroid (given by an oracle) is representable over GF(2). Note that being representable over GF(2) is equivalent to the non-existence of U_2^4 minor, which can be expressed in MSO logic. This result generalizes for all finite fields and holds even when restricted to matroids of bounded branch-width. This subsequently implies the intractability of deciding MSO formulas on general matroids of bounded branch-width. See [12] for more details on matroid representability from the computational point of view. Besides MSO properties, algorithmic aspects of first order properties have also been studied [11].

Two width parameters have been proposed to circumvent the restriction of tractability results to matroids representable over finite fields: decomposition width [13, 14] and another width parameter based on 2-sums of matroids [22]. The latter allows the input matroid to be split only along 2-separations, making it of little use for 3-connected matroids. On the other hand, though the first one can split the matroid along more complex separations, it does not correspond to any natural "gluing" operation on matroids. In this work, we present a matroid parameter, called amalgam-width, that has neither of these two disadvantages and still allows proving corresponding algorithmic results. An input matroid can be split along complex separations and the parts of the decomposed matroid can be glued together using the so-called amalgamation [19], which is a well-established matroid operation.

2 Notation

We now introduce definitions and concepts relevant to this work. The reader is referred to the monograph [19] for a detailed treatment of matroid theory.

A *matroid* M is a tuple $(E, \mathcal{I})$ where $\mathcal{I} \subseteq 2^E$. The set E is called *the ground set* and its elements are *the elements of* M. Sets from $\mathcal{I}$ are referred to as *independent sets*. If a set is not independent, we call it *dependent*. The ground set of a particular matroid M is denoted by $E(M)$. The set $\mathcal{I}$ is required (1) to contain the set $\emptyset$, (2) to contain as elements all subsets of any independent set, and (3) to satisfy *the exchange axiom*: if F and F' are independent sets with $|F| < |F'|$, then there exists $x \in F'$ such that $F \cup \{x\} \in \mathcal{I}$. A minimal depedent set is called *a circuit*. The set of all circuits of the matroid, denoted by $\mathcal{C}(M)$, uniquely determines the matroid.

Examples of matroids include *graphic matroids* and *vector matroids*. The former are derived from graphs in the following way: their elements are edges and a set of edges is independent if it does not span a cycle. Vector matroids have vectors as their elements and a set of vectors is independent if the vectors in the set are linearly independent. A matroid M is called representable over a field $\mathbf{F}$ if there exists a vector matroid over $\mathbf{F}$ isomorphic to M. Finally, a matroid is binary if it is representable over the binary field and it is regular if it is representable over any field.

The rank $r(F)$ of a set $F \subseteq E(M)$ is the size of the largest independent subset of F. *The closure operator* $\mathrm{cl}(F)$ acting on subsets of $E(M)$ is defined as $\mathrm{cl}(F) := \{x : r(F \cup \{x\}) = r(F)\}$. It can be shown that $r(\mathrm{cl}(F)) = r(F)$. A set F such that $\mathrm{cl}(F) = F$ is called *a flat*.

By $M \setminus F$ we denote the matroid resulting from deleting the elements of $F \subseteq E(M)$: the elements of $M \setminus F$ are those not contained in F, with $F' \subset E(M \setminus F)$ being independent in $M \setminus F$ if and only if it is independent in M. For $F \subseteq E$, we define *the restriction* $M|F$ as $M \setminus (E \setminus F)$. A *loop* of M is an element e of M with $r(\{e\}) = 0$. A *separation* (A, B) is a bipartition of $E(M)$ into sets A and B. We call a separation (A, B) a k-*separation* if $r(A) + r(B) - r(M) \leq k - 1$.

A *branch-decomposition of a matroid* $M = (E, \mathcal{I})$ is a tree T, in which

- the leaves of T are in one-to-one correspondence with the elements of E and
- all inner nodes have degree three.

The width of an edge e *of* T is defined as $r(E_1) + r(E_2) - r(E) + 1$, where E_1 and E_2 are the subsets of $E(M)$ corresponding to the leaves of the two components of $T \setminus e$. Thus, the width of an edge e is the smallest k such that the induced bipartition (E_1, E_2) is a k-separation of M. *The width of the branch-decomposition* T *is a maximum width of an edge* $e \in T$. *The branch-width* $bw(M)$ *of a matroid is the minimum width of a branch-decomposition of* M.

The question of constructing a branch decomposition of a small width was positively settled in [16, 17] for general matroids (given by an oracle).

Theorem 2. *[17, Corollary 7.2] For each k, there is an $\mathcal{O}(n^4)$ algorithm constructing a decomposition of width at most $3k - 1$ or outputting a true statement that the matroid has branch-width at least $k + 1$.*

Moreover, for matroids representable over a fixed finite field, an efficient algorithm for constructing a branch decomposition of optimal width is given in [10].

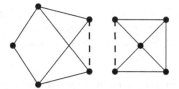 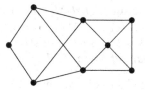

Fig. 1. The underlying graphs of matroids M_1, M_2 (with edges p_1, p_2 being dashed) and the underlying graph of the graphic matroid $M_1 \odot_{p_1, p_2} M_2$

Let M_1 and M_2 be two matroids satisfying $p_i \in E(M_i)$, for $i \in \{1, 2\}$. Then, the 2-sum $M_1 \odot_{p_1, p_2} M_2$ is defined as the matroid with the set of circuits below:

$$\mathcal{C} = \mathcal{C}(M_1 \setminus p_1) \cup \ \mathcal{C}(M_2 \setminus p_2) \cup$$
$$\{(C_1 \setminus p_1) \cup (C_2 \setminus p_2) : p_i \in C_i \in \mathcal{C}(M_i) \text{ for } i \in \{1, 2\}\}.$$

An example of a 2-sum of a pair of graphic matroids can be found in Figure 1.

We say that an algorithm runs in *linear time* if it always finishes in $\mathcal{O}(n)$ steps, where n is the length of the input in an appropriate encoding. Similarly, an algorithm runs in *polynomial time* if it always finishes in $\mathcal{O}(n^k)$ steps, for $k \in \mathbf{N}$. When a part of the algorithm's input is given by an oracle (e.g., a rank-oracle specifying an input matroid), the time the oracle spent computing the answer is not counted towards the number of steps the main algorithm took – only the time spent on constructing the input for the oracle and reading its output is accounted for in the overall runtime.

A *monadic second order (MSO) formula* ψ for a matroid M can contain the logical connectives $\vee, \wedge, \neg, \Rightarrow$, the equality predicate $=$, quantifications $\exists x$ over elements of $E(M)$ and subsets of $E(M)$ (we refer x as the element or set variable, respectively), the predicate $\in$ of containment of an element in a set, and, finally, the independence predicate $\mathrm{ind}(\cdot)$ determining whether a subset of $E(M)$ is independent. The independence predicate encodes the input matroid.

Deciding MSO properties of matroids is NP-hard in general, since, for example, the property that a graph is hamiltonian can be determined by deciding the following formula on the graphic matroid corresponding to the input graph:

$$\exists H \exists e \big(\mathrm{is\_circuit}(H) \wedge \mathrm{is\_base}(H \setminus \{e\}) \big),$$

where H is a set variable, e an element variable, and $\mathrm{is\_circuit}(\cdot)$ and $\mathrm{is\_base}(\cdot)$ are predicates testing the property of being a circuit and a base, respectively. These can be defined in MSO logic as follows:

$$\mathrm{is\_circuit}(H) \equiv \big(\neg \mathrm{ind}(H) \big) \wedge \big(\forall e : (e \in H) \Rightarrow \mathrm{ind}(H \setminus \{e\}) \big),$$
$$\mathrm{is\_base}(H) \equiv \neg \big(\exists e : \mathrm{ind}(H \cup \{e\}) \big).$$

3   Matroid Amalgams

In this section we define the operation of *a generalized parallel connection*, which plays a key role in the definition of an amalgam decomposition. We begin by introducing matroid amalgams and modular flats.

Definition 1. *Let M_1 and M_2 be two matroids. Let $E = E(M_1) \cup E(M_2)$ and $T = E(M_1) \cap E(M_2)$. Suppose that $M_1|T = M_2|T$. If M is a matroid with the ground set E such that $M|E_1 = M_1$ and $M|E_2 = M_2$, we say that M is an amalgam of M_1 and M_2.*

An amalgam of two matroids does not necessarily exist, even if the matroids coincide on the intersection of their ground sets. Our aim is to investigate a condition on matroids sufficient for the existence of an amalgam. To do so, we introduce the notions of *free amalgams* and *proper amalgams*.

Definition 2. *Let M_0 be an amalgam of M_1 and M_2. We say that M_0 is the free amalgam of M_1 and M_2 if for every amalgam M of M_1 and M_2 every set independent in M is also independent in M_0.*

The definition of a more restrictive *proper* amalgam is more involved.

Definition 3. *Let M_1 and M_2 be two matroids with rank functions r_1 and r_2, respectively, and independent sets coinciding on $E_1 \cap E_2$. First, define functions η and ζ on subsets of $E := E_1 \cup E_2$ as follows.*

$$\eta(X) := r_1(X \cap E_1) + r_2(X \cap E_2) - r(X \cap T),$$

$$\zeta(X) := \min\{\eta(Y) : Y \supseteq X\},$$

where $T := E_1 \cap E_2$ and r is the rank function of the matroid $N := M_1|T = M_2|T$. (Note that η provides an upper bound on the rank of the set X in a supposed amalgam of M_1 and M_2, while ζ is the least of these upper bounds.) If ζ is submodular on 2^E, we say that the matroid on $E_1 \cup E_2$ with ζ as its rank function is the proper amalgam of M_1 and M_2.

It can be verified that if the proper amalgam of two matroids exists then it is a free amalgam. The next lemma provides a necessary and sufficient condition for an amalgam to be the proper amalgam of two given matroids.

Lemma 1. *Let M_1 and M_2 be two matroids and M one of their amalgams. M is the proper amalgam of M_1 and M_2 if and only if it holds for every flat F of M that*

$$r(F) = r(F \cap E_1) + r(F \cap E_2) - r(F \cap T).$$

However, Lemma 1 says nothing about the existence of the proper amalgam of M_1 and M_2. Below, we give a condition that guarantees it.

Definition 4. *A flat $X = cl(T)$ of a matroid M is modular if for any flat Y of M the following holds:*

$$r(X \cup Y) = r(X) + r(Y) - r(X \cap Y).$$

Furthermore, we say that T is a modular semiflat if $cl(T)$ is a modular flat in M and every element of $cl(T)$ is either in T, a loop, or parallel to some other element of T.

For example, the set of all elements, the set of all loops, and any flat of rank one are modular flats. Each single-element set is a modular semiflat.

Theorem 3. *[19] Suppose that M_1 and M_2 are two matroids with a common restriction $N := M_1|T = M_2|T$, where $T = E(M_1) \cap E(M_2)$. If T is a modular semiflat in M_1, then the proper amalgam of M_1 and M_2 exists.*

We are now ready to introduce the operation of a generalized parallel connection, which can be used to glue matroids.

If M_1 and M_2 satisfy the assumptions of Theorem 3, then the resulting proper amalgam is called *the generalized parallel connection* of M_1 and M_2 and denoted by $M_1 \oplus_N M_2$, where $N := M_1|(E(M_1) \cap E(M_2))$. If we use $M_1 \oplus_N M_2$ without specifying N in advance, then N refers to the unique intersection of the two matroids. The generalized parallel connection satisfies the following properties.

Lemma 2. *If the generalized parallel connection of matroids M_1 and M_2 exists, $cl(E_2)$ is a modular semiflat in $M_1 \oplus_N M_2$.*

Lemma 3. *[19, p. 446] Let M_1 and M_2 be two matroids, $T = E(M_1) \cap E(M_2)$, N the matroid $M_1|T = M_2|T$, and $M = M_1 \oplus_N M_2$. For $X \subseteq E(M_1) \cup E(M_2)$, let $X_i = cl_i(X \cap E_i) \cup X$. It holds that*

$$cl_M(X) = cl_1(X_2 \cap E_1) \cup cl_2(X_1 \cap E_2), \text{ and}$$

$$r_M(X) = r_{M_1}(X_2 \cap E_1) + r_{M_2}(X_1 \cap E_2) - r(T \cap (X_1 \cup X_2)).$$

The operation of generalized parallel connection also commutes:

Lemma 4. *Let K, M_1 and M_2 be matroids such that $M_1|T_1 = K|T_1$ and $M_2|T_2 = K|T_2$. If T_1 is a modular semiflat in M_1 and T_2 is a modular semiflat in M_2, then*

$$M_2 \oplus_{N_2} (M_1 \oplus_{N_1} K) = M_1 \oplus_{N_1} (M_2 \oplus_{N_2} K).$$

3.1   Amalgam Width

Recall that the class of graphs of bounded tree-width can be introduced as the set of all subgraphs of a k-tree, where a k-tree is a graph that can be obtained by glueing two smaller k-trees along a clique of size k. Similarly, matroids of bounded branch-width can be introduced in terms of an operation taking two matroids of bounded branch-width and producing a larger matroid of bounded branch-width by glueing them along a low-rank separation. The amalgam-width is also defined using a glueing operation. Analogously to the definition of tree-width, where some elements of the clique can be effectively removed after glueing takes place, the operation includes a set of elements to be deleted. A typical situation when applying the glueing operation is illustrated on Figure 2.

Definition 5. *Suppose we are given matroids M_1, M_2, and K such that $E(M_1) \cap E(M_2) \subseteq E(K)$. Furthermore, suppose we are also given a set $D \subseteq E(K)$. Let $J_i := E(M_i) \cap E(K), i \in \{1,2\}$ and assume the two conditions below hold:*

- $M_i|J_i = K|J_i, i \in \{1,2\}$,
- J_1 and J_2 are both modular semiflats in K.

Then, the matroid $M_1 \oplus^{K,D} M_2$ is defined as follows:

$$M_1 \oplus^{K,D} M_2 := \left((K \oplus_{J_1} M_1) \oplus_{J_2} M_2\right) \setminus D.$$

We also say that the matroid $M_1 \oplus^{K,D} M_2$ is a result of glueing of M_1 and M_2 along K and removing the elements D.

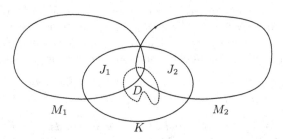

Fig. 2. M_1, M_2 are the matroids being combined, K is a small matroid used to glue them together, and D is a set of elements that are subsequently removed

Note that Theorem 3 guarantees the matroid $M_1 \oplus^{K,D} M_2$ to be well defined. We are now ready to introduce our width parameter.

Definition 6. *Matroid M has amalgam-width at most $k \in \mathbf{N}$ if $|E(M)| \leq 1$, or there are matroids M_1 and M_2 of amalgam width at most k, a matroid K satisfying $|E(K)| \leq k$, and a choice of $D \subseteq E(K)$ such that*

$$M = M_1 \oplus^{K,D} M_2.$$

Note that the first condition can be weakened to $|E(M)| \leq k$ without affecting the definition. Every finite matroid M has an amalgam width at most $|E(M)|$. The amalgam-width of M is the smallest k such that M has amalgam-width at most k. The definition above naturally yields a tree-like representation of the construction of the matroid in question:

Definition 7. *Assume that M is a matroid with amalgam width k. Any rooted tree $\mathcal{T}$ satisfying either of the following statements is called an amalgam decomposition of M of width at most k:*

- *$|E(M)| \leq 1$ and $\mathcal{T}$ is a trivial tree containing precisely one node,*
- *$M = M_1 \oplus^{K,D} M_2$ and $\mathcal{T}$ has a root r with children r_1 and r_2 such that the subtrees of $\mathcal{T}$ rooted at r_1 and r_2 are amalgam decompositions of M_1 and M_2 of width at most k.*

The above definition leads to a natural assignment of matroids to the nodes of $\mathcal{T}$: whenever a glueing operation is performed, we assign the resulting matroid to the node. We use $M^{\mathcal{T}}(v)$ to refer to this matroid and say that the node v

represents $M^{\mathcal{T}}(v)$. For an internal node $v \in \mathcal{T}$, we use $M_1^{\mathcal{T}}(v)$, $M_2^{\mathcal{T}}(v)$, $K^{\mathcal{T}}(v)$, $D^{\mathcal{T}}(v)$, $J_1^{\mathcal{T}}(v)$ and $J_2^{\mathcal{T}}(v)$ to denote the corresponding elements appearing in the glueing operation used to obtain $M^{\mathcal{T}}(v) = M_1^{\mathcal{T}}(v) \oplus^{K^{\mathcal{T}}(v), D^{\mathcal{T}}(v)} M_2^{\mathcal{T}}(v)$. If v is a leaf of a decomposition $\mathcal{T}$, we let $M_1^{\mathcal{T}}(v)$ and $M_2^{\mathcal{T}}(v)$ be matroids with empty groundsets, $K^{\mathcal{T}}(v) := M^{\mathcal{T}}(v)$, and $D^{\mathcal{T}}(v) := \emptyset$. Finally, we denote by $J^{\mathcal{T}}(v) \subseteq K(v)$ the set of elements used to glue $M(v)$ to its parent. More formally, we set $J^{\mathcal{T}}(v) := J_i^{\mathcal{T}}(u)$, where $i \in \{1, 2\}$ is chosen depending on whether v is a left or right child of u. Since the decomposition under consideration is typically clear from context, we usually omit the upper index $\mathcal{T}$.

Strozecki [22] introduces a similar parameter that uses the operation of a matroid 2-sum instead of the generalized parallel connection. However, its applicability is limited since it allows to join matroids only using separations of size at most 2 and thus a corresponding decomposition of a 3-connected matroid M has a width of $|E(M)|$. The next proposition implies that the latter is able to express the 2-sum operation as a special case. Therefore, the amalgam-width is a more general parameter than the one from [22].

Proposition 1. *A 2-sum of matroids M_1 and M_2 can be replaced by finitely many operations of generalized parallel connections and deletions.*

Furthermore, the amalgam width is a generalization of the branch-width parameter for finitely representable matroids in the sense that a bound on the value of branch-width implies a bound on the amalgam-width:

Proposition 2. *If M is a matroid with branch-width k and M is representable over a finite field $\mathbf{F}$, then the amalgam width of M is at most $|\mathbf{F}|^{3k/2}$.*

4 Algorithms

As a main result of this section, we show that the problem of deciding monadic second order properties is computationally tractable for matroids of bounded amalgam width:

Theorem 4. *MSO properties can be decided in linear time for matroids with amalgam width bounded by k (assuming the corresponding amalgam decomposition $\mathcal{T}$ of the matroid is given explicitly as a part of the input).*

For the purpose of induction used in the proof of Theorem 4, we need to generalize the considered problem by introducing free variables:

INPUT:

- an MSO formula ψ with p free variables,
- amalgam decomposition $\mathcal{T}$ of a matroid M with width at most k,
- a function Q defined on the set $\{1, \ldots, p\}$ assigning the i-th free variable its value; specifically, $Q(i)$ is equal to an element of $E(M)$ if x_i is an element variable, and it is a subset of $E(M)$ if x_i is a set variable.

OUTPUT:

- *ACCEPT* if ψ is satisfied on M with the values prescribed by Q to its free variables.
- *REJECT* otherwise.

The resulting problem is referred to as *the MSO-DECIDE problem*. To simplify notation, let us assume that if ψ is a formula with free variables, we use x_i for the i-th variable if it appears in ψ as an element variable and X_i if it appears as a set variable. We prove the following generalization of Theorem 4:

Theorem 5. *The problem MSO-DECIDE can be solved in linear time for matroids with amalgam width bounded by k (assuming the corresponding amalgam decomposition $\mathcal{T}$ of the matroid is given as a part of the input).*

Our aim in the proof of Theorem 5 is to construct a linear time algorithm based on deterministic bottom-up tree automatons. Let us introduce such automatons.

Definition 8. *A finite tree automaton is a 5-tuple $(S, S_A, \delta, \Delta, \Sigma)$, where*

- *S is a finite set of states containing a special initial state 0,*
- *$S_A \subseteq S$ is a non-empty set of accepting states,*
- *Σ is a finite alphabet,*
- *$\delta : S \times \Sigma \to S$ is set of transition rules that determine a new state of the automaton based on its current state and the information, represented by Σ, contained in the current node of the processed tree, and*
- *$\Delta : S \times S \to S$ is a function combining the states of two children into a new state.*

Let us also establish the following simple notation.

Definition 9. *Consider an instance of an MSO-DECIDE problem. In particular, let Q be the variable-assignment function as in the definition of our generalized problem. For $F \subseteq E(M)$, we define the local view of Q at F to be the following function:*

$$Q_F(i) := \begin{cases} Q(i) \cap F & \text{if the } i\text{-th variable is a set variable,} \\ Q(i) & \text{if the } i\text{-th variable is an element variable and } Q(i) \in F, \\ \boxtimes & \text{otherwise,} \end{cases}$$

where $\boxtimes$ is a special symbol that is not an element of the input matroid.

The symbol $\boxtimes$ stands for values outside of F. We simplify the notation by writing $Q_v(x_i)$ instead of $Q_{E(K(v))}(i)$, where v is a node of $\mathcal{T}$ from the problem's input.

The alphabet Σ of the automaton we construct will correspond to the set of all possible "configurations" at a node v in an amalgam decomposition of width at most k. A finite tree automaton processes a tree (in our case $\mathcal{T}$) from its leaves to the root, assigning states to each node based on the information read in the node and on the states of its children. When processing a node whose two children were already processed the automaton calculates the state $s := \Delta(s_1, s_2)$, where s_1 and s_2 are the states of the children, and moves to the state $\delta(s, q)$, where $q \in \Sigma$ represents the information contained in that node of the tree. If the state eventually assigned to the root of the tree is contained in the set S_A, we say that the automaton accepts. It rejects otherwise.

As a final step of our preparation for the proof of Theorem 5, we slightly alter the definition of an MSO formula by replacing the use of $\text{ind}(X)$ predicate with the use of $x_1 \in \text{cl}(X_2)$, where $\text{cl}(\cdot)$ is the closure function of M. The predicate $\text{ind}(X)$ can be expressed while adhering to the altered definition as follows:

$$\text{ind}(X) \equiv \neg(\exists e \in X : \text{cl}(X) = \text{cl}(X \setminus \{e\})).$$

Proof (of Theorem 5). We proceed by induction on the complexity of the formula ψ, starting with simple formulas such as $x_1 = x_2$ or $x_1 \in X_2$. In each step of the induction, we design a tree automaton processing the amalgam decomposition tree $\mathcal{T}$ and correctly solving the corresponding MSO-DECIDE problem. As already mentioned, the alphabet will encode all possible non-isomorphic choices of the matroid $K(v)$, sets $J(v)$, $J_1(v)$, $J_2(v)$, and $D(v)$ combined with all possible local views of Q at v, allowing this information to be read when processing the corresponding node. Note that if k is bounded, the size of the set Σ of such configurations is bounded. Since the automaton size does not depend on n and the amount of information read in each node of $\mathcal{T}$ is bounded by a constant (assuming bounded amalgam-width), we will be able to conclude that the running time of our algorithm, which will just simulate the tree automaton, is linear in the size of $\mathcal{T}$.

To start the induction, we first consider the case $\psi = x_1 \in X_2$. Such instances of MSO-DECIDE can be solved by the automaton given in Figure 3. This automaton stays in its original state if x_1 is assigned $\boxtimes$ by the local view of Q at $E(K(v))$. Otherwise, it moves to designated ACCEPT and REJECT states based on whether $Q_v(x_1) \in Q_v(X_2)$ holds. The set S_A is defined to be $\{\text{ACCEPT}\}$. The function $\Delta : S \times S \to S$ assigns the ACCEPT state to any tuple containing an ACCEPT state. Similarly for the REJECT state. We are guaranteed not to encounter the situation where one child node is in the ACCEPT state and the other in the REJECT state, since the free variable assignment function Q maps x_1 precisely to one element of $E(M)$. It is clear that this tree automaton correctly propagates the information of whether $x_1 \in X_2$ or not from the leaf representing the value of x_1 to the root of $\mathcal{T}$.

The cases of formulas $x_1 = x_2$ and $X_1 = X_2$ can be handled similarly. For formulas of the form $\psi_1 \vee \psi_2$, we construct the automaton by taking the Cartesian product of the automatons $A_i = (S^i, S_A^i, \delta^i, \Delta^i, \Sigma^i), i \in \{1, 2\}$ for the partial formulas ψ_i. Specifically, $\Sigma = \Sigma^1 \times \Sigma^2$, $S = S^1 \times S^2$, $S_A = (S_A^1 \times S^2) \cup (S^1 \times S_A^2)$, $\Delta((x, y)) = (\Delta^1(x), \Delta^2(y))$, $\delta((x, y), (q, r)) = (\delta^1(x, q), \delta^2(y, r))$. Informally, the two automatons run in parallel and the new automaton accepts precisely if at least one of the two is in an accepting state. A formula of the form

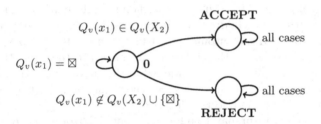

$Q_v(x_1) \in Q_v(X_2)$

ACCEPT

all cases

$Q_v(x_1) = \boxtimes$ $\mathbf{0}$

all cases

$Q_v(x_1) \notin Q_v(X_2) \cup \{\boxtimes\}$

REJECT

Fig. 3. The states and transition rules δ of the tree automaton for the formula $x_1 \in X_2$. Here, v is the currently processed node of the amalgam decomposition. The names of the states are typed using bold font.

$\neg\psi$ can be processed by the same automaton as ψ, except we change the set accepting states to their complement. The connectives $\wedge, \Rightarrow, \ldots$ can be expressed using $\vee$ and $\neg$ by a standard reduction.

The special properties of amalgam decompositions come into play when constructing the automaton deciding $x_1 \in \mathrm{cl}(X_2)$. When processing a node $v \in \mathcal{T}$, we see the elements of $K(v)$, can query the independent sets on $E(K(v))$, and see local view of $Q(X_2)$ at $E(K(v))$. Our strategy will be to compute $\mathrm{cl}_M(X_2)$ restricted to $E(K(v))$ and determine whether x_1 is contained in it. However, the state at v does not encode necessary information about the remaining part of M. The matroid $M(v)$ is joined to this part by a generalized parallel connection using $J(v)$. Lemma 3 says that the remaining part of M can influence the restriction of the closure of X_2 on $E(K(v))$ only through forcing some of the elements of this modular flat into the closure. Since $|J(v)|$ is bounded, we can precompute the behavior of the resulting closure for all possible cases. This information is encoded in the state of the finite automaton passed to the parent node. The parent node can then use the information encoded in the states corresponding to its children when precomputing its intersection with $\mathrm{cl}_M(X_2)$. We formalize this approach using the following definition.

Definition 10. *Let v be a node of an amalgam decomposition $\mathcal{T}$ of M and X be a subset of $E(M)$. A map f_v^X from $2^{J(v)} \to 2^{J(v)}$ satisfying*

$$f_v^X(Y) = \mathrm{cl}_{M(v)}\big(\big(X \cap E(M(v))\big) \cup Y\big) \cap J(v)$$

is called the type of a node v with respect to X.

When processing a node v, we can assume we are given the types f_1^X and f_2^X of the children of v and we want to determine the type of v. The type is then encoded into the state of the finite automaton (along with the information for which choices of $Y \subseteq J(v)$ the formula ψ holds) and is passed to the parent node. This information is then reused to determine the type of the parent node, etc. This process is captured by the following definition.

Definition 11. *Let v be a node of an amalgam decomposition $\mathcal{T}$ of a matroid M, v_1 and v_2 the children of v, and X a subset of $E(M)$. If $f_{v_1}^X$ is the type of*

v_1 with respect to X and $f_{v_2}^X$ is a type of v_2 with respect to X, we say that the type $f_{v_1}^X +_{K(v)} f_{v_2}^X$ of v is the join of $f_{v_1}^X$ and $f_{v_2}^X$ if for every subset Y of $J(v)$ it holds that $f_{v_1}^X +_{K(v)} f_{v_2}^X = Z \cap J(v)$, where Z is the smallest subset of $E(K(v))$ such that (1) $f_{v_1}^X(Z \cap J_1(v)) = Z \cap J_1(v)$, (2) $f_{v_2}^X(Z \cap J_2(v)) = Z \cap J_2(v)$, (3) $Z \supseteq Y \cup (X \cap E(K(v)))$.

Lemma 3 implies that $f_{v_1}^X +_{K(v)} f_{v_2}^X$ is the type of the node v with respect to X. Observe that the type $f_1^X +_{K(v)} f_2^X$ in the above definition is determined by $f_{v_1}^X, f_{v_2}^X, K(v)$ and $X \cap E(K(v))$ – each of which has bounded size. This implies that the computation of the type $f_1^X +_{K(v)} f_2^X$ can be wired in the transition function of the automaton. Deciding if $Q(x_1) \in \mathrm{cl}(X_2) \cap J(v)$ is then reduced to verifying if $Q(x_1) \in f_v^{X_2}(Y)$ for a particular choice of Y.

The case of a formula $\exists x : \psi$ is solved by a standard argument of taking the finite tree automaton recognizing ψ and transforming it to a non-deterministic finite tree automaton that tries to guess the value of x (in our case, the automaton also checks if this guessed value of x lies in the set $D(v)$ of deleted elements). This automaton can in turn be simulated using a *deterministic* finite tree automaton with up to an exponential blow-up of the number of states. The case $\exists X : \psi$ is solved analogously. Since the algorithm simulating the automaton on $\mathcal{T}$ spends $\mathcal{O}(1)$ time in each of the nodes of $\mathcal{T}$, there exists a linear time algorithm solving the problem from the statement of the theorem.

We can also show that the Tutte polynomial can be computed and evaluated in an FPT way. We omit the proof due to space constraints.

Theorem 6. *For a fixed k, there exists a polynomial-time algorithm that computes the coefficients of the Tutte polynomial of M from its amalgam decomposition $\mathcal{T}$ of width at most k. The degree of the polynomial in the running time estimate of the algorithm is independent of k.*

5   Conclusion

The Theorem 5 includes the assumption that the amalgam decomposition is given as a part of the input. This can be removed for matroids representable over a fixed finite field, since the proof of Proposition 2 gives a linear time algorithm constructing an amalgam decomposition from a branch decomposition. Therefore, we can use a polynomial-time algorithm [7, 15] for constructing a branch decomposition and then convert it to an amalgam decomposition of width bounded by a constant multiple of the original branch-width. Similarly, it can be shown that the branch decomposition of a representable matroid can be obtained from an amalgam decomposition in a natural way. However, we have not been able to settle the complexity of constructing amalgam decomposition of (approximately) optimal width for a general oracle-given matroid.

Acknowledgment. The authors would like to thank Dan Král' for valuable insights into the problem.

References

1. Bui-Xuan, B.-M., Telle, J.A., Vatshelle, M.: Boolean-width of graphs. In: Chen, J., Fomin, F.V. (eds.) IWPEC 2009. LNCS, vol. 5917, pp. 61–74. Springer, Heidelberg (2009)
2. Courcelle, B.: The monadic second-order logic of graph I. Recognizable sets of finite graphs. Inform. and Comput. 85, 12–75 (1990)
3. Courcelle, B.: The expression of graph properties and graph transformations in monadic second-order logic. In: Rozenberg, G. (ed.) Handbook of Graph Grammars and Computing by Graph Transformations. Foundations, vol. 1, pp. 313–400. World Scientific (1997)
4. Courcelle, B., Olariu, S.: Upper bounds to the clique width of graphs. Discrete Appl. Math. 101, 77–114 (2000)
5. Courcelle, B., Makowsky, J.A., Rotics, U.: On the fixed parameter complexity of graph enumeration problems definable in monadic second-order logic. Discrete Appl. Math. 108, 23–52 (2001)
6. Hliněný, P., Whittle, G.: Matroid tree-width. European J. Combin. 27, 1117–1128 (2003)
7. Hliněný, P.: A parametrized algorithm for matroid branch-width. SIAM J. Computing 35, 259–277 (2005)
8. Hliněný, P.: Branch-width, parse trees, and monadic second-order logic for matroids. J. Combin. Theory Ser. B 96, 325–351 (2006)
9. Hliněný, P.: On matroid representatibility and minor problems. In: Královič, R., Urzyczyn, P. (eds.) MFCS 2006. LNCS, vol. 4162, pp. 505–516. Springer, Heidelberg (2006)
10. Hliněný, P., Oum, S.: Finding branch-decomposition and rank-decomposition. SIAM J. Computing 38, 1012–1032 (2008)
11. Gavenčiak, T., Král, D., Oum, S.-I.: Deciding first order logic properties of matroids. In: Czumaj, A., Mehlhorn, K., Pitts, A., Wattenhofer, R. (eds.) ICALP 2012, Part II. LNCS, vol. 7392, pp. 239–250. Springer, Heidelberg (2012)
12. Král', D.: Computing representations of matroids of bounded branch-width. In: Thomas, W., Weil, P. (eds.) STACS 2007. LNCS, vol. 4393, pp. 224–235. Springer, Heidelberg (2007)
13. Král', D.: Decomposition width of matroids. Discrete Appl. Math. 160, 913–923 (2012)
14. Král', D.: Decomposition width of matroids. In: Abramsky, S., Gavoille, C., Kirchner, C., Meyer auf der Heide, F., Spirakis, P.G. (eds.) ICALP 2010. LNCS, vol. 6198, pp. 55–66. Springer, Heidelberg (2010)
15. Oum, S., Seymour, P.D.: Approximating clique-width and branch-width. J. Combin. Theory Ser. B 96, 514–528 (2006)
16. Oum, S., Seymour, P.D.: Certifying large branch-width. In: Proc. of SODA, pp. 810–813 (2006)
17. Oum, S., Seymour, P.D.: Approximating clique-width and branch-width. J. Combin. Theory Ser. B 96, 514–528 (2006)
18. Oum, S., Seymour, P.D.: Testing branch-width. J. Combin. Theory Ser. B 97, 385–393 (2007)
19. Oxley, J.: Matroid Theory, 2nd edn. Oxford University Press (2011)
20. Robertson, N.S., Seymour, P.D.N.: Graph minors. X. Obstructions to tree-decomposition. J. Combin. Theory Ser. B 52, 153–190 (1992)
21. Seymour, P.: Recognizing graphic matroids. Combinatorica 1, 75–78 (1981)
22. Strozecki, Y.: Monadic second-order model-checking on decomposable matroids. Discrete Appl. Math. 159, 1022–1039 (2011)

On the Parameterized Complexity
of Reconfiguration Problems

Amer E. Mouawad[1,*], Naomi Nishimura[1,*], Venkatesh Raman[2],
Narges Simjour[1,*], and Akira Suzuki[3,**]

[1] David R. Cheriton School of Computer Science
University of Waterloo, Waterloo, Ontario, Canada
{aabdomou,nishi,nsimjour}@uwaterloo.ca
[2] The Institute of Mathematical Sciences
Chennai, India
vraman@imsc.res.in
[3] Graduate School of Information Sciences, Tohoku University
Aoba-yama 6-6-05, Aoba-ku, Sendai, 980-8579, Japan
a.suzuki@ecei.tohoku.ac.jp

Abstract. We present the first results on the parameterized complexity of reconfiguration problems, where a reconfiguration version of an optimization problem Q takes as input two feasible solutions S and T and determines if there is a sequence of *reconfiguration steps* that can be applied to transform S into T such that each step results in a feasible solution to Q. For most of the results in this paper, S and T are subsets of vertices of a given graph and a reconfiguration step adds or deletes a vertex. Our study is motivated by recent results establishing that for most NP-hard problems, the classical complexity of reconfiguration is PSPACE-complete.

We address the question for several important graph properties under two natural parameterizations: k, the size of the solutions, and ℓ, the length of the sequence of steps. Our first general result is an algorithmic paradigm, the *reconfiguration kernel*, used to obtain fixed-parameter algorithms for the reconfiguration versions of VERTEX COVER and, more generally, BOUNDED HITTING SET and FEEDBACK VERTEX SET, all parameterized by k. In contrast, we show that reconfiguring UNBOUNDED HITTING SET is $W[2]$-hard when parameterized by $k+\ell$. We also demonstrate the $W[1]$-hardness of the reconfiguration versions of a large class of maximization problems parameterized by $k + \ell$, and of their corresponding deletion problems parameterized by ℓ; in doing so, we show that there exist problems in FPT when parameterized by k, but whose reconfiguration versions are $W[1]$-hard when parameterized by $k + \ell$.

1 Introduction

The reconfiguration version of an optimization problem asks whether it is possible to transform a source feasible solution S into a target feasible solution T by

* Research supported by the Natural Science and Engineering Research Council of Canada.
** Research supported by JSPS Grant-in-Aid for Scientific Research, Grant Number 24.3660.

G. Gutin and S. Szeider (Eds.): IPEC 2013, LNCS 8246, pp. 281–294, 2013.

a (possibly minimum-length) sequence of *reconfiguration steps* such that every intermediate solution is also feasible. Reconfiguration problems model dynamic situations in which we seek to transform a solution into a more desirable one, maintaining feasibility during the process. The study of reconfiguration yields insights into the structure of the solution space of the underlying optimization problem, crucial for the design of efficient algorithms.

Motivated by these facts, there has been a lot of recent interest in studying the complexity of reconfiguration problems. Problems for which reconfiguration has been studied include VERTEX COLOURING [1–5], LIST EDGE-COLOURING [6], INDEPENDENT SET [7, 8], SET COVER, MATCHING, MATROID BASES [8], SATISFIABILITY [9], SHORTEST PATH [10, 11], and DOMINATING SET [12, 13]. Most work has been limited to the problem of determining the existence of a reconfiguration sequence between two given solutions; for most NP-complete problems, this problem has been shown to be PSPACE-complete.

As there are typically exponentially many feasible solutions, the length of a reconfiguration sequence can be exponential in the size of the input instance. It is thus natural to ask whether reconfiguration problems become tractable if we allow the running time to depend on the length of the sequence. In this work, we explore reconfiguration in the framework of parameterized complexity [14] under two natural parameterizations: k, a bound on the size of feasible solutions, and ℓ, the length of the reconfiguration sequence. One of our key results is that for most problems, the reconfiguration versions remain intractable in the parameterized framework when we parameterize by ℓ. It is important to note that when k is not bounded, the reconfiguration problems we study become easy.

We present fixed-parameter algorithms for problems parameterized by k by modifying known parameterized algorithms for the problems. The paradigms of bounded search tree and kernelization typically work by exploring minimal solutions. However, a reconfiguration sequence may necessarily include non-minimal solutions. Any kernel that removes solutions (non-minimal or otherwise) may render finding a reconfiguration sequence impossible, as the missing solutions might appear in every reconfiguration sequence. To handle these difficulties, we introduce a general approach for parameterized reconfiguration problems. We use a *reconfiguration kernel*, showing how to adapt Bodlaender's cubic kernel [15] for FEEDBACK VERTEX SET, and a special kernel by Damaschke and Molokov [16] for BOUNDED HITTING SET (where the cardinality of each input set is bounded) to obtain polynomial reconfiguration kernels, with respect to k. These results can be considered as interesting applications of kernelization, and a general approach for other similar reconfiguration problems.

As a counterpart to our result for BOUNDED HITTING SET, we show that reconfiguring UNBOUNDED HITTING SET or DOMINATING SET is $W[2]$-hard parameterized by $k + \ell$ (Section 4). Finally, we show a general result on reconfiguration problems of hereditary properties and their 'parametric duals', implying the $W[1]$-hardness of reconfiguring INDEPENDENT SET, INDUCED FOREST, and BIPARTITE SUBGRAPH parameterized by $k + \ell$ and VERTEX COVER, FEEDBACK VERTEX SET, and ODD CYCLE TRANSVERSAL parameterized by ℓ.

2   Preliminaries

Unless otherwise stated, we assume that each input graph G is a simple, undirected graph on n vertices with vertex set $V(G)$ and edge set $E(G)$. To avoid confusion, we refer to *nodes* in reconfiguration graphs (defined below), as distinguished from *vertices* in the input graph. We use the modified big-Oh notation O^* that suppresses all polynomially bounded factors.

Our definitions are based on optimization problems, each consisting of a polynomial-time recognizable set of valid instances, a set of feasible solutions for each instance, and an objective function assigning a nonnegative rational value to each feasible solution.

Definition 1. *The* reconfiguration graph $R_Q(I, \mathrm{adj}, k)$, *consists of a node for each feasible solution to instance I of optimization problem Q, where the size of each solution is at least k for Q a maximization problem (of size at most k for Q a minimization problem, respectively), for positive integer k, and an edge between each pair of nodes corresponding to solutions in the binary adjacency relation* adj *on feasible solutions.*

We define the following *reconfiguration problems*, where S and T are feasible solutions for I: Q RECONFIGURATION determines if there is a path from S to T in $R_Q(I, \mathrm{adj}, k)$; the *search variant* returns a *reconfiguration sequence*, the sequence of feasible solutions associated with such a path; and the *shortest path variant* returns the reconfiguration sequence associated with a path of minimum length.

Using the framework developed by Downey and Fellows [14], a *parameterized reconfiguration problem* includes in the input a positive integer ℓ (an upper bound on the length of the reconfiguration sequence) and a parameter p (typically k or ℓ). For a parameterized problem Q with inputs of the form (x, p), $|x| = n$ and p a positive integer, Q is *fixed-parameter tractable* (or in *FPT*) if it can be decided in $f(p)n^c$ time, where f is an arbitrary function and c is a constant independent of both n and p. Q has a *kernel* of size $f(p)$ if there is an algorithm A that transforms the input (x, p) to (x', p') such that A runs in polynomial time (with respect to $|x|$ and p) and (x, p) is a yes-instance if and only if (x', p') is a yes-instance, $p' \le g(p)$, and $|x'| \le f(p)$. Each problem in *FPT* has a kernel, possibly of exponential (or worse) size. The main hierarchy of parameterized complexity classes is $FPT \subseteq W[1] \subseteq W[2] \subseteq \ldots \subseteq XP$, where W-hardness, shown using *FPT reductions*, is the analogue of NP-hardness in classical complexity. The reader is referred to [17, 18] for more on parameterized complexity.

We introduce the notion of a *reconfiguration kernel*; it follows from the definition that a reconfiguration problem that has such a kernel is in *FPT*.

Definition 2. *A* reconfiguration kernel *of an instance $(x, p) = (Q, \mathrm{adj}, S, T, k, \ell, p)$ of a parameterized reconfiguration problem is a set of $h(p)$ instances, for an arbitrary function h, such that for $1 \le i \le h(p)$:*

- *for each instance in the set, $(x_i, p_i) = (Q, \mathrm{adj}, S_i, T_i, k_i, \ell_i, p_i)$, the values of S_i, T_i, k_i, ℓ_i, and p_i can all be computed in polynomial time,*

- *the size of each x_i is bounded by $j(p)$, for an arbitrary function j, and*
- *(x, p) is a yes-instance if and only if at least one (x_i, p_i) is a yes-instance.*

Most problems we consider can be defined using graph properties, where a *graph property* π is a collection of graphs, and is *non-trivial* if it is non-empty and does not contain all graphs. A graph property is *polynomially decidable* if for any graph G, it can be decided in polynomial time whether G is in π. For a subset $V' \subseteq V$, $G[V']$ is the *subgraph of G induced on V'*, with vertex set V' and edge set $\{\{u, v\} \in E \mid u, v \in V'\}$. The property π is *hereditary* if for any $G \in \pi$, any induced subgraph of G is also in π. It is well-known [19] that every hereditary property π has a forbidden set $\mathcal{F}_\pi$, in that a graph has property π if and only if it does not contain any graph in $\mathcal{F}_\pi$ as an induced subgraph.

For a graph property π, we define two reconfiguration graphs, where solutions are sets of vertices and two solutions are adjacent if they differ by the addition or deletion of a vertex. The *subset reconfiguration graph of G with respect to π*, $R^\pi_{\mathrm{SUB}}(G, k)$, has a node for each $S \subseteq V(G)$ such that $|S| \geq k$ and $G[S]$ has property π, and the *deletion reconfiguration graph of G with respect to π*, $R^\pi_{\mathrm{DEL}}(G, k)$, has a node for each $S \subseteq V(G)$ such that $|S| \leq k$ and $G[V(G) \setminus S]$ has property π. We can obtain $R^\pi_{\mathrm{DEL}}(G, |V(G)| - k)$ by replacing the set corresponding to each node in $R^\pi_{\mathrm{SUB}}(G, k)$ by its (setwise) complement.

Definition 3. *For any graph property π, graph G, positive integer k, $S \subseteq V(G)$, and $T \subseteq V(G)$, we define the following decision problems: π-DELETION(G, k): Is there $V' \subseteq V(G)$ such that $|V'| \leq k$ and $G[V(G) \setminus V'] \in \pi$?*
π-SUBSET(G, k): Is there $V' \subseteq V(G)$ such that $|V'| \geq k$ and $G[V'] \in \pi$?
π-DEL-RECONF(G, S, T, k, ℓ): For $S, T \in V(R^\pi_{\mathrm{DEL}}(G, k))$, is there a path of length at most ℓ between S and T in $R^\pi_{\mathrm{DEL}}(G, k)$?
π-SUB-RECONF(G, S, T, k, ℓ): For $S, T \in V(R^\pi_{\mathrm{SUB}}(G, k))$, is there a path of length at most ℓ between S and T in $R^\pi_{\mathrm{SUB}}(G, k)$?

We say that π-DELETION(G, k) and π-SUBSET(G, k) are *parametric duals* of each other. We refer to π-DEL-RECONF(G, S, T, k, ℓ) and π-SUB-RECONF(G, S, T, k, ℓ) as *π-reconfiguration problems*.

Due to the page limitation, some proofs (marked with an asterisk) have been omitted and can be found in the full version of the paper [20].

3   Fixed-Parameter Tractability Results

For an instance (G, S, T, k, ℓ) of a π-reconfiguration problem, we partition $V(G)$ into the sets $C = S \cap T$, $S_D = S \setminus C$, $T_A = T \setminus C$, and $O = V(G) \setminus (S \cup T) = V(G) \setminus (C \cup S_D \cup T_A)$ (all other vertices). Furthermore, we can partition C into two sets C_F and $C_M = C \setminus C_F$, where a vertex is in C_F if and only if it is in every feasible solution of size bounded by k. In any reconfiguration sequence, each vertex in S_D must be deleted and each vertex in T_A must be added. We say that a vertex v is *touched* if v is either added or deleted in at least one reconfiguration step. The following fact is a consequence of the definitions above, the fact that π is hereditary, and the observations that $G[S_D]$ and $G[O]$ are both subgraphs of $G[V(G) \setminus T]$, and $G[T_A]$ and $G[O]$ are both subgraphs of $G[V(G) \setminus S]$.

Fact 1. *For an instance π-DEL-RECONF(G, S, T, k, ℓ) of a reconfiguration problem for hereditary property π, $G[O]$, $G[S_D]$, and $G[T_A]$ all have property π.*

In the next section, we show that for most hereditary properties, reconfiguration problems are hard when parameterized by ℓ. Here, we prove the parameterized tractability of reconfiguration for certain superset-closed k-subset problems when parameterized by k, where a k-*subset problem* is a parameterized problem Q whose solutions for an instance (I, k) are all subsets of size at most k of a domain set, and is *superset-closed* if any superset of a solution of Q is also a solution of Q.

Theorem 4. *If a k-subset problem Q is superset-closed and has an FPT algorithm to enumerate all its minimal solutions, the number of which is bounded by a function of k, then Q RECONFIGURATION parameterized by k is in FPT, as well as the search and shortest path variants.*

Proof. By enumerating all minimal solutions of Q, we compute the set M of all elements v of the domain set such that v is in a minimal solution to Q. For (I, S, T, k, ℓ) an instance of Q RECONFIGURATION, we show that there exists a reconfiguration sequence from S to T if and only if there exists a reconfiguration sequence from $S \cap M$ to $T \cap M$ that uses only subsets of M.

Each set U in the reconfiguration sequence from S to T is a solution, hence contains at least one minimal solution in $U \cap M$; $U \cap M$ is a superset of the minimal solution and hence also a solution. Moreover, since any two consecutive solutions U and U' in the sequence differ by a single element, $U \cap M$ and $U' \cap M$ differ by at most a single element. By replacing each subsequence of identical sets by a single set, we obtain a reconfiguration sequence from $S \cap M$ to $T \cap M$ that uses only subsets of M.

The reconfiguration sequence from $S \cap M$ to $T \cap M$ using only subsets of M can be extended to a reconfiguration sequence from S to T by transforming S to $S \cap M$ in $|S \setminus M|$ steps and transforming $T \cap M$ to T in $|T \setminus M|$ steps. In this sequence, each vertex in $C \setminus M$ is removed from S to form $S \setminus M$ and then readded to form T from $T \setminus M$. For each vertex $v \in C \setminus M$, we can choose instead to add v to each solution in the sequence, thereby decreasing ℓ by two (the steps needed to remove and then readd v) at the cost of increasing by one the capacity used in the sequence from $S \cap M$ to $T \cap M$. This choice can be made independently for each of these $\mathcal{E} = |C \setminus M|$ vertices.

Consequently, (I, S, T, k, ℓ) is a yes-instance for Q RECONFIGURATION if and only if one of the $\mathcal{E} + 1$ reduced instances $(I, S \cap M, T \cap M, k - e, \ell - 2(\mathcal{E} - e))$, for $0 \leq e \leq \mathcal{E}$ and $\mathcal{E} = |C \setminus M|$, is a yes-instance for Q' RECONFIGURATION: we define Q' as a k-subset problem whose solutions for an instance (I, k) are solutions of instance (I, k) of Q that are contained in M. To show that Q' RECONFIGURATION is in FPT, we observe that the number of nodes in the reconfiguration graph for

Q' is bounded by a function of k: each solution of Q' is a subset of M, yielding at most $2^{|M|}$ nodes, and $|M|$ is bounded by a function of k.   □

Corollary 5. BOUNDED HITTING SET RECONFIGURATION, FEEDBACK VERTEX SET IN TOURNAMENTS RECONFIGURATION, *and* MINIMUM WEIGHT SAT IN BOUNDED CNF FORMULAS RECONFIGUATION *parameterized by* k *are in FPT.*

For BOUNDED HITTING SET, the proof of Theorem 4 can be strengthened to develop a polynomial reconfiguration kernel. In fact, we use the ideas in Theorem 4 to adapt a special kernel that retains all minimal k-hitting sets in the reduced instances [16].

Theorem 6. BOUNDED HITTING SET RECONFIGURATION *parameterized by* k *has a polynomial reconfiguration kernel.*

Proof. We let (G, S, T, k, ℓ) be an instance of BOUNDED HITTING SET RECONFIGURATION: G is a family of sets of vertices of size at most r and each of S and T is a hitting set of size at most k, that is, a set of vertices intersecting each set in G. We form a reconfiguration kernel using the reduction algorithm $\mathcal{A}$ of Damaschke and Molokov [16]: $G' = \mathcal{A}(G)$ contains all minimal hitting set solutions of size at most k, and is of size at most $(r-1)k^r + k$.

$V(G')$ includes all minimal k-hitting sets, and the k-hitting sets for G' are actually those k-hitting sets for G that are completely included in $V(G')$. Therefore, as in the proof of Theorem 4, (G, S, T, k, ℓ) is a yes-instance for BOUNDED HITTING SET RECONFIGURATION if and only if one of the $\mathcal{E}+1$ reduced instances $(G', S \cap V(G'), T \cap V(G'), k - e, \ell - 2(\mathcal{E} - e))$, for $0 \le e \le \mathcal{E}$, is a yes-instance for BOUNDED HITTING SET RECONFIGURATION.

Notice that unlike in the proof of Theorem 4, here the set containing all minimal solutions can be computed in polynomial time, whereas Theorem 4 guarantees only a fixed-parameter tractable procedure.   □

BOUNDED HITTING SET generalizes any deletion problem for a hereditary property with a finite forbidden set:

Corollary 7. *If* π *is a hereditary graph property with a finite forbidden set, then* π-DEL-RECONF(G, S, T, k, ℓ) *parameterized by* k *has a polynomial reconfiguration kernel.*

Corollary 7 does not apply to FEEDBACK VERTEX SET, for which the associated hereditary graph property is the collection of all forests; the forbidden set is the set of all cycles and hence is not finite. Indeed, Theorem 4 does not apply to FEEDBACK VERTEX SET either, since the number of minimal solutions exceeds $f(k)$ if the input graph includes a cycle of length $f(k) + 1$, for any function f. While it may be possible to adapt the compact enumeration of minimal feedback vertex sets [21] for reconfiguration, we develop a reconfiguration kernel for feedback vertex set by modifying a specific kernel for the problem.

We are given an undirected graph and two feedback vertex sets S and T of size at most k. We make use of Bodlaender's cubic kernel for FEEDBACK VERTEX SET [15], modifying reduction rules (shown in italics in the rules below) to allow the reconfiguration sequence to use non-minimal solutions, and to take into account the roles of C, S_D, T_A, and O. In some cases we remove vertices from O only, as others may be needed in a reconfiguration sequence.

The reduction may introduce multiple edges, forming a multigraph. Bodlaender specifies that a double edge between vertices u and v consists of two edges with u and v as endpoints. Since we preserve certain degree-two vertices, we extend the notion by saying that there is a *double edge* between u and v if either there are two edges with u and v as endpoints, one edge between u and v and one path from u to v in which each internal vertex is of degree two, or two paths (necessarily sharing only u and v) from u to v in which each internal vertex is of degree two. Following Bodlaender, we define two sets of vertices, a feedback vertex set A of size at most $2k$ and the set B containing each vertex with a double edge to at least one vertex in A. A *piece* is a connected component of $G[V \setminus (A \cup B)]$, the *border* of a piece with vertex set X is the set of vertices in $A \cup B$ adjacent to any vertex in X, and a vertex v in the border *governs* a piece if there is a double edge between v and each other vertex in the border. We introduce $\mathcal{E}$ to denote how much capacity we can "free up" for use in the reduced instance by removing vertices and then readding them.

Bodlaender's algorithm makes use of a repeated initialization phase in which an approximate solution A is found and B is initialized; for our purposes, we set $A = C \cup S_D \cup T_A$ in the first round and thereafter remove vertices as dictated by the application of reduction rules. Although not strictly necessary, we preserve this idea in order to be able to apply Bodlaender's counting arguments. In the following rules, v, w, and x are vertices.

Rule 1. If v has degree 0, remove v from G. *If v is in $S_D \cup T_A$, subtract 1 from ℓ. If v is in C, increment $\mathcal{E}$ by 1.*

Rule 2. If v has degree 1, remove v and its incident edge from G. *If v is in $S_D \cup T_A$, subtract 1 from ℓ. If v is in C, increment $\mathcal{E}$ by 1.*

Rule 3. If there are three or more edges $\{v, w\}$, remove all but two.

Rule 4. If v has degree 2 *and v is in O*, remove v and its incident edges from G and add an edge between its neighbours w and x; add w (respectively, x) to B if a double edge is formed, w (respectively, x) is not in $A \cup B$, and x (respectively, w) is in A.

Rule 5. If v has a self-loop, remove v and all incident edges and decrease k by 1, then restart the initialization phase.

Rule 6. If there are at least $k + 2$ vertex-disjoint paths between $v \in A$ and any w and there is no double edge between v and w, add two edges between v and w, and if $w \notin A \cup B$, add w to B.

Rule 7. If for $v \in A$ there exist at least $k + 1$ cycles such that each pair of cycles has exactly $\{v\}$ as the intersection, remove v and all incident edges and decrease k by 1, then restart the initialization phase.

Rule 8. If v has at least $k + 1$ neighbours with double edges, remove v and all incident edges and decrease k by 1, then restart the initialization phase.

Rule 9. If $v \in A \cup B$ governs a piece with vertex set X and has exactly one neighbour w in X, then remove the edge $\{v, w\}$.

Rule 10. If $v \in A \cup B$ governs a piece with vertex set X and has at least two neighbours in X, then remove v and all incident edges and decrease k by 1, then restart the initialization phase. *Replaced by the following rule: If a piece with vertex set X has a border set Y such that there is a double edge between each pair of vertices in Y, remove X.*

Lemma 8. *The instance (G, S, T, k, ℓ) is a yes-instance if and only if one of the $\mathcal{E} + 1$ reduced instances $(G', S', T', k - e, \ell - 2(\mathcal{E} - e))$, for $0 \le e \le \mathcal{E}$, is a yes-instance.*

Proof. We show that no modification of a reduction rule removes possible reconfiguration sequences. This is trivially true for Rules 3 and 6.

The vertices removed by Rules 1, 2, and 4 play different roles in converting a reconfiguration sequence for a reduced instance to a reconfiguration sequence for the original instance. As there is no cycle that can be destroyed only by a vertex removed from O by Rule 1, 2, or 4, none of these vertices are needed. To account for the required removal (addition) of each such vertex in S_D (T_A), we remove all d such vertices and decrease ℓ by d. We can choose to leave a $v \in C_M$ in each solution in the sequence (with no impact on ℓ) or to remove and then readd v to free up extra capacity, at a cost of incrementing ℓ by two; in the reduced instance we thus remove v and either decrement k or subtract two from ℓ. Since this choice can be made for each of these vertices, $\mathcal{E}$ in total, we try to solve any of $\mathcal{E} + 1$ versions $(G', S', T', k - e, \ell - 2(\mathcal{E} - e))$ for $0 \le e \le \mathcal{E}$.

For each of Rules 5, 7, and 8, we show that the removed vertex v is in C_F; since the cycles formed by v must be handled by each solution in the sequence, the instance can be reduced by removing v and decrementing k. For Rule 5, $v \in C_F$ since every feedback arc set must contain v. For Rules 7 and 8, $v \in C_F$, since any feedback vertex set not containing v would have to contain at least $k + 1$ vertices, one for each cycle.

For Rule 9, Bodlaender's Lemma 8 shows that the removed edge has no impact on feedback vertex sets.

For Rule 10, we first assume that Rule 9 has been exhaustively applied, and thus each vertex in the border has two edges to X. By Fact 1 for π the set of acyclic graphs, there cannot be a cycle in $G[O \cup \{v\}]$ for any $v \in S_D \cup T_A \cup O$, and hence each member of the border is in C. Lemma 9 in Bodlaender's paper shows that there is a minimum size feedback vertex set containing v: even if all the neighbours of v in the border are included in a feedback vertex set, at least one more vertex is required to break the cycle formed by v and X. There is no gain in capacity possible by replacing v in the reconfiguration sequence, and hence this particular piece is of no value in finding a solution. □

We first present the key points and lemmas in Bodlaender's counting argument and then show that, with minor modifications, the same argument goes through for our modified reduction rules and altered definition of *double edge*. In Bodlaender's proof, the size of the reduced instance is bounded by bounding the sizes of A and B (Lemma 10), bounding the number of pieces (Lemma 12), and bounding the size of each piece. Crucial to the proof of Lemma 12 is Lemma 11, as the counting associates each piece with a pair of vertices in its border that are not connected by a double edge and then counts the number of pieces associated with each different type of pair. We use Lemma 9 in the discussion below.

Lemma 9. *[15] Suppose $v \in A \cup B$ governs a piece with vertex set X. Suppose there are at least two edges with one endpoints v and one endpoint in X. Then there is a minimum size feedback vertex set in G that contains v.*

Lemma 10. *[15] In a reduced instance, there are at most $2k$ vertices in A and at most $2k^2$ vertices in B.*

Lemma 11. *[15] Suppose none of the Rules 1–10 can be applied to G. Suppose $Y \subseteq V$ is the border of a piece in G. Then there are two disjoint vertices $v, w \in Y$ such that $\{v, w\}$ is not a double edge.*

Lemma 12. *[15] Suppose we have a reduced instance. There are at most $8k^3 + 9k^2 + k$ pieces.*

Lemma 13. *Each reduced instance has $O(k^3)$ vertices and $O(k^3)$ edges, and can be obtained in polynomial time.*

Proof. Our modifications to Rules 1–3 and 5–9 do not have an impact on the size of the kernel. Although our Rule 4 preserves some vertices in A of degree two, due to the initialization of A to be $C \cup S_D \cup T_A$, and hence of size at most $2k$, the bound on B and hence Lemma 10 follows from Rule 8. In essence, our extended definition of double edges handles the degree-two vertices that in Bodlaender's constructions would have been replaced by an edge.

To claim the result of Lemma 12, it suffices to show that Lemma 11 holds for our modified rules. Bodlaender shows that if there is a piece such that each pair of vertices in the border set is connected by a double edge, Rule 10 along with Rule 9 can be applied repeatedly to remove vertices from the border of the piece and thereafter Rules 2 and 1 to remove the piece entirely.

To justify Rule 10, Bodlaender shows in Lemma 9 that if $v \in A \cup B$ governs a piece with vertex set X and there are at least two edges between v and X, then there is a minimum size feedback vertex set in G that contains v. For our purposes, however, since there may be non-minimum size feedback vertex sets used in the reconfiguration sequence, we wish to retain v rather than removing it. Our modification to Rule 10 allows us to retain v, handling all the removals from the piece without changing the border, and thus establishing Lemma 11, as needed to prove Lemma 12. In counting the sizes of pieces, our modifications result in extra degree-two vertices. Rule 4 removes all degree-two vertices in O, and hence the number of extra vertices is at most $2k$, having no effect on the asymptotic count. □

Theorem 14. FEEDBACK VERTEX SET RECONFIGURATION *and the search variant parameterized by k are in* FPT.

Proof. Since the number of reduced instances is $\mathcal{E} + 1 \leq |C| + 1 \leq k + 1$, as a consequence of Lemmas 8 and 13, we have a reconfiguration kernel, proving the first result. For the search version, we observe that we can generate the reconfiguration graph of the reduced yes-instance and use it to extract a reconfiguration sequence. We demonstrate that we can form a reconfiguration sequence for (G, S, T, k, ℓ) from the reconfiguration sequence σ for the reduced yes-instance $(G', S', t', k - e, \ell - 2(\mathcal{E} - e))$. We choose an arbitrary partition of the vertices removed from G by Rules 1 and 2 into two sets, K (the ones to keep) of size e and M (the ones to modify) of size $\mathcal{E} - e$. We can modify σ into a sequence σ' in which all vertices in K are added to each set; clearly no set will have size greater than k. Our reconfiguration sequence then consists of $\mathcal{E} - e$ steps each deleting an element of M, the sequence σ', and $\mathcal{E} - e$ steps each adding an element of M, for a length of at most $(\mathcal{E} - e) + (\ell - (\mathcal{E} - e)) + (\mathcal{E} - e) \leq \ell$, as needed. □

4   Hardness Results

The reductions presented in this section make use of the forbidden set characterization of heredity properties. A π-*critical graph* H is a (minimal) graph in the forbidden set $\mathcal{F}_\pi$ that has at least two vertices; we use the fact that $H \notin \pi$, but the deletion of any vertex from H results in a graph in π. For convenience, we will refer to two of the vertices in a π-critical graph as *terminals* and the rest as *internal vertices*. We construct graphs from multiple copies of H. For a positive integer c, we let H_c^* be the ("star") graph obtained from each of c copies H_i of H by identifying an arbitrary terminal v_i, $1 \leq i \leq c$, from each H_i; in H_c^* vertices v_1 through v_c are replaced with a vertex w, the *gluing vertex of* v_1 *to* v_c, to form a graph with vertex set $\cup_{1 \leq i \leq c}(V(H_i) \setminus \{v_i\}) \cup \{w\}$ and edge set $\cup_{1 \leq i \leq c}\{\{u, v\} \in E(H_i) \mid v_i \notin \{u, v\}\} \cup \cup_{1 \leq i \leq c}\{\{u, w\} \mid \{u, v_i\} \in E(H_i)\}$. A terminal is *non-identified* if it is not used in forming a gluing vertex. In Figure 1, H is a K_3 with terminals marked black and gray; H_4^* is formed by identifying all the gray terminals to form w.

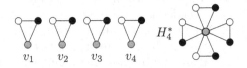

Fig. 1. An example H_c^*

Theorem 15. *Let π be any hereditary property satisfying the following:*

- *For any two graphs G_1 and G_2 in π, their disjoint union is in π.*
- *There exists an $H \in \mathcal{F}_\pi$ such that if H_c^* is the graph obtained from identifying a terminal from each of c copies of H, then $R = H_c^*[V(H_c^*) \setminus \{u_1 \ldots u_c\}]$ is in π, where $u_1 \ldots u_c$ are the non-identified terminals in the c copies of H.*

Then each of the following is at least as hard as π-SUBSET(G, k):

1. π-DEL-RECONF(G, S, T, k, ℓ) *parameterized by ℓ, and*
2. π-SUB-RECONF(G, S, T, k, ℓ) *parameterized by $k + \ell$.*

Proof. Given an instance of π-SUBSET(G, k) and a π-critical graph H satisfying the hypothesis of the lemma, we form an instance of π-DEL-RECONF$(G', S, T, |V(G)| + k, 4k)$, with G', S, and T defined below. The graph G' is the disjoint union of G and a graph W formed from k^2 copies of H, where $H_{i,j}$ has terminals $\ell_{i,j}$ and $r_{i,j}$. We let a_i, $1 \leq i \leq k$, be the gluing vertex of $\ell_{i,1}$ through $\ell_{i,k}$, and let b_j, $1 \leq j \leq k$, be the gluing vertex of $r_{1,j}$ through $r_{k,j}$, so that there is a copy of H joining each a_i and b_j. We let $A = \{a_i \mid 1 \leq i \leq k\}$, $B = \{b_j \mid 1 \leq j \leq k\}$, $S = V(G) \cup A$, and $T = V(G) \cup B$. Clearly $|V(G')| = |V(G)| + 2k + k^2(|V(H)| - 2)$ and $|S| = |T| = |V(G)| + k$. Moreover, each of $V(G') \setminus S$ and $V(G') \setminus T$ induce a graph in π, as each consists of k disjoint copies of H_k^* with one of the terminals removed from each H in H_k^*.

Suppose the instance of π-DEL-RECONF$(G', S, T, |V(G)| + k, 4k)$ is a yes-instance. As there is a copy of H joining each vertex of A to each vertex of B, before deleting $a \in A$ from S the reconfiguration sequence must add all of B to ensure that the complement of each intermediate set induces a graph in π. Otherwise, the complement will contain at least one copy of H as a subgraph and is therefore not in π. The capacity bound of $|V(G)| + k$ implies that the reconfiguration sequence must have deleted from S a subset $S' \subseteq V(G)$ of size at least k such that $V(G') \setminus (S \setminus S') = S' \cup B$ induces a subgraph in π. Thus, $G[S'] \in \pi$, and hence π-SUBSET(G, k) is a yes-instance.

Conversely if the instance of π-SUBSET(G, k) is a yes-instance, then there exists $V' \subseteq V(G)$ such that $|V'| = k$ and $G[V'] \in \pi$. We form a reconfiguration sequence between S and T by first deleting all vertices in V' from S to yield a set of size $|V(G)|$. $G'[V(G') \setminus (S \setminus V')]$ consists of the union of $G'[V'(G) \setminus S]$ and $G'[V'] = G[V']$, both of which are in π. Next we add one by one all vertices of B, then delete one by one all vertices of A and then add back one by one each vertex in the set V' resulting in a reconfiguration sequence of length $k + k + k + k = 4k$. It is clear that in every step, the complement of the set induces a graph in π.

Thus we have showed that π-SUBSET(G, k) is a yes-instance if and only if there is a path of length at most $4k$ between S and T in $R_{\mathrm{DEL}}^{\pi}(G', |V(G)| + k)$. Since $|V(G')| - (|V(G)| + k) = k + k^2(|V(H)| - 2))$, this implies that π-SUBSET(G, k) is a yes-instance if and only if there is a path of length at most $4k$ between $V(G') \setminus S$ and $V(G') \setminus T$ in $R_{\mathrm{SUB}}^{\pi}(G', k + k^2(|V(H)| - 2))$. Therefore, π-SUB-RECONF(G, S, T, k, ℓ) parameterized by $k + \ell$ is at least as hard as π-SUBSET(G, k), proving the second part.   □

It is easy to see that for π the collection of all edgeless graphs, or all forests, or all bipartite graphs, the hypothesis of Theorem 15 is satisfied. Since the π-SUBSET(G, k) problem is $W[1]$-hard for these properties [22], it follows that:

Corollary 16. * VERTEX COVER RECONFIGURATION, FEEDBACK VERTEX SET RECONFIGURATION, *and* ODD CYCLE TRANSVERSAL RECONFIGURATION

parameterized by ℓ are all $W[1]$-hard and INDEPENDENT SET RECONFIGURA-
TION, FOREST RECONFIGURATION, *and* BIPARTITE SUBGRAPH RECONFIGU-
RATION *parameterized by $k + \ell$ are all $W[1]$-hard.*

We obtain further results for properties not covered by Theorem 15. Lemma 17
handles the collection of all cliques, which does not satisfy the first condition of the theorem and the collection of all *cluster graphs* (disjoint unions of
cliques), which satisfies the first condition but not the second. Moreover, as π-
SUBSET(G, k) is in *FPT* for π the collection of all cluster graphs [22], Theorem 15
provides no lower bounds.

Lemma 17. * CLIQUE RECONFIGURATION *and* CLUSTER SUBGRAPH RECON-
FIGURATION *parameterized by $k + \ell$ are $W[1]$-hard.*

As neither DOMINATING SET nor its parametric dual is a hereditary graph
property, Theorem 15 is inapplicable; we instead use a construction specific to
the problem in Lemma 18.

Lemma 18. * DOMINATING SET RECONFIGURATION *and* (UNBOUNDED) HIT-
TING SET RECONFIGURATION *parameterized by $k + \ell$ are $W[2]$-hard.*

5   Conclusions and Directions for Further Work

Our results constitute the first study of the parameterized complexity of reconfiguration problems. We give a general paradigm, the reconfiguration kernel, for proving fixed-parameter tractability, and provide hardness reductions
that apply to problems associated with hereditary graph properties. Our result
on cluster graphs (Lemma 17) demonstrates the existence of a problem that
is fixed-parameter tractable [22], but whose reconfiguration version is W-hard
when parameterized by k. It remains open whether there exists an NP-hard
problem for which the reconfiguration version is in *FPT* if parameterized by ℓ.

Our *FPT* algorithms for reconfiguration of BOUNDED HITTING SET and
FEEDBACK VERTEX SET have running times of $O^*(2^{O(k \lg k)})$. Further work is
needed to determine whether the running times can be improved to $O^*(2^{O(k)})$,
or whether these bounds are tight under the *Exponential Time Hypothesis*.

We observe connections to another well-studied paradigm, local search [23],
where the aim is to find an *improved solution* at distance ℓ of a given solution
S. Not surprisingly, as in local search, the problems we study turn out to be
hard even in the parameterized setting when parameterized by ℓ. Other natural
directions to pursue (as in the study of local search) are the parameterized
complexity of reconfiguration problems in special classes of graphs and of non-
graph reconfiguration problems, as well as other parameterizations.

Acknowledgements. The second author wishes to thank Marcin Kamiński for
suggesting the examination of reconfiguration in the parameterized setting.

References

1. Bonamy, M., Bousquet, N.: Recoloring bounded treewidth graphs. In: Proc. of Latin-American Algorithms Graphs and Optimization Symposium (2013)
2. Bonsma, P.S., Cereceda, L.: Finding paths between graph colourings: PSPACE-completeness and superpolynomial distances. Theor. Comput. Sci. 410(50), 5215–5226 (2009)
3. Cereceda, L., van den Heuvel, J., Johnson, M.: Connectedness of the graph of vertex-colourings. Discrete Mathematics 308(56), 913–919 (2008)
4. Cereceda, L., van den Heuvel, J., Johnson, M.: Mixing 3-colourings in bipartite graphs. European Journal of Combinatorics 30(7), 1593–1606 (2009)
5. Cereceda, L., van den Heuvel, J., Johnson, M.: Finding paths between 3-colorings. Journal of Graph Theory 67(1), 69–82 (2011)
6. Ito, T., Kamiński, M., Demaine, E.D.: Reconfiguration of list edge-colorings in a graph. Discrete Applied Mathematics 160(15), 2199–2207 (2012)
7. Hearn, R.A., Demaine, E.D.: PSPACE-completeness of sliding-block puzzles and other problems through the nondeterministic constraint logic model of computation. Theor. Comput. Sci. 343(1-2), 72–96 (2005)
8. Ito, T., Demaine, E.D., Harvey, N.J.A., Papadimitriou, C.H., Sideri, M., Uehara, R., Uno, Y.: On the complexity of reconfiguration problems. Theor. Comput. Sci. 412(12-14), 1054–1065 (2011)
9. Gopalan, P., Kolaitis, P.G., Maneva, E.N., Papadimitriou, C.H.: The connectivity of boolean satisfiability: computational and structural dichotomies. SIAM J. Comput. 38(6), 2330–2355 (2009)
10. Bonsma, P.: The complexity of rerouting shortest paths. In: Rovan, B., Sassone, V., Widmayer, P. (eds.) MFCS 2012. LNCS, vol. 7464, pp. 222–233. Springer, Heidelberg (2012)
11. Kamiński, M., Medvedev, P., Milanič, M.: Shortest paths between shortest paths. Theor. Comput. Sci. 412(39), 5205–5210 (2011)
12. Haas, R., Seyffarth, K.: The k-Dominating Graph, arXiv:1209.5138 (2012)
13. Suzuki, A., Mouawad, A.E., Nishimura, N.: Reconfiguration of dominating sets (submitted)
14. Downey, R.G., Fellows, M.R.: Parameterized complexity. Springer, New York (1997)
15. Bodlaender, H.L.: A cubic kernel for feedback vertex set. In: Thomas, W., Weil, P. (eds.) STACS 2007. LNCS, vol. 4393, pp. 320–331. Springer, Heidelberg (2007)
16. Damaschke, P., Molokov, L.: The union of minimal hitting sets: Parameterized combinatorial bounds and counting. Journal of Discrete Algorithms 7(4), 391–401 (2009)
17. Flum, J., Grohe, M.: Parameterized complexity theory. Springer, Berlin (2006)
18. Niedermeier, R.: Invitation to fixed-parameter algorithms. Oxford University Press (2006)
19. Lewis, J.M., Yannakakis, M.: The node-deletion problem for hereditary properties is NP-complete. Journal of Computer and System Sciences 20(2), 219–230 (1980)
20. Mouawad, A.E., Nishimura, N., Raman, V., Simjour, N., Suzuki, A.: On the parameterized complexity of reconfiguration problems, arXiv:1308.2409 (2013)

21. Guo, J., Gramm, J., Hüffner, F., Niedermeier, R., Wernicke, S.: Compression-based fixed-parameter algorithms for feedback vertex set and edge bipartization. Journal of Computer and System Sciences 72(8), 1386–1396 (2006)
22. Khot, S., Raman, V.: Parameterized complexity of finding subgraphs with hereditary properties. Theor. Comput. Sci. 289(2), 997–1008 (2002)
23. Fellows, M.R., Rosamond, F.A., Fomin, F.V., Lokshtanov, D., Saurabh, S., Villanger, Y.: Local search: is brute-force avoidable? In: Proc. of the 21st International Joint Conference on Artifical Intelligence, pp. 486–491 (2009)

FPT Algorithms for Consecutive Ones Submatrix Problems

N.S. Narayanaswamy and R. Subashini

Department of Computer Science and Engineering
Indian Institute of Technology Madras, India
{swamy,rsuba}@cse.iitm.ac.in

Abstract. A binary matrix M has the Consecutive Ones Property (COP) if there exists a permutation of columns that arranges the ones consecutively in all the rows. We consider the parameterized complexity of d-COS-R (Consecutive Ones Submatrix by Row deletions) problem [8]: Given a matrix M and a positive integer d, decide whether there exists a set of at most d rows of M whose deletion results in a matrix with the COP. The closely related Interval Deletion problem has recently been shown to be FPT [5]. In this work, we describe a recursive depth-bounded search tree algorithm in which the problems at the leaf-level of the recursion tree are solved as instances of Interval Deletion. Therefore, we show that d-COS-R is fixed-parameter tractable and has the current best run-time of $O^*(10^d)$, which is associated with the Interval Deletion problem. We then consider a closely related optimization problem, called MIN-ICPIA, and prove that it is computationally equivalent to the Vertex Cover problem.

Keywords: Consecutive Ones Property, Consecutive Ones Submatrix, Parameterized complexity.

1 Introduction

A binary matrix has the *Consecutive Ones Property (COP)*, if there is a permutation of its columns that places the ones consecutively in all the rows. Testing the COP is a classical algorithmic problem and has various applications that include information retrieval [13], physical mapping of DNA [1], recognizing interval graphs, planar graphs and Hamiltonian cubic graphs [4,28]. Problems like INTEGER LINEAR PROGRAMMING and SET COVER, on instances for which the associated constraint matrices have the COP, are polynomial-time solvable [7]. After early attempts at testing for the COP by Robinson [25], the first polynomial-time algorithm was designed by Fulkerson and Gross [10]. They decomposed the given matrix into simpler matrices which have exactly two permutations that guarantee the COP or none. This decomposition into simpler matrices have been repeatedly used in the study of the COP. Using a data structure called PQ-trees, Booth and Lueker [4] came up with a linear-time algorithm to test the COP. PQ-trees exist only for matrices with the COP and it is an empty data structure if the associated matrix does not have the COP. PQR-trees are extensions of PQ-trees introduced by Meidanis, Porto and Telles

G. Gutin and S. Szeider (Eds.): IPEC 2013, LNCS 8246, pp. 295–307, 2013.

[22] to ensure that the data structure is well defined for all matrices. Further, R-nodes in PQR-trees give the forbidden submatrices for the COP. Hsu [18] introduced PC-trees, as a generalization of PQ-trees to test the COP and this algorithm also runs in linear-time. Habib, McConnell, Paul and Viennot [15] used Lex-BFS and partition refinement to test matrices with the COP. An alternate linear time solution (without using any of PQ-trees, PQR-trees, PC-trees) for testing the COP using decomposition technique was also proposed by Hsu [19]. Recently, Raffinot [24] presented an algorithm for testing the COP based on a new partitioning scheme.

Matrices with the COP are characterized based on a set of *forbidden submatrices* and this characterization is due to Tucker [27]. Efficiently finding minimum size forbidden submatrices for the COP is a vibrant line of research. Dom, Guo and Niedermeier [8] first came up with a polynomial-time algorithm for finding these forbidden submatrices. Recently, Blin, Rizzi and Vialette [2] designed a faster algorithm for finding minimum size Tucker forbidden submatrices, thereby improving the running time of Dom et al. [8]. Another fundamentally different characterization of matrices with the COP is due to McConnell [21] based on the bipartiteness of an associated incompatibility graph. An odd cycle of the incompatibility graph serves as a certificate, if the matrix does not have the COP.

Matrices with the COP have been characterized in terms of *interval assignments* to an associated set system such that the intersection cardinalities are preserved [23]. This characterization, implicitly present in the paper by Fulkerson and Gross (Theorem 2.1, [10]), is re-discovered in [23] (Theorem 3, [23]). Let $S = \{S_1, \ldots, S_n\}$ be a family of subsets of $[m]$. An *interval assignment* to the collection S is denoted by a set of ordered pairs, $\{(S_i, I_i) \mid 1 \leq i \leq n\}$, where each interval I_i is a set of consecutive integers from the set $[m]$. An interval assignment to the collection S is an *Intersection Cardinality Preserving Interval Assignment (ICPIA)*, if it satisfies the following conditions: for each i, $|S_i| = |I_i|$ and for every pair of sets S_i and S_j, $|S_i \cap S_j| = |I_i \cap I_j|$. The relevance of an ICPIA associated with a binary matrix M is as follows: for each $1 \leq i \leq n$, let $S_i = \{j \mid M_{ij} = 1\}$, and $S(M) = \{S_i \mid 1 \leq i \leq n\}$. It is easy to verify that if M has the COP then the family $S(M)$ has an ICPIA. The converse is also true, and this is a characterization of matrices with the COP [10,23].

Many approaches are taken to deal with matrices that do not have the COP: *Consecutive Ones Submatrix (COS)*, consecutive ones matrix partition, consecutive ones matrix augmentation and consecutive block minimization (SR14, SR15, SR16, SR17 [11]) are some of the known problems. COS problem considers a natural way of addressing matrices that do not have the COP: Find a minimum set of rows to be deleted such that the resulting matrix has the COP. This problem is referred to as Min-COS-R (COS by Row deletions) [8] and it is known to be NP-complete [3]. However, many variants of Min-COS-R have been looked at in the literature both from parameterized and classical complexity. To highlight the results of [8,16,26], we follow the notation of [26]. The notation (x, y)-matrix denote the matrix consisting of at most x-number of ones in each column and at most y-number of ones in each row and $x = *$ or $y = *$ denote that no bound on the number of ones in columns or rows. It is known from [16] that Min-COS-R on (2,2)-matrices is polynomial-time solvable. The closely related variants of Min-COS-R on (2,3) and (3,2)-matrices are NP-hard [26]. In the parameterized setting, Min-COS-R with the number d of rows to be deleted as the parameter is referred to as d-COS-R from now on. The parameterized complexity of d-COS-R results as established by Dom, Guo and Niedermeier [8] are listed as follows.

◊ $(*, 2)$-matrices have problem kernels with $O(d^2)$ rows and $O(d)$ columns
◊ $(2, *)$-matrices have run-time $O^*(4^d)$
◊ $(*, \Delta)$-matrices have run-time $O^*((\Delta + 1)^d . (2\Delta)^{2d})$

In the above results of d-COS-R, number of ones in the columns/rows of the matrices are bounded and are shown to be FPT. These FPT algorithms are based on a refinement of the forbidden submatrix characterization [27] of matrices with the COP. Dom et al. [8] used this characterization repeatedly to delete the forbidden submatrices, for solving d-COS-R problem. To the best of our knowledge, the parameterized complexity of d-COS-R on general binary matrices is open.

Our Results: We consider two deletion problems with respect to matrices that do not have the COP: Consecutive Ones Submatrix (COS) problem [11] and MIN-ICPIA problem (defined later) posed by us, motivated by the results of (Theorem 2.1, [10]) and (Theorem 3, [23]), which relate the COP and ICPIA. The decision version of Min-COS-R called d-COS-R on arbitrary binary matrices is defined as follows:

---
d-COS-R
Instance: $\langle M, d \rangle$ - A binary matrix $M_{n \times m}$ and an integer $d \geq 0$.
Parameter: d
Question: Does there exist a set of at most d rows of M whose deletion results in a matrix with the COP?

We show that d-COS-R admits an FPT algorithm with run-time $O^*(10^d)$. This is obtained by a recursive branching algorithm in which the instances at the leaf nodes are instances of Interval Deletion. For these instances, we employ the recent $O^*(10^d)$ algorithm for Interval Deletion [5].

Interval Deletion
Instance: $\langle G, d \rangle$ - A graph G and an integer $d \geq 0$.
Parameter: d
Question: Does G have a set V' of at most d vertices such that $G \setminus V'$ is an interval graph?

This is a significant advancement over the current knowledge on this problem, where current FPT results [8] are known only when the number of 1s in the rows or columns are bounded. It is well known that a graph is an interval graph if and only if its clique matrix (vertices versus maximal cliques incidence matrix) has the COP (see Theorem 1). An algorithm for the Interval-deletion problem would solve the d-COS-R problem, if the given matrix M can be recognized as a clique matrix of the associated graph. Our approach, which is a recursive algorithm, processes M such that matrices at the leaves of the recursion tree are clique matrices of the associated graphs. Then, at the leaf nodes we apply the interval-deletion algorithm of Cao and Marx [5] and get the overall running time of d-COS-R as $O^*(10^d)$.

We pose a natural problem MIN-ICPIA, based on the characterization of matrices with the COP in terms of set systems that have an ICPIA [10,23], and is defined as follows.

MIN-ICPIA

Instance: An interval assignment $\mathcal{F} = \{(S_i, I_i) \mid 1 \leq i \leq n\}$, where each $S_i, I_i \subseteq [m]$ satisfying $\bigcup S_i = \bigcup I_i = [m]$ and an integer $d \geq 0$.
Question: Does there exist a set $\mathcal{D} \subseteq \mathcal{F}$ with $|\mathcal{D}| \leq d$ such that $\mathcal{F} \setminus \mathcal{D}$ is an ICPIA?

We show that MIN-ICPIA is NP-complete by a polynomial time reduction from Vertex Cover. We also present a parameter preserving reduction from MIN-ICPIA to Vertex Cover. By using the current best FPT algorithm for Vertex Cover [6] we obtain an algorithm that solves MIN-ICPIA in time $O(1.2738^d + dn)$, where d is the number of ordered pairs to be deleted. Also, the 2-approximation algorithm of Vertex Cover [17] is applicable to MIN-ICPIA.

2   Preliminaries

Throughout this paper we consider only binary matrices. Recall that an $n \times m$ matrix M can be represented as a set system $(U, \mathcal{S}(M))$ where $\mathcal{S}(M) = \{S_1, \ldots, S_n\}$ being the collection of subsets of $U = \{1, \ldots, m\}$ where $S_i = \{j \mid M_{ij} = 1\}$.

Definition 1. *Let $\mathcal{F} = \{S_1, \ldots, S_n\}$, $S_i \subseteq [m]$ be a collection of sets. Let $\mathcal{F}' \subseteq \mathcal{F}$ be a subcollection such that any two sets in $\mathcal{F}'$ have a non-empty intersection. If all such pairwise intersecting subcollections $\mathcal{F}'$, contain at least one common element, then $\mathcal{F}$ is said to have the* Helly *property [9].*

For an $n \times m$ matrix M, let $\mathcal{R}(M) = \{r_1, \ldots, r_n\}$ and $\mathcal{C}(M) = \{c_1, \ldots, c_m\}$ denote the sets of rows and columns, respectively. Here, r_i and c_j denote the binary vectors corresponding to the row r_i and column c_j of M, respectively. The $(i,j)^{th}$ entry in M is denoted as M_{ij}. For a subset $\mathcal{D} \subseteq \mathcal{R}(M)$ of rows, the *submatrix* induced on $\mathcal{D}$ and $\mathcal{R}(M) \setminus \mathcal{D}$ are denoted by $M[\mathcal{D}]$ and $M \setminus \mathcal{D}$, respectively. We follow graph theoretic definitions and notations of [14,29].

Definition 2. *The derived graph* (see Golumbic [14]) *associated with a 0-1 matrix M is $G(M) = (V, E)$ is defined as $V = \{v_i \mid r_i \in \mathcal{R}(M)\}$ and $E = \{\{v_i, v_j\} \mid \exists c_k \in \mathcal{C}(M) \text{ with } M_{ik} = M_{jk} = 1\}$.*

In other words, $G(M)$ is obtained from M by visualizing each column c_k as a clique involving the vertices corresponding to rows which have a 1 entry in c_k. For a column c_k, the set of vertices in $G(M)$ is defined as $\text{vert}(c_k) = \{v_i \mid r_i \in \mathcal{R}(M) \text{ and } M_{ik} = 1\}$. It is easy to see that $G(M)$ is an intersection graph of the set system $(U = [m], \mathcal{S}(M))$ associated with M.

Definition 3. *A graph is called an* interval graph *if its vertices can be assigned intervals such that two vertices are adjacent if and only if their corresponding intervals have a nonempty intersection.*

Let G be a graph on the vertex set $\{v_1, \ldots, v_n\}$ and let $\{Q_1, \ldots, Q_l\}$ be the set of maximal cliques in G. The *clique matrix* M of G is the matrix whose rows and columns correspond to the vertices and the maximal cliques, respectively, in G. The entry $M_{ij} = 1$ if the vertex v_i is in the clique Q_j and it is 0 otherwise. The following characterization by Fulkerson and Gross [10] relates the COP and interval graphs.

Theorem 1. *(Theorem 7.1, [10]) A graph is an* interval graph *if and only if its clique matrix has the COP.*

2.1 COP, ICPIA and Clique-Matrices of Derived Graphs

In this section, we state and prove few results that are crucially used in proving the correctness of our parameterized algorithm described in Section 3. We present the following characterization of matrices with the COP [23], which is essential in proving our claims.

Theorem 2. *(Theorem 3, [23]) A binary matrix M has the COP if and only if $\mathcal{S}(M)$ has an ICPIA.*

We recall an important property of ICPIA, by generalizing the Lemma 1 in Lemma 2.

Lemma 1. *(Lemma 1, [23]) Let $(S_i, I_i), (S_j, I_j), (S_k, I_k)$ be elements of an ICPIA. Then, $|S_i \cap S_j \cap S_k| = |I_i \cap I_j \cap I_k|$.*

Lemma 2. *Let $\mathcal{F} = \{(S_i, I_i) \mid 1 \leq i \leq n\}$ be an ICPIA for $\mathcal{S}(M)$. Then, for any $1 \leq r \leq n$ and for any collection of ordered pairs $\{(S_{i_1}, I_{i_1}), \ldots, (S_{i_r}, I_{i_r})\} \subseteq \mathcal{F}$, $|\bigcap_{j=1}^{r} S_{i_j}| = |\bigcap_{j=1}^{r} I_{i_j}|$.*

Proof. We prove this by induction on r. *Base:* For $r = 3$, the claim is true for any three ordered pairs (by Lemma 1). *Hypothesis:* Let us assume that the claim is true for all collections of r ordered pairs, i.e for each $k \leq r$, $|\bigcap_{j=1}^{k} S_{i_j}| = |\bigcap_{j=1}^{k} I_{i_j}|$. *Induction step:* We now prove the claim for a collection of $r+1$ ordered pairs. For a collection of k ordered pairs, $k \leq r$, the claim is true by the induction hypothesis. For $k = r+1$, we will prove the claim as follows: let $\bigcap_{j=1}^{r+1} S_{i_j} = S_{i_1} \cap (\bigcap_{j=2}^{r} S_{i_j}) \cap S_{i_{r+1}}$ and $\bigcap_{j=1}^{r+1} I_{i_j} = I_{i_1} \cap (\bigcap_{j=2}^{r} I_{i_j}) \cap I_{i_{r+1}}$. Define $S^* = \bigcap_{j=2}^{r} S_{i_j}$ and $I^* = \bigcap_{j=2}^{r} I_{i_j}$. From the induction hypothesis, it is clear that for any k ordered pairs, where $k \leq r$, $|\bigcap_{j=1}^{k} S_{i_j}| = |\bigcap_{j=1}^{k} I_{i_j}|$. Therefore $|S^*| = |I^*|$. Further, it follows that $|S_{i_1} \cap S_{i_{r+1}}| = |I_{i_1} \cap I_{i_{r+1}}|$, $|S_{i_1} \cap S^*| = |I_{i_1} \cap I^*|$, $|S_{i_{r+1}} \cap S^*| = |I_{i_{r+1}} \cap I^*|$ i.e the interval assignment $\{(S_{i_1}, I_{i_1}), (S^*, I^*), (S_{i_{r+1}}, I_{i_{r+1}})\}$ is an ICPIA. Thus, by applying Lemma 1 to the ICPIA $\{(S_{i_1}, I_{i_1}), (S^*, I^*), (S_{i_{r+1}}, I_{i_{r+1}})\}$, we obtain $|S_{i_1} \cap S^* \cap S_{i_{r+1}}| = |I_{i_1} \cap I^* \cap I_{i_{r+1}}|$ and this is equivalent to $|\bigcap_{j=1}^{r+1} S_{i_j}| = |\bigcap_{j=1}^{r+1} I_{i_j}|$. Hence, the lemma is proved. $\square$

We use the following lemma in proving our structural observation in Lemma 4.

Lemma 3. *If M has the COP then $\mathcal{S}(M)$ satisfies the Helly Property.*

Proof. Since M has the COP, let M' be the column permuted matrix obtained from M which has consecutive ones in the rows. For each $1 \leq i \leq n$, let I_i be the interval of column indices corresponding to the ones in row r_i of M'. We know that $\{(S_i, I_i) \mid 1 \leq i \leq n\}$ is an ICPIA for $\mathcal{S}(M)$ (from Theorem 2). From Lemma 2 and the fact that the set of intervals $\{I_1, \ldots, I_n\}$ satisfies the Helly Property, it follows that $\mathcal{S}(M)$ satisfies the Helly Property. □

Augmented Matrix $\tilde{M}$ and its Derived Graph $G(\tilde{M})$: For the given matrix M, the augmented matrix $\tilde{M}$ of order $(m+n) \times m$, is defined as $\left(\begin{smallmatrix} I \\ M \end{smallmatrix}\right)$ where I is the $m \times m$ identity matrix. The main reason for considering $\tilde{M}$ is that in $G(\tilde{M})$, each column corresponds to a maximal clique. This may not necessarily be the case in $G(M)$. We first observe that M and $\tilde{M}$ behave the same with respect to the COP, and the proof of this observation is very easy and is based on the fact that $\tilde{M}$ is obtained from M by *padding* with an identity matrix.

Observation 1. *$\tilde{M}$ has the COP if and only if M has the COP.*

Corollary 1. *Let $\mathcal{D} \subseteq \mathcal{R}(M)$. Then, $M \setminus \mathcal{D}$ has the COP if and only if $\tilde{M} \setminus \mathcal{D}$ has the COP.*

Lemma 4. *If M has the COP, then $G(M)$ is an interval graph. Further, for every maximal clique Q in $G(M)$ there exists a column c_k in M such that $\text{vert}(c_k) = Q$.*

Proof. Consider the columns of M in the order of a permutation σ that results in the COP. Now, for every vertex v_i in $G(M)$ assign the interval $I_i = [j, k]$ where j and k are the minimum and maximum column indices, respectively, with $M_{ij} = M_{ik} = 1$. Now, $\{v_i, v_j\}$ is an edge in $G(M)$ if and only if there is a column r in which $M_{ir} = M_{jr} = 1$, and this happens if and only if r is in both I_i and I_j. Clearly, this is an interval representation of $G(M)$, and therefore, $G(M)$ is an interval graph. Moreover, from Theorem 2 $\{(S_i, I_i) \mid 1 \leq i \leq n\}$ is an ICPIA.

We now prove the second part of the lemma. Let $Q = \{v_1, \ldots, v_q\}$ be a maximal clique in $G(M)$. Let $r_1, \ldots, r_q$ be the rows in M corresponding to the vertices in Q, and let $S_1, \ldots, S_q$ be the sets corresponding to the rows $r_1, \ldots, r_q$, respectively. Since Q is a clique, any two sets in $S_1, \ldots, S_q$ have a non-empty intersection. Consequently, using the fact that M has the COP and by Lemma 3, it follows that $|\bigcap_{i=1}^{q} S_i| > 0$. Let k be an element in $\bigcap_{i=1}^{q} S_i$, then it follows that vertices of Q are in $\text{vert}(c_k)$. Now $\text{vert}(c_k) = Q$, since Q is a maximal clique. Hence the lemma is proved. □

Based on the above results, we prove the following theorem which forms a crucial step in the parameterized algorithm for d-COS-R.

Theorem 3. *If M has the COP, then $G(\tilde{M})$ is an interval graph, and $\tilde{M}$ is the clique matrix of $G(\tilde{M})$.*

Proof. From Observation 1, M has the COP implies that $\widetilde{M}$ has the COP. From Lemma 4 it follows that $G(\widetilde{M})$ is an interval graph, and each maximal clique corresponds to a column in $\widetilde{M}$. In $\widetilde{M}$, each column c_j has a *distinguishing* entry $\widetilde{M}_{ij}$ where there is a 1, and all other entries $\widetilde{M}_{ik}$ with $k \neq j$ in that row r_i are zero. This shows that v_i is adjacent only to vertices $\text{vert}(c_j) \setminus \{v_i\}$ in $G(\widetilde{M})$. In other words, each column corresponds to a maximal clique in $G(\widetilde{M})$. Therefore, $\widetilde{M}$ is the clique matrix of $G(\widetilde{M})$.   □

3   An FPT Algorithm for d-COS-R via Interval Deletion

In this section, we present a parameterized algorithm for solving d-COS-R. The recursive algorithm *COS-R* is called with a binary matrix M, the initial solution set $\mathcal{D} = \emptyset$ and the parameter d as inputs. The basic idea in this algorithm is that we transform the given instance $\langle M, d \rangle$ of d-COS-R into at most 3^d instances $\langle M', d' \rangle$ where M' has the additional property that $\widetilde{M}'$ is the clique matrix of $G(\widetilde{M}')$. This is achieved by applying a branching rule referred to as M_{I_1}-Hitting rule (described below) at each internal node in the depth bounded search tree. This rule is based on Tucker's forbidden submatrix [27] $M_{I_1} = \begin{pmatrix} 1 & 1 & 0 \\ 0 & 1 & 1 \\ 1 & 0 & 1 \end{pmatrix}$. At the leaves of the recursion tree, *COS-R* calls the function Interval-Deletion$(G(\widetilde{M}'), d')$, which determines the existence of a set X of at most d' vertices such that $G(\widetilde{M}') \setminus X$ is an interval graph and returns such a set, if one exists. In $O^*(10^d)$ time, *COS-R(M, $\emptyset$, d)* either returns a set $\mathcal{D}$ of at most d rows such that $M \setminus \mathcal{D}$ has the COP or returns 'NO'. Algorithm *COS-R* is described as follows:

Algorithm COS-R$(M, \mathcal{D}, d)$
Input: An instance $\langle M_{n \times m}, d \rangle$ where M is a binary matrix and $d \geq 0$.
Output: Return a set $\mathcal{D}$ of at most d rows (if one exists) such that $M \setminus \mathcal{D}$ has the COP.
(Step 0) If M has the COP and $d \geq 0$ then Return $\mathcal{D}$.
(Step 1) If $d < 0$ then Return 'NO'/* parameter budget exhausted */
(Step 2) (M_{I_1}-Hitting Rule) If there exists three rows $\{r_1, r_2, r_3\} \subseteq \mathcal{R}(M)$ such that $M[\{r_1, r_2, r_3\}]$ contains the matrix $M_{I_1} = \begin{pmatrix} 1 & 1 & 0 \\ 0 & 1 & 1 \\ 1 & 0 & 1 \end{pmatrix}$, then branch into 3 instances $\mathcal{I}_i = \langle M_i, d_i \rangle$, where $i \in \{1, 2, 3\}$
     Set $\mathcal{D}_i \leftarrow \mathcal{D} \cup \{r_i\}$ and $M_i \leftarrow M \setminus \{r_i\}$
     Update $d_i \leftarrow d - 1$ /* Parameter drops by 1 */
For some $i \in \{1, 2, 3\}$, if COS-R$(M_i, \mathcal{D}_i, d_i)$ returns a solution $\mathcal{D}_i$, then Return $\mathcal{D}_i$, else Return 'NO'
/* INVARIANT: See Lemma 6 */
(Step 3) (Interval Deletion) V'=Interval-Deletion$(G(\widetilde{M}), d)$.
(Step 4) If Interval-Deletion returns 'NO' then Return 'NO'. Otherwise, Return the set $\mathcal{D} = \mathcal{D} \cup \{r_i \in \mathcal{R}(M) \mid v_i \in V'\}$.

We now prove the correctness of M_{I_1}-Hitting rule in the following Lemma.

Lemma 5. *Let* M *be a matrix for which* M_{I_1}*-Hitting rule applies, and* $M[\{r_1, r_2, r_3\}]$ *contains the forbidden matrix* $M_{I_1} = \begin{pmatrix} 1 & 1 & 0 \\ 0 & 1 & 1 \\ 1 & 0 & 1 \end{pmatrix}$, *where* $\{r_1, r_2, r_3\} \subseteq R(M)$. *Then, any solution* D *of* d*-COS-R includes at least one of the rows* r_1, r_2, r_3.

Proof. We prove this by contradiction. Suppose, there exists a solution D that contains none of r_1, r_2 and r_3. Let $M' = M \setminus D$ be the matrix with the COP. This implies that $M[\{r_1, r_2, r_3\}]$ in M' satisfies the COP. This is a contradiction to the fact that $M[\{r_1, r_2, r_3\}]$ contains the forbidden matrix M_{I_1}.   □

The consequence of M_{I_1}-Hitting rule is as follows.

Lemma 6. *Let* M *be a matrix on which* M_{I_1}*-Hitting rule is not applicable. Then, for every maximal clique* Q *in* $G(M)$, *there exists a column* c_k *such that* $Q = \text{vert}(c_k)$. *Further,* $\widetilde{M}$ *is the clique matrix of* $G(\widetilde{M})$.

Proof. Assume on the contrary that Q is a maximal clique in $G(M)$ such that there is no column c_l satisfying the property $Q \subseteq \text{vert}(c_l)$. Further, let Q' be a minimal subset of Q with this property. Since Q' is a minimal subset with this property, then for every strict subset Q'' of Q' there exists a column c_k such that $Q'' \subseteq \text{vert}(c_k)$. Now, Q' is a clique of size at least 3, because for every edge $\{v_i, v_j\}$ in $G(M)$ there exists a column c_k such that $\{v_i, v_j\} \subseteq \text{vert}(c_k)$. Let v_1, v_2, v_3 be 3 vertices in Q'. Let c_1, c_2, c_3 be three distinct columns in M such that $Q' \setminus \{v_1\} \subseteq \text{vert}(c_1)$, $Q' \setminus \{v_2\} \subseteq \text{vert}(c_2)$ and $Q' \setminus \{v_3\} \subseteq \text{vert}(c_3)$, respectively. Let r_1, r_2, r_3 be the rows corresponding to the vertices v_1, v_2, v_3, respectively. From this, it follows that the submatrix formed by rows r_1, r_2, r_3, and columns c_1, c_2, c_3 is a column permutation of the forbidden submatrix $M_{I_1} = \begin{pmatrix} 0 & 1 & 1 \\ 1 & 0 & 1 \\ 1 & 1 & 0 \end{pmatrix}$. Therefore, the matrix M has M_{I_1} as a submatrix, and this contradicts the hypothesis that M_{I_1}-Hitting rule is not applicable. Therefore, our assumption is wrong. This finishes the first claim in the lemma. The example shown in Fig.1. illustrates the above argument.

We prove the second part of the lemma as follows. For every column $c_k, 1 \le k \le m$ in M whose vertices $\text{vert}(c_k)$ is not a maximal clique in $G(M)$, becomes a maximal clique in $G(\widetilde{M})$. This is because $G(\widetilde{M})$ can be viewed as a graph obtained from $G(M)$ by adding a new vertex u_k for each clique $Q_k = \text{vert}(c_k), 1 \le k \le m$, and making u_k adjacent to all the vertices in Q_k. Further, if Q_k is a maximal clique in $G(M)$, then in $G(\widetilde{M})$, Q_k is a maximal clique with one additional vertex. This completes the proof that $\widetilde{M}$ is the clique matrix of $G(\widetilde{M})$.   □

Algorithm *COS-R* explores a recursion tree in which the leaves fall into one of the following cases: (1) M_{I_1}-Hitting rule is not applicable further and the number of rows deleted is less than d (2) Number of rows deleted by M_{I_1}-Hitting rule is more than d. An instance that falls into the first case is a *preprocessed instance*. In the second case, since the parameter is exhausted we abort this branch. Each node in the recursion tree has at most 3 subproblems, and therefore, the tree has at most 3^d leaves. At a leaf, let M' be the resultant matrix obtained after the application of rule and d' is the remaining parameter. Thus, at this leaf, an

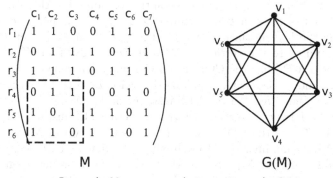

M   G(M)

Row r_i in M corresponds to vertex v_i in G(M)

Fig. 1. For the maximal clique $Q = \{v_1, \ldots, v_6\}$ in G(M), there is no column c_1 in M satisfying the property $Q \subseteq \mathrm{vert}(c_1)$. Also, $Q' = \{v_4, v_5, v_6\}$ is a minimal subset of Q with the same property. The submatrix induced by the rows corresponding to the vertices of Q' i.e $M[\{r_4, r_5, r_6\}]$ contains M_{I_1} (shown in dotted lines).

instance $\langle M, d \rangle$ of d-COS-R is transformed to an instance $\langle G(\tilde{M}'), d' \rangle$ of the interval deletion problem, by Theorem 3. Now we show that, solving d-COS-R on $\tilde{M}$ is equivalent to solving the Interval-Deletion problem on the graph $G(\tilde{M})$.

Theorem 4. *Let $\tilde{M}$ be the clique matrix of $G(\tilde{M})$. Given $\tilde{M}$ and integer $d \geq 0$, there exists a set $\mathcal{D}$ of rows such that $|\mathcal{D}| \leq d$ and $\tilde{M} \setminus \mathcal{D}$ has the COP if and only if $G(\tilde{M})$ has a set of vertices V' such that $|V'| \leq d$ and $G(\tilde{M}) \setminus V'$ is an interval graph.*

Proof. Let $\mathcal{D}$ be a set of rows in $\tilde{M}$, and let V' be the corresponding vertices in $G(\tilde{M})$. From Lemma 4, it follows that $\tilde{M} \setminus \mathcal{D}$ has the COP implies $G(\tilde{M} \setminus \mathcal{D})$ is an interval graph. Further, $G(\tilde{M} \setminus \mathcal{D})$ is the graph obtained by removing V' from $G(\tilde{M})$. This completes the forward direction of the claim. For the reverse direction, let V' be a minimal set of vertices such that $G(\tilde{M}) \setminus V'$ is an interval graph. Due to the minimality of V', no vertex in $G(\tilde{M})$ corresponding to a row of the identity matrix padded to M is an element of V'. Let $\mathcal{D}$ be the set of rows in $\tilde{M}$ corresponding to V'. Since $\tilde{M}$ is the clique matrix of $G(\tilde{M})$, the columns of $\tilde{M} \setminus \mathcal{D}$ are exactly the maximal cliques of $G(\tilde{M}) \setminus V'$. Therefore, $\tilde{M} \setminus \mathcal{D}$ is the clique matrix of $G(\tilde{M}) \setminus V'$. Since $G(\tilde{M}) \setminus V'$ is an interval graph, it follows from Theorem 1 that $\tilde{M} \setminus \mathcal{D}$ has the COP. Hence the theorem is proved. □

The following theorem states our main result. In the theorem, let $\langle G(\tilde{M}_1), d_1 \rangle$, $\ldots, \langle G(\tilde{M}_r), d_r \rangle$ be the instances of interval deletion at the leaf nodes of the depth bounded search tree explored by Algorithm *COS-R*.

Theorem 5. $\langle M, d \rangle$ *is an YES-instance of d-COS-R if and only if $\langle G(\tilde{M}_k), d_k \rangle$, $1 \leq k \leq r$ is an YES-instance of interval deletion.*

Proof. Let $\langle G(\tilde{M}_k), d_k \rangle$ be an YES-instance of interval deletion such that $G(\tilde{M}_k) \setminus V'$ is an interval graph, where V' is a minimal set of vertices in $G(\tilde{M}_k)$

guaranteeing this property. In the recursion tree explored by Algorithm COS-R, let $\mathcal{D}'$ be the set of rows that are removed to get M_k from M. Since d_k is the parameter at the leaf instance, it follows that $|\mathcal{D}'| \le d - d_k$. Let $\mathcal{D}''$ be the rows corresponding to the vertices in V' and $|\mathcal{D}''| \le d_k$. From Theorem 4, it follows that $\widetilde{M}_k \setminus \mathcal{D}''$ has the COP. From Corollary 1 it follows that $M_k \setminus \mathcal{D}''$ has the COP, and consequently it follows that $M \setminus (\mathcal{D}' \cup \mathcal{D}'')$ has the COP and $|\mathcal{D}' \cup \mathcal{D}''| \le d$. Thus, $\langle M, d \rangle$ is an YES-instance of d-COS-R.

To prove the forward direction, let $\langle M, d \rangle$ be an YES-instance of d-COS-R such that $M \setminus \mathcal{D}$ has the COP, where $\mathcal{D}$ is a set of rows in M and $|\mathcal{D}| \le d$. Let $\mathcal{D}' \subseteq \mathcal{D}$ be a minimal set of cardinality d', such that $\mathcal{D}'$ contains a row of the forbidden matrix M_{I_1}. Since Algorithm *COS-R* branches on each row of forbidden submatrix of the form M_{I_1}, one path in the recursion tree will remove all the elements of $\mathcal{D}'$. Consequently the resulting matrix $M' = M \setminus \mathcal{D}'$ does not have an M_{I_1}, and becomes an interval deletion instance at a leaf node in the recursion tree associated with Algorithm *COS-R*. From the hypothesis that $M \setminus \mathcal{D}$ has the COP, it follows that $M' \setminus (\mathcal{D} \setminus \mathcal{D}')$ has the COP, and from Corollary 1, it is clear that $\widetilde{M}' \setminus (\mathcal{D} \setminus \mathcal{D}')$ has the COP. Further, $\widetilde{M}'$ is the clique matrix of $G(\widetilde{M}')$. Therefore, from Theorem 4 it follows that for $\langle G(\widetilde{M}'), d - d' \rangle$ there is a minimal set of vertices V' such that $|V'| \le d - d'$ and $G(\widetilde{M}') \setminus V'$ is an interval graph and $|\mathcal{D}'| + |V'| \le d$. Thus, $G(\widetilde{M}') \setminus V'$ is an YES-instance of interval deletion.   □

As a consequence of Algorithm *COS-R*$(M, \mathcal{D}, d)$, we conclude that d-COS-R is FPT in the parameterized complexity framework. Thus, by employing the recent parameterized algorithm for Interval Deletion [5] at the leaves of the search tree, we get an overall running time of $O^*(10^d)$ for d-COS-R.

3.1   Convex Bipartite Deletion is FPT

Using our algorithm for d-COS-R, we observe that the Convex Bipartite Deletion problem is FPT. Let $G = (V_1, V_2, E)$ be a bipartite graph with $V_1 = \{x_1, \ldots, x_n\}$ and $V_2 = \{y_1, \ldots, y_m\}$. Let M be the half adjacency matrix of G. That is, $M_{ij} = 1$ if and only if $\{x_i, y_j\} \in E$. G is *convex* bipartite graph iff M has the COP [7,27]. The Convex Bipartite Deletion problem is defined as follows.

Convex Bipartite Deletion
Input: A bipartite graph $G = (V_1, V_2, E)$, $|V_1| = n$, $|V_2| = m$ and $d \ge 0$
Parameter: d
Question: Does there exist a set $D \subseteq V_1$ with $|D| \le d$ such that $G[V_1 \setminus D, V_2]$ is a convex bipartite graph?

This problem is known to be NP-complete from [20]. However, *COS-R* algorithm in Section 3 can be used to solve this problem in $O^*(10^d)$ time. Here, the inputs to the algorithm are the half adjacency matrix M of G and the parameter d. The algorithm returns a set $\mathcal{D}$ of at most d rows (if one exists) such that $G[V_1 \setminus D, V_2]$ is convex bipartite where D is the subset of vertices of V_1 corresponding to $\mathcal{D}$.

4   Min-ICPIA is equivalent to Vertex Cover

We first prove that MIN-ICPIA (defined in Section 1) is NP-complete. It is easy
to see that MIN-ICPIA is in NP. To prove that it is NP-hard, we show that
Vertex Cover [11], a classical NP-hard problem, reduces to MIN-ICPIA. Let
$\langle G, d \rangle$ be an instance of Vertex Cover. Let $\{v_1, \ldots, v_n\}$ and $\{e_1, \ldots, e_m\}$ denote
the vertex and edge set, respectively, of G. We construct a MIN-ICPIA instance
$\langle U = [2m + 1], \mathcal{F} = \{(S_i, I_i) \mid 0 \leq i \leq n\}, d + 1 \rangle$ as follows. For each vertex v_i in
G, define $S_i = \{k \mid e_k \in E(G)$ and e_k incident on $v_i\}$ and $I_i = [l_i, r_i]$, where l_i
and r_i are the left and right end points, respectively and $l_i = |\bigcup_{1 \leq j \leq i-1} S_j| + 1$,
$r_i = l_i + |S_i| - 1$. Also, $S_0 = U \setminus \bigcup_{1 \leq i \leq n} S_i$, $I_0 = [2m + 1, 2m + 1]$. From the
definition, it is clear that $|S_0| \neq |I_0|$.

Theorem 6. G *has a vertex cover of size* d *if and only if the corresponding*
MIN-ICPIA *instance* $\langle U = [2m + 1], \mathcal{F} = \{(S_i, I_i) \mid 0 \leq i \leq n\}, d + 1 \rangle$ *has a*
$(d + 1)$-*sized solution.*

Proof. Let X be a vertex cover of size d in G. Let $\mathcal{D} = \{(S_i, I_i) \mid v_i \in X\}$. From
the reduction, it is clear that for any two non-adjacent distinct vertices v_i and
v_j, $S_i \cap S_j = \emptyset$. Further, for any v_i and v_j, $I_i \cap I_j = \emptyset$. For every edge $\{v_i, v_j\}$
in G, $|S_i \cap S_j| \neq |I_i \cap I_j|$. Finally, except for S_0, for all v_i, $|S_i| = |I_i|$. We now
observe that since X is a vertex cover for G it follows that $\mathcal{F} \setminus (\mathcal{D} \cup \{(S_0, I_0)\})$
is an ICPIA. This ICPIA has been obtained by removing a set of size $d + 1$.
Conversely, let $\mathcal{D} \subseteq \mathcal{F}$ be a set of size d such that $\mathcal{F} \setminus \mathcal{D}$ is an ICPIA. Then,
(S_0, I_0) must be in $\mathcal{D}$. Further, since $\mathcal{F} \setminus \mathcal{D}$ is an ICPIA, if (S_i, I_i) and (S_j, I_j)
are in $\mathcal{F} \setminus \mathcal{D}$, then $S_i \cap S_j = \emptyset$ as $I_i \cap I_j = \emptyset$. Therefore, the corresponding two
vertices are not adjacent in G. Consequently the vertices corresponding to $\mathcal{D}$
form a vertex cover of size $d - 1$. Hence the theorem is proved.   □

The following corollary shows that MIN-ICPIA restricted to special class of
inputs remains NP-Complete. It follows from the same reduction described above
to prove that MIN-ICPIA is NP-Complete and the fact that Vertex Cover on
cubic graphs is NP-Complete [12].

Corollary 2. MIN-ICPIA *is NP-Complete on instances* $\langle U = [m], \{(S_i, I_i) \mid 1 \leq i \leq n\}, d \rangle$ *such that* $|S_i| \leq 3$ *and for all* $1 \leq i < j \leq n$, $|S_i \cap S_j| \leq 1$.

4.1   Min-ICPIA is FPT w.r.t Solution Size as the Parameter

In this section, we show that MIN-ICPIA, parameterized by d - the number of
ordered pairs to delete, is FPT by reducing it to Vertex Cover. A MIN-ICPIA
instance $\langle U = [m], \{(S_i, I_i) \mid 1 \leq i \leq n\}, d \rangle$ is transformed to a Vertex Cover
instance $\langle G = (V, E), d \rangle$ as follows:

(1) Discard all the pairs (S_i, I_i) such that $|S_i| \neq |I_i|$, and reduce the parameter
d by as many pairs discarded.
(2) For each remaining pair (S_i, I_i), there is a vertex v_i.
(3) There is an edge between two vertices (v_i, v_j) if and only if the corresponding
ordered pairs (S_i, I_i) and (S_j, I_j), $1 \leq i < j \leq n$ are such that $|S_i \cap S_j| \neq |I_i \cap I_j|$.

Observation 2. *Let* $\langle \mathsf{U} = [\mathsf{m}], \mathcal{F} = \{(S_i, I_i) \mid 1 \le i \le \mathsf{n}\}, \mathsf{d}\rangle$ *be an instance of* MIN-ICPIA. *Let* $\langle \mathsf{G}, \mathsf{d} \rangle$ *be the corresponding instance of Vertex Cover obtained from the above reduction. Then,* $\mathcal{F}$ *has a* d-*sized solution if and only if* G *has* d-*sized Vertex Cover.*

Proof. Let $\mathcal{D} \subseteq \mathcal{F}$ be such that $|\mathcal{D}| = \mathsf{d}$ and $\mathcal{F} \setminus \mathcal{D}$ is an ICPIA. Let X be the set of vertices in G corresponding to $\mathcal{D}$. Clearly, $\mathsf{G}[\mathsf{V} \setminus \mathsf{X}]$ is an independent set, and therefore X is a vertex cover of size d. In the other direction, let X be a vertex cover of G of size at most d. Let $\mathcal{D} \subseteq \mathcal{F}$ be the ordered pairs corresponding to the vertices in X. Since $\mathsf{G}[\mathsf{V} \setminus \mathsf{X}]$ is an independent set, it follows from the construction that $\mathcal{F} \setminus \mathcal{D}$ is an ICPIA. Hence the Lemma is proved. $\qquad\Box$

As a consequence of the above reduction which is both parameter and approximation preserving, it follows that MIN-ICPIA is FPT. In particular, the current best parameterized and approximation results [6,17] of Vertex Cover applies to MIN-ICPIA too. As a result of the reduction from Vertex Cover to MIN-ICPIA, the inapproximability of Vertex Cover also applies to MIN-ICPIA.

Corollary 3. MIN-ICPIA *can be solved in time* $O(1.2738^{\mathsf{d}} + \mathsf{dn})$ *where* d *is the number of ordered pairs to be deleted from* MIN-ICPIA *instance.*

Corollary 4. MIN-ICPIA *has polynomial time 2-approximation algorithm.*

Acknowledgements. The second author would like to thank R. Krithika, Sajin Koroth and G. Ramakrishna for their helpful discussions. We are also very grateful to anonymous reviewers for many helpful remarks.

References

1. Atkins, J.E., Boman, E.G., Hendrickson, B.: A spectral algorithm for seriation and the consecutive ones problem. SIAM Journal on Computing 28(1), 297–310 (1988)
2. Blin, G., Rizzi, R., Vialette, S.: A faster algorithm for finding minimum tucker submatrices. Theory of Computing Systems 51(3), 270–281 (2012)
3. Booth, K.S.: PQ-tree algorithms. Ph.D. thesis, Dept. of Electrical Engineering and Computer Science, University of California, Berkeley, CA (1975)
4. Booth, K.S., Lueker, G.S.: Testing for the consecutive ones property, interval graphs, and graph planarity using PQ-tree algorithms. Journal of Computer and System Sciences 13(3), 335–379 (1976)
5. Cao, Y., Marx, D.: Interval deletion is fixed-parameter tractable. arXiv:1211.5933v2(cs.DS) (2013)
6. Chen, J., Kanj, I.A., Xia, G.: Improved upper bounds for vertex cover. Theoretical Computer Science 411(40-42), 3736–3756 (2010)
7. Dom, M.: Recognition, generation, and application of binary matrices with the consecutive-ones property. Ph.D. thesis, Institut fur Informatik, Friedrich-Schiller-Universitat (2008)
8. Dom, M., Guo, J., Niedermeier, R.: Approximation and fixed-parameter algorithms for consecutive ones submatrix problems. Journal of Computing System Sciences 76(3-4), 204–221 (2010)

9. Dourado, M.C., Protti, F., Szwarcfiter, J.L.: Computational aspects of the Helly property: a survey. Journal of the Brazilian Computer Society 12(1), 7–33 (2006)

10. Fulkerson, D.R., Gross, O.A.: Incidence matrices and interval graphs. Pacific Journal of Mathematics 15(3), 835–855 (1965)

11. Garey, M.R., Johnson, D.S.: Computers and intractability: A guide to the theory of NP-completeness. W.H. Freeman (1979)

12. Garey, M.R., Johnson, D.S., Stockmeyer, L.: Some simplified NP-complete problems. In: Proceedings of the Sixth Annual ACM Symposium on Theory of Computing, pp. 47–63 (1974)

13. Ghosh, S.P.: File organization: The consecutive retrieval property. Communications of the ACM 15(9), 802–808 (1972)

14. Golumbic, M.C.: Algorithmic graph theory and perfect graphs, 2nd edn. Annals of Discrete Mathematics, vol. 57. Elsevier B.V. (2004)

15. Habib, M., McConnell, R.M., Paul, C., Viennot, L.: Lex-BFS and partition refinement, with applications to transitive orientation, interval graph recognition and consecutive ones testing. Theoretical Computer Science 234(1-2), 59–84 (2000)

16. Hajiaghayi, M., Ganjali, Y.: A note on the consecutive ones submatrix problem. Information Processing Letters 83(3), 163–166 (2002)

17. Hochbaum, D.S.: Approximation algorithms for NP-hard problems. PWS Publishing Company (1997)

18. Hsu, W.L.: PC-trees vs. PQ-trees. In: Wang, J. (ed.) COCOON 2001. LNCS, vol. 2108, pp. 207–217. Springer, Heidelberg (2001)

19. Hsu, W.L.: A simple test for the consecutive ones property. Journal of Algorithms 43(1), 1–16 (2002)

20. Lewis, J.M., Yannakakis, M.: The node-deletion problem for hereditary properties is NP-Complete. Journal of Computer and System Sciences 20(2), 219–230 (1980)

21. McConnell, R.M.: A certifying algorithm for the consecutive-ones property. In: Proceedings of the Fifteenth Annual ACM-SIAM Symposium on Discrete Algorithms (SODA 2004), pp. 768–777 (2004)

22. Meidanis, J., Porto, O., Telles, G.P.: On the consecutive ones property. Discrete Applied Mathematics 88(1-3), 325–354 (1998)

23. Narayansaswamy, N.S., Subashini, R.: A new characterization of matrices with the consecutive ones property. Discrete Applied Mathematics 157, 3721–3727 (2009)

24. Raffinot, M.: Consecutive ones property testing: cut or swap. In: Models of Computation in Context - 7th Conference on Computability in Europe, pp. 239–249 (2011)

25. Robinson, W.S.: A method for chronologically ordering archaeological deposits. American Antiquity 16(4), 293–301 (1951)

26. Tan, J., Zhang, L.: The consecutive ones submatrix problem for sparse matrices. Algorithmica 48(3), 287–299 (2007)

27. Tucker, A.C.: A structure theorem for the consecutive ones property. Journal of Combinatorial Theory. Series B 12, 153–162 (1972)

28. Wang, R., Lau, F.C.M., Zhao, Y.C.: Hamiltonicity of regular graphs and blocks of consecutive ones in symmetric matrices. Discrete Applied Mathematics 155(17), 2312–2320 (2007)

29. West, D.B.: Introduction to graph theory, 2nd edn. Prentice Hall (2001)

Upper Bounds on Boolean-Width
with Applications to Exact Algorithms

Yuri Rabinovich[1], Jan Arne Telle[2], and Martin Vatshelle[2]

[1] Department of Computer Science, University of Haifa, Israel
[2] Department of Informatics, University of Bergen, Norway

Abstract. Boolean-width is similar to clique-width, rank-width and NLC-width in that all these graph parameters are constantly bounded on the same classes of graphs. In many classes where these parameters are not constantly bounded, boolean-width is distinguished by its much lower value, such as in permutation graphs and interval graphs where boolean-width was shown to be $O(\log n)$ [1]. Together with FPT algorithms having runtime $O^*(c^{boolw})$ for a constant c this helped explain why a variety of problems could be solved in polynomial-time on these graph classes.

In this paper we continue this line of research and establish non-trivial upper-bounds on the boolean-width and linear boolean-width of *any* graph. Again we combine these bounds with FPT algorithms having runtime $O^*(c^{boolw})$, now to give a common framework of moderately-exponential exact algorithms that beat brute-force search for several independence and domination-type problems, on general graphs.

Boolean-width is closely related to the number of maximal independent sets in bipartite graphs. Our main result breaking the triviality bound of $n/3$ for boolean-width and $n/2$ for linear boolean-width is proved by new techniques for bounding the number of maximal independent sets in bipartite graphs.

1 Introduction

Boolean-width is a recently introduced graph parameter motivated by algorithms [2]. Having small boolean-width is witnessed by a decomposition of the graph into cuts with few different unions of neighborhoods - Boolean sums of neighborhoods - across the cut. This makes the decomposition natural to guide dynamic programming algorithms to solve problems where vertex sets having the same neighborhood across a cut can be treated as equivalent. Such dynamic programming on a given decomposition of boolean-width $boolw$ will for several problems related to independence and domination have runtime $O^*(c^{boolw})$ for a small constant c [2].

Boolean-width is similar to clique-width, rank-width and NLC-width in that all these graph parameters are constantly bounded on the same classes of graphs. However, in many classes where these parameters are not constantly bounded, boolean-width is distinguished by its much lower value. For example, permutation graphs, interval graphs, convex graphs and Dilworth k graphs all have boolean-width $O(\log n)$, and the decompositions are easy to find [1]. Since $O^*(c^{O(\log n)})$ is $n^{O(1)}$ this helps explain why several problems related to independence and domination are polynomial-time solvable on these graph classes.

G. Gutin and S. Szeider (Eds.): IPEC 2013, LNCS 8246, pp. 308–320, 2013.
© Springer International Publishing Switzerland 2013

In this paper we continue this line of research, combining $O^*(c^{boolw})$ dynamic programming for independence and domination problems with new bounds on boolean-width. Rather than giving a framework for polynomial-time algorithms on restricted graph classes, our goal in this paper is a framework for moderately-exponential exact algorithms on general graphs. Our main results are non-trivial upper-bounds of $(1-c)n/3$ on the boolean-width, and $(1-c)n/2$ on the linear boolean-width, of *any* graph, for some $c > 0$ and sufficiently large values of n. This is accompanied by a polynomial-time algorithm computing a decomposition witnessing the non-trivial bound on linear boolean-width.

We combine this with dynamic programming algorithms on decompositions of linear boolean-width k that solve INDEPENDENT SET in time $O^*(2^k)$ and DOMINATING SET, INDEPENDENT DOMINATING SET and TOTAL DOMINATING SET in time $O^*(2^{2k})$. The combination gives moderately-exponential exact algorithms on general graphs solving all these problems, also weighted versions and counting versions, by a runtime beating brute-force search. Note that faster algorithms do exist in the literature, our goal in this paper is mainly to show the viability of this line of research. This is the first time a non-trivial upper bound on the value of a graph width parameter has been shown to hold *for every graph*.

Boolean-width is defined based on branch decompositions of a graph, using as cut function what is called the boolean dimension $bd(H)$ of a (bipartite) graph H. This corresponds to the logarithm (base 2) of the number of maximal independent sets $mis(H)$ of H. Our upper bounds on (linear) boolean-width rely on new techniques for bounding the number of maximal independent sets. This number has received much attention both from the algorithmic and the structural perspectives. While it is known [3] that computing $mis(G)$ is #P-hard even for planar bipartite graphs, approximating it is a much more delicate problem. From the structural point of view, bounding the number of maximal independent sets in special as well as in general graphs leads to interesting hard problems. Let us just mention the entropy-based results about $bd(G) = \log_2 mis(G)$ of d-regular graphs [4], see the references therein for an updated picture of the state of research in this area.

We introduce three techniques for our bounds. The first (Theorem 5) is based on a vertex partition achieved from a packing of paths and goes via $im(H)$, the size of a maximum induced matching in bipartite graph H. The second (Theorem 6) is based on a random partition and also goes via $im(H)$. The third (Theorem 7) is based on Hoeffding's inequality and in contrast to the first two applies also to boolean-width rather than just linear boolean-width. As already mentioned, our goal is to show the viability of this line of research and these various techniques should be helpful in later attempts to improve the bounds.

Our paper is organized as follows. In Section 2 we give all definitions and some preliminary results, for example showing that a non-trivial upper bound on (linear) boolean-width of a graph G will follow from a non-trivial upper bound on the boolean dimension of some balanced partition of G. In Section 3 we aim at an understanding of the structure of bipartite graphs of high boolean dimension. It is well known and easy that $bd(H)$ is at most $n/2$, and the maximum is attained by a size-$n/2$ matching. We tie $bd(H)$ to $im(H)$ and introduce and study the values of $co\text{-}im(H) = n/2 - im(H)$

and co-bd$(H) = n/2 - bd(H)$, the high-end ranges of im(H) and bd(H). We show constant factor approximation algorithms for these values, as well as a stability result showing that the smaller the values are, the closer is the bipartite graph to the size-$n/2$ matching. In Section 4 we turn to general graphs, and show, constructively, by a polynomial-time algorithm, that every graph has a balanced partition where the boolean dimension of the associated bipartite graph beats the triviality bound. Combined with the result from Section 2 this implies a constructive result for linear boolean-width of general graphs, beating the triviality bound of $n/2$. In Section 5 we turn to the standard boolean-width parameter and show also in this case a non-trivial upper bound beating the triviality bound of $n/3$, also constructive, but now by a randomized and low-exponential-time algorithm.

2 Terminology and Preliminaries

We consider undirected unweighted simple graphs $G = (V, E)$ and bipartite graphs $H = (A, B, E)$. We also denote the vertex set V by $V(G)$. For $S \subseteq V$ we denote by $G|_S$ the subgraph induced by S. The *neighborhood* of a vertex $v \in V$ is denoted $N(v)$. The neighborhood of a set $S \subset V$ is $N(S) = \cup_{v \in S} N(v)$. Any $S \subseteq V$ defines a cut $(S, V{-}S)$, and a bipartite graph $G_{S,V{-}S} = (S, V{-}S, \{(u, v) \in E : u \in S \wedge v \in V{-}S\})$.

A *decomposition tree* of a graph $G = (V, E)$ is a pair (T, δ) where T is a ternary tree, i.e. all internal nodes are of degree three, and δ a bijection between the leaves of T and $V(G)$. Removing an edge (a, b) from T results in two subtrees T_a and T_b, and a bipartition of V into V_a and V_b corresponding, respectively, to the δ-labels of leaves of T_a and T_b, and a bipartite graph G_{V_a, V_b}.

Definition 1 (Boolean Dimension, Boolean Width and Linear Boolean Width). *For a bipartite graph $H = (A, B, E)$, let $\mathcal{N}_A = \{N(X) \subseteq B \mid X \subseteq A\}$ be the family of neighborhoods of all sets $X \subseteq A$. The* boolean dimension *of H is defined as* bd$(H) = \log_2 |\mathcal{N}_A|$.

The boolean-width *of a decomposition tree (T, δ) is the maximum value of* bd(G_{V_a, V_b}) *over all edges (a, b) of T. The* boolean-width *of G, denoted* bw(G), *is the minimum boolean-width over all decomposition trees of G.*

The linear boolean-width *of G, denoted* lbw(G), *is the minimum boolean-width over all decomposition trees (T, δ) of G where T is a path on $|V|$ inner nodes, each with an attached leaf, corresponding to a linear arrangement of V.*

Given a graph G there is a $O^*(2.52^n)$ algorithm computing its boolean-width exactly [5] and in FPT time parameterized by $bw(G)$ we can compute a decomposition of boolean-width $2^{2bw(G)}$ using the algorithm for decompositions of optimal rank-width [6]. The boolean-width parameter was originally introduced in [2] in the context of parameterized algorithms. In particular, using a natural dynamic programming approach it was shown there that

Theorem 1. *[2,5] Given a graph G and a decomposition tree of boolean-width k, one can solve weighted and counting versions of* INDEPENDENT SET *in time $O^*(2^{2k})$ and* DOMINATING SET, INDEPENDENT DOMINATING SET *and* TOTAL DOMINATING SET *in time $O^*(2^{3k})$.*

These are dynamic programming algorithms that choose a root of the decomposition tree and traverse it bottom-up, with each node of the tree representing the subgraph induced by vertices corresponding to the leaves of the subtree. In a decomposition tree for linear boolean-width we choose one end of the path of inner nodes as root so that one of the two children of any node will always represent a subgraph on a single vertex. The runtime on linear decompositions can for this reason be improved

Corollary 1. *[2,5] Given a linear arrangement of $V(G)$ of linear boolean-width k, one can solve weighted and counting versions of* INDEPENDENT SET *in time $O^*(2^k)$ and* DOMINATING SET, INDEPENDENT DOMINATING SET *and* TOTAL DOMINATING SET *in time $O^*(2^{2k})$.*

Note that for any bipartite graph $H = (A, B, E)$, we have $|\mathcal{N}_A| = |\mathcal{N}_B|$, see, e.g., [7]. A good combinatorial way to demonstrate this is by establishing a bijection between the elements of $\mathcal{N}_A$ (or $\mathcal{N}_B$) and the set of all maximal independent sets of $H = (A, B, E)$. Here is a sketch of the argument. Given a set $S \in \mathcal{N}_A$, let X be the maximal set in A such that $N(X) = S$. Then X is uniquely defined, and $X \cup B-S$ is maximal independent. In the other direction, given a maximal independent set I, $B-I \in \mathcal{N}_A$. Moreover, if I resulted from S, then S results from I. [1] Hence,

Proposition 1. *Let $\mathrm{mis}(H)$ be the number of maximal independent sets in a bipartite graph H. Then, $\mathrm{bd}(H) = \log_2 \mathrm{mis}(H)$.*

The following simple property of $\mathrm{bd}(H)$ will prove useful; it is an immediate consequence of the definition of bd.

Proposition 2. $\mathrm{bd}(H)$ *is monotone decreasing with respect to vertex removal. Moreover, such removal may decrease $\mathrm{bd}(H)$ by at most 1. Hence, for a bipartite graph $H = (A, B, E)$ we have $\mathrm{bd}(H) \leq \min(|A|, |B|)$.*
More generally, given two bipartite graphs $G = (A, B, E)$ and $H = (A, B, E')$ on the same vertex set and the same two sides, it holds that $\mathrm{bd}(G \cup H) \leq \mathrm{bd}(G) + \mathrm{bd}(H)$.

Proposition 2 implies that $\mathrm{bd}(H) \leq n/2$, and this bound is met when H is a matching of size $n/2$. For a finer study of the structure of sub-extremal graphs H, we shall need the following notions.

Definition 2. *Define $\mathrm{im}(G)$ as the size of a maximum induced matching in G, i.e. a maximum-size set of edges whose endpoints do not induce any other edges in G. Note that $\mathrm{im}(G) \leq n/2$ and this bound is met only by a size-$n/2$ matching. To study the high-end range of $\mathrm{bd}(H)$ and $\mathrm{im}(G)$ we define $\mathrm{co\text{-}im}(G) = \frac{n}{2} - \mathrm{im}(G)$ and the boolean co-dimension $\mathrm{co\text{-}bd}(H) = \frac{n}{2} - \mathrm{bd}(H)$.*

The extremal values of $\mathrm{bw}(G)$ and $\mathrm{lbw}(G)$ are not known. The following proposition provides a preliminary tool for the study of the former.

---

[1] Observed by Nathann Cohen in a course of discussion with the authors. Later we have learned that a similar observation was made in [8].

Proposition 3. *Let* $A \subseteq V$ *be a subset of vertices with* $\frac{1}{3}n \leq |A| \leq \frac{2}{3}n$ *and* $\mathrm{bd}(G_{A,V-A}) = (\frac{1}{3} - \epsilon)n$ *for some* $\frac{1}{4} > \epsilon \geq 0$. *Then one can construct a decomposition tree of boolean-width at most* $(\frac{1}{3} - \frac{\epsilon}{3})n$.

In particular, $\mathrm{bw}(G) \leq n/3$. *We call this the* triviality bound *for* $\mathrm{bw}(G)$.

Proof. Without loss of generality, $|A| \leq n/2$; otherwise we switch to $V-A$. Partition V into three sets $A \cup X, B_1, B_2$ where X is disjoint from A and $|X| \leq \frac{2}{3}\epsilon n$, $|A \cup X| \leq n/2$ and $|B_1|, |B_2| \leq (\frac{1}{3} - \frac{1}{3}\epsilon)n$. Refining this partition to a decomposition tree arbitrarily with the sole restriction that at the upper level $A \cup X$ is split into two approximately equal parts we argue that the proposition holds using Proposition 2. For the top cut $\mathrm{bd}(G_{A \cup X, V-(A \cup X)}) \leq n/3 - \epsilon + |X| \leq (\frac{1}{3} - \frac{\epsilon}{3})n$. For any node representing a subset of B_1 or B_2 the proposition holds since the size of one side will be small. When splitting $A \cup X$ into two approximately equal parts, each part has size at most $n/4$, we have $n/4 \leq n/3 - \epsilon/3$ since $\epsilon < n/4$, and hence the proposition holds. The conclusion about $\mathrm{bw}(G) \leq n/3$ corresponds to choosing an A of size $\frac{1}{3}n$, and $\epsilon = 0$. $\square$

For $\mathrm{lbw}(G)$ one has an analogous statement:

Proposition 4. *Let* $A \subseteq V$ *be subset of vertices of size* $n/2$ *such that* $\mathrm{bd}(G_{A,V-A}) \leq (\frac{1}{2} - \epsilon) \cdot n$ *for some* $\epsilon \geq 0$. *Then one can construct a linear arrangement of the vertices of linear boolean-width at most* $(\frac{1}{2} - \frac{1}{2}\epsilon) \cdot n$.

In particular, $\mathrm{lbw}(G) \leq n/2$. *We call this the* triviality bound *for* $\mathrm{lbw}(G)$.

Proof. Take any linear arrangement whose first $n/2$ elements are precisely A. We claim that it has the desired property. Let A_i denote the set of the first (equivalently, the last) i elements in this arrangement. Let $G_i = G_{A_i, V-A_i}$. Proposition 2 implies that $\mathrm{bd}(G_i) \leq i$. It also implies that $|\mathrm{bd}(G_i) - \mathrm{bd}(G_{i+1})| \leq 1$ for every i, hence $\mathrm{bd}(G_i) \leq \mathrm{bd}(G_{n/2}) + (n/2 - i) \leq n - \epsilon n - i$. Combining the two bounds on $\mathrm{bd}(G_i)$, the statement follows. $\square$

The present paper, besides studying $\mathrm{bd}(H)$, is mostly dedicated to establishing upper bounds on $\mathrm{lbw}(G)$ and $\mathrm{bw}(G)$. Before starting with our toil, let us just mention that the above values can in general be as large as $\Omega(n)$, which is achieved e.g., when G is a constant-degree expander. Indeed, in this case any $G_{A,V-A}$ where both sides are $\geq n/3$ has at least $\Omega(n)$ edges, and hence, due to the constant degree, $\mathrm{im}(G_{A,V-A}) = \Omega(n)$. Since the size of an induced matching is a lower bound on the boolean dimension (see the next section for details), the conclusion follows. The constants obtained along this line of reasoning are, however, quite miserable. The extremal values of $\mathrm{lbw}(G), \mathrm{bw}(G)$ and the structure of the corresponding extremal graphs remain a (highly inspiring) mystery.

3   On Boolean Co-Dimension

3.1   Boolean Dimension vs. the Size of Maximum Induced Matching

We start with a lemma relating the boolean dimension of a bipartite graph $G = (A, B, E)$, $|V(G)| = n$, to the size of the maximum induced matching in G:

Lemma 1. *When* $\text{im}(G) \leq n/4$, *it holds that*

$$\text{im}(G) \;\leq\; \text{bd}(G) \;\leq\; \text{im}(G) \cdot \log_2(n/\text{im}(G)) \cdot \phi(2 \cdot \text{im}(G)/n),$$

where ϕ *is a function which never exceeds* 1.088, *and tends to* 1 *as* $\text{im}(G)$ *tends to* $n/4$.

Proof. The first inequality is obvious, as the boolean dimension is monotone with respect to taking induced subgraphs. For the second inequality, assume w.l.o.g., that $|A| \leq n/2$, and consider the family of neighborhoods $\mathcal{N}_A$ in B. For every $S \in \mathcal{N}_A$, there is a a minimal set $S^* \subseteq A$ such that $N(S^*) = S$. By minimality of S^*, each vertex v^* in it has a neighbour $v \in S$ not seen by the other vertices. Forming a set $S' \subseteq S \subseteq B$ by picking (one) such v for every $v^* \in S^*$, we conclude that the subgraph of G induced by (S^*, S') is an induced matching. In particular, it holds that $|S^*| \leq \text{im}(G)$, and thus any $S \in \mathcal{N}_A$ is a neighbourhood of a subset of A of size $\leq \text{im}(G)$. Consequently, using a standard estimation for the sum of binomial coefficients,

$$2^{\text{bd}(G)} \;=\; |\mathcal{N}_A| \;\leq\; \sum_{i=0}^{\text{im}(G)} \binom{|A|}{i} \;\leq\; \sum_{i=0}^{\text{im}(G)} \binom{n/2}{i} \;\leq\; 2^{n/2 \cdot H(\text{im}(G)/(n/2))}$$

where $H(p) = p \log_2 \frac{1}{p} + (1-p) \log_2 \frac{1}{1-p}$ is the entropy function. We introduce $p \log_2 \frac{1}{p} + p = p \log_2 \frac{2}{p}$ as an approximator of H, and set $\phi(p) = H(p)/(p \log_2 \frac{2}{p})$. Then,

$$\text{bd}(G) \;\leq\; \frac{n}{2} \cdot H\left(\frac{\text{im}(G)}{n/2}\right) \;=\; \text{im}(G) \cdot \log_2\left(\frac{n}{\text{im}(G)}\right) \cdot \phi\left(\frac{\text{im}(G)}{n/2}\right)$$

Through numerical analysis we found that $\phi(0.157) \approx 1.08798$ is the global maximum of ϕ in the range $[0, 0.5]$. $\qquad\square$

Thus, $\text{im}(G)$ is a $\log n$-approximation of $\text{bd}(G)$, and the quality of approximation improves as $\text{im}(G)$ grows. However, when $\text{im}(G) \geq n/4$, the approach of Lemma 1 fails to imply anything beyond the trivial upper bound $\text{bd}(G) \leq n/2$. This makes Lemma 1 inapplicable to the study of co-bd(G) vs. co-im(G). The key result of this subsection is that when co-bd(G) is small, so is co-im(G), and, moreover, co-bd(G) and co-im(G) are linearly related.

We start with a special case:

Lemma 2. *Let* $G = (A, B, E)$ *be a bipartite graph of degree at most* 2. *Then,*

$$\text{co-im}(G) \;\geq\; \text{co-bd}(G) \;\geq\; 0.339 \cdot \text{co-im}(G).$$

Proof. The first inequality was already established in Lemma 1. For the second inequality, observe that both co-bd(G) and co-im(G) are additive with respect to disjoint union of graphs. Thus it suffices to consider connected G's, i.e. G is either C_n, the (even) n-cycle, or P_{n-1}, the path on $(n-1)$ edges. For such graphs both im(G) and bd(G) are tractable. For the maximum induced matching one easily gets $\text{im}(C_n) = \lfloor \frac{n}{3} \rfloor$ and $\text{im}(P_{n-1}) = \lfloor \frac{n+1}{3} \rfloor$. For boolean dimension, recall that by Proposition 1, $\text{bd}(G) = \log_2 \text{mis}(G)$, where mis$(G)$ is the number of maximal independent sets in

G. Let $c(n) = \text{mis}(C_n)$ and $p(n) = \text{mis}(P_{n-1})$. The recurrence formulae for these values are well known (see e.g. [9]). Namely, $c(n) = c(n-2) + c(n-3)$, and $p(n) = p(n-2) + p(n-3)$. The initial conditions are $c(1) = 0$, $c(2) = 2$, $c(3) = 3$ and $p(1) = 1$, $p(2) = 2$, $p(3) = 2$ respectively.

Thus, to compare co-bd(G) to co-im(G) one needs to lower-bound the expressions

$$\frac{n/2 - \log_2(c(n))}{n/2 - \lfloor \frac{n}{3} \rfloor} \quad \text{and} \quad \frac{n/2 - \log_2(p(n))}{n/2 - \lfloor \frac{n+1}{3} \rfloor}.$$

Combining case analysis (according to $n \bmod 3$), numerical computations and an inductive argument, we conclude that the minimum is achieved on the 8-cycle C_8, and its value is $\frac{1}{2}(4 - \log_2 10) \approx 0.339036$. □

We continue with the general case.

Theorem 2. *Let $G = (A, B, E)$ be a bipartite graph. Then,*

$$\text{co-im}(G) \geq \text{co-bd}(G) \geq 0.0698 \cdot \text{co-im}(G) - 4.$$

Proof. As before, we shall be concerned only with the second inequality. Set $\Delta = \text{co-bd}(G)$. Keeping in mind that $\text{mis}(G) = 2^{\text{bd}(G)}$, observe that

$$\text{mis}(G) \leq \text{mis}(G|_{V-\{v\}}) + \text{mis}(G|_{V-\{v\}-N(v)}), \tag{1}$$

where the first term (over-)counts the maximal independent sets not containing v, and the second term counts those containing v. In accordance with inequality, we define a splitting process, or a (weighted) rooted splitting tree T, as follows.

Each inner node x of T is labelled by (G_x, v), where G_x is an induced subgraph of G, and $v \in V(G_x)$ is a vertex of degree 3 or more in G_x. At the root $G_x = G$; the leaves correspond to induced subgraphs of degree at most 2. An inner node x has two children, one corresponding to the graph obtained from G_x by removing v, the other corresponding to the graph obtained from G_x by removing v and all its neighbours (at least 4 vertices removed). The weight of the respective edge is defined as the number of vertices (respectively) removed. The weight of the node x, $w(x)$, is the defined as the sum of weights on the path from the root to x.

In view of (1), it holds that

$$2^{\text{bd}(G)} \leq \sum_{x: \text{ leaf of } T} 2^{\text{bd}(G_x)}. \tag{2}$$

The strategy of proof is as follows. The leaves L of T will be split into $L^+ = \{x \mid w(x) \geq z\}$ and $L^- = \{x \mid w(x) < z\}$ according to a suitably defined threshold value z. Then, it is shown that the leaves in L^+ contribute little to the above sum, while the graphs G_x corresponding to $x \in L^-$ have co-im comparable with Δ. The value of z will be set later, in the course of analysis.

Upper-bounding the contribution of L^+

By Proposition 2, $\text{bd}(G_x) \leq (n - w(x))/2$ for any $x \in T$. Thus,

$$\sum_{x \in L^+} 2^{\text{bd}(G_x)} \leq 2^{n/2} \cdot \sum_{x \in L^+} 2^{-w(x)/2}.$$

It is readily checked that the right-hand side is maximized when T is the complete 1-4 tree, where every inner node has an outgoing edge of weight 1 and an outgoing edge of weight 4. Moreover, for an inner node x and it two children x_1 and x_2 it holds that

$$2^{-w(x_1)/2} + 2^{-w(x_2)/2} = 2^{-w(x)/2} \cdot (2^{-1/2} + 2^{-4/2}) < 2^{-w(x)/2}.$$

Therefore, the right-hand side is maximized when the leaves are immediate descendants of the inner nodes of weight $< z$.

Let $s(i)$ denote the number of nodes of weight i in the complete 1-4 tree. Then, $|L^+| \leq s(z) + s(z+1) + s(z+2) + s(z+3)$. A closer look reveals that $s(0), s(1), s(2) = 1$, $s(3) = 2$, and that for $i \geq 4$, $s(i) = s(i-1) + s(i-4)$, where the first term counts the strings with leading "1", and the second term counts those with leading "4". Moreover, it holds that $s(z) + s(z+1) + s(z+2) + s(z+3) = s(z+6)$. Finding the maximal absolute-value root $\alpha = 1.38028$ of the equation $x^4 = x^3 + 1$, and using the estimation $s(z) \leq \alpha^z$, we conclude that

$$|L^+| \leq s(z+6) \leq 8 \cdot 1.38028^z \leq 8 \cdot 2^{0.4649589\, z},$$

and the total contribution of L^+ to the right-hand side of (2) is bounded from above by

$$2^{n/2} \cdot \sum_{x \in L^+} 2^{-w(x)/2} \leq 8 \cdot 2^{n/2} \cdot 2^{-z/2} \cdot 2^{0.4649589\, z} \leq 8 \cdot 2^{n/2} \cdot 2^{-0.035\, z}.$$

Setting $z = \lceil (\Delta + 4)/0.035 \rceil$, where $\Delta = \text{co-bd}(G) = n/2 - \text{bd}(G)$, ensures that the total contribution of L^+ is at most $0.5 \cdot 2^{n/2} \cdot 2^{-\Delta} = 0.5 \cdot 2^{\text{bd}(G)}$.

Consequently, the total contribution of L^- to the right-hand side of (2) is at least

$$\sum_{x \in L^-} 2^{\text{bd}(G_x)} \geq 0.5 \cdot 2^{\text{bd}(G)} = 2^{n/2} \cdot 2^{-\Delta-1}. \tag{3}$$

Upper bounding the contribution of L^-

To get an estimation from above on $\sum_{x \in L^-} 2^{\text{bd}(G_x)}$, consider $\text{bd}(G_x)$ for a leaf x of T. Since G_x has degree ≤ 2, Lemma 2 implies that:

$$\text{bd}(G_x) = (n - w(x))/2 - \text{co-bd}(G_x) \leq (n - w(x))/2 - 0.339 \cdot \text{co-im}(G_x).$$

Keeping in mind that $w(x) = n - |V(G_x)|$, it follows that $\text{co-im}(G_x) \geq \text{co-im}(G) - w(x)/2$. Substituting this in the previous line yields

$$\text{bd}(G_x) \leq n/2 - 0.33\, w(x) - 0.339 \cdot \text{co-im}(G),$$

and thus,

$$\sum_{x \in L^-} 2^{\text{bd}(G_x)} \leq \sum_{x \in L^-} 2^{\frac{n}{2} - 0.33 w(x) - 0.339 \text{co-im}(G)} = 2^{\frac{n}{2} - 0.339 \text{co-im}(G)} \sum_{x \in L^-} 2^{-0.33 w(x)}$$

As before, the complete 1-4 tree yields the most general (i.e., the weakest possible) upper bound on the sum $\sum_{x \in L^-} 2^{-0.33\, w(x)}$, as it maximizes the number of nodes of any weight i in T. Since this time the contribution of the father node is dominated by that of its sons, it suffices to analyse the case when the leaves of L^- have weights $z - 4, z - 3, z - 2$ or $z - 1$. Arguing as before, we conclude that $|L^-| \leq s(z + 2) \leq 2^{0.4649589(z+2)} < 2^{0.4649589z+1}$. That is,

$$\sum_{x \in L^-} 2^{-0.33 \cdot w(x)} \leq |L^-| \cdot 2^{-0.33(z-4)} \leq 2^{0.4649589z+1} \cdot 2^{-0.33(z-4)} < 2^{0.1349589z+2.32}.$$

Now, $z < (\Delta + 4)/0.035 + 1$, implying $0.1349589z + 2.32 \leq 3.856\Delta + 18$. The bottom line is:

$$\sum_{x \in L^-} 2^{\mathrm{bd}(G_x)} \leq 2^{n/2} \cdot 2^{-0.339 \cdot \mathrm{co\text{-}im}(G)} \cdot 2^{3.856\Delta+18}. \tag{4}$$

We are ready to conclude the proof of Theorem 2. Combining (3) and (4) yields

$$2^{n/2} \cdot 2^{-\Delta-1} \leq \sum_{x \in L^-} 2^{\mathrm{bd}(G_x)} \leq 2^{n/2} \cdot 2^{-0.339 \cdot \mathrm{co\text{-}im}(G)} \cdot 2^{3.856\Delta+18}.$$

Combining the two sides, it follows that $0.339 \cdot \mathrm{co\text{-}im}(G) \leq 4.856\Delta + 19$, and, finally, $\mathrm{co\text{-}im}(G) \leq 14.33\,(\Delta + 4) = 14.33\,(\mathrm{co\text{-}bd}(G) + 4)$.   □

One curious structural implication following at once from Theorem 2 is the following result (for asymptotically tight results see [4]):

Corollary 2. *Let G be a d-regular bipartite graph, $d > 1$. Then $\mathrm{mis}(G) \leq 2^{(\frac{1}{2}-\epsilon)n}$ for some universal $\epsilon > 0$.*

The reason is that by a trivial computation, for such graphs one has $\mathrm{im}(G) \leq \frac{n}{2} \cdot \frac{d}{2d-1}$, and since $\mathrm{co\text{-}bd}(G)$ is proportional to $\mathrm{co\text{-}im}(G)$, the conclusion follows.

3.2   A Constant Factor Polynomial Approximation Algorithm for co-bd(G)

Let us first give a polynomial time constant-factor approximation algorithm for $\mathrm{co\text{-}im}(G)$. As before, G is bipartite.

Approx-CoIm: *Construct (greedily or otherwise) a maximal vertex-disjoint packing $\mathcal{P}$ of P_2's (paths on 2 edges) in G. Remove all the vertices in $\mathcal{P}$. Output $\widetilde{M}$, the set of the remaining edges.*

Theorem 3. *The above algorithm produces an induced matching $\widetilde{M}$ with $n/2 - |\widetilde{M}| \leq 5 \cdot \mathrm{co\text{-}im}(G)$. In particular, it provides a 5-approximation for $\mathrm{co\text{-}im}(G)$.*

Proof. Observe that after the removal of P_2's in $\mathcal{P}$, the remaining induced graph consists of singletons and isolated edges, and thus $\widetilde{M}$ is indeed an induced matching.

Let M^* denote the maximum induced matching of G. Since every P_2 in the packing must contain at least one vertex outside of M^*, the size of $\mathcal{P}$ is at most $n - 2|M^*|$. Now, since each P_2 in $\mathcal{P}$ may hit at most 2 edges of M^*, at least $|M^*| - 2|\mathcal{P}|$ edges of M^* will survive the removal of $\mathcal{P}$. Thus,

$$|\widetilde{M}| \geq |M^*| - 2|\mathcal{P}| \geq |M^*| - 2 \cdot (n - 2|M^*|) = 5|M^*| - 2n\,;$$

$$n/2 - |\widetilde{M}| \leq 5/2\,n - 5|M^*| = 5 \cdot (n/2 - |M^*|) = 5 \cdot \mathrm{co\text{-}im}(G) \qquad □.$$

Theorem 3 combined with Theorem 2 yields a constant factor approximation algorithm for co-bd(G):

Theorem 4. *The co-size of $\widetilde{M}$ produced by* **Appox-CoIm** *on input G, i.e., $n/2 - |\widetilde{M}|$, approximates* co-bd(G) *within a multiplicative factor of $5 \cdot 14.3 < 72$.*

4   Linear Boolean Width: Beyond the Triviality Bound

We show (constructively and efficiently) that every size-n graph has a balanced bipartition of its vertex set such that the boolean dimension of the associated bipartite graph is $\leq (1/2 - c)n$ for some universal $c > 0$. Combined with Proposition 4, this implies that lbw$(G) \leq (1/2 - c/2)n$; the argument therein provides also the corresponding linear arrangement of the vertices.

Two constructions are provided, the first deterministic and somewhat elaborate, the other is just the random uniform bipartition. We start with the former.

GoodBipartition: *Construct (greedily or otherwise) a maximal vertex-disjoint packing $\mathcal{P}$ of P_2's (paths on 2 edges) in G. For each $P_2 \in \mathcal{P}$ mark the middle vertex. Partition the vertices into two equal- (up to ± 2) size sets so that*
   (i) *for every $P_2 \in \mathcal{P}$, the marked and the unmarked vertices lay on different sides;*
   (ii) *no edge (parity permitting) remaining after the removal of $V(\mathcal{P})$ is split.*

Since $\mathcal{P}$ is maximal, removing $V(\mathcal{P})$ one obtains an (induced) graph that consists of isolated edges and vertices. The required partition is obtained by placing the middle vertices of P_2's, one by one, on alternating sides. The same is done for surviving isolated edges and isolated vertices.

Let $H = (A, B, E')$ be the graph defined by this bipartition.

Theorem 5. *It holds that* co-im$(H) \geq \frac{1}{10} n$, *and* bd$(H) \leq (\frac{1}{2} - \frac{1}{143})n + O(1)$.

Proof. Let IM be a set of edges corresponding to a maximum induced matching of H, on vertices $V(IM)$. For each $P_2 \in \mathcal{P}$ we have at least one vertex of this P_2 not belonging to $V(IM)$, let $h(P_2)$ be such a vertex. For each $(u, v) \in IM$ we have either u or v (or both) a vertex of some $P_{uv} \in \mathcal{P}$. Fix for each $(u, v) \in IM$ arbitrarily such a P_{uv} and define a function $f : IM \to V(H) - V(IM)$ by $f((u, v)) = h(P_{uv})$. Since IM is an induced matching this function assigns to each vertex $h(P_{uv})$ at most two edges of IM. Thus there are at least $|IM|/2$ vertices in $V(H) - V(IM)$ and we have $|V(H)| \geq 2|IM| + |IM|/2$ which gives im$(H) \leq \frac{2}{5}n$, and, equivalently, co-im$(H) \geq \frac{1}{2}n - \frac{2}{5}n = \frac{1}{10}n$. By Theorem 2, this implies that co-bd$(H) \geq \frac{1}{10 \cdot 14.3} n = \frac{1}{143} n$. $\square$

Next, we show that a statement similar to that of Theorem 5 holds also for a random uniform bipartition of $V(G)$, when each vertex of V is assigned randomly and independently either to side A or to side B, resulting in $H(A, B, E'')$. While the bound will be weaker, this structural result is of independent interest. The proof is left out of this extended abstract.

Theorem 6. *For H as above, it holds almost surely that* co-im$(H) \geq \frac{1}{28} n - o(n)$. *Consequently,* co-bd$(H) \geq \frac{1}{801} n - o(n)$.

5 Boolean Width: Beyond the Triviality Bound

We turn to the boolean-width of general graphs. Since we currently have much less understanding of boolean dimension of unbalanced partitions than of the balanced ones, in this section we provide an existential argument, which can nevertheless be turned into an exponential time algorithm with a relatively small exponent.

We shall need the following standard estimation for the sum of binomial coefficients, to be called here the Entropy Bound:

$$\sum_{i=0}^{cm} \binom{m}{i} \leq 2^{H(c)\,m} \tag{5}$$

where $H(p) = p \log_2 \frac{1}{p} + (1-p) \log_2 \frac{1}{1-p}$ is the entropy function.

Lemma 3. *Every graph G has $A \subset V(G)$ with $|A| = \frac{n}{3} \pm o(n)$ such that* $\mathrm{bd}(G_{A,V-A}) \leq \frac{n}{3} - \frac{n}{226} + o(n)$.

Proof. If there exists $S \subseteq V(G)$ such that $|S| = \frac{n}{226}$ and $|N(S) \cup S| \leq \frac{n}{3}$ we just take any set A of size $n/3$ containing $N(S) \cup S$. Then, $\mathrm{bd}(G_{A,V-A}) \leq \frac{n}{3} - \frac{n}{226}$, as no vertex in S has neighbours in $V-A$.

Otherwise, every set S of size $\frac{n}{226}$ has $|N(S) - S| \geq \frac{n}{3} - \frac{n}{226}$. The set A will be constructed by a random procedure by choosing every vertex v with probability $\frac{1}{3}$, randomly and independently from the others. We claim that almost surely two following events take place. First, $|A| = n/3 \pm o(n)$, and second, all sets $S \subseteq V-A$ of size $\frac{n}{226}$ have $|N(S) \cap A| > \left(\frac{1}{3} - 0.202 \right) \cdot \left(\frac{n}{3} - \frac{n}{226} \right)$, which we short-cut as αn for the suitable α. Such an A will be called good.

Since $1 - \Pr[X \cap Y] \leq (1 - \Pr[X]) + (1 - \Pr[Y])$, it suffices to show that each of the two events holds almost surely *separately*. To bound the probabilities of failure, we use a suitable Chernoff-Hoeffding Bound.

Let Σ be the number of successes in r i.i.d. 0/1 events, each happening with probability p. Then by [10], $\Pr[\, \Sigma \leq (p-t)r \,] \leq e^{-2t^2 r}$ and $\Pr[\, \Sigma \geq (p+t)r \,] \leq e^{-2t^2 r}$. The desired bound on the probability of the first event follows at once with $o(n)$ standing for any sublinear function majorizing $n^{0.5}$, e.g., $6(n^{0.5+\epsilon})$. For the second event, the analysis is more involved.

Let S be any subset of $V(G)$ of size $\frac{n}{226}$. The probability that S causes a failure is

$$\Pr\left[\{S \subseteq V-A\} \wedge \{|N(S) \cap A| \leq \alpha n\} \right] =$$

$$= \Pr\left[\{|N(S) \cap A| \leq \alpha n\} \mid \{S \subseteq V-A\} \right] \cdot \Pr[S \subset V-A].$$

We start with upper-bounding the first factor in the above product. Then, the set S is fixed and is in $V-A$. The choosing process on the unfixed vertices in $V-S$ remains, however, unaltered. In particular, the vertices in $N(S)-S$ are chosen randomly and independently as before. Recall that by our assumption there are $\geq n/3 - n/226$ vertices in this set. Choosing $t = 0.202$, we get from the above Hoeffding Bound:

$$\Pr\left[|N(S) \cap A| \leq \left(\frac{1}{3} - 0.202 \right) \left(\frac{n}{3} - \frac{n}{226} \right) \mid \{S \subseteq V-A\} \right] < e^{-2 \cdot 0.202^2 \left(\frac{n}{3} - \frac{n}{226} \right)} <$$
$$< e^{-0.02684\,n}. \text{ Thus,}$$

$\Pr[\,S \text{ is bad}\,] < e^{-0.02684\,n} \cdot \Pr[S \subset V{-}A] = e^{-0.02684\,n} \cdot \left(\frac{2}{3}\right)^{n/226} < e^{-0.02863\,n}$.
Next, we apply the union bound summing over all sets S of size $\frac{n}{226}$. As always, the binomial coefficients are upper-bounded using the Entropy Bound from Equation (5):
$\Pr[\,\text{there exists a bad } S\,] \leq e^{-0.02863\,n} \cdot \binom{n}{n/226} < e^{-0.02863\,n} \cdot 2^{H(1/226)\,n} <$
$< e^{-0.00023\,n} = o(1)$. Thus, a random A is good almost surely for a large enough n. We proceed with upper bounding $\mathrm{bd}(G_{A,V{-}A})$ for a good A by counting the sets in $\mathcal{N}_{V{-}A}$, the family of neighbourhoods of subsets of $V-A$ in A.

Recall that $|V-A| \approx 2n/3$. The sets $S \subset V-A$ of size $i < n/226$ may contribute only as many as $\sum_{i=0}^{n/226} \binom{2n/3+o(n)}{i}$ distinct neigbourhoods in A. The contribution of sets $S \subset V-A$ of size $\geq n/226$ may be bounded as follows. In each such S mark an arbitrary subset $X \subseteq S$ of size precisely $\lceil n/226 \rceil$. Call two large S's equivalent if the same X was marked in both of them. Then, since every X sees at least αn vertices n, the contribution of the entire equivalence class of large sets defined X is at most $2^{n/3-\alpha n}$. The number of X's is at most $\binom{2n/3+o(n)}{n/226}$. By plugging in the numerical value of α and using the Entropy Bound from Equation (5) for $\binom{2n/3}{n/226}$, the entire contribution can be bounded by:

$$|\mathcal{N}_{V{-}A}| \leq \sum_{i=0}^{n/226} \binom{2n/3 + o(n)}{i} + \binom{2n/3 + o(n)}{n/226} \cdot 2^{n/3-\alpha n} \leq 2^{0.3286n+o(1)} \leq$$

$$\leq 2^{\frac{n}{3} - \frac{n}{226} + o(1)},$$

and the upper bound on $\mathrm{bd}(G_{A,V{-}A}) = \log_2 |\mathcal{N}_{V{-}A}|$ follows.

As an immediate consequence of Lemma 3 and Proposition 3 we obtain the following result:

Theorem 7. *For any graph G, it holds that* $\mathrm{bw}(G) \leq \frac{n}{3} - \frac{n}{672} + o(n)$.

6 Conclusion

Our results are the first non-trivial upper bounds on the value of a graph width parameter that hold *for every graph*. In this paper we gave three techniques to show such bounds, respectively Theorems 5, 6 and 7. At the moment the first two work only for linear boolean-width and the third is here applied only to boolean-width but it should work also for the linear case. We believe these bounds can be substantially improved. Combining Corollary 1 with Proposition 4 and Theorem 5 we can solve MAXIMUM WEIGHT INDEPENDENT SET and COUNTING INDEPENDENT SETS OF SIZE K in time $O^*(1.4108^n)$, and solve MINIMUM WEIGHT DOMINATING SET, MINIMUM WEIGHT TOTAL DOMINATING SET, MAXIMUM/MINIMUM WEIGHT INDEPENDENT DOMI-NATING SET, and counting versions of these, in time $O^*(1.9904^n)$. These runtimes beat brute-force search but faster algorithms exist in the literature, see [11]. Our goal was mainly to prove the viability of this new line of research by establishing structural qualitative results. The natural directions for further work are to improve the bounds and hence the runtime, and to increase the class of problems handled by Corollary 1.

References

1. Belmonte, R., Vatshelle, M.: Graph classes with structured neighborhoods and algorithmic applications. In: TCS (2013),
   http://dx.doi.org/10.1016/j.tcs.2013.01.011
2. Bui-Xuan, B.-M., Telle, J.A., Vatshelle, M.: Boolean-width of graphs. Theoretical Computer Science 412(39), 5187–5204 (2011)
3. Rödl, V., Duffus, D., Frankl, P.: Maximal independent sets in bipartite graphs obtained from boolean lattices. Eur. J. Comb. 32(1), 1–9 (2011)
4. Fomin, F.V., Kratsch, D.: Exact Exponential Algorithms, 1st edn. Texts in Theoretical Computer Science (2010)
5. Füredi, Z.: The number of maximal independent sets in connected graphs. Journal of Graph Theory 11(4), 463–470 (1987)
6. Hlinený, P., Oum, S.I.: Finding branch-decompositions and rank-decompositions. SIAM J. Comput. 38(3), 1012–1032 (2008)
7. Hoeffding, W.: Probability inequalities for sums of bounded random variables. Journal of the American Statistical Association 58(301), 13–30 (1963)
8. Ilinca, L., Kahn, J.: Counting maximal antichains and independent sets. Order 30(2), 427–435 (2013)
9. Kim, K.H.: Boolean matrix theory and its applications. Monographs and textbooks in pure and applied mathematics. Marcel Dekker (1982)
10. Vadhan, S.P.: The complexity of counting in sparse, regular, and planar graphs. SIAM Journal on Computing 31, 398–427 (1997)
11. Vatshelle, M.: New Width Parameters of Graphs. PhD thesis, University of Bergen (2012) ISBN:978-82-308-2098-8

Speeding Up Dynamic Programming
with Representative Sets*
An Experimental Evaluation of Algorithms
for Steiner Tree on Tree Decompositions

Stefan Fafianie, Hans L. Bodlaender, and Jesper Nederlof

Utrecht University, The Netherlands
{S.Fafianie,H.L.Bodlaender,J.Nederlof}@uu.nl

Abstract. Dynamic programming on tree decompositions is a frequently used approach to solve otherwise intractable problems on instances of small treewidth. In recent work by Bodlaender et al. [5], it was shown that for many connectivity problems, there exist algorithms that use time, linear in the number of vertices, and single exponential in the width of the tree decomposition that is used. The central idea is that it suffices to compute representative sets, and these can be computed efficiently with help of Gaussian elimination.

In this paper, we give an experimental evaluation of this technique for the STEINER TREE problem. A comparison of the classic dynamic programming algorithm and the improved dynamic programming algorithm that employs the table reduction shows that the new approach gives significant improvements on the running time of the algorithm and the size of the tables computed by the dynamic programming algorithm, and thus that the rank based approach from Bodlaender et al. [5] does not only give significant theoretical improvements but also is a viable approach in a practical setting, and showcases the potential of exploiting the idea of representative sets for speeding up dynamic programming algorithms.

Keywords: Experimental evaluation, Algorithmic engineering, Steiner tree, Treewidth, Dynamic programming, Exact algorithms.

1 Introduction

The notion of treewidth provides us with a method of solving many $\mathcal{NP}$-hard problems by means of dynamic programming algorithms on tree decompositions of graphs, resulting in algorithmic solutions which are fixed-parameter tractable in the treewidth of the input graph. For many problems, this gives algorithms that are linear in the number of vertices n, but at least exponential in the width

---

* The third author is supported by the NWO project 'Space and Time Efficient Structural Improvements of Dynamic Programming Algorithms'.

G. Gutin and S. Szeider (Eds.): IPEC 2013, LNCS 8246, pp. 321–334, 2013.

of the tree decomposition on which the dynamic programming algorithm is executed. The dependency of the running time on the width of the tree decomposition has been a point of several investigations. For many problems, algorithms were known whose running time is single exponential on the width, see e.g., [25]. A recent breakthrough was obtained by Cygan et al. [11] who showed for several *connectivity* problems, including HAMILTONIAN CIRCUIT, STEINER TREE, CONNECTED DOMINATING SET (and many other problems) that these can be solved in time, single exponential in the width, but at the cost of introducing randomization and an additional factor in the running time that is polynomial in n. Very recently, Bodlaender et al. [5] introduced a new technique (termed the *rank based approach*) that allows algorithms for connectivity problems that are (i) deterministic, (ii) can handle weighted vertices, and (iii) have a running time of the type $O(c^k n)$ for graphs with a given tree decomposition of width k and n vertices, i.e., the running time is single exponential in the width, and linear in the number of vertices.

The main ideas of the rank based approach are the following. (Many details are abstracted away in the discussion below. See [5] for more details.) Suppose we store during dynamic programming a table T with each entry giving the characteristic of a partial solution. If we have an entry s in T, such that for each extension t of s to a 'full solution', $s \cdot t$, there is an other entry $s' \neq s$ in T, that can be extended in the same way to a full solution $s' \cdot t$, and solution $s' \cdot t$ has a value that is as least as good as the value of $s \cdot t$, then s is not needed for obtaining an optimal solution, and we can delete s from T. This idea leads to the notion of *representativity*, pioneered by Monien in 1985 [23]. Consider the matrix M with rows indexed by partial solutions, and columns indexed by manners to extend partial solutions, with a 1 if the combination gives a full solution, and a 0 otherwise. A table T corresponds to a set of rows in M, with a value associated to each row. (E.g., for the STEINER TREE problem, a row corresponds to the characteristic of a forest in a subgraph, and the value is the sum of the edges in the forest.) It is not hard to see that a maximal subset of linear independent rows of minimal cost (in case of minimization problems, and of maximal value in case of maximization problems) forms a representative set. Now, if we have an explicit basis of M (the characteristics of the columns in a maximal set of independent columns in T) and M has 'small' rank, then we can find a 'small' representative set efficiently, just by performing Gaussian elimination on a submatrix of M. Now, for many connectivity problems, including STEINER TREE, FEEDBACK VERTEX SET, LONG PATH, HAMILTONIAN CIRCUIT, CONNECTED DOMINATING SET, the rank of this matrix M when solving these problems on a tree decomposition is single exponential in the width of the current bag. This leads to the improved dynamic programming algorithm: interleave the steps of the existing DP algorithm with computing representative sets by computing the submatrix of M and then carrying out Gaussian elimination on this submatrix.

The notion of representative sets was pioneered by Monien in 1985 [23]. Using the well known two families theorem by Lovász [21], it is possible to obtain efficient FPT algorithms for several other problems [22,14]. Cygan et al. [10] give an improved bound on the rank as a function of the width of the tree decomposition for problems on finding cycles and paths in graphs of small treewidth, including TSP, HAMILTONIAN CIRCUIT, LONG PATH.

In this paper, we perform an *experimental evaluation* of the rank based approach, targeted at the STEINER TREE problem, i.e., we discuss an implementation of the algorithm, described by Bodlaender et al. [5] for the STEINER TREE problem and its performance. We test the algorithm on a number of graphs from a benchmark for STEINER TREE, and some randomly generated graphs. The results of our experiments are very positive: the new algorithm is considerably faster compared to the classic dynamic programming algorithm, i.e., the time that is needed to reduce the tables with help of Gaussian elimination is significantly smaller than the gain in time caused by the fact that tables are much smaller.

The STEINER TREE problem (of which MINIMUM SPANNING TREE is a special case) is a classic $\mathcal{NP}$-hard problem which was one of Karp's original 21 $\mathcal{NP}$-complete problems [17]. Extensive overviews on this problem and algorithms for it can be found in [16,30]. Applications of STEINER TREE include electronic design automation, very large scale integration (VSLI) of circuits and wire routing. In this paper we consider the weighted variant, i.e., edges have a weight, and we want to find a Steiner tree of minimum weight. It is well known that STEINER TREE can be solved in linear time for graphs of bounded treewidth. In 1983, Wald and Colbourn [27] showed this for graphs of treewidth two. For larger fixed values of k, polynomial time algorithms are obtained as consequence of a general characterization by Bodlaender [4] and linear time algorithms are obtained as consequence of extensions of Courcelles theorem, by Arnborg et al. [2] and Borie et al. [7]. In 1990, Korach and Solel [20] gave an explicit linear time algorithm for STEINER TREE on graphs of bounded treewidth. Inspection shows that the running time of this algorithm is $O(2^{O(k \log k)} n)$; k denotes the width of the tree decomposition. We call this algorithm the *classic* algorithm. Recently, Chimani et al. [8] gave an improved algorithm for STEINER TREE on tree decompositions that uses $O(B_{k+1}^2 \cdot k \cdot n)$ time, where the Bell number B_i denotes the number of partitions of an i element set. Our description of the classic algorithm departs somewhat from the description in Korach and Solel [20], but the underlying technique is essentially the same. We have chosen not to use the coloring schemes from Chimani et al. [8], but instead use hash tables to store information. While the coloring schemes give a better worst case running time, we also spend time with these on 'non-existing table entries', and thus we expect faster computations when using hash tables. Wei-Kleiner [29] gives a tree decomposition based algorithm for STEINER TREE, that particularly aims at instances with a small set of Steiner vertices.

In this paper, we compare three different algorithms:

– The *classic* dynamic programming algorithm (CDP), see the discussion above. On a nice tree decomposition, we build for each node i a table. Tables map partitions of subsets of X_i to values, characterizing the minimum weight of a 'partial solution' that has this partition of a subset as 'fingerprint'.

- RBA: To the classic dynamic programming algorithm, we add a step where we apply the *reduce* algorithm from [5]. With help of Gaussian elimination on a specific matrix (with rows corresponding to entries in the DP table, columns corresponding to a 'basis of the fingerprints of ways of extending partial solutions to Steiner trees', and values 1, if the extension of the column applied to the entry of the row gives a Steiner tree and 0 otherwise), we delete some entries from the table. It can be shown that deleted entries are not needed to obtain an optimal solution, i.e., the step does not affect optimality of the solution. This elimination step is performed each time after the DP algorithm has computed a table for a node of the nice tree decomposition.
- RBC: Similar to RBA, but now the elimination step is only performed for 'large' tables, i.e., tables where the theory tells us that we will delete at least one entry when we perform the elimination step.

Our software is publicly available, can be used under a GNU Lesser General Public Licence, and can be downloaded at:
http://www.staff.science.uu.nl/~bodla101/java/steiner.zip

This paper is organized as follows. Some preliminary definitions are given in Section 2. In Section 3, we briefly describe both the classic dynamic programming algorithm for Steiner Tree on nice tree decompositions, as well as the improvement with the rank based approach as presented in [5]. In Section 4, we describe the setup of our experiments, and in Section 5, we discuss the results of the experiments. Some final conclusions are given in Section 6.

2   Preliminaries

We use standard graph theory notation and quite some additional notation from [5]. For a subset of edges $X \subseteq E$ of an undirected graph $G = (V, E)$, we let $G[X]$ denote the subgraph induced by edges and endpoints of X, i.e. $G[X] = (V(X), X)$.

For two partitions p and q of a set W, we say that p is a coarsening of q (or, q is a refinement of p) if every block of q is contained in a block of p, and we let $p \sqcap q$ denote the finest partition that is a coarsening of p and of q. (In graph terms: take an edge between $v \in W$ and $w \in W$ iff $v \neq w$ and v and w belong to the same block in p or to the same block in q. Now, the classes of $p \sqcap q$ are the connected components of this graph.)

The STEINER TREE problem can be defined as follows.

STEINER TREE
Input: A graph $G = (V, E)$, weight function $\omega : E \to \mathbb{N} \setminus \{0\}$, a terminal set $K \subseteq V$ and a nice tree decomposition $\mathbb{T}$ of G of width **tw**.
Question: The minimum of $\omega(X)$ over all subsets $X \subseteq E$ of G such that $G[X]$ is connected and $K \subseteq V(G[X])$.

Definition 1 (Tree decomposition, [24]). *A tree decomposition of a graph G is a tree $\mathbb{T}$ in which each node x has an assigned set of vertices $B_x \subseteq V$ (called a bag) such that $\bigcup_{x \in \mathbb{T}} B_x = V$ with the following properties:*

 – *for any* $e = (u, v) \in E$, *there exists an* $x \in \mathbb{T}$ *such that* $u, v \in B_x$.
 – *if* $v \in B_x$ *and* $v \in B_y$, *then* $v \in B_z$ *for all* z *on the (unique) path from* x *to* y *in* $\mathbb{T}$.

The treewidth $tw(\mathbb{T})$ of a tree decomposition $\mathbb{T}$ is the size of the largest bag of $\mathbb{T}$ minus one, and the treewidth of a graph G is the minimum treewidth over all possible tree decompositions of G.

Definition 2 (Nice tree decomposition). *A* nice tree decomposition *is a tree decomposition with one special bag* z *called the* root *and in which each bag is one of the following types:*

 – leaf bag: *a leaf* x *of* $\mathbb{T}$ *with* $B_x = \emptyset$.
 – introduce vertex bag: *an internal vertex* x *of* $\mathbb{T}$ *with one child vertex* y *for which* $B_x = B_y \cup \{v\}$ *for some* $v \notin B_y$. *This bag is said to* introduce v.
 – introduce edge bag: *an internal vertex* x *of* $\mathbb{T}$ *labelled with an edge* $e = (u, v) \in E$ *with one child bag* y *for which* $u, v \in B_x = B_y$. *This bag is said to* introduce e.
 – forget vertex bag: *an internal vertex* x *of* $\mathbb{T}$ *with one child bag* y *for which* $B_x = B_y \setminus \{v\}$ *for some* $v \in B_y$. *This bag is said to* forget v.
 – join bag: *an internal vertex* x *with two child vertices* y *and* y' *with* $B_x = B_y = B_{y'}$.

We additionally require that every edge in E *is introduced exactly once.*

Nice tree decompositions were introduced in the 1990s by Kloks [18]. We use here a more recent version that distinguishes *introduce edge* and *introduce vertex* bags [11]. To each bag x we associate the graph $G_x = (V_x, E_x)$, with V_x the union of all B_y with $y = x$ or y a descendant of x, and E_x the set of all edges introduced at bags y with $y = x$ or y a descendant of x. There are also many heuristics for finding a tree decomposition of small width; see [6] for a recent overview. Given a tree decomposition $\mathbb{T}$ of G, a nice tree decomposition rooted at a forget bag can be computed in $n \cdot tw^{\mathcal{O}(1)}$ time by following the arguments given in [18], with the following modification: between a forget bag X_i where we 'forget vertex v' and its child bag X_j, we add a series of introduce edge bags for each edge $e = \{v, w\} \in E$ and $w \in X_j$. We also assume that root bag z is a forget node with $B_x = \emptyset$ and that the vertex that is forgotten at the root bag is a terminal.

3 Dynamic Programming Algorithms for Steiner Tree Parameterized by Treewidth

In this section we briefly sketch both the classic dynamic programming algorithm on (nice) tree decompositions for (the edge weighted version of) STEINER TREE and its variant with the rank based approach. For details, see the full paper [13].

3.1   Classic Dynamic Programming

The classic dynamic programming algorithm computes for each bag x a function A_x. This function is represented by a table, with only trivial entries (e.g., partitions mapping to infinity, as there are no forests corresponding to the partition) not stored.

The function A_x maps a subset $W \subseteq B_x$ to a collection of pairs. Each pair consists of a partition p of W and a weight w. If (p, w) is in the collection associated to W, then w is the minimum weight over all forests F in G_x with the following properties: (1) For all $v \in B_x$, $v \in W$ iff v belongs to F; (2) each terminal in V_x belongs to F; (3) each tree in F contains at least one vertex in W; and (4) two vertices in W belong to the same class in the partition p, iff they belong to the same tree in F. I.e., for each partition p, we store at most one weight; if for set W and a partition p, no forest exists that fulfills the properties, then we have no pair in $A_x(W)$ of the form $(p, \ldots)$.

In the full paper [13] we can find a more formal description with slightly different notation (based on the notation in [5]), and recurrences for A for each of the types (leaf, introduce vertex, introduce edge, join, forget) of nodes in nice tree decompositions. In bottom-up order, we compute for each node x in the nice tree decomposition a table for A_x. The minimum value of a Steiner tree in G can be directly observed given the table for the root node.

In our implementation, we use two levels of hash tables: one with keys the different subsets W of B_x, and for each W with at least one partial solution, we have a hash table storing for each p the value z of the pair $(p, z) \in A_x(W)$, in case such a pair exists.

3.2   Rank Based Table Reductions

The main idea of the rank based approach from [5] is that after we have computed a table for a bag x in the nice tree decomposition, we can carry out a reduction step and possibly remove a number of entries from the table without affecting optimality. A table is transformed thus to a (possibly smaller) table whose weighted partitions are *representative* for the collection of weighted partitions in the earlier table.

The reduction step is performed as follows: for each $W \subseteq B_x$, we do the following. We build a matrix $\mathcal{M}$ with a row for each partition p appearing in a pair in $A_x(W)$, and a column for each partition q of W in two sets, with $\mathcal{M}(p, q) = 1$ if and only if $p \sqcap q = U$. Now, from [5], it follows that it is sufficient to keep a minimum weight basis of rows. With help of Gaussian elimination, we compute such a minimum weight basis (after first sorting the rows with respect to their weights), and then delete all other entries from the table. Correctness follows from the analysis in [5]. In our experiments, we consider both the case where we always apply the reduction step, and the case where we only apply it when $A \geq 2^{|B_x|}$. Both cases give the same guarantees on the size of tables and worst case upper bound on the running time, but the actual running times in experiments differ, as we discuss in later sections.

4   Implementation

In this section, we give some details on our implementation of the algorithms described in the previous section. We have implemented the algorithms in Java. For each of the test graphs, we used the well known (and quite simple and effective, see e.g., [6]) *Greedy Degree* heuristic to find a tree decomposition. These tree decompositions were subsequently transformed into nice tree decompositions, using the procedure which was previously described in Section 2. The algorithms were executed on the thus obtained nice tree decompositions.

The recursions for the different types of nodes were implemented such that we spend linear time per generated entry (before removing double entries, and before the reduction step). For most types, this is trivial. The computation for join bags contains a step, where we are given two partitions, and must compute the partition that is the closure of the combination of the two (i.e., the finest partition that is a coarsening of both). We implemented this step with a breadth first search on the vertices in the bag, with the children of a vertex v all not yet discovered vertices that are in the same block as v in either of the partitions.

Sets $W \subseteq B_x$ are represented by a bitstring. In the computations of join, introduce edge, and forget nodes, it is possible that we generate two or more entries for the same W and partition p of W. Of these duplicate partial solutions, we need to keep only the one with the smallest weight. In order to find such duplicate partial solutions we have represented the partial solution tables in a nested hash-map structure. First we use sets of vertices that where not used in a partial solution as keys, pointing to tables of weighted partitions, effectively grouping partitions consisting of the same base set of vertices together. These weighted partition tables are then represented by another hash-map where the partitions, which are represented as nested sets, are used as keys, pointing to the minimum weight corresponding to the partial solution. This allows us to find and replace any duplicate partial solution in amortized constant time. Java provides hash-codes for sets by adding the hash-codes for all objects contained within a set, which works well enough for the outer hash-table used in our structure. This standard approach breaks down when we use it to calculate hash-codes for partitions however, as it effectively adds all hash-codes of vertices used in the partition together. This results in the same hash-code for all partitions used in the same inner hash-map. To resolve this problem we disrupt this commutative effect by multiplying indexes of vertices contained in each block, and then taking the sum of these values of blocks in order to calculate hash-codes for partitions. We apply the multiplications modulo a prime number to avoid integer overflows. In our experiments, we observed that this approach results in approximately 3% collisions for large tables.

In the implementation of the rank based approach, for each bag, we first compute a table as in the classic algorithm, and then compute the corresponding matrix $\mathcal{M}$, as discussed above. We perform the steps of Gaussian elimination with rows in order of nondecreasing weight. I.e., first we order the rows of M in order of nondecreasing weight. find the first 1 in the row, and now add the values in this row to all later rows with a 1 in the same column (modulo 2).

(This is precisely one step of Gaussian elimination). When a row consists of only 0's, it is linearly dependent on previous processed rows (of smaller weight), and thus safely eliminated. We stop when all partial solutions have been processed, or when we have processed $2^{|W|}$ rows, since all remaining partial solutions are linearly dependent on solutions in $\mathcal{A}$. Any time a partial solution is processed we can eliminate the column containing its leading 1, since all elements in this column are 0.

Chimani et al. [8] give an efficient algorithm for STEINER TREE for graphs given with a tree decomposition, that runs in $O(B_{k+2}^2 kn)$ time, with k the width of the tree decomposition. We have chosen not to use the coloring scheme from Chimani et al. [8], but instead use hash tables (as discussed above) to store the tables. Of course, our choice has the disadvantage that we lose a guarantee on the worst case running time (as we cannot rule out scenarios where many elements are hashed to the same position in the hash table), but gives a simple mechanism which works in practice very well. In fact, if we assume that the expected number of collisions of an element in the hash table is bounded by a constant (which can be observed in practice), then the expected running time of our implementation matches asymptotically the worst case running time of Chimani et al.

5   Experimental Results

In this section, we will report the results for experiments with the algorithms discussed in Section 3. We will denote the classic dynamic programming algorithm as CDP. With RBA, we denote the algorithm where we always apply the reduction step, whereas RBC denotes the algorithm which only applies the reduction step when we have a table whose size is larger than the bound guaranteed by reduction. We will compare the runtime of these three algorithms. Furthermore we will compare the number of partial solutions generated during the execution of these algorithms to illustrate how much work is being saved by reducing the tables.

Each of the three algorithms receives as input the same nice tree decomposition of the input graph; this nice tree decomposition is rooted at a forget bag of a terminal vertex. The experiments where performed on sets of graphs of different origin, spanning a range of treewidth sizes of their tree decompositions, and where possible diversified on the number of vertices, edges and terminals. Our graphs come from benchmarks for algorithms for the STEINER TREE problem and for Treewidth. The graphs from Steiner tree benchmarks can be found in Steinlib [19], a repository for Steiner Tree problems. These are prefixed by b, i080 or es. Graph instances prefixed by b are randomly generated sparse graphs with edge weights between 1 and 10; these were introduced in [3] and were generated following a scheme outlined in [1]. The i080 graph instances are randomly generated sparse graphs with incidence edge weights, introduced in [12]. We have grouped these sparse graphs together in the results. The next set of instances, prefixed by es, were generated by placing random points on a two-dimensional grid, which serve as terminals. By building the grid outlined in [15] they where converted to rectilinear graphs with L1 edge weights and preprocessed with GeoSteiner [28]. The last collection of graphs are often used as benchmarks for algorithms for TREEWIDTH. These come from Bayesian network and graph colouring applications. We transformed these to STEINER TREE instances by adding random edge weights between 1 and 1000, and by selecting randomly a subset of the vertices as terminals (about 20% of the original vertices). These graphs can be found in [26].

All algorithms have been implemented in Java and the computations have been carried out on a Windows-7 operated PC with an Intel Core i5-3550 processor and 16.0 GB of available main memory. We have given each of the algorithms a maximum time of one hour to find a solution for a given instance; in the tables, we marked instances halted due to the use of the maximum time by a *.

Table 1. Runtime in milliseconds for instances from Steinlib (1)

instance	tw($\mathbb{T}$)	$|V|$	$|E|$	$|T|$	CDP	RBA	RBC
b01.stp	4	50	63	9	55	53	17
b02.stp	4	50	63	13	12	30	12
b08.stp	6	75	94	19	171	92	48
b09.stp	6	75	94	38	78	46	31
b13.stp	7	100	125	17	1328	618	408
b14.stp	7	100	125	25	2190	385	275
b15.stp	8	100	125	50	14421	1542	1281
i080-001.stp	9	80	120	6	98617	11270	7953
i080-003.stp	9	80	120	6	144796	12689	10211
i080-004.stp	10	80	120	6	1618531	70192	68930
b06.stp	10	50	100	25	1325669	36986	29082
b05.stp	11	50	100	13	*	270376	207516
i080-005.stp	11	80	120	6	*	936074	840466

In Tables 1 – 3, we have gathered the results for the runtimes of the three algorithms for the aforementioned graph instances. We immediately notice that RBC outperforms RBA in all cases. In Table 4 we give the number of partial solutions (table entries) computed for each of the three algorithms; similar data is available for the other instances in the full paper [13]. If we investigate Table 4 we notice that the number of partial solutions computed during RBA is not significantly smaller compared to the number computed during RBC. From these results we can conclude that it is preferable to use the reductions more sparingly in order to decrease runtime, since applying the reductions when the tables are already smaller than their size guarantee does not seem to have a noteworthy effect.

We also notice that, while RBA outperforms CDP in numerous cases, RBC outperforms CDP in all but one (discussed below). For example, in the case of *i080-004* we see a significant speed-up: the classic DP uses 26 minutes to find the

Table 2. Runtime in milliseconds for instances from Steinlib (2)

instance	tw($\mathbb{T}$)	$|V|$	$|E|$	$|T|$	CDP	RBA	RBC
es90fst12.stp	5	207	284	90	71	120	60
es100fst10.stp	5	229	312	100	105	166	86
es80fst06.stp	6	172	224	80	272	276	151
es100fst14.stp	6	198	253	100	109	160	78
es90fst01.stp	7	181	231	90	250	270	148
es100fst13.stp	7	254	361	100	1223	1200	679
es100fst15.stp	8	231	319	100	2600	1688	1033
es250fst03.stp	8	543	727	250	2904	2010	1251
es100fst08.stp	9	210	276	100	4670	2302	1942
es250fst05.stp	9	596	832	250	24460	15550	9742
es250fst07.stp	10	585	799	250	107150	54605	31729
es500fst05.stp	10	1172	1627	500	124664	47336	31102
es250fst12.stp	11	619	872	250	*	144932	95855
es100fst02.stp	12	339	522	100	*	426078	334785
es250fst01.stp	12	623	876	250	*	332389	246704
es250fst08.stp	13	657	947	250	*	2670464	2251728
es250fst15.stp	13	713	1053	250	*	2120913	1671672

Table 3. Runtime in milliseconds for instances on graphs from TreewidthLib

instance	tw($\mathbb{T}$)	$|V|$	$|E|$	$|T|$	CDP	RBA	RBC
myciel3.stp	5	11	20	2	5	8	4
BN_28.stp	5	24	49	4	8	15	7
pathfinder.stp	6	109	211	21	422	254	155
csf.stp	6	32	94	6	335	198	116
oow-trad.stp	7	33	72	6	766	594	364
mainuk.stp	7	48	198	9	8842	3495	2025
ship-ship.stp	8	50	114	10	10579	4511	2841
barley.stp	8	48	126	9	9281	2410	1473
miles250.stp	9	128	387	25	35369	14423	9382
jean.stp	9	80	254	16	39192	18237	10862
huck.stp	10	74	301	14	*17030*	38486	21050
myciel4.stp	11	23	71	4	1510595	98720	93107
munin1.stp	11	189	366	37	*	460051	372718
pigs.stp	12	441	806	88	*	1431083	1280194
anna.stp	12	138	493	27	*	*	3291591

optimal solution, but RBC uses just 69 seconds. Furthermore we see a strong increase in the runtime difference when the width of the tree decompositions increases. This is further reflected in Table 4, where we see that when the width of the tree decompositions increases, the difference in the number of of generated partial solutions grows significantly.

The *huck* instance is the only example where using the rank based approach does not pay off. Upon further inspection we found that the tree decomposition

Table 4. Number of generated partial solutions for instances of Steinlib (1)

instance	tw($\mathbb{T}$)	$|V|$	$|E|$	$|T|$	CDP	RBA	RBC
b01.stp	4	50	63	9	3141	2854	2854
b02.stp	4	50	63	13	3263	2763	2769
b08.stp	6	75	94	19	39178	11278	11345
b09.stp	6	75	94	38	18970	5177	5449
b13.stp	7	100	125	17	328366	68533	70693
b14.stp	7	100	125	25	400940	35554	40012
b15.stp	8	100	125	50	2294557	84567	94951
i080-001.stp	9	80	120	6	15757284	529805	565777
i080-003.stp	9	80	120	6	18841974	589313	589773
i080-004.stp	10	80	120	6	196513167	2611426	3270334
b06.stp	10	50	100	25	156669926	903700	938800
b05.stp	11	50	100	13	*	6320072	6320264
i080-005.stp	11	80	120	6	*	26653282	31275766

for this instance has only one bag of size 11, while most of the other bags are of size 7 and below. This is also reflected by the difference in the number of generated partial solutions, where the improvement factor is not comparable to the other cases. Conversely we found that the *i080-004* case included 18 bags of treewidth 11 of which 6 where join bags, which explains the extreme difference. In practice, when we run dynamic programming algorithms on tree decompositions, the underlying structure of the decomposition has a big influence on the performance, which is not always properly reflected by the treewidth of a graph. In general however, the rank based approach is more and more advantageous as the treewidth increases, even allowing us to find solutions where CDP does not find any within the time limit. During the execution of the experiments we have also tracked the amount of time spent on filling the cut-matrices and the time spent on performing Gaussian elimination, and found that we spent significantly more time on filling the table whereas the Gaussian elimination step only takes up a small fraction of the time spent on reducing the tables.

6   Discussion and Concluding Remarks

In this paper, we presented an experimental evaluation of the rank based approach by Bodlaender et al. [5], comparing the classic dynamic programming for STEINER TREE and the new versions based on Gaussian elimination. The results are very promising: even for relatively small values of the width of the tree decompositions, the new approach shows a notable speed-up in practice. The theoretical analysis of the algorithm already predicts that the new algorithms are asymptotically faster, but it is good to see that the improvement already is clearly visible at small size benchmark instances.

Overall, the rank based approach is an example of the general technique of representativity: a powerful but so far underestimated paradigmatic improvement to dynamic programming. A further exploration of this concept, both in theory (improving the asymptotic running time for problems) as in experiment and algorithm engineering seems highly interesting. Our current paper gives a clear indication of the practical relevance of this concept.

We end this paper with a number of specific points for further study:

- The rank based approach also promises faster algorithms on tree decompositions for several other problems. The experimental evaluation can be executed for other problems. In particular, for HAMILTONIAN CIRCUIT and similar problems, it would be interesting to compare the use of the basis from [5] with the smaller basis given by Cygan et al. [10].
- How well does the *Cut and Count* method perform? As remarked in [11], it seems advantageous to use polynomial identity testing rather then the isolation lemma to optimize the running time.
- Are further significant improvements on the running time possible by using different data structures or variants of the approach, e.g., by not storing table entries as partitions of subsets by identifying them by their row in the matrix $\mathcal{M}$?
- In what extent do results change if we use normal (instead of nice) tree decompositions?
- What is the effect of the ratio between the number of terminals and the number of vertices on the running times?
- Are running time improvements possible by other forms of reduction of tables (without affecting optimality)? If we exploit the two families theorem by Lovász [21], we obtain a variant of our algorithm, with a somewhat different reduce algorithm [14] (see also [22]); how does the running time of this version compare with the running time of the algorithm we studied?
- Can we use the rank based approach to obtain a faster version of the *tour merging* heuristic for TSP by Cook and Seymour [9]? Also, it would be interesting to try a variant of tour merging for other problems, e.g., 'tree merging' as a heuristic for STEINER TREE.
- For what other problems does the rank based approach give faster algorithms in practical settings?
- Are there good heuristic ways of obtaining small representative sets, even for problems where theory tells us that representative sets are large in the worst case?

References

1. Aneja, Y.P.: An integer linear programming approach to the Steiner problem in graphs. Networks 10, 167–178 (1980)
2. Arnborg, S., Lagergren, J., Seese, D.: Easy problems for tree-decomposable graphs. Journal of Algorithms 12, 308–340 (1991)
3. Beasley, J.E.: An algorithm for the Steiner problem in graphs. Networks 14, 147–159 (1984)
4. Bodlaender, H.L.: Dynamic programming algorithms on graphs with bounded treewidth. In: Lepistö, T., Salomaa, A. (eds.) ICALP 1988. LNCS, vol. 317, pp. 105–119. Springer, Heidelberg (1988)

5. Bodlaender, H.L., Cygan, M., Kratsch, S., Nederlof, J.: Deterministic single exponential time algorithms for connectivity problems parameterized by treewidth. In: Fomin, F.V., Freivalds, R., Kwiatkowska, M., Peleg, D. (eds.) ICALP 2013, Part I. LNCS, vol. 7965, pp. 196–207. Springer, Heidelberg (2013)
6. Bodlaender, H.L., Koster, A.M.C.A.: Treewidth computations I. Upper bounds. Information and Computation 208, 259–275 (2010)
7. Borie, R.B., Parker, R.G., Tovey, C.A.: Automatic generation of linear-time algorithms from predicate calculus descriptions of problems on recursively constructed graph families. Algorithmica 7, 555–581 (1992)
8. Chimani, M., Mutzel, P., Zey, B.: Improved Steiner tree algorithms for bounded treewidth. Journal of Discrete Algorithms 16, 67–78 (2012)
9. Cook, W., Seymour, P.D.: Tour merging via branch-decomposition. INFORMS Journal on Computing 15(3), 233–248 (2003)
10. Cygan, M., Kratsch, S., Nederlof, J.: Fast Hamiltonicity checking via bases of perfect matchings. In: Proceedings of the 45th Annual Symposium on Theory of Computing, STOC 2013, pp. 301–310 (2013)
11. Cygan, M., Nederlof, J., Pilipczuk, M., Pilipczuk, M., van Rooij, J., Wojtaszczyk, J.O.: Solving connectivity problems parameterized by treewidth in single exponential time. In: Proceedings of the 52nd Annual Symposium on Foundations of Computer Science, FOCS 2011, pp. 150–159 (2011)
12. Duin, C.: Steiner Problems in Graphs. PhD thesis, University of Amsterdam, Amsterdam, the Netherlands (1993)
13. Fafianie, S., Bodlaender, H.L., Nederlof, J.: Speeding-up dynamic programming with representative sets — an experimental evaluation of algorithms for Steiner tree on tree decompositions. Report on arXiv 1305.7448 (2013)
14. Fomin, F.V., Lokshtanov, D., Saurabh, S.: Efficient computation of representative sets with applications in parameterized and exact algorithms. Report on arXiv 1304.4626 (2013)
15. Hanan, M.: On Steiner's problem with rectilinear distance. SIAM J. Applied Math. 14, 255–265 (1966)
16. Hwang, F., Richards, D.S., Winter, P.: The Steiner Tree Problem. Annals of Discrete Mathematics, vol. 53. Elsevier (1992)
17. Karp, R.M.: Reducibility among combinatorial problems. In: Miller, R.E., Thatcher, J.W. (eds.) Complexity of Computer Computations, pp. 85–104. Plenum Press (1972)
18. Kloks, T.: Treewidth. LNCS, vol. 842. Springer, Heidelberg (1994)
19. Koch, T., Martin, A., Voß, S.: Steinlib, an updated library on Steiner tree problems in graphs. Technical Report ZIB-Report 00-37, Konrad-Zuse Zentrum für Informationstechnik Berlin (2000), http://elib.zib.de/steinlib
20. Korach, E., Solel, N.: Linear time algorithm for minimum weight Steiner tree in graphs with bounded treewidth. Technical Report 632, Technion, Haifa, Israel (1990)
21. Lovász, L.: Flats in matroids and geometric graphs. In: Combinatorial Surveys. Proceedings 6th Britisch Combinatorial Conference, pp. 45–86. Academic Press (1977)
22. Marx, D.: A parameterized view on matroid optimization problems. Theoretical Computer Science 410, 4471–4479 (2009)
23. Monien, B.: How to find long paths efficiently. Annals of Discrete Mathematics 25, 239–254 (1985)
24. Robertson, N., Seymour, P.D.: Graph minors. II. Algorithmic aspects of tree-width. Journal of Algorithms 7, 309–322 (1986)

25. Telle, J., Proskurowski, A.: Efficient sets in partial k-trees. Discrete Applied Mathematics 44, 109–117 (1993)
26. Treewidthlib (2004), http://www.cs.uu.nl/people/hansb/treewidthlib
27. Wald, J.A., Colbourn, C.J.: Steiner trees, partial 2-trees, and minimum IFI networks. Networks 13, 159–167 (1983)
28. Warme, D., Winter, P., Zachariasen, M.: GeoSteiner, software for computing Steiner trees, http://www.diku.dk/hjemmesider/ansatte/martinz/geosteiner/
29. Wei-Kleiner, F.: Tree decomposition based Steiner tree computation over large graphs. Report on arXiv 1305.5757 (2013)
30. Winter, P.: Steiner problem in networks: A survey. Networks 17, 129–167 (1987)

Completeness Results
for Parameterized Space Classes

Christoph Stockhusen and Till Tantau

Institute for Theoretical Computer Science
Universität zu Lübeck
23538 Lübeck, Germany
{stockhus,tantau}@tcs.uni-luebeck.de

Abstract. The parameterized complexity of a problem is generally considered "settled" once it has been shown to lie in FPT or to be complete for a class in the W-hierarchy or a similar parameterized hierarchy. Several natural parameterized problems have, however, resisted such a classification. At least in some cases, the reason is that upper and lower bounds for their parameterized *space* complexity have recently been obtained that rule out completeness results for parameterized *time* classes. In this paper, we make progress in this direction by proving that the associative generability problem and the longest common subsequence problem are complete for parameterized space classes. These classes are defined in terms of different forms of bounded nondeterminism and in terms of simultaneous time–space bounds. As a technical tool we introduce a "union operation" that translates between problems complete for classical complexity classes and for W-classes.

1   Introduction

Parameterization has become a powerful paradigm in complexity theory, both in theory and practice. Instead of just considering the runtime of an algorithm as a function of the input *length*, we analyse the runtime as a multivariate function depending on a number of different problem *parameters*, the input length being just one of them. While in classical complexity theory instead of "runtime" many other resource bounds have been studied in great detail, in the parameterized world the focus has lain almost entirely on time complexity. This changed when in a number of papers [1–3] it was shown for different natural problems, including the vertex cover problem, the feedback vertex set problem, and the longest common subsequence problem, that their parameterized *space* complexity is of interest. Indeed, the parameterized space complexity of natural problems can explain why some problems in FPT are easier to solve than others (namely, because they lie in much smaller space classes) and why some problems cannot be classified as complete for levels of the weft hierarchy (namely, because upper and lower bounds on their space complexity rule out such completeness results unless unlikely collapses occur).

G. Gutin and S. Szeider (Eds.): IPEC 2013, LNCS 8246, pp. 335–347, 2013.

Our Contributions. In the present paper, we present completeness results of natural parameterized problems for different parameterized space complexity classes. The classes we study are of two kinds: First, parameterized classes of *bounded nondeterminism* and, second, parameterized classes where the *space and time* resources of the machines are bounded *simultaneously.* In both cases, we introduce the classes for systematic reasons, but also because they are needed to classify the complexity of the natural problems that we are interested in.

In the context of bounded nondeterminism, we introduce a general "union operation" that turns any language into a parameterized problem in such a way that completeness of the language for some complexity class C carries over to completeness of the parameterized problem for a class "paraWC," which we will define rigorously later. Building on this result, we show that many union versions of graph problems are complete for paraWNL and paraWL, but the theorem can also be used to show that p-WEIGHTED-SAT is complete for paraWNC[1]. Our technically most challenging result is that the associative generability problem parameterized by the generator set size is complete for the class paraWNL.

Regarding time–space classes, we present different problems that are complete for the class of problems solvable "nondeterministically in fixed-parameter time and slice-wise logarithmic space." Among these problems are the longest common subsequence problem parameterized by the number of strings, but also the acceptance problem for certain cellular automata parameterized by the number of cells.

Related Work. Early work on parameterized space classes is due to Cai et al. [1] who introduced the classes para-L and para-NL, albeit under different names, and showed that several important problems in FPT lie in these classes: the parameterized vertex cover problem lies in para-L and the parameterized k-leaf spanning tree problem lies in para-NL. Later, Flum and Grohe [4] showed that the parameterized model checking problem of first-order formulas on graphs of bounded degree lies in para-L. In particular, standard parameterized graph problems belong to para-L when we restrict attention to bounded-degree graphs. Recently, Guillemot [3] showed that the longest common subsequence problem (LCS) is equivalent under fpt-reductions to the short halting problem for NTMs, where the time and space bounds are part of the input and the space bound is the parameter. Our results differ from Guillemot's insofar as we use weaker reductions (para-L- rather than fpt-reductions) and prove completeness for a class defined using a machine model rather than for a class defined as a reduction closure. The paper [2] by Elberfeld and us is similar to the present paper insofar as it also introduces new parameterized space complexity classes and presents upper and lower bounds for natural parameterized problems. The core difference is that in the present paper we focus on completeness results for natural problems rather than "just" on upper and lower bounds.

Organisation of This Paper. In Section 2 we review the parameterized space classes previously studied in the literature and introduce some new classes that will be needed in the later sections. For some of the classes from the literature we propose new names in order to systematise the naming and to make connections between the different classes easier to spot. In Section 3 we study problems complete for classes defined in terms of bounded nondeterminism, in Section 4 we do the same for time–space classes.

Due to lack of space, all proofs have been omitted. They can be found in the technical report version of this paper [5].

2   Parameterized Space Classes

Before we turn our attention to parameterized *space* classes, let us first review some basic terminology. As in [2], we define a *parameterized problem* as a pair (Q, κ) of a language $Q \subseteq \Sigma^*$ and a parameterization $\kappa \colon \Sigma^* \to \mathbb{N}$ that maps input instances to parameter values and that is computable in logarithmic space.[1] For a classical complexity class C, a parameterized problem (Q, κ) belongs to the *para-class* para-C if there are an alphabet Π, a computable function $\pi \colon \mathbb{N} \to \Pi^*$, and a language $A \subseteq \Sigma^* \times \Pi^*$ with $A \in C$ such that for all $x \in \Sigma^*$ we have $x \in Q \iff \big(x, \pi(\kappa(x))\big) \in A$. The problem is in the *X-class* XC if for every number $w \in \mathbb{N}$ the slice $Q_w = \{\, x \mid x \in Q \text{ and } \kappa(x) = w \,\}$ lies in C. It is immediate from the definition that para-$C \subseteq$ XC holds.

The "popular" class FPT is the same as para-P. In terms of the O-notation, a parameterized problem (Q, κ) is in para-P if there is a function $f \colon \mathbb{N} \to \mathbb{N}$ such that the question $x \in Q$ can be decided within time $f(\kappa(x)) \cdot |x|^{O(1)}$. By comparison, (Q, κ) is in para-L if $x \in Q$ can be decided within space $f(\kappa(x)) + O(\log |x|)$; and for para-PSPACE the space requirement is $f(\kappa(x)) \cdot |x|^{O(1)}$. The class XP is in wide use in parameterized complexity theory; the logarithmic space classes XL and XNL have previously been studied by Chen et al. [4, 8].

To simplify the notation, let us write f_x for $f\big(\kappa(x)\big)$ and n for $|x|$ in the following. Then the time bound for para-P can be written as $f_x n^{O(1)}$ and the space bound for para-L as $f_x + O(\log n)$.

Parameterized logspace reductions (para-L-reductions) are the natural restriction of fpt-reductions to logarithmic space: A para-L-reduction from a parameterized problem (Q_1, κ_1) to (Q_2, κ_2) is a mapping $r \colon \Sigma_1^* \to \Sigma_2^*$ such that

1. for all $x \in \Sigma_1^*$ we have $x \in Q_1 \iff r(x) \in Q_2$,
2. $\kappa_2\big(r(x)\big) \leq g\big(\kappa_1(x)\big)$ for some computable function g, and,
3. r is para-L-computable with respect to κ_1 (that is, there is a Turing machine that outputs $r(x)$ on input x and needs space at most $f(\kappa_1(x)) + O(\log |x|)$ for some computable function f).

Using standard arguments one can show that all classes in this paper are closed with respect to para-L-reductions; with the possible exception of paraWNC[1], a class we encounter in Theorem 3.2. Throughout this paper, all completeness and hardness results are meant with respect to para-L-reductions.

---

[1] In the classical definition, Downey and Fellows [6] just require the parameterization to be computable, while Flum and Grohe [7] require it to be computable in polynomial time. Whenever the parameter is part of the input, it is certainly computable in logarithmic space.

2.1   Parameterized Bounded Nondeterminism

While the interplay of nondeterminism and parameterized space may seem to be simple at first sight (NL is closed under complement and NPSPACE is even equal to PSPACE, so only XNL and para-NL appear interesting), a closer look reveals that useful and interesting new classes arise when we bound the amount of nondeterminism used by machines in dependence on the parameter. For this, it is useful to view nondeterministic computations as deterministic computations using "choice tapes" or "tapes filled with nondeterministic bits." These are extra tapes for a deterministic Turing machine, and an input word is accepted if there is at least one bitstring that we can place on this extra tape at the beginning of the computation such that the Turing machine accepts. It is well known that NP and NL can be defined in this way using deterministic polynomial-time or logarithmic-space machines, respectively, that have *one-way* access to a choice tape. (For NP it makes no difference whether we have one- or two-way access, but logspace DTMs with access to a two-way choice tape can accept all of NP.)

Classes of *bounded* nondeterminism arise when we restrict the length of the bitstrings on the choice tape. For instance, the classes β^h for $h \geq 1$, see [9] and also [10] for variants, are defined in the same way as NP above, only the length of the bitstring on the choice tape may be at most $O(\log^h n)$. Classes of *parameterized bounded* nondeterminism arise when we restrict the length the bitstring on the choice tape in dependence not only on the input length, but also of the parameter. Furthermore, in the context of bounded space computations, it also makes a difference whether we have one-way or two-way access to the choice tapes.

Definition 2.1. *Let C be a complexity class defined in terms of a deterministic Turing machine model (like L or P). We define para$\exists^{\leftrightarrow}C$ as the class of parameterized problems (Q, κ) for which there exists a C-machine M, an alphabet Π, and a computable function $\pi \colon \mathbb{N} \to \Pi^*$ such that: For every $x \in \Sigma^*$ we have $x \in Q$ if, and only if, there exists a bitstring $b \in \{0,1\}^*$ such that M accepts with $(x, \pi(\kappa(x)))$ on its input tape and b on the two-way choice tape. We define para$\exists^{\rightarrow}C$ similarly, only access to the choice tape is now one-way.*

We define para$\exists^{\leftrightarrow}_{f\log}C$ and para$\exists^{\rightarrow}_{f\log}C$ in the same way, but the length of b may be at most $|\pi(\kappa(x))| \cdot O(\log n)$.

Observe that, as argued earlier, para$\exists^{\leftrightarrow}$L = para$\exists^{\leftrightarrow}$P = para$\exists^{\rightarrow}$P = para-NP and para$\exists^{\rightarrow}$L = para-NL. Also observe that para$\exists^{\leftrightarrow}_{f\log}$P = para$\exists^{\rightarrow}_{f\log}$P = W[P] by one of the many possible definitions of W[P].

The above definition can easily be extended to the case where a universal quantifier is used instead of an existential one and where *sequences* of quantifiers are used. This is interpreted in the usual way as having a choice tape for each quantifier and the different "exists ... for all"-conditions must be met in the order the quantifiers appear. For instance, for problems in para$\exists^{\leftrightarrow}_{f\log}\exists^{\rightarrow}$L we have $x \in Q$ if, and only if, there exists a bitstring of length $f_x \log_2 n$ for the first, two-way-readable choice tape for which an NL-machine accepts. The classes para-NL[$f \log$], para-L-cert, and para-NL-cert introduced in an ad hoc manner by

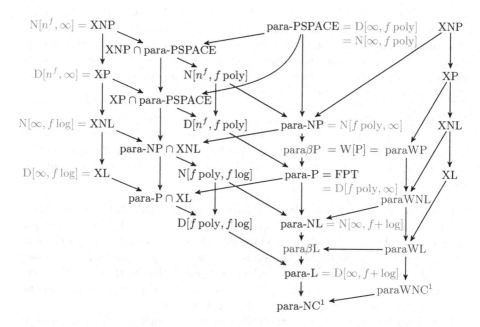

Fig. 1. In this inclusion diagram bounded nondeterminism classes are shown in red and time–space classes in blue. The X-classes are shown twice to keep the diagram readable. All known inclusions are indicated, where $C \to D$ means $C \supseteq D$.

Elberfeld et al. in [2] can now be represented systematically: They are $\text{para}\exists^{\to}_{f\log}L$, $\text{para}\exists^{\leftrightarrow}_{f\log}L$, and $\text{para}\exists^{\leftrightarrow}_{f\log}\exists^{\to}L$, respectively.

In order to make the notation more useful in practice, instead of "$\exists^{\to}$" let us write "N" and instead of "$\exists^{\to}_{f\log}$" we write "β" as is customary. As a new notation, instead of "$\exists^{\leftrightarrow}_{f\log}$" and "$\forall^{\leftrightarrow}_{f\log}$" we write "W" and "$W_\forall$," respectively. The three classes of [2] now become $\text{para}\beta L$, paraWL, and paraWNL.

Our reasons for using "W" to denote $\exists^{\leftrightarrow}_{f\log}$ will be explained fully in Section 3; for the moment just observe that $W[P] = \text{paraWP}$ holds. To get a better intuition on the W-operator, note that it provides machines with "$f_x \log_2 n$ bits of nondeterministic information" or, equivalently, with "f_x many nondeterministic positions in the input" and these bits are provided as part of the input. This allows us to also apply the W-operator to classes like NC^1 that are not defined in terms of Turing machines.

The right half of Figure 1 depicts the known inclusions between the introduced classes, the left half shows the classes introduced next.

2.2 Parameterized Time–Space Classes

In classical complexity theory, the major complexity classes are either defined in terms of time complexity (P, NP, EXP) or in terms of space complexity (L, NL, PSPACE), but not both at the same time: by the well-known inclusion chain

$L \subseteq NL \subseteq P \subseteq NP \subseteq PSPACE = NPSPACE \subseteq EXP$ space and time are intertwined in such a way that bounding either automatically bounds the other in a specific way (at least for the major complexity classes). In the parameterized world, interesting new classes arise when we restrict time and space simultaneously: namely whenever the time is "para-restricted" while space is "X-restricted" or *vice versa*.

Definition 2.2. *For a space bound s and a time bound t, both of which may depend on a parameter k and the input length n, let* $D[t, s]$ *denote the class of all parameterized problems that can be accepted by a deterministic Turing machine in time* $t(\kappa(x), |x|)$ *and space* $s(\kappa(x), |x|)$. *Let* $N[t, s]$ *denote the corresponding nondeterministic class.*

Four cases are of interest: First, $D[f\text{poly}, f\log]$, meaning that $t(k, n) = f(k) \cdot n^{O(1)}$ and $s(k, n) = f(k) \cdot O(\log n)$, contains all problems that are "fixed parameter tractable via a machine needing only slice-wise logarithmic space," and, second, the nondeterministic counterpart $N[f\text{poly}, f\log]$. The two other cases are $D[n^f, f\text{poly}]$ and $N[n^f, f\text{poly}]$, which contain problems that are "in slice-wise polynomial time via machines that need only fixed parameter polynomial space." See Figure 1 for the trivial inclusions between the classes.

In Section 4 we will see that these classes are not only of scholarly interest. Rather, we will show that LCS parameterized by the number of input strings is complete for $N[f\text{poly}, f\log]$.

3   Complete Problems for Bounded Nondeterminism

In this section we present new natural problems that are complete for paraWNL and paraWL. Previously, it was only known that the following "colored reachability problem" [2] is complete for paraWNL: We are given an edge-colored graph, two vertices s and t, and a parameter k. The question is whether there is a path from s to t that uses only k colors. Our key tool for proving new completeness results will be the introduction of a "union operation," which turns P-, NL-, and L-complete problems into paraWP-, paraWNL-, and paraWL-complete problems, respectively. Building on this, we prove the parameterized associative generability problem to be complete for paraWNL. Note that the underlying classical problem is well-known to be NL-complete and, furthermore, if we drop the requirement of associativity, the parameterized and classical versions are known to be complete for paraWP and P, respectively.

At this point, we remark that Guillemot, in a paper [3] on parameterized *time* complexity, uses "WNL" to denote a class different from the class paraWNL defined in this paper. Guillemot chose the name because his definition of the class is derived from one possible definition of W[1] by replacing a time by a space constraint. Nevertheless, we believe that our definition of a "W-operator" yields the "right analogue" of W[P]: First, there is the above pattern that parameterized version of problems complete P, NL, and L tend to be complete for paraWP, paraWNL, and paraWL, respectively. Furthermore, in Section 4 we show that the class WNL defined and studied by Guillemot is exactly the fpt-reduction closure of the time–space class $N[f \text{ poly}, f \log]$.

Union Problems. For numerous problems studied in complexity theory the input consists of a string in which some positions can be "selected" and the objective is to select a "good" subset of these positions. For instance, for the satisfiability problem we must select some variables such that setting them to true makes a formula true; for the circuit satisfiability problem we must select some input gates such that when they are set to 1 the circuit evaluates to 1; and for the exact cover problem we must select some sets from a family of sets so that they form a partition of the union of the family. In the following, we introduce some terminology that allows us to formulate all of these problems in a uniform way and to link them to the W-operator.

Let Σ be an alphabet that contains none of the three special symbols ?, 0, and 1. We call a word $t \in (\Sigma \cup \{?\})^*$ a *template*. We call a word $s \in (\Sigma \cup \{0,1\})^*$ an *instantiation of t* if s is obtained from t by replacing exactly the ?-symbols arbitrarily by 0- or 1-symbols. Given instantiations $s_1, \ldots, s_k$ of the same template t, their *union s* is the instantiation of t that has a 1 exactly at those positions i where at least one s_j has a 1 at position i (the union is the "bitwise or" of the instantiated positions and is otherwise equal to the template).

Given a language $A \subseteq (\Sigma \cup \{0,1\})^*$, we define three different kinds of union problems for A. Each of them is a parameterized problem where the parameter is k. As we will see in a moment, the first kind is linked to the W-operator while the last kind links several well-known languages from classical complexity theory to well-known parameterized problems. We will also see that the three kinds of union problems for a language A often all have the same complexity.

1. The input for p-FAMILY-UNION-A are a template $t \in (\Sigma \cup \{?\})^*$ and a family $(S_1, \ldots, S_k)$ of k sets of instantiations of t. The question is whether there are $s_i \in S_i$ for $i \in \{1, \ldots, k\}$ such that the union of $s_1, \ldots, s_k$ lies in A.
2. The input for p-SUBSET-UNION-A are a template $t \in (\Sigma \cup \{?\})^*$, a set S of instantiations of t, and a number k. The question is whether there exists a subset $R \subseteq S$ of size $|R| = k$ such that the union of R's elements lies in A.
3. The input for p-WEIGHTED-UNION-A are a template $t \in (\Sigma \cup \{?\})^*$ and a number k. The question is whether there exists an instantiation s of t containing exactly k many 1-symbols such that $s \in A$?

To get an intuition for these definitions, think of instantiations as words written on transparencies with 0 rendered as an empty box and 1 as a checked box. Then for the family union problem we are given k heaps of transparencies and the task is to pick one transparency from each heap such that "stacking them on top of each other" yields an element of A. For the subset union problem, we are only given one stack and must pick k elements from it. We call the weighted union problem a "union" problem partly in order to avoid a clash with existing terminology and partly because the weighted union problem is the same as the subset union problem for the special set S containing all instantiations of the template of weight 1.

Concerning the promised link between well-known languages and parameterized problems, consider $A = $ CIRCUIT-VALUE-PROBLEM (CVP) where we use Σ to encode a circuit and use 0's and 1's solely to describe an assignment to the input gates. Then the input for p-WEIGHTED-UNION-CVP are a circuit with ?-symbols instead of a concrete assignment together with a number k, and the question is whether we can replace exactly k of the ?-symbols by 1's (and the other by 0's) so that the resulting instantiation lies in CVP. Clearly, p-WEIGHTED-UNION-CVP is exactly the W[P]-complete problem p-CIRCUIT-SAT, which asks whether there is a satisfying assignment for a given circuit that sets exactly k input gates to 1.

Concerning the promised link between the union problems and the W-operator, recall that the operator provides machines with f_x nondeterministic indices as part of the input. In particular, a W-machine can mark f_x different "parts" of the input – like one element from each of f_x many sets in a family, like the elements of a size-f_x subset of some set, or like f_x many positions in a template. With this observation it is not difficult to see that if $A \in C$, then all union versions of A lie in paraWC. A much deeper observation is that the union versions are also often *complete* for these classes. In the next theorem, which states this claim precisely, the following definition of a *compatible logspace projection* p from a language A to a language B is used: First, p must be a logspace reduction from A to B. Second, p is a projection, meaning that each symbol of $p(x)$ depends on at most one symbol of x. Third, for each word length n there is a single template t_n such for all $x \in \Sigma^n$ the word $p(x)$ is an instantiation of t_n.

Theorem 3.1. *Let $C \in \{\mathrm{NC}^1, \mathrm{L}, \mathrm{NL}, \mathrm{P}\}$. Let A be complete for C via compatible logspace projections. Then p-FAMILY-UNION-A is complete for* paraWC *under para-L-reductions.*[2]

Parameterized Satisfiability Problems. Recall that the problem p-WEIGHTED-UNION-CVP equals p-CIRCUIT-SAT. Since one can reduce p-FAMILY-UNION-CVP to p-WEIGHTED-UNION-CVP (via essentially the same reduction as that used in the proof of Theorem 3.2 below), Theorem 3.1 provides us with a direct proof that p-CIRCUIT-SAT $= p$-WEIGHTED-UNION-CVP is complete for paraWP. We get an even more interesting result when we apply the theorem to BF, the propositional formula evaluation problem. We encode pairs of formulas and assignments in the straightforward way by using 0 and 1 solely for the assignment. Since BF is complete for NC^1 under compatible logspace reductions, see [11, 12], p-FAMILY-UNION-BF is complete for paraWNC^1 by Theorem 3.1. By further reducing the problem to p-WEIGHTED-UNION-BF, we obtain:

Theorem 3.2. *p-WEIGHTED-UNION-BF is para-L-complete[3] for* paraWNC^1.

By definition, W[SAT] is the fpt-reduction closure of p-WEIGHTED-SAT $= p$-WEIGHTED-UNION-BF. Thus, by the theorem, W[SAT] is also the fpt-reduction

---

[2] The proof shows that the theorem actually also holds for any "reasonable" class C and any "reasonable" weaker reduction.

[3] As in Theorem 3.1 one can also use weaker reductions.

closure of paraWNC^1 – a result that may be of independent interest. For example, it shows that NC^1 = P implies W[SAT] = W[P]. Note that we do not claim W[SAT] = paraWNC^1 since paraWNC^1 is presumably not closed under fpt-reductions.

Graph Problems. In order to apply Theorem 3.1 to standard graph problems like REACH or CYCLE, we encode graphs using adjacency matrices consisting of 0- and 1-symbols. Then a template is always a string of n^2 many ?-symbols for n vertex graphs. The "colored reachability problem" mentioned at the beginning of this section equals p-SUBSET-UNION-REACH.[4] Note that any reduction to a union problem for this encoding is automatically compatible as long as the number of vertices in the reduction's output depends only on the length of its input.

Applying Theorem 3.1 to standard L- or NL-complete problems yields that their family union versions are complete for paraWL and paraWNL, respectively. By reducing the family versions further to the subset union version, we get the following:

Theorem 3.3. *For $A \in \{$REACH, DAG-REACH, CYCLE$\}$, p-SUBSET-UNION-A is complete for* paraWNL, *while for $B \in \{$UNDIRECTED-REACH, TREE, FOREST, UNDIRECTED-CYCLE$\}$, p-SUBSET-UNION-B is complete for* paraWL.

Associative Generability. The last union problem we study is based on the GENERATORS problem, which contains tuples $(U, \circ, x, G)$ where U is a set, $\circ \colon U^2 \to U$ is (the table of) a binary operation, $x \in U$, and $G \subseteq U$ is a set. The question is whether the closure of G under $\circ$ (the smallest superset of G closed under $\circ$) contains x. A restriction of this problem is ASSOCIATIVE-GENERATOR, where $\circ$ must be associative. By two classical results, GENERATORS is P-complete [13] and ASSOCIATIVE-GENERATOR is NL-complete [14].

In order to apply the union operation to generator problems, we encode $(U, \circ, x, G)$ as follows: U, $\circ$, and x are encoded in some sensible way using the alphabet Σ. To encode G, we add a 1 after the elements of U that are in G and we add a 0 after *some* elements of U that are not in G. This means that in the underlying templates we get the freedom to specify that only some elements of U may be chosen for G. Now, p-WEIGHTED-UNION-GENERATORS equals the problem known as p-GENERATORS in the literature: Given $\circ$, a subset $C \subseteq U$ of generator candidates, a parameter k, and a target element x, the question is whether there exists a set $G \subseteq C$ of size $|G| = k$ such that the closure of G under $\circ$ contains x. Flum and Grohe [7] have shown that p-GENERATORS is complete for W[P] = paraWP (using a slightly different problem definition that has the same complexity, however). Similarly, p-WEIGHTED-UNION-ASSOCIATIVE-GENERATOR is also known as p-AGEN and we show:

Theorem 3.4. *p-AGEN is complete for* paraWNL.

---

[4] For exact equality, in the colored reachability problem we must allow edges to have several colors, but this is does not change the problem complexity.

With the machinery introduced in this section, this result may not seem surprising: ASSOCIATIVE-GENERATORS is known to be complete for NL via compatible logspace reductions and, thus, by Theorem 3.1, p-FAMILY-UNION-ASSOCI-ATIVE-GENERATORS is complete for paraWNL. To prove Theorem 3.4 we "just" need to further reduce to the weighted union version. However, unlike for satisfiability and graph problems, this reduction turns out to be technically difficult.

4   Problems Complete for Time–Space Classes

The classes para-P = FPT and XL appear to be incomparable: Machines for the first class may use $f_x n^{O(1)}$ time and as much space as they want (which will be at most $f_x n^{O(1)}$), while machines for the second class may use $f_x \log n$ space and as much time as they want (which will be at most n^{f_x}). A natural question is which problems are in the intersection para-P $\cap$ XL or – even better – in the class D[f poly, f log], which means that there is a *single* machine using only fixed-parameter time and slice-wise logarithmic space simultaneously.

It is not particularly hard to find artificial problems that are complete for the different time–space classes introduced in Section 2.2; we present such problems at the beginning of this section. We then move on to automata problems, but still some ad hoc restrictions are needed to make the problems complete for time–space classes. The real challenge lies in finding problems together with *natural* parameterization that are complete. We present one such problem: the longest common subsequence problem parameterized by the number of strings.

Resource-Bounded Machine Acceptance. A good starting point for finding complete problems for new classes is typically some variant of Turing machine acceptance (or halting). Since we study machines with simultaneous time–space limitations, it makes sense to start with the following "time and space bounded computation" problems: For DTSC the input is a single-tape DTM M together with two numbers s and t given in unary. The question is whether M accepts the empty string making at most t steps and using at most s tape cells. The problem NTSC is the nondeterministic variant. As observed by Cai et al. [15], the fpt-reduction closure of p_t-NTSC (that is, the problem parameterized by t) is exactly W[1]. In analogy, Guillemot [3] proposed the name "WNL" for the fpt-reduction closure of p_s-NTSC (now parameterized by s rather than t). As pointed out in Section 3, we believe that this name should be reserved for the class resulting from applying the operator $\exists^{\leftrightarrow}_{f \log}$ to the class NL. Furthermore, the following theorem shows that p_s-NTSC is better understood in terms of time–space classes:

Theorem 4.1. *The problems p_s-DTSC and p_s-NTSC are complete for the classes* D[f poly, f log] *and* N[f poly, f log]*, respectively.*

Automata. A classical result of Hartmanis [16] states that L contains exactly the languages accepted by finite multi-head automata. In [2], Elberfeld et al. used this to show that p_{heads}-MDFA (the multi-head automata acceptance problem parameterized by the number of heads) is complete for XL. It turns out that multi-head automata can also be used to define a (fairly) natural complete problem for D[f poly, f log]: A DAG-*automaton* is an automaton whose transition graph is a topologically sorted DAG (formally, the states must form the set $\{1, \ldots, |Q|\}$ and the transition function must map each state to a strictly greater state). Clearly, a DAG-automata will never need more than $|Q|$ steps to accept a word, which allows us to prove the following theorem:

Theorem 4.2. *The problems p_{heads}-DAG-MDFA and p_{heads}-DAG-MNFA are complete for D[f poly, f log] and N[f poly, f log], respectively.*

Instead of DAG-automata, we can also consider a "bounded time version" of MDFA and MNFA, where we ask whether the automaton accepts within s steps (s being given in unary). Both versions are clearly equivalent: The number of nodes in the DAG bounds the number of steps the automaton can make and cyclic transitions graphs can be made acyclic by making s layered copies.

Another, rather natural kind of automata are *cellular automata,* where there is one instance of the automaton (called a *cell*) for each input symbol. The cells perform individual synchronous computations, but "see" the states of the two neighbouring cells (we only consider one-dimensional automata, but the results hold for any fixed number of dimensions). Formally, the transition function of such an automaton is a function $\delta\colon Q^3 \to Q$ (for the cells at the left and right end this has to be modified appropriately). The "input" is just a string $q_1 \ldots q_k \in Q^*$ of states and the question is whether k cells started in the states q_1 to q_k will arrive at a situation where one of them is in an accepting state (one can also require all to be in an accepting state, this makes no difference).

Let DCA be the language $\{(C, q_1 \ldots q_k) \mid C$ is a deterministic cellular automaton that accepts $q_1 \ldots q_k\}$. Let NCA denote the nondeterministic version and let DAG-DCA and DAG-NCA be the versions where C is required to be a DAG-automaton (meaning that δ must always output a number strictly larger than all its inputs). The following theorem states the complexity of the resulting problems when we parameterize by k (number of cells):

Theorem 4.3. *The problems p_{cells}-DCA and p_{cells}-NCA are complete for XL and XNL, respectively. The problems p_{cells}-DAG-DCA and p_{cells}-DAG-NCA are complete for D[$f$poly, f log] and N[$f$poly, f log], respectively.*

We remark that, for once, the nondeterministic cases need special arguments.

Longest Common Subsequence. The input for the longest common subsequence problem LCS is a set S of strings over some alphabet Σ together with a number l. The question is whether there is a string $c \in \Sigma^l$ that is a *subsequence* of all strings in S, meaning that for all $s \in S$ just by removing symbols from s we arrive at c.

There are several natural parameterization of LCS: We can parameterize by the number of strings in S, by the size of the alphabet, by the length l, or any combination thereof. Guillemot has shown [3] that $p_{\text{strings,length}}$-LCS is fpt-complete for W[1], while p_{strings}-LCS is fpt-equivalent to p_s-NTSC. Hence, by Theorem 4.1, both problems are complete under fpt-reductions for the fpt-reduction closure of

N[f poly, f log]. We tighten this in Theorem 4.6 below (using a weaker reduction is more than a technicality: N[f poly, f log] is presumably not even closed under fpt-reduction, while it *is* closed under para-L-reductions).

As a preparation for the proof of Theorem 4.6, we first present a simpler-to-prove result: Let LCS-INJECTIVE denote the restriction of LCS where all input words must be *p-sequences* [17], which are words containing any symbol at most once (the function mapping word indices to word symbols is injective).

Theorem 4.4. LCS-INJECTIVE *is* NL-*complete and this holds already under the restriction* $|S| \leq 4$.

Although we do not prove this, we remark that NL-completeness already holds for $|S| = 3$, while for $|S| = 2$ the complexity appears to drop significantly.

Corollary 4.5. p_{strings}-LCS-INJECTIVE *is para-L-complete for* para-NL.

Theorem 4.6. p_{strings}-LCS *is para-L-complete for* N[$f$poly, f log].

5 Conclusion

Bounded nondeterminism plays a key role in parameterized complexity theory since it lies at the heart of the definition of important classes like W[P], but also of W[1]. In the present paper we introduced a "W-operator" that cannot only be applied to P, yielding paraWP, but also to classes like NL or NC^1. We showed that "union versions" of problems complete for P, NL, and L tend to be complete for paraWP, paraWNL, and paraWL. Several important problems studied in parameterized complexity turn out to be union problems, including p-CIRCUIT-SAT and p-WEIGHTED-SAT, and we could show that the latter problem is complete for paraWNC^1. For the associative generability problem p-AGEN, which is also a union problem, we established its paraWNL-completeness. An interesting open problem is determining the complexity of the "universal" version of AGEN, where the question is whether *all* size-k subsets of the universe are generators. Possibly, this problem is complete for paraW$_\forall$NL.

We showed that different problems are complete for the time–space class N[f poly, f log]. We shied away from presenting complete problem for the classes D[n^f, f poly] and N[n^f, f poly] because in their definition we need restrictions like "the machine may make at most n^k steps where k is the parameter." Such artificial parameterizations have been studied, though: In [7, Theorem 2.25] Flum and Grohe show that "p-EXP-DTM-HALT" is complete for XP. Adding a unary upper bound on the number of steps to the definition of the problem yields a problem easily seen to be complete for D[n^f, f poly]. Finding a *natural* problem complete for the latter class is, however, an open problem.

Acknowledgements. We would like to thank Michael Elberfeld for helping us with the proof of Theorem 3.4 and the anonymous reviewers for their comments.

References

1. Cai, L., Chen, J., Downey, R.G., Fellows, M.R.: Advice classes of parameterized tractability. Annals of Pure and Applied Logic 84(1), 119–138 (1997)
2. Elberfeld, M., Stockhusen, C., Tantau, T.: On the Space Complexity of Parameterized Problems. In: Thilikos, D.M., Woeginger, G.J. (eds.) IPEC 2012. LNCS, vol. 7535, pp. 206–217. Springer, Heidelberg (2012)
3. Guillemot, S.: Parameterized complexity and approximability of the longest compatible sequence problem. Discrete Optimization 8(1), 50–60 (2011)
4. Flum, J., Grohe, M.: Describing parameterized complexity classes. Information and Computation 187, 291–319 (2003)
5. Stockhusen, C., Tantau, T.: Completeness results for parameterized space classes. Technical Report arXiv:1308.2892, ArXiv e-prints (2013)
6. Downey, R.G., Fellows, M.R.: Parameterized Complexity. Springer (1999)
7. Flum, J., Grohe, M.: Parameterized Complexity Theory. Springer (2006)
8. Chen, Y., Flum, J., Grohe, M.: Bounded nondeterminism and alternation in parameterized complexity theory. In: Proceedings of the 18th IEEE Annual Conference on Computational Complexity, CCC 2003, pp. 13–29 (2003)
9. Kintala, C.M.R., Fischer, P.C.: Refining nondeterminism in relativized polynomial-time bounded computations. SIAM Journal on Computing 1(9), 46–53 (1980)
10. Buss, J., Goldsmith, J.: Nondeterminism within P. SIAM Journal on Computing, 560–572 (1993)
11. Buss, S.R.: The Boolean formula value problem is in ALOGTIME. In: Proceedings of the 19th Annual ACM Symposium on Theory of Computing, STOC 1987, pp. 123–131. ACM (1987)
12. Buss, S., Cook, S., Gupta, A., Ramachandran, V.: An optimal parallel algorithm for formula evaluation. SIAM Journal on Computing 21(4), 755–780 (1992)
13. Jones, N.D., Laaser, W.T.: Complete problems for deterministic polynomial time. Theoretical Computer Science 3(1), 105–117 (1976)
14. Jones, N.D., Lien, Y.E., Laaser, W.T.: New problems complete for nondeterministic log space. Mathematical Systems Theory 10(1), 1–17 (1976)
15. Cai, L., Chen, J., Downey, R.G., Fellows, M.R.: On the parameterized complexity of short computation and factorization. Archive for Mathematical Logic 36(4-5), 321–337 (1997)
16. Hartmanis, J.: On non-determinancy in simple computing devices. Acta Informatica 1, 336–344 (1972)
17. Fellows, M., Hallett, M., Stege, U.: Analogs & duals of the MAST problem for squences & trees. Journal of Algorithms 49(1), 192–216 (2003)

Treewidth and Pure Nash Equilibria

Antonis Thomas[1] and Jan van Leeuwen[2]

[1] Institute of Theoretical Computer Science, ETH Zurich,
8092 Zurich, Switzerland
athomas@inf.ethz.ch
[2] Department of Information and Computing Sciences, Utrecht University,
3584 CC Utrecht, The Netherlands
J.vanLeeuwen1@uu.nl

Abstract. We consider the complexity of w-PNE-GG, the problem of computing pure Nash equilibria in graphical games parameterized by the treewidth w of the underlying graph. It is well-known that the problem of computing pure Nash equilibria is NP-hard in general, but in polynomial time when restricted to games of bounded treewidth. We now prove that w-PNE-GG is $W[1]$-hard. Next we present a dynamic programming approach, which in contrast to previous algorithms that rely on reductions to other problems, directly attacks w-PNE-GG. We show that our algorithm is in FPT for games with strategy sets of bounded cardinality. Finally, we discuss the implications for solving games of $O(\log n)$ treewidth, the existence of polynomial kernels for w-PNE-GG, and constructing a sample or a maximum-payoff pure Nash equilibrium.

1 Introduction

The computation of solution concepts of finite games is a fundamental class of problems arising in algorithmic game theory. The computation of Nash equilibria is a case in point. Several recent breakthroughs have settled the complexity of computing approximate mixed Nash equilibria [9,6]. Such equilibria are guaranteed to exist, but are very fragile as models of behavior and rationality. On the other hand, *pure Nash equilibria* are more intuitive but they do not exist in every game.

Games are commonly represented in normal form, i.e. with the payoff of each player defined by a matrix with one column for each combination of all players' actions. Lately, it is widely noted that more succinct representations for multi-party game theory are essential, since most large games of any practical interest have highly structured payoff functions. A prime example is the *graphical games* representation, introduced by Kearns *et al.* [19]. A graphical game consists of a graph and a collection of matrices -one for each player; a player is represented by a vertex in the input graph and her payoff is determined entirely by her action and that of her neighbors.

We focus in this paper on the computational aspects of pure Nash equilibria for graphical games and the role of treewidth in such computations. We treat the

G. Gutin and S. Szeider (Eds.): IPEC 2013, LNCS 8246, pp. 348–360, 2013.

problem from the viewpoint of parameterized complexity, when the parameter is the treewidth of the input graph. First we prove that computing pure Nash equilibria for graphical games is $W[1]$-hard for the parameter treewidth (Section 3). Then, we develop a direct dynamic programming method to compute pure Nash equilibria (Section 4). Our algorithm decides the existence of pure Nash equilibria in $O(\alpha^w \cdot n \cdot |\mathcal{M}|)$ time, where α is the size of the largest strategy set, w is the treewidth of the input graph and $|\mathcal{M}|$ is the size of the description of the input matrices. As a consequence, the problem is fixed-parameter tractable when the cardinality of the strategy sets is bounded. Finally, we treat the existence of a polynomial kernel for w-PNE-GG, and discuss the implications of our algorithm for games of $O(\log n)$ treewidth and for constructing a sample or the maximum-payoff equilibrium (Section 5).

Related Work. The computational complexity of computing pure Nash equilibria for graphical games was proved to be NP-complete by Gottlob *et al.* in [13], even in the restricted case of neighborhoods of size at most 3 and a fixed number of actions. On the other hand, they prove that the problem is tractable for games with graphs of bounded hypertreewidth and in extension bounded treewidth. This is proved by mapping graphical games to Constraint Satisfaction problems while maintaining pure Nash equilibria as solutions of the resulting instance. The time complexity of the suggested procedure is $O(||\mathcal{G}||^{w+1} \cdot \log ||\mathcal{G}||)$, exponential in treewidth w, where $||\mathcal{G}||$ is the size of the description of the game instance. Moreover, Marx shows that the algorithm for solving CSP is essentially optimal under the Exponential Time Hypothesis [20] and thus, faster algorithms are not expected using this approach.

A different approach was provided by Daskalakis and Papadimitriou in [7] where they attacked the problem by providing a reduction from graphical games to Markov random fields. Their result yields a unified proof to the previously known tractable cases with time complexity $O(n \cdot |M_p|^{w+1}) = O(n \cdot \alpha^{\Delta \cdot (w+1)})$, where n is the number of players, p is the player with the largest neighborhood (of size Δ) and M_p its local game matrix (cf section 2). It additionally implies that the class of games with $O(\log n)$ treewidth is tractable. Furthermore, Jiang and Leyton-Brown provide an algorithm for another class of succinctly represented games, namely action graph games, that is polynomial for symmetric action graph games of bounded treewidth [16]. It is known that any graphical game can be mapped to a non-symmetric action graph game [18]. For bounded cardinality strategy sets this mapping keeps the treewidth bounded. However, computing pure Nash equilibria for non-symmetric action-graph games is NP-complete even when the treewidth is 1 [8].

Greco and Scarcello build on [13] and provide a dynamic programming approach that decides, in polynomial time, the existence of constrained pure Nash equilibria for graphical games of bounded treewidth and with bounded number of constraints [15]. Their approach is based on a non-deterministic algorithm, implicitly provided in [19], that associates pure with approximate mixed equilibria. Finally, in [17] it is shown that every recursively enumerable class of graphical games of bounded in-degree that is in FPT must be in P, with the representational size of the graph as the parameter and assuming $FPT \neq W[1]$. Observe that none of the known methods implies fixed-parameter tractability with treewidth of the input graph as the parameter.

2   Preliminaries

In a graphical game with graph $G = (V, E)$ we have $n = |V|$ players and each player $p \in V$ has a finite set of strategies, each strategy $St(p)$ being a finite set of actions with $|St(p)| \geq 2$. The cardinality of the largest strategy set is denoted with $\alpha = \max_{p \in V} |St(p)|$. For a non-empty set of players $P \subseteq V$ a joint strategy or *configuration* C is a set containing exactly one strategy for each player $p \in P$. The set of all joint strategies of players in P is denoted as $St(P)$ and thus we write $C \in St(P)$. For a player p, C_p denotes the strategy of player p with respect to configuration C and C_{-p} denotes the configuration resulting from removing the strategy suggested for p in C. Additionally, for every $a_p \in St(p)$ and $C_{-p} \in St(V \backslash \{p\})$ we denote by $(C_{-p}; a_p)$ the configuration in which p plays a_p and every other player $p' \neq p$ plays according to C. Abusing notation, we use $C \cup \{a_p\}$ to denote the configuration resulting by adding strategy $a_p \in St(p)$ to configuration $C \in St(P')$ where $p \notin P'$. A configuration C is termed *global* if it is over the set of all players ($C \in St(V)$). The global configurations are the possible outcomes of the game. We define the neighborhood of player $p \in V$ as $\mathcal{N}(p) = \{u \in V | (p, u) \in E\}$.

Definition 1 ([19]). *A graphical game is a pair $(G, \mathcal{M})$, where $G = (V, E)$ is an undirected graph and $\mathcal{M}$ is a set of $n = |V|$ local matrices. For any joint strategy C, the local game matrix $M_p \in \mathcal{M}$ specifies the payoff $M_p(C)$ for player $p \in V$, which depends only on the actions taken by p and the players in $\mathcal{N}(p)$.*

Note that for graphical games on undirected graphs, players' interests are necessarily symmetric, i.e. for any pair of players p_1 and p_2, $p_1 \in \mathcal{N}(p_2)$ if and only if $p_2 \in \mathcal{N}(p_1)$. Let the size of the collection of matrices be $|\mathcal{M}| = \sum_{p \in V} |M_p|$.

Definition 2. *The best response function of a player p is a function $\beta_p : St(\mathcal{N}(p)) \to 2^{St(p)}$ such that:*

$$\beta_p(C) = \{a_p | a_p \in St(p) \text{ and } \forall a'_p \in St(p) : M_p(C_{-p}; a_p) \geq M_p(C_{-p}; a'_p)\}$$

Intuitively, $\beta_p(C)$ is the set of strategies that maximize the payoff of player p when the players in p's neighborhood play according to C. Consequently, a pure Nash equilibrium (PNE for short) is a global configuration C such that for every player $p \in V$, $C_p \in \beta_p(C_{-p})$. Alternatively:

Definition 3. *A global configuration C is a pure Nash equilibrium if for every player p and strategy $a_p \in St(p)$ we have $M_p(C) \geq M_p(C_{-p}; a_p)$.*

We end this section with the definitions of treewidth and of some basic concepts from the theory of parameterized complexity [10,21].

Definition 4 ([22]). *A* tree decomposition *of a graph* $G = (V, E)$ *is a pair* $(\{X_i | i \in I\}, T = (I, F))$, *where* T *is a tree and each node* $i \in I$ *has associated to it a subset of vertices* $X_i \subseteq V$, *called the* bag *of i, such that:*

1. *Each vertex belongs to at least one bag,* $\cup_{i \in I} X_i = V$;
2. $\forall \{v, u\} \in E$, $\exists i \in I$ *with* $v, u \in X_i$;
3. $\forall v \in V$, *the set of nodes* $\{i \in I | v \in X_i\}$ *induces a subtree of* T.

The width *of a tree decomposition* T *is* $\max_{i \in I} |X_i| - 1$. *The* treewidth *of a graph* G *is the minimum width over all tree decompositions of* G.

Definition 5 ([21]). *A* parameterized problem *is a language* $L \subseteq \Sigma^* \times \Sigma^*$, *where* Σ *is a finite alphabet. The second component is called the* parameter *of the problem.*

The only parameters we consider here are nonnegative integers, hence we write $L \in \Sigma^* \times \mathbb{N}$ from now on. For $(x, k) \in L$, the two dimensions of parameterized complexity are the input size n, $n = |(x, k)|$, and the parameter value k.

Definition 6 ([21]). *A* parameterized problem L *is* fixed-parameter tractable *if, for all* (x, k), *it can be determined in* $f(k) \cdot n^{O(1)}$ *time whether* $(x, k) \in L$, *where* f *is a computable function depending only on* k.

The class of parameterized problems of the form (x, k), that are solvable in time $f(k) \cdot n^{O(1)}$, is denoted as FPT. In order to prove hardness for parameterized problems we also need a reducibility concept.

Definition 7 ([21]). *Let* (Q, k) *and* (Q', k') *be parameterized problems over the alphabets* Σ *and* Σ'. *An* fpt-reduction *is a mapping* $R : \Sigma^* \to (\Sigma')^*$ *such that*

- $\forall x \in \Sigma^*$ *we have* $(x \in Q \Leftrightarrow R(x) \in Q')$;
- R *is computable in* FPT *time (with respect to* k);
- $\exists$ *computable function* $g : \mathbb{N} \to \mathbb{N}$ *such that* $k' \leq g(k)$.

Fixed-parameter intractability beyond FPT is captured in the W-*hierarchy* (cf. [10,21]). A parameterized problem is $W[1]$-hard if WEIGHTED 3SAT is reducible to it by an fpt-reduction. It is currently open whether $FPT \subset W[1]$.

3   W[1]-Hardness

Treewidth plays an important role in the study of pure Nash equilibria for graphical games (cf. [13]). However, none of the previous results implies the existence of a fixed-parameter tractable algorithm with respect to the treewidth of the input graph. Here we argue that this is not surprising. Consider the following parameterized problem:

w-PNE-GG
Input: $\mathcal{G} = (G, \mathcal{M})$, T a tree decomposition of G.
Parameter: w - the width of T.
Question: Does $\mathcal{G}$ admit a PNE?

We will prove that w-PNE-GG is $W[1]$-hard. For this, a reduction from the $W[1]$-hard problem k-MULTICOLOR CLIQUE will be used. The input of this problem is a graph $G = (V, E)$ and a vertex coloring $c : V \to \{1, \dots, k\}$, k is the parameter and the question is whether G contains a clique with vertices of all k colors. Hardness follows follows easily by reduction from k-CLIQUE [12].

Before proceeding to the reduction, we introduce some useful notation. Let G be the input graph, and $c : V \to \{1, \dots, k\}$ a k-coloring of G. We let V_a denote the vertices colored a, i.e. $V_a = \{v \in V | c(v) = a\}$, and we let E_{c_i, c_j} be the set of edges $(u, v) \in E$ such that $\{c(u), c(v)\} = \{c_i, c_j\}$. Observe that w.l.o.g we can assume that the input coloring is *proper*, i.e. for any color c, $E_{c,c} = \emptyset$, as any such edge can be removed from G [12]. W.l.o.g. we can also assume that the color classes of G, and the edge sets between them, have uniform sizes, i.e $|V_c| = N$ for all c and $|E_{c_i, c_j}| = M$ for all $c_i < c_j$.

Theorem 1. *w-PNE-GG is $W[1]$-hard.*

Proof. Given an instance of MULTICOLOR CLIQUE, graph $G = (V, E)$ with k-coloring c, we construct an instance $\mathcal{G} = (G' = (P, E'), \mathcal{M})$ of PNE-GG as follows: The players of $\mathcal{G}$ are separated in two distinct sets, the *colorful* P_c and the *auxiliary* P_a players, $P = P_c \cup P_a$. Every $c \in P_c$ is connected to all the other colorful players $c' \in P_c$, through an auxiliary vertex $a \in P_a$. Thus, G' arises by taking a k-clique and adding one auxiliary player on each edge. By construction, the treewidth of G' is exactly k and thus the parameter is preserved.

The strategy sets are defined in the following manner: For a player $c \in P_c$, the possible strategies are all the vertices of G that are colored c plus an extra NA strategy, that stands for non-adjacent. Formally, $St(c) = \{v \in V | c(v) = c\} \cup \{NA\}$. An auxiliary player $a \in P_a$ has only two possible strategies, $St(p) = \{A, NA\}$, that stand for adjacent and non-adjacent respectively. Observe that G' is built such that all colorful vertices neighbor only with auxiliary vertices and each auxiliary vertex is neighbor to exactly 2 colorful ones. An example reduction can be found in Figure 1.

Let x be a global configuration. For an auxiliary player $a \in P_a$ let i, j be the two neighboring colorful players, i.e. $i, j \in \mathcal{N}(a)$. Then, the utility function u_a is such that:

1. $u_a(x) = 1$ if a plays A and i, j play v, u such that $(v, u) \in E$ or at least one of i, j plays NA;
2. $u_a(x) = 1$ if a plays NA and i, j play v, u such that $(v, u) \notin E$ and neither of i, j plays NA;
3. $u_a(x) = 0$ in all other cases.

For each player $c \in P_c$, her utility function u_c is such that:

4. $u_c(x) = 1$ if c plays a strategy in $St(c) \backslash \{NA\}$, and all of her neighbors play A;
5. $u_c(x) = 1$ if c plays NA and at least one of her neighbors plays NA;
6. $u_c(x) = 0$ in all other cases.

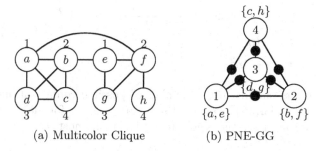

(a) Multicolor Clique   (b) PNE-GG

Fig. 1. An example of the reduction, where the numbers correspond to different colors. In (b) the strategy sets are shown in curly brackets (omitting NA) and the auxiliary players are represented as black vertices.

In the following paragraphs we will show that G has a clique including all k colors if and only if $\mathcal{G}$ has a pure Nash equilibrium. Let $(v_1, \ldots, v_k)$ be a k-clique of G that contains all k colors. Consider the global strategy x where each player $c \in P_c$ plays the strategy that corresponds to vertex v_c (the vertex from the clique that is colored c) and each auxiliary vertex plays A. Observe that in this case all players receive payoff 1 which is the maximum they can receive and thus x is a pure Nash equilibrium.

To prove the opposite direction of the claim we will first argue that there is no pure Nash equilibrium of $\mathcal{G}$ where there is an auxiliary vertex that plays NA. Assume that x is a PNE and $\exists a \in P_a$ that plays NA, with neighbor $j \in \mathcal{N}(a)$. Then, j would have an incentive to play NA and get payoff 1 rather than a strategy in $St(j) \backslash \{NA\}$. Consequently, a would prefer A over NA which contradicts our assumption that x is a PNE.

Now, let x be a global configuration and a pure Nash equilibrium of $\mathcal{G}$. From the previous paragraph, every $a \in P_a$ plays A and thus every $c \in P_c$ plays a strategy in $St(j) \backslash \{NA\}$. Consider the set of vertices $K = (v_1, \ldots, v_k)$ where each v_c corresponds to the strategy of player $c \in P_c$. Since each auxiliary vertex plays A, it means that all vertices in K are pairwise connected to each other and therefore form a clique. In addition, they all belong to a different color class because of the construction of $\mathcal{G}$. Therefore, K is a multicolored k-clique of G.

To conclude our proof we need to show that the reduction takes at most time of the form $f(k) \cdot p(|G|, k)$ for some computable function f and polynomial $p(X)$. The time of the construction is dominated by the computation of the matrix collection $\mathcal{M}$, whose size is the summation of the sizes of the individual matrices $|\mathcal{M}| = \sum_{c \in P_c} |M_c| + \sum_{a \in P_a} |M_a|$.

As mentioned earlier, we assume that the color classes of the MULTICOLOR CLIQUE instance have uniform size N and thus $N = \frac{n}{k}$ and for $c \in P_c$, $|St(c)| = N + 1$. In addition, observe that $|P_c| = k$, that each player $c \in P_c$ has $k - 1$ auxiliary neighbors with 2 available strategies each, and that $|P_a| = \frac{k(k-1)}{2}$ since we have one auxiliary vertex for each edge of the k-clique. Then the above summation can be rewritten as

$$k \cdot ((N+1) \cdot 2^{k-1}) + \frac{k(k-1)}{2} 2 \cdot (N+1)^2 \leq$$
$$2^{k-1} \cdot (n+k) + k^2 \cdot (N+1)^2 \leq$$
$$2^k \cdot n + 4n^2$$

The 2^{k-1} term corresponds to the number of possible configurations over the neighborhood of a colorful player, i.e. $|St(\mathcal{N}(c))| = 2^{k-1}$, for all $c \in P_c$. Therefore, the time we need for the whole reduction is at most $f(k) \cdot p(|G|)$ which concludes our proof.   □

We conclude that w-PNE-GG does not admit a fixed-parameter tractable algorithm, unless $FPT = W[1]$. Nevertheless, in the next section we will demonstrate an algorithm that becomes FPT for games with a bounded number of available strategies per player.

4   Fixed-Parameter Tractability

When the input graph of a graphical game is a tree, a relatively simple algorithm can answer the question of existence of a PNE in time linear in the input. The idea is that every vertex is able to compute the best response(s) for each configuration of its children, while ignoring its parent. Then, visiting the vertices in a bottom-up manner the parent will be taken into account in a subsequent step. The details of the algorithm and the proof of the result below are omitted.

Proposition 1. *Given a graphical game $(T, \mathcal{M})$, where T is a tree, one can compute a PNE in time $O(|\mathcal{M}|)$.*

The idea of the tree algorithm will now be generalized to tree decompositions; the problem under consideration is w-PNE-GG as defined in the previous section. The intuition is to go through all possible configurations for each bag of the tree, which count to α^{w+1}. Then we put together this information on the tree decomposition in polynomial time. The analysis we provide, is based on a *nice tree decomposition*. In such a decomposition, one node in $\mathcal{T}$ is considered to be the root and each node $i \in I$ is one of the following four types:
- Leaf: node i is a leaf of $\mathcal{T}$ and $|X_i| = 1$;
- Join: node i has exactly two children, say j_1, j_2 and $X_i = X_{j_1} = X_{j_2}$;
- Introduce: node i has exactly one child, say j, and $\exists v \in V$ with $X_i = X_j \cup \{v\}$;
- Forget: node i has exactly one child, say j, and $\exists v \in V$ with $X_j = X_i \cup \{v\}$.

It is known that if a graph $G = (V, E)$ has a tree decomposition with width at most w, then it also has a *nice* tree decomposition of width at most w and $O(|V|)$ nodes. A given tree decomposition can be turned into a nice one in linear time [3].

4.1   A Dynamic Programming Approach

Suppose we are given an instance of a graphical game; a graph $G = (V, E)$, a collection of matrices $\mathcal{M}$ -one matrix M_p for each node $p \in V$- and a tree decomposition $\mathcal{T}$. We assume w.l.o.g. that the tree decomposition $(\{X_i | i \in I\}, \mathcal{T} = (I, F))$ is nice. Each node $i \in I$ is associated to a graph $G_i = (V_i, E_i)$. V_i is the union of all bags X_j, with j equaling i or a descendant of i in $\mathcal{T}$, and $E_i = E \cap (V_i \times V_i)$. In other words, G_i is the subgraph of G induced by V_i.

A table A_i is to be computed for each node $i \in I$ and contains an integer value for each possible configuration $C \in St(X_i)$. Therefore, when the treewidth is w, table A_i contains at most $\alpha^{|X_i|} \leq \alpha^{w+1}$ values. Given configuration $C \in St(X_i)$, the table value $A_i(C)$ corresponds to the (maximum) number of players in best-response in G_i w.r.t. C. Thus, $A_i(C) = |V_i|$ if and only if $\exists C' \in St(V_i)$ such that $C' \supseteq C$ and $\forall p \in V_i$, $C'_p \in \beta_p(C'_{-p})$. Note that the strategy for the players in $V_i - X_i$ is not explicitly mentioned at this point (where the algorithm is treating bag X_i) but has been treated at an earlier time of the execution of the algorithm. Table A_i is computed for all nodes $i \in I$ in bottom-up order; for each non-leaf node we use the tables of its children to compute table A_i.

In addition, we have a $0, 1$-table F_p for each $p \in V$ that has the same number of entries as matrix M_p. Initially, F_p has the value 1 at all entries. For the sake of simplicity, we will assume that $|F_p| = |M_p|$ (same description size) for all $p \in V$. Essentially, F_p is where we mark which joint strategies are allowed at PNE over the neighborhood of p, with respect to the neighbors of p that have been forgotten (through a forget node). It follows that F tables will be updated at forget nodes -when a player is forgotten, her neighbors will update their F tables. At introduce nodes the F table will be examined -when a player is introduced, her neighbors will check their F tables for joint strategies that are allowed with regards to their forgotten neighbors. Formally: Let $i \in \mathcal{T}$ be a forget node of the tree decomposition. Then, after i is examined and corresponding F tables updated, we have that: For every $p \in V_i$ and $C \in St(\mathcal{N}(p) \cup \{p\})$, $F_p(C)$ has the value 0 if and only if $\exists u \in \mathcal{N}(p) \cap V_i$ such that $C_u \notin \beta_u(C_{-u})$.

The algorithm presented here might use the best response function with input a configuration for only a subset of the neighbors of the player under consideration. Let $p \in V$, P' be subset of the neighborhood of p, $P' \subset \mathcal{N}(p)$, and C a configuration over the players in P', $C \in St(P')$. In this case, $\beta_p(C)$ contains $a_p \in St(p)$ if and only if $\exists C' \in St(\mathcal{N}(p))$ such that $C' \supset C$ and $a_p \in \beta_p(C')$. Similarly, let $C \in St(P' \cup \{p\})$. Then, by $F_p(C)$ we mean all entries $F_p(C')$ such that $C' \supset C$. Also, if the configuration given as input includes strategies for players that are not in $\mathcal{N}(p)$, these strategies are ignored. A case analysis based on the type of the node under examination follows.

Leaf Nodes. Suppose node i is a leaf of $\mathcal{T}$ with $X_i = \{p\}$. Then, table A_i has only $|St(p)|$ entries. The value 1 will be attributed to these entries since a single player can be in PNE, no matter what strategy it follows, when there is no other player to compete with. Hence, for each configuration C over the vertices of X_i (in this case $St(X_i) = St(p)$) we set $A_i(C) = 1$.

Forget Nodes. Suppose i is a forget node of $\mathcal{T}$ with child j. In this case, G_i and G_j is the same graph but X_i and X_j differ by one vertex. Suppose this vertex is $p \in X_j - X_i$. To compute the tables of a forget node we use the procedure suggested by Lemma 1. For each of the $\alpha^{|X_i|}$ possible configurations we perform a number of α steps for a total of $O(\alpha^{|X_i|+1})$.

Lemma 1. *Let* $C \in St(X_i)$, $A_i(C) = \max_{a_p \in St(p)} A_j(C \cup \{a_p\})$.

In the case of a forget node we additionally have to update the F_u table for each $u \in X_i \cap \mathcal{N}(p)$. While computing the maximizing values for the procedure suggested by Lemma 1 we encounter all the possible combinations of joint strategies over the players in X_j (the bag including p). For each $C \in St(X_i)$ and $a_p \in St(p)$, if $a_p \notin B_p(C)$ then we set $F_u(C \cup \{a_p\}) = 0$. Information about the preferences of forgotten players propagates this way.

For each $u \in X_j$ we read the table M_p (to conclude if $a_p \notin B_p(C)$) and table F_j once (to update it). Since X_i is a forget node we have $|X_i| \leq w$. Thus, the time needed to compute the values of the table A_i and to update the tables F is $O(\alpha^w \cdot (|M_p| + \sum_{u \in \mathcal{N}(p) \cap X_j} |M_u|))$.

Introduce Nodes. Suppose i is an introduce node of $\mathcal{T}$ with child j and that $X_i = X_j \cup \{p\}$. It is known that there is no vertex $u \in V_j - X_j$ such that $\{p, u\} \in E$ [3]. Hence, G_i is formed from G_j by adding p and zero or more edges from p to vertices in X_j.

Lemma 2. *Let* $C \in St(X_j)$. *If* $\forall u \in X_j$, $\{p, u\} \notin E$, *then* $A_i(C \cup \{a_p\}) = A_j(C) + 1$ *for all strategies* $a_p \in St(p)$.

In the case above, p is not connected to any vertex in G_i. For the other case we have to be more elaborate. Assume that there is $u \in X_j$ such that $\{p, u\} \in E$. We use Algorithm 1 which proceeds in the following manner: Given $C \in St(X_j)$ and a best response for $p \in X_i - X_j$, $a_p \in B_p(C)$, for each player $u \in X_j \cap \mathcal{N}(p)$ it checks if u is in best-response with respect to configuration $C \cup \{a_p\}$. If $C_u \in B_u(C \cup \{a_p\})$ it also checks that $F_u(C \cup \{a_p\}) = 1$ and thus that the suggested joint strategy $C \cup \{a_p\}$ is allowed from the perspective of u with regards to her forgotten neighbors. In the positive case it adds player u to the set P_i. In the end of the iteration if all players in $\mathcal{N}(p) \cap X_j$ are also in P_i it means that all players in G_i connected to p are in best-response with respect to the current configuration. Hence, for $A_i(C \cup \{a_p\})$ we take the value $A_j(C) + 1$.

Lemma 3. *Given an introduce node* $i \in \mathcal{T}$ *with child* $j \in \mathcal{T}$ *such that* $p \in X_i - X_j$, *we can compute* $A_i(C)$ *for all configurations* $C \in St(X_i)$ *in time* $O(\alpha^{|X_i|} \cdot (|M_p| + \sum_{u \in \mathcal{N}(p) \cap X_j} |M_u|))$.

Proof. Before we start the procedure we compute the set $\mathcal{N}(p) \cap X_j$ in at most $|X_j|$ steps. This happens only once for each introduce node. In the case $\nexists u \in X_j$ such that $\{p, u\} \in E$ we compute the table value for each configuration in constant time and thus the total time needed is $O(\alpha^{|X_i|})$.

Algorithm 1. IntroNode

Input: $C \in St(X_j)$, $p \in X_i - X_j$, $a_p \in \beta_p(C)$
Output: $A_i(C \cup \{a_v\})$
1: Initiate set $P_i = \emptyset$
2: **for** $u \in \mathcal{N}(p) \cap X_j$ **do**
3:     **if** $C_u \in \beta_u(C \cup \{a_p\})$ **and** $F_u(C \cup \{a_p\}) = 1$ **then**
4:        $P_i \leftarrow P_i \cup \{u\}$
5:     **end if**
6: **end for**
7: **if** $P_i = \mathcal{N}(p) \cap X_j$ **then**
8:     $A_i(C \cup \{a_p\}) \leftarrow A_j(C) + 1$
9: **else**
10:     $A_i(C \cup \{a_p\}) \leftarrow A_j(C)$
11: **end if**

In the other case, we use Algorithm 1 for each configuration $C \in St(X_j)$. Computing $\beta_p(C)$ takes at most $|M_p|$ steps[1] The loop at lines 2-6 is through all vertices $u \in \mathcal{N}(p) \cap X_j$ and for each vertex u, $\beta_u(C \cup \{a_p\})$ is computed once and table F_u is checked once at line 3 in at most $2 \cdot |M_u|$ steps. The operation at line 7 takes one step. For each of the $\alpha^{|X_j|} = \alpha^{|X_i|-1}$ configurations Algorithm 1 has to be run at most α times (for each $a_p \in St(p)$). The lemma follows. □

The algorithm and lemma above show us how to compute the A_i table for an introduce node using information found in the table of the child node. For the introduced vertex p, we have that the matrix M_p and at most other $|X_i| - 1$ matrices are read once for each configuration. Note that adjacency is only checked once since it does not change for different entries.

Join Nodes. Suppose i is a join node of $\mathcal{T}$ with children j_1 and j_2. Remember that $X_i = X_{j_1} = X_{j_2}$. Then, G_i can be interpreted as a union of G_{j_1} and G_{j_2}. What we need to capture here, is that a configuration C for the players of X_i may be part of a PNE for G_i if and only if it also is for both G_{j_1} and G_{j_2}. Given a configuration C the computation of $A_i(C)$ for a join node takes only constant time as described by Lemma 4. Therefore, the computation of the whole table for a join node takes place in time $O(\alpha^{|X_i|})$.

Lemma 4. Let $C \in St(X_i)$, $A_i(C) = A_{j_1}(C) + A_{j_2}(C) - |X_i|$.

4.2 Combining the Tables

The algorithm proposed in the previous section is a bottom-up tree walk that finds partial configurations for each bag of the tree decomposition, that could be part of a PNE configuration. Then, these configurations are synthesized together on every step of the tree walk and an answer can be achieved when the root of the tree decomposition is reached.

---

[1] For a justification, see [23], p. 14.

Lemma 5. *Graphical game $(G, \mathcal{M})$ has a PNE if and only if $\exists C \in St(X_r)$ such that $A_r(C) = |V|$.*

In addition, note that during the bottom-up tree walk, if there exists a bag X_i such that $\forall C \in St(X_i): A_i(C) < |V_i|$, then we can stop the execution of the algorithm and reply NO. The tables of all bags of the tree decomposition have to be computed to verify a YES instance since any partial PNE configuration might be jeopardized by a newly introduced vertex. Upper time bounds of the algorithm suggested in this section are provided by the theorem below.

Theorem 2. *Given a graphical game $(G, \mathcal{M})$ and a tree decomposition $\mathcal{T}$ of width w, there is an algorithm that determines the existence of a PNE in $O(\alpha^w \cdot n \cdot |\mathcal{M}|)$ time.*

Proof. As discussed above, the computationally expensive nodes are the forget and introduce nodes. Both types have asymptotically the same upper bound. Thus, here we assume that every node $i \in \mathcal{T}$ is an introduce node[2] with child j and vertex $p \in X_i - X_j$. We derive the following upper bound:

$$\alpha^{w+1} \cdot \sum_{i \in \mathcal{T}} \left(|M_p| + \sum_{u \in \mathcal{N}(p) \cap X_j} |M_u| \right) \le \alpha^{w+1} \cdot (|\mathcal{M}| + (n-1)|\mathcal{M}|) \quad (1)$$

The summation over all matrices $|M_p|$ gives $|\mathcal{M}|$. In addition, the second summation is over at most $n - 1$ elements which in turn are upper bounded by $|\mathcal{M}|$ because of the first summation. The theorem follows.   □

Our algorithm improves significantly on the previous known bounds, since the base of the exponent is only the number of available strategies and not the whole game description. For example, if we assume that the number of available strategies is bounded by a constant, our algorithm becomes fixed-parameter tractable. Furthermore, it is known that for a game $\mathcal{G}$, $hwd \le w + 1$, where hwd and w are the hypertree width and the treewidth of $\mathcal{G}$, respectively [13]. Combining this result with Theorem 2, we obtain as a corollary the fixed-parameter tractability of computing a PNE for graphical games with bounded cardinality strategy sets when the parameter is hypertreewidth. Determining whether the treewidth of a given graph G is at most w, and if so, find a tree decomposition of width at most w is fixed-parameter tractable [2]. However, the same problem is $W[2]$-hard for hypertreewidth [14].

Corollary 1. *Given a graphical game $(G, \mathcal{M})$ and a hypertree decomposition $\mathcal{T}$ of width hwd, there is an algorithm that determines the existence of a PNE in $O(\alpha^{hwd} \cdot n \cdot |\mathcal{M}|)$ time.*

Finally, we address the question of the existence of a kernelization algorithm. A *polynomial kernel* means that given an instance of w-PNE-GG, we can obtain in

---

[2] Observe that in the case of a clique all the nodes of $\mathcal{T}$ but one are introduce nodes. The treewidth of an n-clique is $n - 1$.

polynomial time an equivalent instance whose size is bounded polynomially to w. The theorem below states that the existence of a polynomial kernel for w-PNE-GG is rather unlikely. It is derived from using the method of AND-Distillation described in [4]. There, AND-Distillation was formulated as a conjecture, but recently, Drucker proved that AND-Distillation holds unless $NP \subseteq coNP/poly$ [11]. It is easy to prove that w-PNE-GG is AND-Compositional by taking the disjoint union of m instances of the problem. The resulting graph has treewidth w and there exists a PNE if and only if all m instances have a PNE. The relevant definitions can be found in [4].

Theorem 3. *w-PNE-GG does not admit a polynomial kernel, unless $NP \subseteq coNP/poly$.*

5 Final Remarks

Daskalakis and Papadimitriou [7] proved that deciding the existence of a PNE is in P for all classes of games with $O(\log n)$ treewidth, bounded number of strategies and bounded neighborhood size. Our algorithm improves on their results in the following ways: First, it is polynomial for graphical games of $O(\log n)$ treewidth and bounded number of strategies, even without the bounded neighborhood size assumption. Second, if the size of the neighborhood is bounded we achieve an upper bound that is polynomial[3] in n. Our bound improves on the time complexity of the algorithms presented in [7] by removing $\Delta = \max_{p \in V} |\mathcal{N}(p)|$ from the exponent. Details can be found in [23], § 7.6.

Theorem 4. *Given a graphical game with $O(\log n)$ treewidth and bounded number of strategies, there is an algorithm that decides the existence of a PNE in time polynomial in the description of the game. Moreover, if the size of the neighborhood is bounded the algorithm becomes polynomial in the number of players.*

Finally, we prove that constructing a sample or the maximum-payoff PNE does not require additional computational effort. Details can be found in [23], § 7.7.

Theorem 5. *Given a graphical game $(G, \mathcal{M})$ and a tree decomposition $\mathcal{T}$ of width w, there is an algorithm that constructs a sample or maximum-payoff PNE, if one exists, or answers NO otherwise in $O(\alpha^w \cdot n \cdot |\mathcal{M}|)$ time. Moreover, the same algorithm computes a succinct description of all PNE.*

References

1. Becker, A., Geiger, D.: A sufficiently fast algorithm for finding close to optimal clique trees. Artif. Intell. 125(1-2), 3–17 (2001)
2. Bodlaender, H.L.: A linear-time algorithm for finding tree-decompositions of small treewidth. SIAM J. Comput 26(6), 1305–1317 (1996)

---

[3] Note that if the degree of the graph is bounded, then the description of the graphical game is polynomial in the number of players.

3. Bodlaender, H.L., Koster, A.M.C.A.: Combinatorial optimization on graphs of bounded treewidth. Comput. J. 51(3), 255–269 (2008)
4. Bodlaender, H.L., Downey, R.G., Fellows, M.R., Hermelin, D.: On problems without polynomial kernels. J. Comput. and Syst. Sciences 75(8), 423–434 (2009)
5. Bodlaender, H.L., Thomassé, S., Yeo, A.: Kernel bounds for disjoint cycles and disjoint paths. In: Fiat, A., Sanders, P. (eds.) ESA 2009. LNCS, vol. 5757, pp. 635–646. Springer, Heidelberg (2009)
6. Chen, X., Deng, X., Teng, S.-H.: Settling the complexity of computing two-player Nash equilibria. J. ACM 56(3) (2009)
7. Daskalakis, C., Papadimitriou, C.H.: Computing pure Nash equilibria in graphical games via markov random fields. In: Feigenbaum, J., et al. (eds.) ACM Conference on Electronic Commerce, pp. 91–99 (2006)
8. Daskalakis, C., Schoenebeck, G., Valiant, G., Valiant, P.: On the complexity of Nash equilibria of action-graph games. In: ACM-SIAM SODA, pp. 710–719 (2009)
9. Daskalakis, C., Goldberg, P.W., Papadimitriou, C.H.: The complexity of computing a Nash equilibrium. SIAM J. Comput. 39(1), 195–259 (2009)
10. Downey, R.G., Fellows, M.R.: Parameterized Complexity. Springer, NY (1999)
11. Drucker, A.: New limits to classical and quantum instance compression. In: FOCS 2012, pp. 609–618 (2012)
12. Fellows, M.R., Hermelin, D., Rosamond, F.A., Vialette, S.: On the parameterized complexity of multiple-interval graph problems. Theor. Comput. Sci. 410(1), 53–61 (2009)
13. Gottlob, G., Greco, G., Scarcello, F.: Pure Nash equilibria: hard and easy games. J. Artif. Intell. Res. 24, 357–406 (2005)
14. Gottlob, G., Grohe, M., Musliu, N., Samer, M., Scarcello, F.: Hypertree decompositions: structure, algorithms, and applications. In: Kratsch, D. (ed.) WG 2005. LNCS, vol. 3787, pp. 1–15. Springer, Heidelberg (2005)
15. Greco, G., Scarcello, F.: On the complexity of constrained Nash equilibria in graphical games. Theor. Comput. Sci. 410(38-40), 3901–3924 (2009)
16. Jiang, A.X., Leyton-Brown, K.: Computing pure Nash equilibria in symmetric action graph games. In: Cohn, A. (ed.) AAAI 2007, vol. 1, pp. 79–85. AAAI Press (2007)
17. Jiang, A.X., Safari, M.A.: Pure Nash equilibria: complete characterization of hard and easy graphical games. In: van der Hoek, W., et al. (eds.) AAMAS 2010, pp. 199–206 (2010)
18. Jiang, A.X., Leyton-Brown, K., Bhat, N.A.R.: Action-graph games. Games and Economic Behavior 71(1), 141–173 (2011)
19. Kearns, M.J., Littman, M.L., Singh, S.P.: Graphical models for game theory. In: Breeses, J.S., Koller, D. (eds.) UAI 2001, pp. 253–260. Morgan Kaufmann (2001)
20. Marx, D.: Can you beat treewidth? Theory of Computing 6(1), 85–112 (2010)
21. Niedermeier, R.: Invitation to Fixed-Parameter Algorithms. Oxford Univ. Press, NY (2006)
22. Robertson, N., Seymour, P.D.: Graph minors II: Algorithmic aspects of tree-width. J. Algorithms 30, 309–322 (1986)
23. Thomas, A.: Games on Graphs - The Complexity of Pure Nash Equilibria, Technical Report UU-CS-2011-024, Dept. of Information and Computing Sciences, Utrecht University (2011)

Algorithms for k-Internal Out-Branching

Meirav Zehavi

Department of Computer Science, Technion - Israel Institute of Technology,
Haifa 32000, Israel
meizeh@cs.technion.ac.il

Abstract. The *k-Internal Out-Branching (k-IOB)* problem asks if a given *directed* graph has an *out-branching* (i.e., a spanning tree with exactly one node of in-degree 0) with at least k internal nodes. The *k-Internal Spanning Tree (k-IST)* problem is a special case of k-IOB, which asks if a given *undirected* graph has a spanning tree with at least k internal nodes. We present an $O^*(4^k)$ time randomized algorithm for k-IOB, which improves the O^* running times of the best known algorithms for both k-IOB and k-IST. Moreover, for graphs of bounded degree Δ, we present an $O^*(2^{(2-\frac{\Delta+1}{\Delta(\Delta-1)})k})$ time randomized algorithm for k-IOB. Both our algorithms use polynomial space.

1 Introduction

In this paper we study the *k-Internal Out-Branching (k-IOB)* problem. The input for k-IOB consists of a *directed* graph $G = (V, E)$ and a parameter $k \in \mathbb{N}$, and the objective is to decide if G has an *out-branching* (i.e., a spanning tree with exactly one node of in-degree 0, that we call the root) with at least k internal nodes (i.e., nodes of out-degree ≥ 1). The k-IOB problem is of interest in database systems [2].

A special case of k-IOB, called *k-Internal Spanning Tree (k-IST)*, asks if a given *undirected* graph $G = (V, E)$ has a spanning tree with at least k internal nodes. A possible application of k-IST, for connecting cities with water pipes, is given in [14].

The k-IST problem is NP-hard even for graphs of bounded degree 3, since it generalizes the Hamiltonian path problem for such graphs [5]; thus k-IOB is also NP-hard for such graphs. In this paper we present parameterized algorithms for k-IOB. Such algorithms are an approach to solve NP-hard problems by confining the combinatorial explosion to a parameter k. More precisely, a problem is *fixed-parameter tractable (FPT)* with respect to a parameter k if an instance of size n can be solved in $O^*(f(k))$ time for some function f [10].[1]

Related Work: Nederlof [9] gave an $O^*(2^{|V|})$ time and polynomial space algorithm for k-IST. For graphs of bounded degree Δ, Raible et al. [14] gave an $O^*(((2^{\Delta+1} - 1)^{\frac{1}{\Delta+1}})^{|V|})$ time and exponential space algorithm for k-IST.

---

[1] O^* hides factors polynomial in the input size.

G. Gutin and S. Szeider (Eds.): IPEC 2013, LNCS 8246, pp. 361–373, 2013.
© Springer International Publishing Switzerland 2013

Table 1. Known parameterized algorithms for k-IOB and k-IST

Reference	Variation	Time Complexity	The Topology of G
Priesto et al. [12]	k-IST	$O^*(2^{O(k \log k)})$	General
Gutin al. [6]	k-IOB	$O^*(2^{O(k \log k)})$	General
Cohen et al. [1]	k-IOB	$O^*(49.4^k)$	General
Fomin et al. [4]	k-IOB	$O^*(16^{k+o(k)})$	General
Fomin et al. [3]	k-IST	$O^*(8^k)$	General
Raible et al. [14]	k-IST	$O^*(2.1364^k)$	$\Delta = 3$
This paper	**k-IOB**	$\mathbf{O^*(4^k)}$	**General**
	k-IOB	$O^*(2^{(2-\frac{\Delta+1}{\Delta(\Delta-1)})k})$	$\mathbf{\Delta = O(1)}$

Table 2. Some concrete figures for the running time of the algorithm Δ-IOB-Alg

Δ	3	4	5	6
Time complexity	$O^*(2.51985^k)$	$O^*(2.99662^k)$	$O^*(3.24901^k)$	$O^*(3.40267^k)$

Table 1 presents a summary of known parameterized algorithms for k-IOB and k-IST. In particular, the algorithms having the best known O^* running times for k-IOB and k-IST are due to [4], [3] and [14]. Fomin et al. [4] gave an $O^*(16^{k+o(k)})$ time and polynomial space randomized algorithm for k-IOB, and an $O^*(16^{k+o(k)})$ time and $O^*(4^{k+o(k)})$ space deterministic algorithm for k-IOB. Fomin et al. [3] gave an $O^*(8^k)$ time and polynomial space deterministic algorithm for k-IST. For graphs of bounded degree 3, Raible et al. [14] gave an $O^*(2.1364^k)$ time and polynomial space deterministic algorithm for k-IST.

Further information on k-IOB, k-IST and variants of these problems is given in surveys [11,15].

Our Contribution: We present an $O^*(4^k)$ time and polynomial space randomized algorithm for k-IOB, that we call IOB-Alg. Our algorithm IOB-Alg improves the O^* running times of the best known algorithms for both k-IOB and k-IST.

For graphs of bounded degree Δ, we present an $O^*(2^{(2-\frac{\Delta+1}{\Delta(\Delta-1)})k})$ time and polynomial space randomized algorithm for k-IOB, that we call Δ-IOB-Alg. Some concrete figures for the running time of Δ-IOB-Alg are given in Table 2.

Techniques: Our algorithm IOB-Alg is based on two reductions as follows. We first reduce k-IOB to a new problem, that we call (k, l)-*Tree*, by using an observation from [1]. This reduction allows us to focus our attention on finding a tree whose size depends on k, rather than a spanning tree whose size depends on $|V|$. We then reduce (k, l)-Tree to the t-*Multilinear Detection (t-MLD)* problem, which concerns multivariate polynomials and has an $O^*(2^t)$ time randomized algorithm [7,17]. We note that reductions to t-MLD have been used to solve several problems quickly (see, e.g., [8]). IOB-Alg is another proof of the applicability of this new tool.

Our algorithm Δ-IOB-Alg, though based on the same technique as IOB-Alg, requires additional new non-trivial ideas and is more technical. In particular, we now use a proper coloring of the graph G when reducing (k,l)-Tree to t-MLD. This idea might be useful in solving other problems.

Organization: Section 2 presents our algorithm IOB-Alg. Specifically, Section 2.1 defines (k,l)-*Tree*, and presents an algorithm that solves k-IOB by using an algorithm for (k,l)-Tree. Section 2.2 defines t-MLD, and reduces (k,l)-Tree to t-MLD. Then, Section 2.3 presents our algorithm for (k,l)-Tree, and thus concludes IOB-Alg. Section 3 presents our algorithm Δ-IOB-Alg. Specifically, Section 3.1 modifies the algorithm presented in Section 2.1, Section 3.2 modifies the reduction presented in Section 2.2, and Section 3.3 modifies the algorithms presented in Section 2.3. Finally, Section 4 presents a few open questions.

2 An $O^*(4^k)$-time k-IOB Algorithm

2.1 The (k,l)-Tree Problem

We first define a new problem, that we call (k,l)-*Tree*.

(k,l)-**Tree**

- Input: A directed graph $G = (V,E)$, a node $r \in V$, and parameters $k,l \in \mathbb{N}$.
- Goal: Decide if G has an *out-tree* (i.e., a tree with exactly one node of in-degree 0) rooted at r with exactly k internal nodes and l leaves.

We now show that we can focus our attention on solving (k,l)-Tree. Let $A(G,r,k,l)$ be a $t(G,r,k,l)$ time and $s(G,r,k,l)$ space algorithm for (k,l)-Tree.

Algorithm 1. IOB-Alg[A](G,k)

1: **for all** $r \in V$ **do**
2:     **if** G has no out-branching T rooted at r **then** Go to the next iteration. **end if**
3:     **for** $l = 1,2,...,k$ **do**
4:         **if** $A(G,r,k,l)$ accepts **then** Accept. **end if**
5:     **end for**
6: **end for**
7: Reject.

The following observation immediately implies the correctness of IOB-Alg[A] (see Algorithm 1).

Observation 1 ([1]). *Let $G = (V,E)$ be a directed graph, and $r \in V$ such that G has an out-branching rooted at r.*

- *If G has an out-branching rooted at r with at least k internal nodes, then G has an out-tree rooted at r with exactly k internal nodes and at most k leaves.*
- *If G has an out-tree rooted at r with exactly k internal nodes, then G has an out-branching with at least k internal nodes.*

By Observation 1, and since Step 2 can be performed in $O(|E|)$ time and $O(|V|)$ space (e.g., by using DFS), we have the following result.

Lemma 1. IOB-Alg[A] *is an* $O(\sum_{r \in V}(|E| + \sum_{1 \leq l \leq k} t(G, r, k, l)))$ *time and* $O($
$|V| + \max_{r \in V, 1 \leq l \leq k} s(G, r, k, l))$ *space algorithm for* k-*IOB.*

2.2 A Reduction from (k, l)-Tree to t-MLD

We first give the definition of t-MLD [7].

t-MLD

– Input: A polynomial P represented by an arithmetic circuit C over a set of variables X, and a parameter $t \in \mathbb{N}$.
– Goal: Decide if P has a multilinear monomial of degree at most t.

Let (G, r, k, l) be an input for (k, l)-Tree. We now construct an input $f(G, r, k, l) = (C_{r,k,l}, X, t)$ for t-MLD. We introduce an indeterminate x_v for each $v \in V$, and define $X = \{x_v : v \in V\}$ and $t = k + l$.

The idea behind the construction is to let each monomial represent a pair of an out-tree $T = (V_T, E_T)$ and a function $h : V_T \to V$, such that if $(v, u) \in E_T$, then $(h(v), h(u)) \in E$ (i.e., h is a homomorphism). The monomial is $\prod_{v \in V_T} x_{h(v)}$. We get that the monomial is multilinear iff $\{h(v) : v \in V_T\}$ is a set (then $h(T) = (\{h(v) : v \in V_T\}, \{(h(v), h(u)) : (v, u) \in E_T\})$ is an out-tree).

Towards presenting $C_{r,k,l}$, we inductively define an arithmetic circuit $C_{v,k',l'}$ over X, for all $v \in V, k' \in \{0, ..., k\}$ and $l' \in \{1, ..., l\}$. Informally, the multilinear monomials of the polynomial represented by $C_{v,k',l'}$ represent out-trees of G rooted at v that have exactly k' internal nodes and l' leaves.

Base Cases:

1. If $k' = 0$ and $l' = 1$: $C_{v,k',l'} = x_v$.
2. If $k' = 0$ and $l' > 1$: $C_{v,k',l'} = 0$.

Steps:

1. If $k' > 0$ and $l' = 1$: $C_{v,k',l'} = \sum_{u \text{ s.t.} (v,u) \in E} x_v C_{u,k'-1,l'}$.
2. If $k' > 0$ and $l' > 1$: $C_{v,k',l'} =$
   $\sum_{u \text{ s.t.} (v,u) \in E}(x_v C_{u,k'-1,l'} + \sum_{1 \leq k^* \leq k'} \sum_{1 \leq l^* \leq l'-1} C_{v,k^*,l^*} \cdot C_{u,k'-k^*,l'-l^*})$.

The following order shows that when computing an arithmetic circuit $C_{v,k',l'}$, we only use arithmetic circuits that have been already computed.

Order:

1. For $k' = 0, 1, ..., k$:
   (a) For $l' = 1, 2, ..., l$:
       i. $\forall v \in V$: Compute $C_{v,k',l'}$.

Denote the polynomial that $C_{v,k',l'}$ represents by $P_{v,k',l'}$.

Lemma 2. (G, r, k, l) *has a solution iff* $(C_{r,k,l}, X, t)$ *has a solution.*

Proof. By using induction, we first prove that if G has an out-tree $T = (V_T, E_T)$ rooted at v with exactly k' internal nodes and l' leaves, then $P_{v,k',l'}$ has the (multilinear) monomial $\prod_{w \in V_T} x_w$.

The claim is clearly true for the base cases, and thus we next assume that $k' > 0$, and the claim is true for all $v \in V$, $k^* \in \{0, ..., k'\}$ and $l^* \in \{1, ..., l'\}$, such that $(k^* < k'$ or $l^* < l')$.

Let $T = (V_T, E_T)$ be an out-tree of G, that is rooted at v and has exactly k' internal nodes and l' leaves. Also, let u be a neighbor of v in T. Denote by $T_v = (V_v, E_v)$ and $T_u = (V_u, E_u)$ the two out-trees of G in the forest $F = (V_T, E_T \setminus \{(v, u)\})$, such that $v \in V_v$. We have the following cases.

1. If $|V_v| = 1$: T_u has $k' - 1$ internal nodes and l' leaves. By the induction hypothesis, $P_{u,k'-1,l'}$ has the monomial $\prod_{w \in V_u} x_w$. Thus, by the definition of $C_{v,k',l'}$, $P_{v,k',l'}$ has the monomial $x_v \prod_{w \in V_u} x_w = \prod_{w \in V_T} x_w$.
2. Else: Denote the number of internal nodes and leaves in T_v by k_v and l_v, respectively. By the induction hypothesis, P_{v,k_v,l_v} has the monomial $\prod_{w \in V_v} x_w$, and $P_{u,k'-k_v,l'-l_v}$ has the monomial $\prod_{w \in V_u} x_w$. By the definition of $C_{v,k',l'}$, $P_{v,k',l'}$ has the monomial $\prod_{w \in V_v} x_w \prod_{w \in V_u} x_w = \prod_{w \in V_T} x_w$.

Now, by using induction, we prove that if $P_{v,k',l'}$ has the (multilinear) monomial $\prod_{w \in U} x_w$, for some $U \subseteq V$, then G has an out-tree $T = (V_T, E_T)$ rooted at v with exactly k' internal nodes and l' leaves, such that $V_T = U$. This claim implies that any multilinear monomial of $P_{v,k',l'}$ is of degree exactly $k' + l'$.

The claim is clearly true for the base cases, and thus we next assume that $k' > 0$, and the claim is true for all $v \in V$, $k^* \in \{0, ..., k'\}$ and $l^* \in \{1, ..., l'\}$, such that $(k^* < k'$ or $l^* < l')$.

Let $\prod_{w \in U} x_w$, for some $U \subseteq V$, be a monomial of $P_{v,k',l'}$. By the definition of $C_{v,k',l'}$, there is u such that $(v, u) \in E$, for which we have the following cases.

1. If $P_{u,k'-1,l'}$ has a monomial $\prod_{w \in U \setminus \{v\}} x_w$: By the induction hypothesis, G has an out-tree $T_u = (V_u, E_u)$ rooted at u with exactly $k' - 1$ internal nodes and l' leaves, such that $V_u = U \setminus \{v\}$. By adding v and (v, u) to T_u, we get an out-tree $T = (V_T, E_T)$ of G that is rooted at v, has exactly k' internal nodes and l' leaves, and such that $V_T = U$.
2. Else: There are $k^* \in \{1, ..., k'\}$, $l^* \in \{1, ..., l' - 1\}$ and $U^* \subseteq U$, such that P_{v,k^*,l^*} has the monomial $\prod_{w \in U^*} x_w$, and $P_{u,k'-k^*,l'-l^*}$ has the monomial $\prod_{w \in U \setminus U^*} x_w$. By the induction hypothesis, G has an out-tree $T_v = (V_v, E_v)$ rooted at v with exactly k^* internal nodes and l^* leaves, such that $V_v = U^*$. Moreover, G has an out-tree $T_u = (V_u, E_u)$ rooted at u with exactly $k' - k^*$ internal nodes and $l' - l^*$ leaves, such that $V_u = U \setminus U^*$. Thus, we get that the out-tree $T = (U, E(T_v) \cup E(T_u) \cup (v, u))$ of G is rooted at v, and has exactly k' internal nodes and l' leaves.

We get that G has an out-tree rooted at r of exactly k internal nodes and l leaves iff $P_{r,k,l}$ has a mutlilinear monomial of degree at most t. $\qquad\square$

The definition of $(C_{r,k,l}, X, t)$ immediately implies the following observation.

Observation 2. *We can compute $(C_{r,k,l}, X, t)$ in polynomial time and space.*

2.3   The Algorithm IOB-Alg[Tree-Alg]

Koutis et al. [7,17] gave an $O^*(2^t)$ time and polynomial space randomized algorithm for t-MLD. We denote this algorithm by MLD-Alg, and use it to get an algorithm for (k, l)-Tree (see Algorithm 2).

Algorithm 2. Tree-Alg(G, r, k, l)

1: Compute $f(G, r, k, l) = (C_{r,k,l}, X, t)$.
2: Accept iff MLD-Alg$(C_{r,k,l}, X, t)$ accepts.

By Lemmas 1 and 2, and Observation 2, we have the following theorem.

Theorem 1. IOB-Alg[Tree-Alg] *is an* $O^*(4^k)$ *time and polynomial space randomized algorithm for k-IOB.*

3   A k-IOB Algorithm for Graphs of Bounded Degree Δ

3.1   A Modification of the Algorithm IOB-Alg[A]

We first prove that in Step 3 of IOB-Alg[A] (see Section 2.1), we can iterate over less than k values for l.

Given an out-tree $T = (V_T, E_T)$ and $i \in \mathbb{N}$, denote the number of degree-i nodes in T by n_i^T.

Observation 3 ([14]). *If* $|V_T| \geq 2$, *then* $2 + \sum_{3 \leq i}(i - 2)n_i^T = n_1^T$.

Observation 4. *An out-tree T of G with exactly k internal nodes contains an out-tree with exactly k internal nodes and at most $k - \frac{k-2}{\Delta-1}$ leaves.*

Proof. As long as T has an internal node v with at least two out-neighbors that are leaves, delete one of these leaves and its adjacent edge from T. Denote the resulting out-tree by T', and denote the tree that we get after deleting all the leaves in T' by T''. Note that T' has exactly k internal nodes, and that T' and T'' have the same number of leaves. Since T'' has k nodes and bounded degree Δ, Observation 3 implies that if $n_1^{T''} + n_\Delta^{T''} = k$, then $n_1^{T''} = k - \frac{k-2}{\Delta-1}$, and if $n_1^{T''} + n_\Delta^{T''} < k$, then $n_1^{T''} < k - \frac{k-2}{\Delta-1}$. We have that $n_1^{T''} \leq k - \frac{k-2}{\Delta-1}$, and thus we conclude that T' has exactly k internal nodes and at most $k - \frac{k-2}{\Delta-1}$ leaves.   □

Thus, in Step 3 of IOB-Alg[A], we can iterate only over $l = 1, 2, ..., k - \lceil \frac{k-2}{\Delta-1} \rceil$. We add some preprocessing steps to IOB-Alg[A], and thus get Δ-IOB-Alg[A] (see Algorithm 3). These preprocessing steps will allow us to assume, when presenting algorithm A, that the underlying undirected graph of G is a connected graph that is neither a cycle nor a clique. This assumption will allow us to compute a proper Δ-coloring of the underlying undirected graph of G (see Section 3.3), which we will use in the following Section 3.2.

Algorithm 3. Δ-IOB-Alg[A](G, k)

1: **if** $k \geq |V|$ or the underlying undirected graph of G is not connected **then**
2:    Reject.
3: **else if** the underlying undirected graph of G is a cycle **then**
4:    **if** $k = |V| - 1$ **then** Accept iff G has a hamiltonian path. **else** Accept iff there is at most one node of out-degree 2 in G. **end if**
5: **else if** the underlying undirected graph of G is a clique **then**
6:    Accept.
7: **end if**
8: **for all** $r \in V$ **do**
9:    **if** G has no out-branching T rooted at r **then** Go to the next iteration. **end if**
10:    **for** $l = 1, 2, ..., k - \lceil \frac{k-2}{\Delta-1} \rceil$ **do**
11:       **if** A(G, r, k, l) accepts **then** Accept. **end if**
12:    **end for**
13: **end for**
14: Reject.

We can clearly perform the new preprocessing steps in $O(|E|)$ time and $O(|V|)$ space. Steps 2 and 4 are clearly correct. Since a tournament (i.e., a directed graph obtained by assigning a direction for each edge in an undirected complete graph) has a hamiltonian path [13], we have that Step 6 is also correct.

We have the following lemma.

Lemma 3. Δ-IOB-Alg[A] *is an* $O(\sum_{r \in V}(|E| + \sum_{1 \leq l \leq k - \lfloor \frac{k-2}{\Delta-1} \rfloor} t(G, r, k, l)))$ *time and* $O(|V| + \max_{r \in V, 1 \leq l \leq k - \lceil \frac{k-2}{\Delta-1} \rceil} s(G, r, k, l))$ *space algorithm for k-IOB.*

3.2 A Modification of the Reduction f

In this section assume that we have a proper Δ-coloring $col : V \to \{c_1, ..., c_\Delta\}$ of the underlying undirected graph of G. Having such col, we modify the reduction f (see Section 2.2) to construct a "better" input for t-MLD (i.e., an input in which $t < k + l$).

The Idea Behind the Modification: Recall that in the previous construction, we let each monomial represent a certain pair of an out-tree $T = (V_T, E_T)$ and a function $h : V_T \to V$. The monomial included indeterminates representing *all* the nodes to which the nodes in V_T are mapped. We can now select some color $c \in \{c_1, ..., c_\Delta\}$, and ignore some occurrences of indeterminates that represent nodes whose color is c and whose degree in $h(T)$ is Δ. We thus construct monomials with smaller degrees, and have an input for t-MLD in which $t < k + l$.

More precisely, the monomial representing T and h is $\prod_{v \in U} x_{h(v)}$, where U is V_T, excluding nodes mapped to nodes whose color is c and whose degree in T is Δ (except the root). We add constraints on T and h to garauntee that nodes in V_T that are mapped to the same node do not have common neighbors in T.

The correctness is based on the following observation. Suppose that there is an indeterminate x_v that occurs more than once in the original monomial representing T and h, but not in the new monomial representing them. Thus the color of v is c. Moreover, there are different nodes $u, w \in V_T$ such that $h(u) = h(w) = v$, and the degree of u in T is Δ. We get that u has a neighbor u' in T and w has a *different* neighbor w' in T, such that $h(u') = h(w')$ and the color of $h(u')$ is not c. Thus $x_{h(u')}$ occurs more than once in the new monomial representing T and h. This implies that monomials that are not multilinear in the original construction do not become multilinear in the new construction.

The Construction: Let (G, r, k, l) be an input for (k, l)-Tree. We now construct an input $f(G, r, k, l, col) = (C, X, t)$ for t-MLD.

We add a node r' to V and the edge (r', r) to E. We color r' with some $c \in \{c_1, ..., c_\Delta\} \setminus \{col(r)\}$. In the following let $<$ be some order on $V \cup \{nil\}$, such that nil is the smallest element. Define $X = \{x_v : v \in V\}$, and $t = (2 - \frac{\Delta+1}{\Delta(\Delta-1)})k + 8$. Denote $N(v, i, o) = \{u \in V \setminus \{i\} : (v, u) \in E, u > o\}$.

We inductively define an arithmetic circuit $C_{v,k',l'}^{c,i,o,b}$ over X, for all $v \in V$, $k' \in \{0, ..., k\}$, $l' \in \{1, ..., l\}$, $c \in \{c_1, ..., c_\Delta\}$, i such that $(i, v) \in E$, o such that $(v, o) \in E$ or $o = nil$, and $b \in \{F, T\}$. Informally, v, k' and l' play the same role as in the original construction; c indicates that only indeterminates representing nodes colored by c can be ignored; i and o are used for constraining the pairs of trees and functions represented by monomials as noted in "The Idea Behind the Modification"; and b indicates whether the indeterminate of v is ignored.

Base Cases:

1. If $k' = 0$, $l' = 1$ and $b = F$: $C_{v,k',l'}^{c,i,o,b} = x_v$.
2. Else if $[k' = 0]$ or $[N(v, i, o) = \emptyset]$ or $[(|N(v, i, o)| > l'$ or $col(v) \neq c$ or $v = r$ or $|N(v, i, nil)| < \Delta - 1)$ and $b = T]$: $C_{v,k',l'}^{c,i,o,b} = 0$.

Steps: (assume that none of the base cases applies)

1. If $l' = 1$ and $b = F$: $C_{v,k',l'}^{c,i,o,b} = x_v \sum_{u \in N(v,i,o)} (C_{u,k'-1,l'}^{c,v,nil,F} + C_{u,k'-1,l'}^{c,v,nil,T})$.
2. Else if $b = F$:
$$C_{v,k',l'}^{c,i,o,b} = \sum_{u \in N(v,i,o)} [x_v C_{u,k'-1,l'}^{c,v,nil,F} + x_v C_{u,k'-1,l'}^{c,v,nil,T} +$$
$$\sum_{1 \leq k^* \leq k'} \sum_{1 \leq l^* \leq l'-1} C_{v,k^*,l^*}^{c,i,u,b} (C_{u,k'-k^*,l'-l^*}^{c,v,nil,F} + C_{u,k'-k^*,l'-l^*}^{c,v,nil,T})].$$
3. If $b = T$ and there is exactly one node u in $N(v, i, o)$: $C_{v,k',l'}^{c,i,o,b} = C_{u,k'-1,l'}^{c,v,nil,F}$.
4. Else if $b = T$:
   (a) Denote $u = \min(N(v, i, o))$.
   (b) $C_{v,k',l'}^{c,i,o,b} = \sum_{1 \leq k^* \leq k'} \sum_{1 \leq l^* \leq l'-1} C_{v,k^*,l^*}^{c,i,u,b} C_{u,k'-k^*,l'-l^*}^{c,v,nil,F}$.

The following order shows that when computing an arithmetic circuit $C_{v,k',l'}^{c,i,o,b}$, we only use arithmetic circuits that have been already computed.

Order:

1. For $k' = 0, 1, ..., k$:
   (a) For $l' = 1, 2, ..., l$:
      i. $\forall v \in V, c \in \{c_1, ..., c_\Delta\}$, i s.t. $(i, v) \in E$, o s.t. $(v, o) \in E$ or $o = nil$, $b \in \{F, T\}$: Compute $C_{v,k',l'}^{c,i,o,b}$.

Define $C = \sum_{c \in \{c_1, ..., c_\Delta\}} C_{r,k,l}^{c,r',nil,F}$.

Denote the polynomial that $C_{v,k',l'}^{c,i,o,b}$ (resp. C) represents by $P_{v,k',l'}^{c,i,o,b}$ (resp. P).

Correctness: We need the next two definitions, which we illustrate in Fig. 1.

Definition 1. *Let* $v \in V$, $k' \in \{0, ..., k\}$, $l' \in \{1, ..., l\}$, $c \in \{c_1, ..., c_\Delta\}$, i *such that* $(i, v) \in E$, o *such that* $(v, o) \in E$ *or* $o = nil$. *Given a subgraph* $T = (V_T, E_T)$ *of* G, *we say that*

1. T *is a* (v, k', l', c, i, o, F)-*tree if*
    (a) T *is an out-tree rooted at* v *with exactly* k' *internal nodes and* l' *leaves.*
    (b) *Every out-neighbor of* v *in* T *belongs to* $N(v, i, o)$.
2. T *is a* (v, k', l', c, i, o, T)-*tree if*
    (a) $col(v) = c$, $v \neq r$, *and* $|N(v, i, nil)| = \Delta - 1$.
    (b) *Every node in* $N(v, i, o)$ *is an out-neighbor of* v *in* T, *and* $N(v, i, o) \neq \emptyset$.
    (c) *There is at most one node* $i' \in V_T$ *such that* $(i', v) \in E_T$.
        i. *If such an* i' *exists:* $(v, i') \notin E_T$, *and* $T' = (V_T, E_T \setminus \{(i', v)\})$ *is an out-tree rooted at* v.
        ii. *Else:* T *is a* (v, k', l', c, i, o, F)-*tree.*

Definition 2. *Given a* (v, k', l', c, i, o, b)-*tree* $T = (V_T, E_T)$, *define* $I(T) =$

$$\{u \in V_T : [u \neq v \wedge (col(u) \neq c \vee u \text{ has less than } (\Delta - 1) \text{ out} - \text{neighbors in } T)]$$

$$\vee [u = v \wedge (b = F \vee v \text{ has an in} - \text{neighbor in } T)]\}.$$

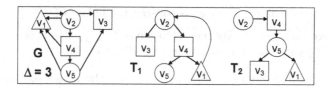

Fig. 1. Assume that $r = v_1 < v_2 < v_3 < v_4 < v_5$, and that shapes represent colors. We have that T_1 is a $(v_2, k', l', O, v_1, nil, T)$-tree for any k' and l', and $I(T_1) = \{v_1, v_2, v_3, v_4, v_5\}$. Moreover, T_2 is a $(v_2, 3, 2, O, v_1, v_3, T)$-tree, and $I(T_2) = \{v_1, v_3, v_4\}$.

Observation 5. *Let* $T = (V_T, E_T)$ *be a* (v, k', l', c, i, o, b)-*tree of* G, *such that there is no* $i' \in V_T$ *for which* $(i', v) \in E_T$. *Then,* $P_{v,k',l'}^{c,i,o,b}$ *has the (multilinear) monomial* $\prod_{w \in I(T)} x_w$.

Proof. We prove the claim by using induction on the construction. The claim is clearly true for the base cases. Next consider a (v, k', l', c, i, o, b)-tree $T = (V_T, E_T)$ of G, such that $C_{v,k',l'}^{c,i,o,b}$ is not constructed in the base cases. Assume that the claim is true for all $(\tilde{v}, \tilde{k}, \tilde{l}, \tilde{c}, \tilde{i}, \tilde{o}, \tilde{b})$ such that $C_{\tilde{v},\tilde{k},\tilde{l}}^{\tilde{c},\tilde{i},\tilde{o},\tilde{b}}$ is constructed before $C_{v,k',l'}^{c,i,o,b}$. Denote by u the smallest out-neighbor of v in T.

Denote by $T_v = (V_v, E_v)$ and $T_u = (V_u, E_u)$ the two out-trees of G in the forest $F = (V_T, E_T \setminus \{(v, u)\})$, such that $v \in V_v$. If $u \notin I(T)$ (this is not the case if $b = T$, since then $col(u) \neq c$), then denote $b' = T$, and note that the set of out-neighbors of u in T_u contains all of the neighbors of u in G, excluding v; else denote $b' = F$. We have the following cases.

1. If $|V_v| = 1$: T_u is a $(u, k'-1, l'c, v, nil, b')$-tree of G. If $b = F$, then $I(T_u) = I(T) \setminus \{v\}$; else $I(T_u) = I(T_v)$. By the induction hypothesis $C_{u,k'-1,l'}^{c,v,nil,b'}$ has the monomial $\prod_{w \in I(T_u)} x_w$. Thus, by the definition of $C_{v,k',l'}^{c,i,o,b}$, $P_{v,k',l'}^{c,i,o,b}$ has the required monomial.

2. Else: Denote the number of internal nodes and leaves in T_v by k_v and l_v, respectively. Note that $1 \leq k_v \leq k'$, $1 \leq l_v < l'$, T_v is a $(v, k_v, l_v, c, i, u, b)$-tree of G, and T_u is a $(u, k'-k_v, l'-l_v, c, v, nil, b')$-tree of G. Moreover, $I(T_v)$ and $I(T_u)$ are disjoint sets whose union is $I(T)$. By the induction hypothesis, $P_{v,k_v,l_v}^{c,i,u,b}$ has the monomial $\prod_{w \in I(T_v)} x_w$, and $P_{u,k'-k_v,l'-l_v}^{c,v,nil,b'}$ has the monomial $\prod_{w \in I(T_u)} x_w$. By the definition of $C_{v,k',l'}^{c,i,o,b}$, $P_{v,k',l'}^{c,i,o,b}$ has the monomial $\prod_{w \in I(T_v)} x_w \prod_{w \in I(T_u)} x_w = \prod_{w \in I(T)} x_w$.

□

Observation 6. If $P_{v,k',l'}^{c,i,o,b}$ has a (multilinear) monomial $\prod_{w \in U} x_w$, for some $U \subseteq V$, then G has a (v, k', l', c, i, o, b)-tree T such that $I(T) = U$.

Proof. We prove the claim by using induction on the construction. The claim is clearly true for the base cases. Let $\prod_{w \in U} x_w$, for some $U \subseteq V$, be a monomial of $P_{v,k',l'}^{c,i,o,b}$, such that $C_{v,k',l'}^{c,i,o,b}$ is not constructed in the base cases. Assume that the claim is true for all $C_{\tilde{v},\tilde{k},\tilde{l}}^{\tilde{c},\tilde{i},\tilde{o},b}$ that is constructed before $C_{v,k',l'}^{c,i,o,b}$.

First suppose that $b = F$. By the definition of $C_{v,k',l'}^{c,i,o,b}$, there are $u \in N(v, i, o)$ and $b' \in \{F, T\}$ such that one of the next conditions is fulfilled.

1. $C_{u,k'-1,l'}^{c,v,nil,b'}$ has the monomial $\prod_{w \in U \setminus \{v\}} x_w$. By the induction hypothesis, G has a $(u, k'-1, l', c, v, nil, b')$-tree $T_u = (V_u, E_u)$, such that $I(T_u) = U \setminus \{v\}$. Suppose that there is $i' \in V_u$ such that $(i', u) \in E_u$. In this case $b' = T$; thus $v \notin V_u$ and the set of out-neighbors of u in T_u contains all the neighbors of u in G, excluding v. We get that i' is an out-neighbor of u in T_u, which a contradiction. Thus, by adding v and (v, u) to T_u, we get a (v, k', l', c, i, o, b)-tree T such that $I(T) = U$ (since $I(T) = I(T_u) \cup \{v\}$).

2. There are $k^* \in \{1, ..., k'\}$, $l^* \in \{1, ..., l'-1\}$ and $U^* \subseteq U$, such that $P_{v,k^*,l^*}^{c,i,u,b}$ has the monomial $\prod_{w \in U^*} x_w$, and $P_{u,k'-k^*,l'-l^*}^{c,v,nil,b'}$ has the monomial $\prod_{w \in U \setminus U^*} x_w$. By the induction hypothesis, G has a $(v, k^*, l^*, c, i, u, b)$-tree $T_v = (V_v, E_v)$ such that $I(T_v) = U^*$, and a $(u, k'-k^*, l'-l^*, c, v, nil, b')$-tree $T_u = (V_u, E_u)$ such that $I(T_u) = U \setminus U^*$. Consider the following cases.

   (a) If $v \in V_u$: $v \notin I(T_u)$ (since $v \in I(T_v)$). Thus $col(v) = c$ and v has $\Delta - 1$ out-neighbors in T_u. Note that v is not an out-neighbor of u in T_u, and thus u is an out-neighbor of v in T_u. Therefore $b' = T$, and thus $col(u) = c$, which is a contradiction (since col is a proper coloring).

(b) If there is $w \in (V_v \cap V_u) \setminus \{v, u\} \neq \emptyset$: Since $I(T_v) \cap I(T_u) = \emptyset$, we get that $col(w) = c$ and (w has Δ neighbors in T_v or T_u). Thus there is w' that is a neighbor of w in both T_v and T_u, such that $col(w') \neq c$. We get that $w' \in I(T_v) \cap I(T_u) = \emptyset$, which is a contradiction.

(c) If $u \in V_v$: u is not an out-neighbor of v in T_v. Therefore u has less than $\Delta - 1$ out-neighbors in T_v, and thus $u \in I(T_v)$. We get that $u \notin I(T_u)$, which implies that the set of out-neighbors of u in T_u contains all the neighbors of u in G, excluding v. Thus u has a neighbor, which is not v, in both T_v and T_u, and we have a contradiction according to Case 2b.

We get that $V_v \cap V_u = \emptyset$. If there is $i' \in V_u$ such that $(i', u) \in E_u$, then we get a contradiction in the same manner as in Case 1. We get that $T = (V_v \cup V_u, E_v \cup E_u \cup \{(v, u)\})$ is an out-tree of G. It is straightforward to verify that T is a (v, k', l', c, i, o, b)-tree of G such that $I(T) = I(T_v) \cup I(T_u)$ (and thus $I(T) = U$).

Now suppose that $b = T$. Denote by u the smallest node in $N(v, i, o)$. By the definition of $C_{v,k',l'}^{c,i,o,b}$, one of the next conditions is fulfilled.

1. If $N(v, i, o) = \{u\}$: $P_{u,k'-1,l'}^{c,v,nil,F}$ has the monomial $\prod_{w \in U} x_w$. By the induction hypothesis, G has a $(u, k' - 1, l', c, v, nil, F)$-tree T_u such that $I(T_u) = U$. Since v is not an out-neighbor of u in T_u, by adding v and (v, u) to T_v, we get a (v, k', l', c, i, o, b)-tree T of G (which may not be an out-tree), such that $I(T) = I(T_u) = U$.

2. Else: There are $k^* \in \{1, ..., k'\}$, $l^* \in \{1, ..., l' - 1\}$ and $U^* \subseteq U$, such that $P_{v,k^*,l^*}^{c,i,u,b}$ has the monomial $\prod_{w \in U^*} x_w$, and $P_{u,k'-k^*,l'-l^*}^{c,v,nil,F}$ has the monomial $\prod_{w \in U \setminus U^*} x_w$. By the induction hypothesis, G has a $(v, k^*, l^*, c, i, u, b)$-tree $T_v = (V_v, E_v)$ such that $I(T_v) = U^*$, and a $(u, k' - k^*, l' - l^*, c, v, nil, F)$-tree $T_u = (V_u, E_u)$ such that $I(T_u) = U \setminus U^*$. Consider the following cases.

   (a) If there is $w \in (V_v \cap V_u) \setminus \{v, u\} \neq \emptyset$: We get a contradiction in the same manner as in the previous Case 2b.

   (b) If $u \in V_v$: Since $col(u) \neq c$, we get that $u \in I(T_v) \cup I(T_u) = \emptyset$, which is a contradiction.

   We get that $V_v \cap V_u \setminus \{v\} = \emptyset$. Denote $T = (V_T = (V_v \cup V_u), E_T = (E_v \cup E_u \cup \{(v, u)\}))$. Suppose, by way of contradiction, that there are two nodes $i_1, i_2 \in V_T$ such that $(i_1, v), (i_2, v) \in E_T$. Since T_v is a $(v, k^*, l^*, c, i, u, b)$-tree and T_u is an out-tree, we can assume WLOG that $i_1 \in V_v$ and $i_2 \in V_u$. We get that $v \in I(T_v)$, and thus $v \notin I(T_u)$. Therefore v has $\Delta - 1$ out-neighbors in T_u; but since T_u is an out-tree rooted at u, and v is not an out-neighbor of u in T_u, we have a contradiction. Thus we get that T is a (v, k', l', c, i, o, b)-tree of G such that $I(T) = I(T_v) \cup I(T_u)$ (and thus $I(T) = U$).   □

Observation 7. *If (G, r, k, l) has a solution, then P has a multilinear monomial of degree at most t.*

Proof. Let $T = (V_T, E_T)$ be a solution. Denote $n(T, c) = \{v \in V_T : col(v) = c, v \text{ has } \Delta \text{ neighbors in } T\}$, and $c^* = \text{argmax}_{c \in \{c_1, ..., c_\Delta\}} \{|n(T, c)|\}$. By Observation 4 and the pseudocode of Δ-IOB-Alg[A] (see Section 3.1), we get that

1. $2 + \sum_{3 \leq i \leq \Delta} (i-2)n_i^T = n_1^T$.
2. $\sum_{1 \leq i \leq \Delta} n_i^T = k + l$.
3. $n_1^T - 1 \leq l \leq k - \frac{k-2}{\Delta-1}$.
4. $|n(T, c^*)| \geq n_\Delta^T / \Delta$.

These conditions imply that $k + l - |n(T, c^*)| \leq (2 - \frac{\Delta+1}{\Delta(\Delta-1)})k + 7$. Since T is an $(r, k, l, c^*, r', nil, F)$-tree, the definition of C and Observation 5 imply that P has the (multilinear) monomial $\prod_{w \in I(T)} x_w$. Note that $|I(T)| \leq k + l - |n(T, c^*)| + 1$, and thus we get the observation.   □

Since Observation 6 implies that if P has a multilinear monomial, then (G, r, k, l) has a solution, and by Observation 7, we get the following lemma.

Lemma 4. (G, r, k, l) has a solution iff (C, X, t) has a solution.

The definition of (C, X, t) immediately implies the following observation.

Observation 8. We can compute (C, X, t) in polynomial time and space.

3.3   The Algorithm Δ-IOB-Alg[Δ-Tree-Alg]

Skulrattanakulchai [16] gave a linear-time algorithm that computes a proper Δ-coloring of an undirected connected graph of bounded degree Δ, which is not an odd cycle or a clique. In Δ-**Tree-Alg** (see Algorithm 4), we assume that the underlying undirected graph of G is connected, and that it is not a cycle or a clique, since these cases are handled in the preprocessing steps of Δ-IOB-Alg[A].

Algorithm 4. Δ-Tree-Alg(G, r, k, l)

1: Use the algorithm in [16] to get a proper Δ-coloring col of the underlying undirected graph of G.
2: Compute $f(G, r, k, l, col) = (C, X, t)$.
3: Accept iff MLD-Alg(C, X, t) accepts.

By Lemmas 3 and 4, and Observation 8, we have the following theorem.

Theorem 2. Δ-IOB-Alg[Δ-Tree-Alg] is an $O^*(2^{(2 - \frac{\Delta+1}{\Delta(\Delta-1)})k})$ time and polynomial space randomized algorithm for k-IOB.

4   Open Questions

In this paper we have presented an $O^*(4^k)$ time algorithm for k-IOB, which improves the previous best known O^* running time for k-IOB. However, our algorithm is randomized, while the algorithm that has the previous best known O^* running time is deterministic. Can we obtain an $O^*(4^k)$ time deterministic algorithm for k-IOB? Moreover, can we further reduce the $O^*(4^k)$ and $O^*(2^{(2 - \frac{\Delta+1}{\Delta(\Delta-1)})k})$ running times for k-IOB presented in this paper?

References

1. Cohen, N., Fomin, F.V., Gutin, G., Kim, E.J., Saurabh, S., Yeo, A.: Algorithm for finding k-vertex out-trees and its application to k-internal out-branching problem. J. Comput. Syst. Sci. 76(7), 650–662 (2010)
2. Demers, A., Downing, A.: Minimum leaf spanning tree. US Patent no. 6,105,018 (August 2013)
3. Fomin, F.V., Gaspers, S., Saurabh, S., Thomassé, S.: A linear vertex kernel for maximum internal spanning tree. J. Comput. Syst. Sci. 79(1), 1–6 (2013)
4. Fomin, F.V., Grandoni, F., Lokshtanov, D., Saurabh, S.: Sharp separation and applications to exact and parameterized algorithms. Algorithmica 63(3), 692–706 (2012)
5. Garey, M.R., Johnson, D.S., Stockmeyer, L.: Some simplified NP-complete problems. In: Proc. STOC, pp. 47–63 (1974)
6. Gutin, G., Razgon, I., Kim, E.J.: Minimum leaf out-branching and related problems. Theor. Comput. Sci. 410(45), 4571–4579 (2009)
7. Koutis, I.: Faster algebraic algorithms for path and packing problems. In: Aceto, L., Damgård, I., Goldberg, L.A., Halldórsson, M.M., Ingólfsdóttir, A., Walukiewicz, I. (eds.) ICALP 2008, Part I. LNCS, vol. 5125, pp. 575–586. Springer, Heidelberg (2008)
8. Koutis, I., Williams, R.: Limits and applications of group algebras for parameterized problems. In: Albers, S., Marchetti-Spaccamela, A., Matias, Y., Nikoletseas, S., Thomas, W. (eds.) ICALP 2009, Part I. LNCS, vol. 5555, pp. 653–664. Springer, Heidelberg (2009)
9. Nederlof, J.: Fast polynomial-space algorithms using mobius inversion: improving on steiner tree and related problems. In: Albers, S., Marchetti-Spaccamela, A., Matias, Y., Nikoletseas, S., Thomas, W. (eds.) ICALP 2009, Part I. LNCS, vol. 5555, pp. 713–725. Springer, Heidelberg (2009)
10. Niedermeier, R.: Invitation to fixed-parameter algorithms. Oxford University Press (2006)
11. Ozeki, K., Yamashita, T.: Spanning trees: A survey. Graphs and Combinatorics 27(1), 1–26 (2011)
12. Prieto, E., Sloper, C.: Reducing to independent set structure – the case of k-internal spanning tree. Nord. J. Comput. 12(3), 308–318 (2005)
13. Rédei, L.: Ein kombinatorischer satz. Acta Litteraria Szeged 7, 39–43 (1934)
14. Raible, D., Fernau, H., Gaspers, D., Liedloff, M.: Exact and parameterized algorithms for max internal spanning tree. Algorithmica 65(1), 95–128 (2013)
15. Salamon, G.: A survey on algorithms for the maximum internal spanning tree and related problems. Electronic Notes in Discrete Mathematics 36, 1209–1216 (2010)
16. Skulrattanakulchai, S.: Delta-list vertex coloring in linear time. Inf. Process. Lett. 98(3), 101–106 (2006)
17. Williams, R.: Finding paths of length k in $O^*(2^k)$ time. Inf. Process. Lett. 109(6), 315–318 (2009)

Author Index